Modern Fuzzy Control
Systems and Applications

Modern Fuzzy Control
Systems and Applications

Sofia Ruiz Suarez

Kruger Brentt
Publishers
2025

Kruger Brentt Publishers UK. LTD.
Company Number 9728962

Regd. Office: 68 St Margarets Road, Edgware, Middlesex HA8 9UU

ISBN: 978-1-78715-414-8

For information on all our publications visit our website at http://krugerbrentt.com/

PREFACE

Fuzzy control systems have emerged as powerful tools for modeling and controlling complex and uncertain systems in various engineering and industrial applications. Rooted in fuzzy logic, these systems offer a flexible and intuitive framework for representing human knowledge and reasoning under uncertainty, making them particularly well-suited for applications where conventional control methods may struggle to provide satisfactory performance. From robotics and automation to process control and decision-making, modern fuzzy control systems have revolutionized numerous fields, enabling robust and adaptive control solutions in diverse real-world scenarios. Understanding the principles, methodologies, and applications of modern fuzzy control systems is essential for engineers, researchers, and practitioners seeking to harness the full potential of this versatile technology.

This handbook, titled "Modern Fuzzy Control: Systems and Applications," is a comprehensive exploration of the principles, methodologies, and applications of fuzzy control systems. Drawing upon the expertise of control engineers, mathematicians, and researchers, this volume offers theoretical insights, practical guidance, and case studies for understanding and utilizing fuzzy control techniques in various applications.

From the fundamentals of fuzzy logic and membership functions to the latest advancements in fuzzy modeling, inference methods, and control strategies, the "Modern Fuzzy Control: Systems and Applications" covers a wide range of topics essential for mastering the field of fuzzy control. Through clear explanations, mathematical derivations, and simulation examples, readers will gain a deep understanding of the principles and techniques used to design and implement fuzzy control systems.

Moreover, this handbook explores the applications of fuzzy control systems across different industries and domains, including robotics, automotive systems, power electronics, biomedical engineering, and finance. By examining case studies, research findings, and practical implementations, we highlight the diverse ways in which fuzzy control systems are used to address complex control problems, enhance system performance, and improve decision-making processes.

As we continue to push the boundaries of automation, artificial intelligence, and robotics, the importance of fuzzy control systems in enabling adaptive and robust control solutions has never been greater. It is our hope that the "Modern Fuzzy Control: Systems and Applications" handbook will serve as a valuable resource for researchers, engineers, and practitioners in the field of control systems and automation. Whether you are developing new control algorithms, implementing fuzzy control solutions, or applying fuzzy logic to real-world problems, this handbook invites you to explore the capabilities and opportunities of modern fuzzy control systems and contribute to the advancement of control engineering knowledge and practice.

Sofia Ruiz Suarez

CONTENTS

1
Chapter

OUTPUT FEEDBACK IS EMPLOYED TO STABILIZE FUZZY CONTROL

1. INTRODUCTION: AN OVERVIEW

In contrast to linear frameworks, nonlinear systems are too intricate to be encapsulated by a singular mathematical approach; hence, a universal method for designing a controller applicable to all nonlinear systems has yet to be established. The Takagi-Sugeno (TS) fuzzy technique serves as an alternative to nonlinear system models, leveraging the benefits of effective linear approximations of sector nonlinearities. The TS fuzzy model employs local system dynamics using linear models, while the nonlinear system is characterized by a set of fuzzy rules. Currently, TS model-based fuzzy control methodologies are rapidly and effectively employed in nonlinear control frameworks. Consequently, many stability analysis requirements [2–5] and control design methodologies have been established for TS fuzzy systems [6–9], primarily dependent on the feasibility of a corresponding set of linear matrix inequalities (LMI) [10]. The design issue constitutes a conventional feasibility problem involving several Linear Matrix Inequalities (LMIs), maybe accompanied with a matrix equality to address the bilinearity challenge. Consequently, state and output feedback management utilizing fuzzy Takagi-Sugeno systems models is predominantly implemented within structures that may be built by numerical methods based on Linear Matrix Inequalities (LMIs). The TS fuzzy model-based state control is based on an implicit assumption that all states are available for measurement. If it is impossible, TS fuzzy observers are used to estimate the unmeasurable state variables, and the state controller exploits the system state variable estimate values [11–14]. The nonlinear output feedback design is so formulated as two LMI set problem, and treated by the two-stage procedure using the separation principle, that is, dealing with a set of LMIs for the observer parameters at first and then solving another set of LMIs for the controller parameters [15]. Since, the fuzzy output control does not require the measurement of system state variables and can be formulated as a one LMI set problem, such structure of feedback control is preferred, of course, if the system is stabilizable.

From a relatively wide range of problems associated with the fuzzy output feedback control design for the continuous-time nonlinear MIMO systems approximated by a TS model, the chapter deals with the techniques incorporating the slack matrix application and fuzzy membership-relaxed approaches. The central idea of the TS fuzzy model-based control design, that is, to derive control rules so as to compensate each rule of a fuzzy system and construct the control strategy based on the parallel-distributed compensators (PDC), is reflected in the approach of output control. Motivated by the above mentioned observations, the proposed design method respects the results presented in Refs. [16, 17], and is constructed on an enhanced form of quadratic Lyapunov function. Comparing with the approaches based only on quadratic Lyapunov matrix [18], which are particular in the case of large number of rules, that are very conservative as a common symmetric positive definite matrix, is used to verify all Lyapunov inequalities, presented principle naturally extends the affine TS model properties using slack matrix variables to decouple Lyapunov matrix and the system matrices in LMIs, and gives substantial reducing of conservativeness. Moreover, extra quadratic constraints are included to incorporate fuzzy membership functions relaxes [19, 20] and applied for static as well as dynamic TS fuzzy output controllers design. Note, other constraints with respect to, for example, to decay rate and closed-loop pole clustering can be utilized to extend the proposed design procedures.

The remainder of this chapter is organized as follows. In Section 2, the structure of TS model for considered class of nonlinear systems is briefly described, and some of its properties are outlined. The output feedback control design problem for systems with measurable promise variables is given in Section 3, where the design conditions that guarantees global quadratic stability are formulated and proven. To complete the solutions, Section 4 formulate the static decoupling principle in static TS fuzzy output control, and the method is reformulated in Section 5 in defined criteria for TS fuzzy dynamic output feedback control design. Section 6 gives the numerical examples to illustrate the effectiveness of the proposed approach, and to confirm the validity of the control scheme. The last section, Section 7, draws conclusions and some future directions.

Throughout the chapter, the following notations are used: $\mathbf{x}^T$, $\mathbf{X}^T$ denotes the transpose of the vector $\mathbf{x}$ and matrix $\mathbf{X}$, respectively, for a square matrix $\mathbf{X} = \mathbf{X}^T > 0$ (respectively, $\mathbf{X} = \mathbf{X}^T < 0$) means that $\mathbf{X}$ is a symmetric positive definite matrix (respectively, negative definite matrix), the symbol $\mathbf{I}_n$ represents the n-th order unit matrix, IR denotes the set of real numbers, and $IR^{n \times r}$ denotes the set of all $n \times r$ real matrices.

2. TAKAGI-SUGENO FUZZY MODELS

The systems under consideration are from one class of multi-input and multi-output (MIMO) dynamic systems, which are nonlinear in sectors and represented by TS fuzzy model. Constructing the set of membership functions $h_i(\theta(t))$, $i = 1, 2, \ldots, s$, where

$$\boldsymbol{\theta}(t) = \begin{bmatrix} \theta_1(t) & \theta_2(t) & \cdots & \theta_q(t) \end{bmatrix}, \tag{1}$$

is the vector of premise variables, the final states of the systems are inferred in the TS fuzzy system model as follows

$$\dot{\boldsymbol{q}}(t) = \sum_{i=1}^{s} h_i(\boldsymbol{\theta}(t))(\boldsymbol{A}_i\boldsymbol{q}(t) + \boldsymbol{B}_i\boldsymbol{u}(t)), \tag{2}$$

with the output given by the relation

$$\boldsymbol{y}(t) = \boldsymbol{C}\boldsymbol{q}(t), \tag{3}$$

where $\mathbf{q}(t) \in IR^n$, $\mathbf{u}(t) \in IR^r$, $\mathbf{y}(t) \in IR^m$ are vectors of the state, input, and output variables, $\mathbf{A}_i \in IR^{n\times n}$, $\mathbf{B}_i \in IR^{n\times r}$, $\mathbf{C} \in IR^{m\times n}$ are real finite values matrix, and where $h_i(\theta(t))$ is the averaging weight for the i-th rule, representing the normalized grade of membership (membership function).

By definition, the membership functions satisfy the following convex sum properties.

$$0 \le h_i(\boldsymbol{\theta}(t)) \le 1, \quad \sum_{i=1}^{s} h_i(\boldsymbol{\theta}(t)) = 1 \quad \forall i \in \langle 1, \ldots, s \rangle. \tag{4}$$

It is assumed that the premise variable is a system state variable or a measurable external variable, and none of the premise variables depends on the inputs $\mathbf{u}(t)$.

It is evident that a general fuzzy model is achieved by fuzzy amalgamation of the linear system models. Using a TS model, the conclusion part of a single rule consists no longer of a fuzzy set [21], but determines a function with state variables as arguments, and the corresponding function is a local function for the fuzzy region that is described by the premise part of the rule. Thus, using linear functions, a system state is described locally (in fuzzy regions) by linear models, and at the boundaries between regions an interpolation is used between the corresponding local models.

Note, the models, Eqs. (2) and (3), are mostly considered for analysis, control, and state estimation of nonlinear systems.

Assumption 1 *Each triplet* ($\mathbf{A}_i$, $\mathbf{B}_i$, $\mathbf{C}$) *is locally controllable and observable, the matrix* $\mathbf{C}$ *is the same for all local models.*

It is supposed in the next that the aforementioned model does not include parameter uncertainties or external disturbances, and the premise variables are measured.

3. STATIC FUZZY OUTPUT CONTROLLER

The subsequent design of the fuzzy output controller employs the principle of parallel-distributed compensation, wherein the fuzzy controller utilizes identical sets of normalized membership functions akin to the TS fuzzy system model.

Definition 1 *Considering* Eqs. (2) *and* (3), *and using the same set of normalized membership function* Eq. (4), *the fuzzy static output controller is defined as*

$$u(t) = \sum_{j=1}^{s} h_j(\theta(t))K_j y(t) = \sum_{j=1}^{s} h_j(\theta(t))K_j C q(t). \tag{5}$$

Note that the fuzzy controller Eq. (5) is in general nonlinear.

Considering the system, Eqs. (2) and (3), and the control law, Eq. (5), yields

$$\dot{q}(t) = \sum_{i=1}^{s}\sum_{j=1}^{s} h_i(\theta(t))h_j(\theta(t))\left(A_i + B_i K_j C\right)q(t) = \sum_{i=1}^{s}\sum_{j=1}^{s} h_i(\theta(t))h_j(\theta(t))A_{cij}q(t), \tag{6}$$

$$A_{cij} = A_i + B_i K_j C, \quad A_{cji} = A_j + B_j K_i C. \tag{7}$$

Proposition 1 (*standard design conditions*). *The equilibrium of the fuzzy system* Eqs. (2) *and* (3), *controlled by the fuzzy controller* Eq. (5), *is global asymptotically stable if there exist a positive definite symmetric matrix* $\mathbf{W} \in IR^{n\times n}$ *and matrices* $\mathbf{Y}_j \in IR^{r\times m}$, $\mathbf{H} \in IR^{m\times m}$ *such that*

$$W = W^T > 0, \tag{8}$$

$$A_i W + W A_i^T + B_i Y_i C + C^T Y_i^T B_i^T < 0, \tag{9}$$

$$\frac{A_i W + W A_i^T}{2} + \frac{A_j W + W A_j^T}{2} + \frac{B_i Y_j C + C^T Y_j^T B_i^T}{2} + \frac{B_j Y_i C + C^T Y_i^T B_j^T}{2} < 0, \tag{10}$$

$$CW = HC \tag{11}$$

for $i = 1, 2, \ldots, s$ as well as $i = 1, 2, \ldots, s - 1, j = i + 1, i + 2, \ldots, s$, and $h_i(\theta(t))h_j(\theta(t)) \neq 0$.

When the above conditions hold, the control law gain matrices are given as

$$K_i = Y_i H^{-1}. \tag{12}$$

Proof. (compare, for example, Ref. [16]) Prescribing the Lyapunov function candidate of the form

$$v(q(t)) = q^T(t)Pq(t) > 0, \tag{13}$$

where $\mathbf{P} \in IR^{n\times n}$ is a symmetric positive definite matrix, the time derivative of Eq. (13) along the system trajectory is

$$\dot{v}(q(t)) = \dot{q}^T(t)Pq(t) + q^T(t)P\dot{q}(t) < 0. \tag{14}$$

Inserting Eq. (6) into Eq. (14), it has to be satisfied

$$\dot{v}(q(t)) = \sum_{i=1}^{s}\sum_{j=1}^{s} h_i(\theta(t))h_j(\theta(t))q^T(t)P_{cij}q(t) < 0, \tag{15}$$

$$P_{cij} = PA_{cij} + A_{cij}^T P. \tag{16}$$

Since $\mathbf{P}$ is positive definite, the state coordinate transform can be defined as

$$q(t) = Wp(t), \quad W = P^{-1}, \tag{17}$$

and subsequently, Eqs. (15) and (16) can be rewritten as

$$\dot{v}(p(t)) = \sum_{i=1}^{s}\sum_{j=1}^{s} h_i(\theta(t))h_j(\theta(t))p^T(t)W_{cij}p(t) < 0, \tag{18}$$

$$W_{cij} = A_{cij}W + WA_{cij}^T. \tag{19}$$

Permuting the subscripts *i* and *j* in Eq. (18), also it can write

$$\dot{v}(q(t)) = \sum_{i=1}^{s}\sum_{j=1}^{s} h_i(\theta(t))h_j(\theta(t))p^T(t)W_{cji}p(t) < 0, \tag{20}$$

$$W_{cji} = A_{cji}W + WA_{cji}^T. \tag{21}$$

Thus, adding Eqs. (17) and (19), it yields

$$2\dot{v}(p(t)) = \sum_{i=1}^{s}\sum_{j=1}^{s} h_i(\theta(t))h_j(\theta(t))p^T(t)\left(W_{cij} + W_{cji}\right)p(t) < 0 \tag{22}$$

and subsequently,

$$\dot{v}(p(t)) = \sum_{i=1}^{s} h_i^2(\theta(t))p^T(t)W_{cii}p(t) + 2\sum_{i=1}^{s-1}\sum_{j=i+1}^{s} h_i(\theta(t))h_j(\theta(t))p^T(t)\frac{W_{cij} + W_{cji}}{2}p(t) < 0, \tag{23}$$

which leads to the set of inequalities.

$$(A_i + B_iK_iC)W + W(A_i + B_iK_iC)^T < 0, \tag{24}$$

$$\frac{(A_i + B_iK_jC)W}{2} + \frac{(A_j + B_jK_iC)W}{2} + \frac{W(A_i + B_iK_jC)^T}{2} + \frac{W(A_j + B_jK_iC)^T}{2} < 0 \tag{25}$$

for i = 1, 2, ..., s as well as i = 1, 2, ..., s - 1, j = 1 + 1, i + 2, ..., s and $h_i(\theta(t))h_j(\theta(t)) \neq 0$. Thus, setting here

$$K_jCW = K_jHH^{-1}CW, \tag{26}$$

where **H** is a regular square matrix of appropriate dimension and defining

$$H^{-1}C = CW^{-1}, \quad Y_j = K_jH, \tag{27}$$

the LMI forms of Eqs. (9) and (10) are obtained from Eqs. (24) and (25), respectively, and Eq. (27) implies Eq. (11). This concludes the proof.

Trying to minimize the number of LMIs owing to the limitation of solvers, Proposition 1 is presented in the structure, in which the number of stabilization conditions, used in fuzzy controller design, is equal to $N = (s^2 + s)/2 + 1$. Evidently, the number of stabilization conditions is substantially reduced if s is large.

Proposition 2 (*enhanced design conditions*). *The equilibrium of the fuzzy system* Eqs. (2) *and* (3), *controlled by the fuzzy controller* Eq. (5), *is global asymptotically stable if for given a positive* $\delta \in IR$ *there exist positive definite symmetric matrices* **V**, $\mathbf{S} \in IR^{n\times n}$, *and matrices* $\mathbf{Y}_j \in IR^{r\times m}$, $\mathbf{H} \in IR^{m\times m}$ *such that*

$$S = S^T > 0, \quad V = V^T > 0, \tag{28}$$

$$\begin{bmatrix} A_iS + SA_i^T + B_iY_iC + C^TY_i^TB_i^T & * \\ V - S + \delta A_iS + \delta B_iY_iC & -2\delta S \end{bmatrix} < 0, \tag{29}$$

$$\begin{bmatrix} \Phi_{ij} & * \\ V - S + \delta\frac{A_iS + A_jS}{2} + \delta\frac{B_iY_j + B_jY_i}{2}C & -2\delta S \end{bmatrix} < 0, \tag{30}$$

$$CS = HC, \tag{31}$$

for i = 1, 2, ..., *s, as well as i* = 1, 2, ..., s - 1, *j* = 1 + 1, *i* + 2, ..., *s*, $h_i(\theta(t))h_j(\theta(t)) \neq 0$, *and*

$$\Phi_{ij} = \frac{A_iS + SA_i^T}{2} + \frac{A_jS + SA_j^T}{2} + \frac{B_iY_jC + C^TY_j^TB_i^T}{2} + \frac{B_jY_iC + C^TY_i^TB_j^T}{2}. \tag{32}$$

When the above conditions hold, the control law gain matrices are given as

$$K_i = Y_i H^{-1}. \tag{33}$$

Here and hereafter, $*$ denotes the symmetric item in a symmetric matrix.

Proof. Writing Eq. (6) in the form

$$\sum_{i=1}^{s}\sum_{j=1}^{s} h_i(\theta(t))h_j(\theta(t))\left(A_{cij}q(t) - \dot{q}(t)\right) = 0, \tag{34}$$

then with an arbitrary symmetric positive definite matrix $\mathbf{S} \in R^{n\times n}$ and a positive scalar $\delta \in R$ it yields

$$\sum_{i=1}^{s}\sum_{j=1}^{s} \mathrm{h}_i(\theta(t))\mathrm{h}_j(\theta(t))\left(q^T(t)\mathbf{S} + \delta\dot{q}^T(t)\mathbf{S}\right)\left(A_{cij}q(t) - \dot{q}(t)\right) = 0. \tag{35}$$

Since $\mathbf{S}$ is positive definite, the new state variable coordinate system can be introduced so that

$$p(t) = Sq(t), \quad \dot{p}(t) = S\dot{q}(t), \quad V = S^{-1}PS^{-1}. \tag{36}$$

Therefore, Eq. (14) can be rewritten as

$$\dot{v}(p(t)) = \dot{p}^T(t)Vp(t) + p^T(t)V\dot{p}(t) < 0 \tag{37}$$

and Eq. (35) takes the form

$$\sum_{i=1}^{s}\sum_{j=1}^{s} \mathrm{h}_i(\theta(t))\mathrm{h}_j(\theta(t))\left(p^T(t) + \delta\dot{p}^T(t)\right)\left(A_{cij}Sp(t) - S\dot{p}(t)\right) = 0. \tag{38}$$

Thus, adding Eq. (38) as well as the transposition of Eq. (38) to Eq. (37), it yields

$$\begin{aligned}
\dot{v}(p(t)) &= \dot{p}^T(t)Vp(t) + p^T(t)V\dot{p}(t) \\
&+ \sum_{i=1}^{s}\sum_{j=1}^{s} h_i(\theta(t))\mathrm{h}_j(\theta(t))\left(p^T(t) + \delta\dot{p}^T(t)\right)\left(A_{cij}Sp(t) - S\dot{p}(t)\right) \\
&+ \sum_{i=1}^{s}\sum_{j=1}^{s} h_i(\theta(t))\mathrm{h}_j(\theta(t))\left(A_{cij}Sp(t) - S\dot{p}(t)\right)^T\left(p(t) + \delta\dot{p}(t)\right) < 0.
\end{aligned} \tag{39}$$

Using the notation

$$p_c^T(t) = \begin{bmatrix} p^T(t) & \dot{p}^T(t) \end{bmatrix}, \tag{40}$$

the inequality Eq. (39) can be written as

$$\dot{v}(\boldsymbol{p}_c(t)) = \sum_{i=1}^{s}\sum_{j=1}^{s} h_i(\boldsymbol{\theta}(t))\mathrm{h}_j(\boldsymbol{\theta}(t))\boldsymbol{p}_c^T(t)\boldsymbol{S}_{cij}\boldsymbol{p}_c(t) < 0, \tag{41}$$

$$\boldsymbol{S}_{cij} = \begin{bmatrix} (\boldsymbol{A}_i + \boldsymbol{B}_i\boldsymbol{K}_j\boldsymbol{C})\boldsymbol{S} + \boldsymbol{S}(\boldsymbol{A}_i + \boldsymbol{B}_i\boldsymbol{K}_j\boldsymbol{C})^T & * \\ \boldsymbol{V} - \boldsymbol{S} + \delta(\boldsymbol{A}_i + \boldsymbol{B}_i\boldsymbol{K}_j\boldsymbol{C})\boldsymbol{S} & -2\delta\boldsymbol{S} \end{bmatrix} < 0. \tag{42}$$

Permuting the subscripts i and j in Eq. (41), and following the way used above, analogously it can obtain

$$\dot{v}(\boldsymbol{p}_c(t)) = \sum_{i=1}^{s}\mathrm{h}_i^2(\boldsymbol{\theta}(t))\boldsymbol{p}_c^T(t)\boldsymbol{S}_{cii}\boldsymbol{p}_c(t) + 2\sum_{i=1}^{s-1}\sum_{j=i+1}^{s}\mathrm{h}_i(\boldsymbol{\theta}(t))\mathrm{h}_j(\boldsymbol{\theta}(t))\boldsymbol{p}_c^T(t)\frac{\boldsymbol{S}_{cij}+\boldsymbol{S}_{cji}}{2}\boldsymbol{p}_c(t) < 0. \tag{43}$$

Since $r = m$, it is now possible to set

$$\boldsymbol{K}_j\boldsymbol{C}\boldsymbol{S} = \boldsymbol{K}_j\boldsymbol{H}\boldsymbol{H}^{-1}\boldsymbol{C}\boldsymbol{S}, \tag{44}$$

where **H** is a regular square matrix of appropriate dimension and introducing

$$\boldsymbol{I}^{-1}\boldsymbol{C} = \boldsymbol{C}\boldsymbol{S}^{-1}, \quad \boldsymbol{Y}_j = \boldsymbol{K}_j\boldsymbol{H} \tag{45}$$

then Eqs. (42) and (45) imply Eqs. (29)–(31). This concludes the proof.

Note, Eq. (42) leads to the set of LMIs only if δ is a prescribed constant. (δ can be considered as a tuning parameter). Considering δ as a LMI variable, Eq. (42) represents the set of bilinear matrix inequalities (BMI).

Theorem 1 (*enhanced relaxed design conditions*). *The equilibrium of the fuzzy system* Eqs. (2) *and* (3), *controlled by the fuzzy controller* Eq. (5), *is global asymptotically stable if for given a positive* $\delta \in IR$ *there exist positive definite symmetric matrices* $\mathbf{V}, \mathbf{S} \in IR^{n\times n}$, *the matrices* $X_{ij}=X^T_{ji} \in IR^{r\times n}$, *and* $\mathbf{Y}_j \in R^{r\times m}$, $\mathbf{H} \in IR^{m\times m}$ *such that*

$$\boldsymbol{S} = \boldsymbol{S}^T > 0, \quad \boldsymbol{V} = \boldsymbol{V}^T > 0, \quad \begin{bmatrix} \boldsymbol{X}_{11} & \boldsymbol{X}_{12} & \cdots & \boldsymbol{X}_{1s} \\ \boldsymbol{X}_{21} & \boldsymbol{X}_{22} & \cdots & \boldsymbol{X}_{2s} \\ \vdots & \vdots & \ddots & \vdots \\ \boldsymbol{X}_{s1} & \boldsymbol{X}_{s2} & \cdots & \boldsymbol{X}_{ss} \end{bmatrix} > 0, \tag{46}$$

$$\begin{bmatrix} \boldsymbol{A}_i\boldsymbol{S} + \boldsymbol{S}\boldsymbol{A}_i^T + \boldsymbol{B}_i\boldsymbol{Y}_i\boldsymbol{C} + \boldsymbol{C}^T\boldsymbol{Y}_i^T\boldsymbol{B}_i^T + \boldsymbol{X}_{ii} & * \\ \boldsymbol{V} - \boldsymbol{S} + \delta\boldsymbol{A}_i\boldsymbol{S} + \delta\boldsymbol{B}_i\boldsymbol{Y}_i\boldsymbol{C} & -2\delta\boldsymbol{S} \end{bmatrix} < 0, \tag{47}$$

$$\begin{bmatrix} \boldsymbol{\Phi}_{ij} & * \\ \boldsymbol{V} - \boldsymbol{S} + \delta\dfrac{\boldsymbol{A}_i\boldsymbol{S} + \boldsymbol{A}_j\boldsymbol{S}}{2} + \delta\dfrac{\boldsymbol{B}_i\boldsymbol{Y}_j + \boldsymbol{B}_j\boldsymbol{Y}_i}{2}\boldsymbol{C} & -2\delta\boldsymbol{S} \end{bmatrix} < 0, \tag{48}$$

$$CS = HC, \tag{49}$$

for i = 1, 2, ..., s, as well as i = 1, 2, ..., s - 1, j = 1 + 1, i + 2, ..., s, $h_i(\theta(t))h_j(\theta(t)) \neq 0$ *and*

$$\Phi_{ij} = \frac{A_iS + SA_i^T}{2} + \frac{A_jS + SA_j^T}{2} + \frac{B_iY_jC + C^TY_j^TB_i^T}{2} + \frac{B_jY_iC + C^TY_i^TB_j^T}{2} + \frac{X_{ij} + X_{ji}}{2}. \tag{50}$$

When the above conditions hold, the control law gain matrices are given as

$$K_i = Y_iH^{-1}. \tag{51}$$

Proof. Introducing the positive real term

$$v_v(\theta(t)) = q^T(t)Z(\theta(t))q(t) > 0, \tag{52}$$

$$Z(\theta(t)) = Z^T(\theta(t)) = \sum_{i=1}^{s}\sum_{j=1}^{s} h_i(\theta(t))h_j(\theta(t))Z_{ij} > 0, \tag{53}$$

where $Z_{ij}=Z^T_{ji} \in IR^{n\times n}$, *i, j* = 1, 2, ..., *s* is the set of associated matrices and using the state coordinate transform Eq. (36), then Eq. (53) can be rewritten as

$$v_v(p(t)) = \sum_{i=1}^{s}\sum_{j=1}^{s} h_i(\theta(t))h_j(\theta(t))p^T(t)X_{ij}p(t) > 0, \quad X_{ij} = S^{-1}Z_{ij}S^{-1} = X_{ji}^T, \tag{54}$$

where

$$Z(\theta(t)) = \begin{bmatrix} h_1(\theta(t)p(t) & h_2(\theta(t)p(t) & \cdots & h_s(\theta(t)p(t) \end{bmatrix} \begin{bmatrix} X_{11} & X_{12} & \cdots & X_{1s} \\ X_{21} & X_{22} & \cdots & X_{2s} \\ \vdots & \vdots & \ddots & \vdots \\ X_{s1} & X_{s2} & \cdots & X_{ss} \end{bmatrix} \begin{bmatrix} h_1(\theta(t)p(t) \\ h_2(\theta(t)p(t) \\ \vdots \\ h_s(\theta(t)p(t) \end{bmatrix} \tag{55}$$

is symmetric, an positive definite if Eq. (46) is satisfied. Then, in the sense of the Krasovskii theorem (see, for example, Ref. [22]), it can be set up in Eq. (39)

$$\begin{aligned} \dot{v}(p(i)) &= \dot{p}^T(t)Vp(t) + p^T(t)V\dot{p}(t) \\ &+ \sum_{i=1}^{s}\sum_{j=1}^{s} h_i(\theta(t))h_j(\theta(t))\left(p^T(t) + \delta\dot{p}^T(t)\right)\left(A_{cij}Sp(t) - S\dot{p}(t)\right) \\ &+ \sum_{i=1}^{s}\sum_{j=1}^{s} h_i(\theta(t))h_j(\theta(t))\left(A_{cij}Sp(t) - S\dot{p}(t)\right)^T\left(p(t) + \delta\dot{p}(t)\right) \\ &< -\sum_{i=1}^{s}\sum_{j=1}^{s} h_i(\theta(t))h_j(\theta(t))p^T(t)X_{ij}p(t) \\ &< 0, \end{aligned} \tag{56}$$

which in the consequence, modifies Eq. (42) as follows

$$S_{cij} = \begin{bmatrix} (A_i + B_iK_jC)S + S(A_i + B_iK_jC)^T + X_{ij} & * \\ V - S + \delta(A_i + B_iK_jC)S & -2\delta S \end{bmatrix} < 0. \tag{57}$$

Following the same way as in the proof of Proposition 2, then Eqs. (47) and (48) can be derived from Eq. (57), while Eq. (55) implies Eq. (46). This concludes the proof.

This principle naturally exploits the affine TS model properties. Introducing the slack matrix variable **S** into the LMIs, the system matrices are decoupled from the equivalent Lyapunov matrix **V**. Note, to respect the conditions $X1_j = X^T_{ji,}$ the set of inequalities Eqs. (47) and (48) have to be constructed. In the opposite case, constructing a set on s^2 LMIs, the constraint conditions have to be set as $X1_j = X^T_{ij} > 0$, that is, the weighting matrices have to be symmetric positive definite.

Corollary 1 *Prescribing* **S** = **V** *and using the Schur complement property, then* Eq. (57) *implies*

$$A_{cij}S + SA_{cij}^T + X_{ij} + 0.5\delta SA_{cij}^T\delta^{-1}S^{-1}\delta A_{cij}S < 0 \tag{58}$$

and for δ = 0 evidently, it has to be

$$A_{cij}S + SA_{cij}^T + X_{ij} < 0. \tag{59}$$

Evidently, then Eqs. (47) and (48) imply

$$S(A_i + B_iK_iC)^T + (A_i + B_iK_iC)S + X_{ii} < 0, \tag{60}$$

$$\frac{(A_i + B_iK_jC)S}{2} + \frac{(A_j + B_jK_iC)S}{2} + \frac{S(A_i + B_iK_jC)^T}{2} + \frac{S(A_j + B_jK_iC)^T}{2} + \frac{X_{ij} + X_{ji}}{2} < 0. \tag{61}$$

Considering **S** = **W** and comparing with Eqs. (23) and (24), then Eqs. (60) and (61) are the extended set of inequalities Eqs. (23) and (24). The result is that the equilibrium of the fuzzy system Eqs. (2) and (3), controlled by the fuzzy controller Eq. (5), is global asymptotically stable if there exist a positive definite symmetric matrices $\mathbf{S} \in IR^{n \times n}$, the matrices $X_{1j} = X^T_{ji} \in R^{r \times n}$, and $\mathbf{Y}_j \in IR^{r \times n}$, $\mathbf{H} \in IR^{m \times m}$ such that

$$S = S^T > 0, \quad \begin{bmatrix} X_{11} & X_{12} & \cdots & X_{1s} \\ X_{21} & X_{22} & \cdots & X_{2s} \\ \vdots & \vdots & \ddots & \vdots \\ X_{s1} & X_{s2} & \cdots & X_{ss} \end{bmatrix} > 0, \tag{62}$$

$$A_iS + SA_i^T + B_iY_iC + C^TY_i^TB_i^T + X_{ii} < 0, \tag{63}$$

$$\frac{A_iS+SA_i^T}{2}+\frac{A_jS+SA_j^T}{2}+\frac{B_iY_jC+C^TY_j^TB_i^T}{2}+\frac{B_jY_iC+C^TY_i^TB_j^T}{2}\frac{X_{ij}+X_{ji}}{2}<0, \qquad (64)$$

$$CS=HC, \qquad (65)$$

for i = 1, 2, ..., s, as well as i = 1, 2, ..., s - 1, j = 1 + 1, i + 2, ..., s and $h_i(\theta(t))h_j(\theta(t)) \neq 0$. Subsequently, if this set of LMIs is satisfied, the set of control law gain matrices is given as

$$K_i=Y_iH^{-1}. \qquad (66)$$

These LMIs form relaxed design conditions.

Note the derived results are linked to some existing finding when the design problem involves additive performance requirements and the relaxed quadratic stability conditions of fuzzy control systems (see, e.g., Refs. [11, 19]) are equivalently steered.

4. FORCED MODE IN STATIC OUTPUT CONTROL

In practice, the plant with $r = m$ (square plants) is often encountered, since in this case, it is possible to associate with each output signal as a reference signal, which is expected to influence this wanted output. Such mode, reflecting nonzero set working points, is called the forced regime.

Definition 2 *A forced regime for the TS fuzzy system* Eqs. (2) *and* (3) *with the TS fuzzy static output controller* Eq. (5) *is foisted by the control policy*

$$u(t)=\sum_{j=1}^{s}h_j(\theta(t))\mathbf{K}_jy(t)+\sum_{i=1}^{s}\sum_{j=1}^{s}h_i(\theta(t))h_j(\theta(t))\mathbf{W}_{ij}w(t), \qquad (67)$$

where $r = m$, $\mathbf{w}(i) \in IR^{m}$ is desired output signal vector, and $\mathbf{W}_{ij} \in IR^{m\times m}$, i, j = 1, 2, ... s is the set of the signal gain matrices.

Lemma 1. *The static decoupling challenge is solvable if* $(\mathbf{A}_i, \mathbf{B}_i)$ *is stabilizable and*

$$rank\begin{bmatrix} A_i & B_i \\ C & 0 \end{bmatrix}=n+m. \qquad (68)$$

Proof. If $(\mathbf{A}_i, \mathbf{B}_i)$ is stabilizable, it is possible to find $\mathbf{K}_j$ such that matrices $\mathbf{A}_{cij} = \mathbf{A}_i + \mathbf{B}_i\mathbf{K}_j\mathbf{C}$ are Hurwitz. Assuming that for such $\mathbf{K}_j$, it yields

$$\text{rank}\begin{bmatrix} A_i & B_i \\ C & 0 \end{bmatrix} = \text{rank}\begin{bmatrix} A_i & B_i \\ C & 0 \end{bmatrix}\begin{bmatrix} I_n & 0 \\ K_jC & I_m \end{bmatrix} = \text{rank}\begin{bmatrix} A_i + B_iK_jC & B_i \\ C & 0 \end{bmatrix}, \tag{69}$$

$$\text{rank}\begin{bmatrix} A_i + B_iK_jC & B_i \\ C & 0 \end{bmatrix} = \text{rank}\begin{bmatrix} I_n & 0 \\ -C(A_i + B_iK_jC)^{-1} & I_m \end{bmatrix}\begin{bmatrix} A_i + B_iK_jC & B_i \\ C & 0 \end{bmatrix}, \tag{70}$$

respectively, then

$$\begin{bmatrix} A_i & B_i \\ C & 0 \end{bmatrix} = \text{rank}\begin{bmatrix} A_i + B_iK_jC & B_i \\ 0 & -C(A_i + B_iK_jC)^{-1}B_i \end{bmatrix} = n + m, \tag{71}$$

since rank($\mathbf{A}_i + \mathbf{B}_i\mathbf{K}_j\mathbf{C}$) = n, and rank$\mathbf{B}_i$ = m.

Thus, evidently, it has to be satisfied

$$\text{rank}\left(C(A_i + B_iK_jC)^{-1}B_i\right) = m. \tag{72}$$

This concludes the proof.

Theorem 2. *To reach a forced regime for the TS fuzzy system* Eqs. (2) *and* (3) *with the TS fuzzy control policy* Eq. (67), *the signal gain matrices have to take the forms*

$$W_{ij} = \left(C(A_i + B_iK_jC)^{-1}B_i\right)^{-1}, \tag{73}$$

where $\mathbf{W}_{ij} \in IR^{m\times m}$, $i, j = 1, 2, \ldots s$.

Proof. In a steady state, which corresponds to q.(t)=0*q*.*t*=0, the equality $\mathbf{y}_o = \mathbf{w}_o$ must hold, where $\mathbf{q}_o \in IR^n$, $\theta_o \in IR^q$, $\mathbf{y}_o$, $\mathbf{w}_o \in IR^m$ are the vectors of steady state values of $\mathbf{q}(t)$, $\theta(t)$, $\mathbf{y}(t)$, $\mathbf{w}(t)$, respectively.

Substituting Eq. (67) in Eq. (2) yields the expression

$$\sum_{i=1}^{s}\sum_{j=1}^{s} \text{h}_i(\theta_o)\text{h}_j(\theta_o)\left((A_i + B_iK_jC)q_o + B_iW_{ij}w_o\right) = 0, \tag{74}$$

$$-\sum_{i=1}^{s}\sum_{j=1}^{s} \text{h}_i(\theta_o)\text{h}_j(\theta_o)q_o = -q_o = \sum_{i=1}^{s}\sum_{j=1}^{s} \text{h}_i(\theta_o)\text{h}_j(\theta_o)(A_i + B_iK_jC)^{-1}B_iW_{ij}w_o, \tag{75}$$

respectively, and it can be set

$$y_o = Cq_o = -\sum_{i=1}^{s}\sum_{j=1}^{s} \text{h}_i(\theta_o)\text{h}_j(\theta_o)C(A_i + B_iK_jC)^{-1}B_iW_{ij}w_o = I_mw_o. \tag{76}$$

Thus, Eq. (76) gives the solution

$$\mathbf{W}_{ij}^{-1} = -\mathbf{C}\left(\mathbf{A}_i + \mathbf{B}_i\mathbf{K}_j\mathbf{C}\right)^{-1}\mathbf{B}_i, \tag{77}$$

which implies Eq. (68). Hence, declaredly,

$$\mathrm{rank}\mathbf{W}_j = \mathrm{rank}\left(\mathbf{C}\left(\mathbf{A}_i + \mathbf{B}_i\mathbf{K}_j\mathbf{C}\right)^{-1}\mathbf{B}_i\right) = m. \tag{78}$$

This concludes the proof.

The forced regime is basically designed for constant references and is very closely related to shift of origin. If the command value **w**(t) is changed "slowly enough," the above scheme can do a reasonable job of tracking, that is, making **y**(t) follow **w**(t) [23].

5. BI-PROPER DYNAMIC OUTPUT CONTROLLER

The full order biproper dynamic output controller is defined by the equation

$$\dot{\mathbf{p}}(t) = \sum_{j=1}^{s} \mathrm{h}_j(\boldsymbol{\theta}(t))\left(\mathbf{J}_j\mathbf{p}(t) + \mathbf{L}_j\mathbf{y}(t)\right), \tag{79}$$

$$\mathbf{u}(t) = \sum_{j=1}^{s} \mathrm{h}_j(\boldsymbol{\theta}(t))\left(\mathbf{M}_j\mathbf{p}(t) + \mathbf{N}_j\mathbf{y}(t)\right), \tag{80}$$

where **p**(t) ∈ IR^{h} is the vector of the controller state variables and the parameter matrix

$$\mathbf{K}_j^{\circ} = \begin{bmatrix} \mathbf{J}_j & \mathbf{L}_j \\ \mathbf{M}_j & \mathbf{N}_j \end{bmatrix}, \tag{81}$$

$K^{\circ}_j \in IR^{(n+r)\times(h+m)}$, is considered in this block matrix structure with respect to the matrices $\mathbf{J}_j \in IR^{h\times h}$, $\mathbf{L}_j \in IR^{h\times m}$, $\mathbf{M}_j \in IR^{r\times h}$, and $\mathbf{N}_j \in IR^{r\times m}$. For simplicity, the full order $p = n$ controller is considered in the following.

To analyze the stability of the closed-loop system structure with the dynamic output controller, the closed-loop system description implies the following form

$$\dot{\mathbf{q}}^{\circ}(t) = \sum_{i=1}^{s}\sum_{j=1}^{s} h_i(\boldsymbol{\theta}(t))h_j(\boldsymbol{\theta}(t))\mathbf{A}_{cij}^{\circ}\mathbf{q}^{\circ}(t), \tag{82}$$

$$\mathbf{y}^{\circ}(t) = \mathbf{I}^{\circ}\mathbf{C}^{\circ}\mathbf{q}^{\circ}(t), \tag{83}$$

where

$$q^{\circ T}(t) = \begin{bmatrix} q^T(t) & p^T(t) \end{bmatrix}, \tag{84}$$

$$A^{\circ}_{cij} = \begin{bmatrix} A_i + B_i N_j C & B_i M_j \\ L_j C & N_j \end{bmatrix}, \quad I^{\circ} = \begin{bmatrix} 0 & I_m \end{bmatrix}, \quad C^{\circ} = \begin{bmatrix} 0 & I_n \\ C & 0 \end{bmatrix} \tag{85}$$

and $A^{\circ}_{cij} \in IR^{2n\times 2n}$, $\mathbf{I}^{\circ} \in IR^{m\times(n+m)}$, $\mathbf{C}^{\circ} \in IR^{(n+m)\times 2n}$.

Introducing the notations

$$A^{\circ}_i = \begin{bmatrix} A_i & 0 \\ 0 & 0 \end{bmatrix}, \quad B^{\circ}_i = \begin{bmatrix} 0 & B_i \\ I_n & 0 \end{bmatrix}, \tag{86}$$

where $A^{\circ}_i \in R^{2n\times 2n}$, $B^{\circ}_i \in R^{2n\times(n+r)}$, the closed-loop system matrices take the equivalent forms

$$A^{\circ}_{cij} = A^{\circ}_i + B^{\circ}_i K^{\circ}_j C^{\circ}. \tag{87}$$

In the sequel, it is supposed that $(A^{\circ}_i, B^{\circ}_i)$ is stabilizable, $(A^{\circ}_i, C^{\circ}_i)$ is detectable [24]. Note this kind of controllers can be preferred in fault tolerant control (FTC) structures with virtual actuators [25].

Theorem 3 (*relaxed design conditions*). *The equilibrium of the fuzzy system* Eqs. (2) *and* (3) *controlled by the fuzzy dynamic output controller* Eqs. (79) *and* (80) *is global asymptotically stable if there exist a positive definite symmetric matrix* $\mathbf{S}^{\circ} \in IR^{2n\times 2n}$, *symmetric matrices* $X^{\circ}_{ij} = X^{\circ}_{ji} \in IR^{2n\times 2n}$, *a regular matrix* $\mathbf{H}^{\circ} \in {}^{(n+m)\times(n+m)}$ *and matrices* $Y^{\circ}_j \in IR^{(n+r)\times(n+m)}$ *such that*

$$S^{\circ} = S^{\circ T} > 0, \quad \begin{bmatrix} X^{\circ}_{11} & X^{\circ}_{12} & \cdots & X^{\circ}_{1s} \\ X^{\circ}_{21} & X^{\circ}_{22} & \cdots & X^{\circ}_{2s} \\ \vdots & \vdots & \ddots & \vdots \\ X^{\circ}_{s1} & X^{\circ}_{s2} & \cdots & X^{\circ}_{ss} \end{bmatrix} > 0, \tag{88}$$

$$A^{\circ}_i S^{\circ} + S^{\circ} A^{\circ T}_i + B^{\circ}_i Y^{\circ}_i C^{\circ} + C^{\circ T} Y^{\circ T}_i B^{\circ T}_i + X^{\circ}_{ii} < 0, \tag{89}$$

$$\frac{A^{\circ}_i + A^{\circ}_j}{2} S^{\circ} + S^{\circ} \frac{A^{\circ T}_i + A^{\circ T}_j}{2} + \frac{B^{\circ}_i Y^{\circ}_j + B^{\circ}_j Y^{\circ}_i}{2} C^{\circ} + C^{\circ T} \frac{Y^{\circ T}_j B^{\circ T}_i + Y^{\circ T}_i B^{\circ T}_j}{2} + \frac{X^{\circ}_{ij} + X^{\circ}_{ji}}{2} < 0, \tag{90}$$

$$C^{\circ} S^{\circ} = H^{\circ} C^{\circ}, \tag{91}$$

for all $i \in \langle 1, 2, \dots s\rangle$, $i < j \le s$, $i, j \in \langle 1, 2, \dots s\rangle$, *respectively, and* $h_i(\theta(t))h_j(\theta(t)) \neq 0$.

When the above conditions hold, the set of control law gain matrices are given as

$$K^{\circ}_j = Y^{\circ}_j (H^{\circ})^{-1}, \quad j = 1, 2, \dots, s \tag{92}$$

Proof. Defining the Lyapunov function as follows

$$v(q^{\circ}(t)) = q^{\circ T}(t)P^{\circ}q^{\circ}(t) > 0, \tag{93}$$

where $\mathbf{P}^{\circ} \in IR^{2n \times 2n}$ is a positive definite matrix, then the time derivative of $v(\mathbf{q}(t))$ along a closed-loop system trajectory is

$$\dot{v}(q^{\circ}(t)) = \dot{q}^{\circ T}(t)P^{\circ}q^{\circ}(t) + q^{\circ T}(t)P^{\circ}\dot{q}^{\circ}(t) < 0. \tag{94}$$

Substituting Eq. (87), then Eq. (94) implies

$$\dot{v}(q^{\circ}(t)) = \sum_{i=1}^{s}\sum_{j=1}^{s} h_i(\theta(t))h_j(\theta(t))q^{\circ T}(t)P^{\circ}_{cij}q^{\circ}(t) < 0, \tag{95}$$

$$P^{\circ}_{cij} = P^{\circ}A^{\circ}_{cij} + A^{\circ T}_{cij}P^{\circ}. \tag{96}$$

Since $\mathbf{P}^{\circ}$ is positive definite, the state coordinate transform can now be defined as

$$q^{\circ}(t) = S^{\circ}p^{\circ}(t), \quad S^{\circ} = (P^{\circ})^{-1}, \tag{97}$$

and subsequently Eqs. (95) and (96) can be rewritten as

$$\dot{v}(p^{\circ}(t)) = \sum_{i=1}^{s}\sum_{j=1}^{s} h_i(\theta(t))h_j(\theta(t))p^{\circ T}(t)S^{\circ}_{cij}p^{\circ}(t) < 0, \tag{98}$$

$$S^{\circ}_{cij} = A^{\circ}_{cij}S^{\circ} + S^{\circ}A^{\circ T}_{cij}. \tag{99}$$

Introducing, analogously to Eqs. (54) and (55), the positive term

$$v_v(p^{\circ}(t)) = p^{\circ T}(t)Z^{\circ}(\theta(t))p^{\circ}(t) > 0, \tag{100}$$

defined by the set of matrices $\{X^{\circ}_{ij} = X^{\circ T}_{ji} \in \mathbb{R}^{n \times n},\ i, j = 1, 2, \ldots, s\}$ in the structure Eq. (88) such that

$$Z^{\circ}(\theta(t)) = Z^{\circ T}(\theta(t)) = \sum_{i=1}^{s}\sum_{j=1}^{s} h_i(\theta(t))h_j(\theta(t))X^{\circ}_{ij} > 0, \tag{101}$$

then, in the sense of Krasovskii theorem, it can be set up

$$\dot{v}(p^{\circ}(t)) = \sum_{i=1}^{s}\sum_{j=1}^{s} h_i(\theta(t))h_j(\theta(t))p^{\circ T}(t)S^{\circ}_{cij}p^{\circ}(t) < 0, \tag{102}$$

where

$$S^{\circ}_{cij} = A^{\circ}_{cij}S^{\circ} + S^{\circ}A^{\circ T}_{cij} + X^{\circ}_{ij}. \tag{103}$$

Therefore, Eq. (102) can be factorized as follows

$$\dot{v}(p^{\circ}(t)) = \sum_{i=1}^{s} h_i^2(\theta(t))p^{\circ T}(t)S_{cii}^{\circ}p^{\circ}(t) + 2\sum_{i=1}^{s-1}\sum_{j=i+1}^{s} h_i(\theta(t))h_j(\theta(t))p^{\circ T}(t)\frac{S_{cij}^{\circ} + S_{cji}^{\circ}}{2}p^{\circ}(t) < 0, \quad (104)$$

which, using Eq. (87), leads to the following sets of inequalities

$$A_i^{\circ}S^{\circ} + S^{\circ}A_i^{\circ T} + B_i^{\circ}K_j^{\circ}C^{\circ}S^{\circ} + S^{\circ}C^{\circ T}K_j^{\circ T}B_i^{\circ T} + X_{ij}^{\circ} < 0, \quad (105)$$

$$\frac{\left(A_i^{\circ} + B_i^{\circ}K_j^{\circ}C^{\circ}\right)S^{\circ}}{2} + \frac{\left(A_j^{\circ} + B_j^{\circ}K_i^{\circ}C^{\circ}\right)S^{\circ}}{2} + \frac{S^{\circ}\left(A_i^{\circ} + B_i^{\circ}K_j^{\circ}C^{\circ}\right)^T}{2} + \frac{S^{\circ}\left(A_j^{\circ} + B_j^{\circ}K_i^{\circ}C^{\circ}\right)^T}{2} + \frac{X_{ij}^{\circ} + X_{ji}^{\circ}}{2} < 0, \quad (106)$$

for $i = 1, 2, \dots, s$, as well as $i = 1, 2, \dots, s - 1, j = 1 + 1, i + 2, \dots, s$, and $h_i(\theta(t))h_j(\theta(t)) \neq 0$. Analyzing the product $B^{\circ}_i K^{\circ}_j C^{\circ} S^{\circ}$, it can set

$$B_i^{\circ}K_j^{\circ}C^{\circ}S^{\circ} = B_i^{\circ}K_j^{\circ}H^{\circ}(H^{\circ})^{-1}C^{\circ}S^{\circ} = B_i^{\circ}Y_j^{\circ}C^{\circ}, \quad (107)$$

where

$$K_j^{\circ}H^{\circ} = Y_j^{\circ}, \quad (H^{\circ})^{-1}C^{\circ} = C^{\circ}(S^{\circ})^{-1} \quad (108)$$

and $\mathbf{H}^{\circ} \in R^{(m+n)\times(m+n)}$ is a regular square matrix. Thus, with Eq. (108), then Eqs. (105) and (106) implies Eqs. (89) and (90) and Eq. (108) gives Eq. (91). This concludes the proof.

This theorem provides the sufficient condition under LMIs and LME formulations for the synthesis of the dynamic output controller reflecting the membership function properties.

For the same reasons as in Theorem 1, the following theorem is proven.

Theorem 4 (*enhanced relaxed design conditions*). *The equilibrium of the fuzzy system* Eqs. (2) *and* (3) *controlled by the fuzzy dynamic output controller* Eqs. (79) *and* (80) *is global asymptotically stable if for given a positive* $\delta \in R$ *there exist positive definite symmetric matrices* $\mathbf{V}^{\circ}, \mathbf{S}^{\circ} \in R^{n\times n}$, *and matrices* $Y^{\circ}_j \in R^{r\times n}$, $\mathbf{H}^{\circ} \in R^{m\times m}$ *such that*

$$S^{\circ} = S^{\circ T} > 0, \quad V^{\circ} = V^{\circ T} > 0, \quad \begin{bmatrix} X_{11}^{\circ} & X_{12}^{\circ} & \cdots & X_{1s}^{\circ} \\ X_{21}^{\circ} & X_{22}^{\circ} & \cdots & X_{2s}^{\circ} \\ \vdots & \vdots & \ddots & \vdots \\ X_{s1}^{\circ} & X_{s2}^{\circ} & \cdots & X_{ss}^{\circ} \end{bmatrix} > 0, \quad (109)$$

$$\begin{bmatrix} A_i^{\circ}S^{\circ} + S^{\circ}A_i^{\circ T} + B_i^{\circ}Y_i^{\circ}C^{\circ} + C^{\circ T}Y_i^{\circ T}B_i^{\circ T} & * \\ V^{\circ} - S^{\circ} + \delta A_i^{\circ}S^{\circ} + \delta B_i^{\circ}Y_i^{\circ}C^{\circ} & -2\delta S^{\circ} \end{bmatrix} < 0, \quad (110)$$

$$\begin{bmatrix} \Phi^{\circ}_{ij} & * \\ V^{\circ} - S^{\circ} + \delta\frac{A^{\circ}_i S^{\circ} + A^{\circ}_j S^{\circ}}{2} + \delta\frac{B^{\circ}_i Y^{\circ}_j + B^{\circ}_j Y^{\circ}_i}{2} C^{\circ} & -2\delta S^{\circ} \end{bmatrix} < 0, \tag{111}$$

$$C^{\circ} S^{\circ} = H^{\circ} C^{\circ}, \tag{112}$$

for $i = 1, 2, \ldots, s$, *as well as* $i = 1, 2, \ldots, s-1, j = 1+1, i+2, \ldots, s$, $h_i(\theta(t))h_j(\theta(t)) \neq 0$ *and*

$$\Phi^{\circ}_{ij} = \frac{A^{\circ}_i S^{\circ} + S^{\circ} A^{\circ T}_i}{2} + \frac{A^{\circ}_j S^{\circ} + S^{\circ} A^{\circ T}_j}{2} + \frac{B^{\circ}_i Y^{\circ}_j C^{\circ} + C^{\circ T} Y^{\circ T}_j B^{\circ T}_i}{2} + \frac{B^{\circ}_j Y^{\circ}_i C^{\circ} + C^{\circ T} Y^{\circ T}_i B^{\circ T}_j}{2}. \tag{113}$$

When the above conditions hold, the control law gain matrices are given as

$$K^{\circ}_i = Y^{\circ}_i (H^{\circ})^{-1}. \tag{114}$$

Proof. Since Eq. (82), Eq. (87) takes formally the same structure as Eqs. (6) and (7), following the same way as in the proof of Theorem 1, the conditions given in Theorem 4 can be obtained. From this reason, the proof is omitted. Compare, for example, Ref. [17].

Following the presented results, it is evident that the standard as well as the enhanced conditions for biproper dynamic output controller design can be derived from Theorem 3 and Theorem 4 in a simple way.

6. EXAMPLE

The nonlinear dynamics of the system is represented by TS model with s = 3 and the system model parameters [20]

$$A_1 = \begin{bmatrix} -1.0522 & -1.8666 & 0.5102 \\ -0.4380 & -5.4335 & 0.9205 \\ -0.5522 & 0.1334 & -0.4898 \end{bmatrix}, A_2 = \begin{bmatrix} -1.0565 & -1.8661 & 0.5116 \\ -0.4380 & -5.4359 & 0.9214 \\ -0.5565 & 0.1339 & -0.4884 \end{bmatrix},$$

$$A_3 = \begin{bmatrix} -1.0602 & -1.8657 & 0.5133 \\ -0.4381 & -5.4353 & 0.9216 \\ -0.5602 & 0.1343 & -0.4867 \end{bmatrix}, B = \begin{bmatrix} -0.1765 & 0.0000 \\ 0.0000 & 0.0000 \\ 0.1176 & 0.4721 \end{bmatrix}, C = \begin{bmatrix} 1 & 0 & 0 \\ 0 & 1 & 0 \end{bmatrix}.$$

To the state vector $\mathbf{q}(t)$ are associated the premise variables and the membership functions as follows

$$\theta(t) = \begin{bmatrix} \theta_1(t) \\ \theta_2(t) \\ \theta_3(t) \end{bmatrix}, \theta_i(t) = \begin{cases} \theta_1(t) \text{ if } q_1(t) \text{ is about } 2.5, & h_1(\theta_2(t)) = 1 - \frac{1}{2.5}|\theta_2(t) - 2.5| \\ \theta_2(t) \text{ if } q_1(t) \text{ is about } 0, & h_2(\theta_1(t)) = 1 - \frac{1}{2.5}|\theta_1(t)| \\ \theta_3(t) \text{ if } q_1(t) \text{ is about } -2.5, & h_3(\theta_3(t)) = 1 - \frac{1}{2.5}|\theta_3(t) + 2.5| \end{cases}$$

while the generalized premise variable is $\theta(t) = q_1(t)$.

Thus, solving Eqs. (46)–(49) for prescribed δ = 1.2 with respect to the LMI matrix variables **S, V H,** $\mathbf{Y}_i$, j = 1, 2, 3, and $\mathbf{X}_{ij}$, i, j = 1, 2, 3 using Self–Dual–Minimization (SeDuMi) package for Matlab [26], then the feedback gain matrix design problem was feasible with the results

$$S = \begin{bmatrix} 0.3899 & -0.0102 & -0.0000 \\ -0.0102 & 0.1596 & -0.0000 \\ -0.0000 & -0.0000 & 0.4099 \end{bmatrix}, V = \begin{bmatrix} 0.9280 & 0.1235 & -0.1525 \\ 0.1235 & 1.1533 & -0.3979 \\ -0.1525 & -0.3979 & 0.7574 \end{bmatrix},$$

$$H = \begin{bmatrix} 0.3899 & -0.0102 \\ -0.0102 & 0.1596 \end{bmatrix},$$

$$X = \begin{bmatrix} 0.4567 & 0.0983 & -0.0517 & 0.0694 & 0.0463 & -0.0174 & 0.0694 & 0.0463 & -0.0174 \\ 0.0983 & 0.7153 & -0.1118 & 0.0463 & 0.1906 & -0.0441 & 0.0463 & 0.1905 & -0.0440 \\ -0.0517 & -0.1118 & 0.1883 & -0.0175 & -0.0442 & 0.0143 & -0.0175 & -0.0442 & 0.0142 \\ 0.0694 & 0.0463 & -0.0175 & 0.4573 & 0.0981 & -0.0515 & 0.0695 & 0.0463 & -0.0174 \\ 0.0463 & 0.1906 & -0.0442 & 0.0981 & 0.7154 & -0.1115 & 0.0463 & 0.1905 & -0.0440 \\ -0.0174 & -0.0441 & 0.0143 & -0.0515 & -0.1115 & 0.1876 & -0.0175 & -0.0441 & 0.0142 \\ 0.0694 & 0.0463 & -0.0175 & 0.0695 & 0.0463 & -0.0175 & 0.4578 & 0.0978 & -0.0514 \\ 0.0463 & 0.1905 & -0.0442 & 0.0463 & 0.1905 & -0.0441 & 0.0978 & 0.7152 & -0.1111 \\ -0.0174 & -0.0440 & 0.0142 & -0.0174 & -0.0440 & 0.0142 & -0.0514 & -0.1111 & 0.1868 \end{bmatrix},$$

$$Y_1 = \begin{bmatrix} 0.5607 & -0.4590 \\ 0.1544 & -0.1191 \end{bmatrix}, Y_2 = \begin{bmatrix} 0.5558 & -0.4577 \\ 0.1579 & -0.1207 \end{bmatrix}, Y_3 = \begin{bmatrix} 0.5518 & -0.4566 \\ 0.1606 & -0.1222 \end{bmatrix},$$

Substituting the above parameters into Eq. (51) to solve the controller parameters, the following gain matrices are obtained

$$K_1 = \begin{bmatrix} 1.3653 & -2.7895 \\ 0.3772 & -0.7224 \end{bmatrix}, K_2 = \begin{bmatrix} 1.3530 & -2.7823 \\ 0.3860 & -0.7318 \end{bmatrix}, K_3 = \begin{bmatrix} 1.3428 & -2.7761 \\ 0.3925 & -0.7406 \end{bmatrix},$$

$$A_{c22} = \begin{bmatrix} -1.2953 & -1.3750 & 0.5116 \\ -0.4380 & -5.4359 & 0.9214 \\ -0.2152 & -0.5388 & -0.4884 \end{bmatrix}, A_{c31} = \begin{bmatrix} -1.3012 & -1.3734 & 0.5133 \\ -0.4381 & -5.4353 & 0.9216 \\ -0.2216 & -0.5348 & -0.4867 \end{bmatrix},$$

For simplicity, other closed-loop matrices of subsystem dynamics are not listed here.

Since the diagonal elements of $\mathbf{A}_{cij}$, i,j = 1, 2, 3, are dominant, in terms of Gerschgorin theorem [27, 28], all eigenvalues of $\mathbf{A}_{cij}$ are real, resulting in the aperiodic dynamics, that is,

$$\tilde{n}(\mathbf{A}_{c11}) = \{-0.6751, \ -1.0816, \ -5.4598\}, \quad \tilde{n}(\mathbf{A}_{c21}) = \{-0.6756, \ -1.0842, \ -5.4620\},$$
$$\tilde{n}(\mathbf{A}_{c31}) = \{-0.6757, \ -1.0861, \ -5.4613\}, \quad \tilde{n}(\mathbf{A}_{c12}) = \{-0.6745, \ -1.0805, \ -5.4593\},$$
$$\tilde{n}(\mathbf{A}_{c22}) = \{-0.6750, \ -1.0831, \ -5.4615\}, \quad \tilde{n}(\mathbf{A}_{c32}) = \{-0.6751, \ -1.0851, \ -5.4609\},$$
$$\tilde{n}(\mathbf{A}_{c13}) = \{-0.6742, \ -1.0795, \ -5.4588\}, \quad \tilde{n}(\mathbf{A}_{c23}) = \{-0.6748, \ -1.0820, \ -5.4610\},$$
$$\tilde{n}(\mathbf{A}_{c33}) = \{-0.6748, \ -1.0840, \ -5.4604\}.$$

Figure 1 gives the associated TS fuzzy static output control structure in a forced mode.

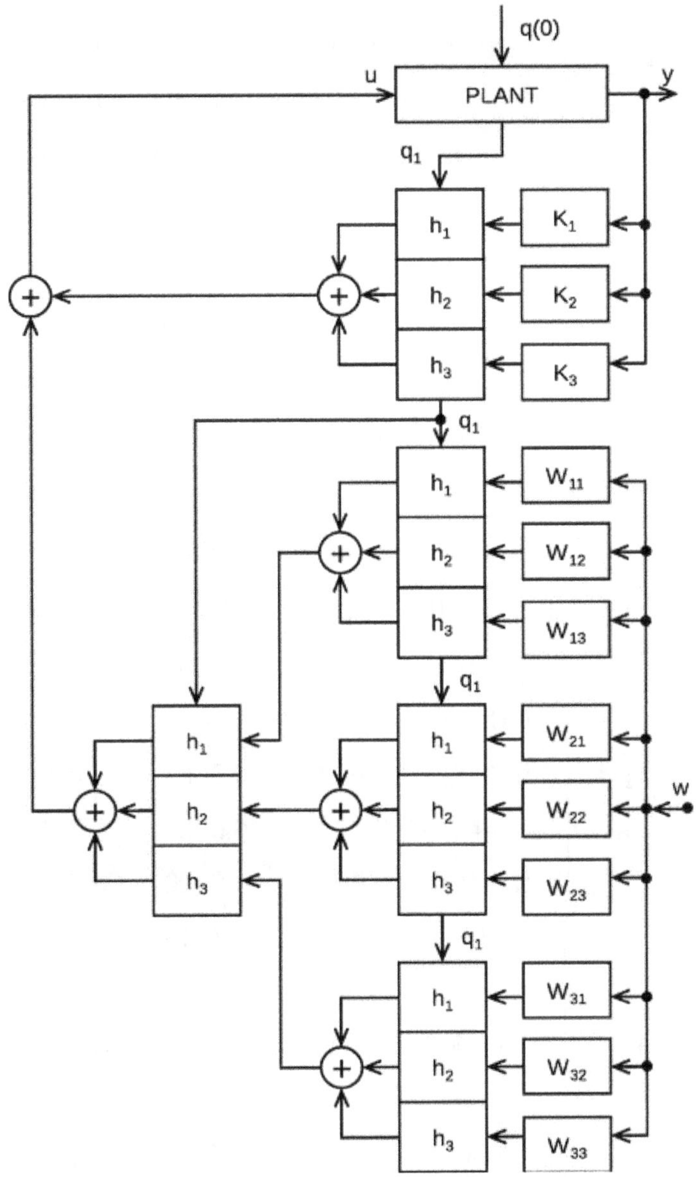

Figure 1. TS fuzzy static output control structure in a forced mode.

For Eqs. (88)–(91), it can find the following feasible solutions by using the given design procedure

$$S^{\circ} = \begin{bmatrix} 0.6194 & -0.0614 & 0.0000 & 0.0000 & 0.0000 & 0.0000 \\ -0.0614 & 0.1305 & 0.0000 & 0.0000 & 0.0000 & 0.0000 \\ 0.0000 & 0.0000 & 0.8724 & 0.0000 & 0.0000 & 0.0000 \\ 0.0000 & 0.0000 & 0.0000 & 0.7066 & 0.0000 & 0.0000 \\ 0.0000 & 0.0000 & 0.0000 & 0.0000 & 0.7066 & 0.0000 \\ 0.0000 & 0.0000 & 0.0000 & 0.0000 & 0.0000 & 0.7066 \end{bmatrix},$$

$$H^{\circ} = \begin{bmatrix} 0.7066 & 0.0000 & 0.0000 & 0.0000 & 0.0000 \\ 0.0000 & 0.7066 & 0.0000 & 0.0000 & 0.0000 \\ 0.0000 & 0.0000 & 0.7066 & 0.0000 & 0.0000 \\ 0.0000 & 0.0000 & 0.0000 & 0.6808 & -0.0614 \\ 0.0000 & 0.0000 & 0.0000 & 0.4889 & 0.0691 \end{bmatrix},$$

$$Y_1 = \begin{bmatrix} -0.5668 & 0.0000 & 0.0000 & 0.0000 & 0.0000 \\ 0.0000 & -0.5668 & 0.0000 & -0.0001 & 0.0000 \\ 0.0000 & 0.0000 & -0.5667 & 0.0000 & 0.0000 \\ 0.0000 & 0.0000 & 0.0000 & 0.3612 & -0.2783 \\ 0.0000 & 0.0000 & 0.0000 & 0.7396 & -1.1397 \end{bmatrix},$$

$$Y_2 = \begin{bmatrix} -0.5668 & 0.0000 & 0.0000 & 0.0000 & 0.0000 \\ 0.0000 & -0.5668 & 0.0000 & -0.0001 & 0.0000 \\ 0.0000 & 0.0000 & -0.5667 & 0.0000 & 0.0000 \\ 0.0000 & 0.0000 & 0.0000 & 0.3615 & -0.2784 \\ 0.0000 & 0.0000 & 0.0000 & 0.7397 & -1.1397 \end{bmatrix},$$

$$Y_3 = \begin{bmatrix} -0.5667 & -0.0001 & 0.0000 & 0.0000 & 0.0000 \\ -0.0001 & -0.5667 & 0.0000 & -0.0001 & 0.0000 \\ 0.0000 & 0.0000 & -0.5668 & 0.0000 & 0.0000 \\ 0.0000 & 0.0000 & 0.0000 & 0.3519 & -0.2859 \\ -0.0001 & 0.0000 & -0.0001 & 0.7486 & -1.1421 \end{bmatrix}$$

and, computing the biproper dynamic output controller parameters, then

$$J_1 = \begin{bmatrix} -0.8022 & 0.0000 & 0.0000 \\ 0.0000 & -0.8021 & 0.0000 \\ 0.0000 & 0.0000 & -0.8021 \end{bmatrix}, \quad L_1 = 10^{-3}\begin{bmatrix} 0.0394 & 0.0048 \\ -0.3143 & 0.3318 \\ 0.0221 & -0.0508 \end{bmatrix},$$

$$M_1 = 10^{-4}\begin{bmatrix} -0.2041 & 0.1504 & -0.0600 \\ -0.5318 & 0.0915 & -0.3275 \end{bmatrix}, \quad N_1 = \begin{bmatrix} 2.0889 & -2.1701 \\ 7.8914 & -9.4765 \end{bmatrix},$$

$$J_2 = \begin{bmatrix} -0.8022 & 0.0001 & 0.0000 \\ 0.0001 & -0.8022 & 0.0000 \\ 0.0000 & 0.0000 & -0.8021 \end{bmatrix}, \quad L_2 = 10^{-3}\begin{bmatrix} -0.2009 & 0.3531 \\ 0.0453 & -0.2017 \\ -0.1903 & 0.1765 \end{bmatrix},$$

$$M_2 = 10^{-4}\begin{bmatrix} -0.2022 & 0.1779 & 0.1796 \\ -0.4575 & 0.0413 & 0.2985 \end{bmatrix}, \quad N_2 = \begin{bmatrix} 2.0897 & -2.1707 \\ 7.8915 & -9.4766 \end{bmatrix},$$

$$J_3 = \begin{bmatrix} -0.8020 & -0.0001 & 0.0000 \\ -0.0001 & -0.8021 & 0.0000 \\ 0.0000 & 0.0000 & -0.8022 \end{bmatrix}, \quad L_3 = 10^{-3}\begin{bmatrix} -0.0641 & 0.0139 \\ -0.1516 & 0.2382 \\ -0.2116 & 0.2630 \end{bmatrix},$$

$$M_3 = 10^{-4}\begin{bmatrix} -0.0135 & 0.0020 & -0.0238 \\ -0.0917 & 0.0218 & -0.1102 \end{bmatrix}, \quad N_3 = \begin{bmatrix} 2.1286 & -2.2445 \\ 7.9148 & -9.4907 \end{bmatrix}.$$

It is evident that all matrices $\mathbf{J}_i$, i = 1.2.3 are Hurwitz, which rise up a TS fuzzy stable dynamic output controller, and based on the solutions obtained, the TS fuzzy dynamic controller can be designed via the concept of PDC.

Verifying the closed-loop stability, it can compute the eigenvalue spectra as follows

$$\tilde{n}(A_{c11}) = \{-0.8022,\ -0.8021,\ -0.8021,\ -4.3774,\ -1.2919 \pm 0.2804i\},$$

$$\tilde{n}(A_{c21}) = \{-0.8022,\ -0.8021,\ -0.8021,\ -4.3774,\ -1.2919 \pm 0.2804i\},$$

$$\tilde{n}(A_{c21}) = \{-0.8022,\ -0.8021,\ -0.8021,\ -4.3774,\ -1.2919 \pm 0.2804i\},$$

$$\tilde{n}(A_{c12}) = \{-0.8022,\ -0.8021,\ -0.8021,\ -4.3774,\ -1.2919 \pm 0.2805i\},$$

$$\tilde{n}(A_{c22}) = \{-0.8022,\ -0.8021,\ -0.8021,\ -4.3774,\ -1.2919 \pm 0.2805i\},$$

$$\tilde{n}(A_{c32}) = \{-0.8022,\ -0.8021,\ -0.8021,\ -4.3788,\ -1.2946 \pm 0.2963i\},$$

$$\tilde{n}(A_{c13}) = \{-0.8020,\ -0.8023,\ -0.8023,\ -4.3713,\ -1.2919 \pm 0.2797i\},$$

$$\tilde{n}(A_{c23}) = \{-0.8020,\ -0.8023,\ -0.8023,\ -4.3713,\ -1.2919 \pm 0.2797i\},$$

$$\tilde{n}(A_{c33}) = \{-0.8020,\ -0.8023,\ -0.8023,\ -4.3728,\ -1.2945 \pm 0.2958i\}.$$

7. CONCISE SUMMARY

This chapter presents a novel method for designing static and dynamic output feedback control, considering the affine characteristics of the TS fuzzy model structure. The implementation of fuzzy output control schemes associated with parallel-distributed output compensators demonstrates a way that markedly diminishes the conservativeness in control design constraints. The adequate existence conditions for both output controller realizations, influencing the global stability of the system, necessitate a parallel decentralized control framework that stabilizes the nonlinear system in accordance with Lyapunov's theory. The design of controller parameters, derived directly from these conditions, constitutes a viable numerical challenge. A further advantage of the method is that controllers utilize minimal feedback information concerning the desired system output, and the approach is sufficiently adaptable to accommodate additional design criteria. The approach's validity and usefulness are illustrated through numerical design examples.

AT A GLANCE

This chapter introduces novel conditions appropriate for the design of stabilizing static and dynamic output controllers for a category of continuous-time nonlinear systems characterized by Takagi-Sugeno models. Considering the affine characteristics of the TS model framework and implementing the fuzzy control methodology associated with parallel-distributed output compensators, the adequate design criteria are delineated in terms of linear matrix inequalities. According to the proposed methodologies, the Lyapunov matrix may be detached from the system parameter matrices by linear matrix inequality approaches, or a fuzzy-relaxed approach can be utilized to enhance the speed of closed-loop dynamics. Numerical examples exemplify the design techniques and showcase the efficacy of the offered design methods.

REFERENCES

1. Takagi T, Sugeno M. Fuzzy identification of systems and its applications to modeling and control. IEEE Transactions on Systems, Man, and Cybernetics. 1985;15(1):116–132. DOI: 10.1109/TSMC.1985.6313399.

2. Wang HO, Tanaka K, Griffin MF. An approach to fuzzy control of nonlinear systems: Stability and design issues. IEEE Transactions on Fuzzy Systems. 1996;4(1):14–23. DOI: 10.1109/91.481841.

3. Johansson M, Rantzer A, Arzen KE. Piecewise quadratic stability of fuzzy systems. Transactions on Fuzzy Systems. 1999;7(6):713–722. DOI: 10.1109/91.811241

4. Tanaka K, Wang HO. Fuzzy Control Systems Design and Analysis. A Linear Matrix Inequality Approach. New York: John Wiley & Sons; 2001:309 p.

5. Lam HK. Polynomial Fuzzy Model-Based Control Systems. Stability Analysis and Control Synthesis Using Membership Function-Dependent Techniques. Cham: Springer-Verlag; 2016:307 p.

6. Michels K, Klawonn F, Kruse R, Nürnberger N. Fuzzy Control. Fundamentals, Stability and Design of Fuzzy Controllers. Berlin: Springer-Verlag; 2006:416 p.

7. Abdelmalek I, Golea N, Hadjili ML. A new fuzzy Lyapunov approach to non-quadratic stabilization of Takagi-Sugeno fuzzy models. International Journal of Applied Mathematics and Computer Science. 2007;17(1):39–51. DOI: 10.2478/v10006-007-0005-4.

8. Krokavec D, Filasová A. Optimal fuzzy control for a class of nonlinear systems. Mathematical Problems in Engineering. 2012;2012. 29 p. DOI: 10.1155/2012/481942.

9. Pan J, Fei S, Ni Y, Xue M. New approaches to relaxed stabilization conditions and H-infinity control designs for T-S fuzzy systems. Journal of Control Theory and Applications. 2012;10(1):82–91. DOI: 10.1007/s11768-012-0088-9.

10. Boyd B, El Ghaoui L, Peron E, Balakrishnan V. Linear Matrix Inequalities in System and Control Theory. Philadelphia: SIAM;1994:205 p. DOI: http://dx.doi.org/10.1137/1.9781611970777.

11. Liu X, Zhang Q. New approaches to H_∞ controller designs based on fuzzy observers for T-S fuzzy systems via LMI. Automatica. 2003;39(9):1571-1582. http://dx.doi.org/10.1016/S0005-1098(03)00172-9.

12. Nguang SK, Shi P. H_∞ fuzzy output feedback control design for nonlinear systems: An LMI approach. IEEE Transactions on Fuzzy Systems. 2003;11(3):331–340. DOI: 10.1109/TFUZZ.2003.812691.

13. Kau SW, Lee HJ, Yang CM, Lee CH, Hong L, Fang CH. Robust H_∞ fuzzy static output feedback control of T-S fuzzy systems with parametric uncertainties. Fuzzy Sets and Systems. 2007;158(2):135–146. DOI: 10.1016/j.fss.2006.09.010.

14. Tognetti ES, Oliveira RCLF, Peres PLD. Improved stabilization conditions for Takagi-Sugeno fuzzy systems via fuzzy integral Lyapunov functions. In Proceedings of the 2011 American Control Conference; 29 Jun–01 Jul 2011; San Francisco, USA, pp. 4970–4975.

15. Chen BS, Tseng CS, Uang HJ. Mixed H_2/H_∞ fuzzy output feedback control design for nonlinear dynamic systems: An LMI approach. IEEE Transactions on Fuzzy Systems. 2000;8(3):249–265. DOI: 10.1109/91.855915.

16. Krokavec D, Filasová A. Stabilizing fuzzy output control for a class of nonlinear systems. Advances in Fuzzy Systems. 2013;2013:9 p. DOI:10.1155/2013/294971.
17. Krokavec D, Filasová A, Liščinský P. Dynamic output control of nonlinear systems described by Takagi-Sugeno models. In: Proceedings of the 2014 IEEE Multi-conference on Systems and Control MSC 2014; 8-10 October 2014; Antibes, France, pp. 959–964. DOI: 10.1109/CCA.2014.6981460.
18. Huang D, Nguang SK. Static output feedback controller design for fuzzy systems: An ILMI approach. Information Sciences. 2007;177(14):3005–3015. DOI: 10.1016/j.ins.2007.02.014.
19. Kim E, Lee H. New approaches to relaxed quadratic stability condition of fuzzy control systems. IEEE Transactions on Fuzzy Systems. 2000;8(5):523–534. DOI: 10.1109/91.873576.
20. Krokavec D, Filasová A. Relaxed design conditions for Takagi-Sugeno unknown input observers. In: Proceedings of the 15th IEEE International Symposium on Applied Machine Intelligence and Informatics SAMI 2017; 26-28 January 2017; Herlany, Slovakia:pp. 61–66.
21. Passino KM, Yurkovich S. Fuzzy Control. Berkeley: Addison-Wesley Longman; 1998:522 p.
22. Haddad WM, Chellaboina V. Nonlinear Dynamical Systems and Control. A Lyapunov-Based Approach. Princeton: Princeton University Press; 2008:976 p.
23. Kailath T. Linear Systems. Englewood Cliffs: Prentice-Hall; 1980:704 p.
24. Doyle JC, Glover K, Khargonekar PP, Francis BA. State-space solutions to standard H_2 and H_∞ control problems. IEEE Transactions on Automatic Control. 1989;34(8):831–847. DOI: 10.1109/9.29425.
25. Krokavec D, Filasová A, Serbák V FTC structures with virtual actuators and dynamic output controllers. IFAC-PapersOnLine. 2015;48(21):511–516. DOI: 10.1016/j.ifacol.2015.09.577.
26. Peaucelle D, Henrion D, Labit Y, Taitz K. User's Guide for SeDuMi Interface 1.04. Toulouse: LAAS-CNRS; 2002:36 p.
27. Gerschgorin S. On the limitation of the eigenvalues of a matrix. Bulletin of the Academy of Sciences of the USSR, Section of Mathematical and Natural Sciences. 1931;1931(6):749–754. (in German).
28. Feingold DG, Varga RS. Block diagonally dominant matrices and generalizations of the Gerschgorin circle theorem. Pacific Journal of Mathematics. 1962;12(4):1241–1250. DOI: 10.2140/pjm.1962.12.1241.

2 Chapter

GUARANTEED COST CONTROL OF NON-FRAGILE NONLINEAR SYSTEMS WITH VARIED STATE AND INPUT DELAYS UTILIZING T-S FUZZY LOCAL BILINEAR MODELS

1. INTRODUCTION: AN OVERVIEW

Recently, T-S (Takagi-Sugeno) model-based fuzzy control has garnered significant interest, primarily due to the fuzzy model's efficacy and adaptability in managing nonlinear systems [1–8]. Utilizing the sector nonlinearity approach, local approximation in fuzzy partition spaces, or several other approximation methods, T-S fuzzy models will be employed to estimate or accurately describe a nonlinear system within a limited set of state variables. The advantage of the model lies in the fact that the consequent component of a fuzzy rule constitutes a linear dynamic subsystem, hence enabling the application of established linear systems theory to nonlinear systems. Additionally, the comprehensive fuzzy model will be derived through the use of the fuzzy inference approach. A fuzzy controller is developed using the method known as 'parallel distributed compensation (PDC)' [3–6], which fundamentally involves implementing a corresponding linear controller for each linear subsystem. The comprehensive nonlinear controller is derived from the fuzzy integration of each distinct linear controller. The T-S fuzzy model has been extensively researched, yielding numerous results [1–8]. Time delay frequently arises in various dynamic systems, including biological and network systems, in practical applications. The presence of delays typically leads to instability and a decline in system performance [3–8]. Typically, the computation of delay-dependent results is complicated by the introduction of the inner product between two vectors. To mitigate this issue, several model modifications were employed in numerous studies; nonetheless, this may lead to the emergence of inequality, leading in potential conservatism. Conversely, various circumstances, including limited word length, truncation errors in numerical computations, and variations in electronic component parameters, will cause alterations in the controller's parameters, resulting in imprecision in its implementation. In this instance, minor variations in the coefficients of the controllers may render the planned controllers sensitive and potentially destabilize the closed-loop control system [9]. The subject of non-fragile control has become significant. Recently, non-fragile control research has garnered significant attention, resulting in a series of findings [10–13]. Bilinear models are extensively utilized in many physical systems, biotechnology, socioeconomics, and dynamic processes across multiple engineering disciplines [14, 15]. The bilinear model is a specific type of nonlinear model, characterized by a nonlinear component that comprises the bilinear function of the state and input. Bilinear models possess two primary advantages over linear models. The bilinear model can more accurately mimic a nonlinear system.. Another is that because of nonlinearity of it, many real physical processes may be appropriately modeled as bilinear systems. A famous example of a bilinear system is the population of biological species, which can be showed by $\frac{d\theta}{dt} = \theta v$. In this equation, v is the birth rate minus death rate, and θ denotes the population. Obviously, the equation cannot be approximated by a linear model [14].

Most of the existing results focus on the stability analysis and synthesis based on T-S fuzzy model with linear local model. However, when a nonlinear system has of complex nonlinearities, the constructed T-S model will consist of a number of fuzzy local models. This will lead to very heavy computational burden. According to the advantages of bilinear systems and T-S fuzzy control, so many researchers paid their attentions to the T-S fuzzy models with bilinear rule consequence [16–18]. From these papers, it is evident that the T-S fuzzy bilinear model may be suitable for some classes of nonlinear plants. In Ref. [16], a nonlinear system was transformed into a bilinear model via Taylor's series expansion, and the stability of T-S fuzzy bilinear model was studied. Moreover, the result was stretched into the complex fuzzy system with state time delay [17]. Ref. [18] presented robust stabilization for a class of discrete-time fuzzy bilinear system. Very recently, a class of nonlinear systems is described by T-S fuzzy models with nonlinear local models in Ref. [19], and in this paper, the scholars put forward a new fuzzy control scheme with local nonlinear feedbacks, the advantage of which over the existing methods is that a fewer fuzzy rules and less computational burden. The non-fragile guaranteed cost controller was designed for a class of T-S discrete-time fuzzy bilinear systems in Ref. [20]. However, in Refs. [19, 20], the time-delay effects on the system is not considered. Ref. [17] is only considered the fuzzy system with the delay in the state and the derivatives of time-delay, $\dot{d}(t)<1$ is required. Refs. [21–23] dealt with the uncertain fuzzy systems with time-delay in different ways. It should be pointed out that all the aforementioned works did not take into account the effect of the control input delays on the systems. The results therein are not applicable to systems with input delay. Recently, some controller design approaches have been presented for systems with input delay, see [2, 3, 4, 18, 24–32] for fuzzy T-S systems and [8, 15, 33, 34] for non-fuzzy systems and the references therein. All of these results are required to know the exact delay values in the implementation. T-S fuzzy stochastic systems with state time-vary or distributed delays were studied in Refs. [35–39]. The researches of fractional order T-S fuzzy systems on robust stability, stability analysis about "$0 < \alpha < 1$", and decentralized stabilization in multiple time delays were presented in Refs. [40–42], respectively. For different delay types, the corresponding adaptive fuzzy controls for nonlinear systems were proposed in Refs. [33, 43, 44]. In Refs. [45, 46], to achieve small control amplitude, a new T-S fuzzy hyperbolic model was developed, moreover, Ref. [46] considered the input delay of the novel model. In Ref. [25, 47], the problems of observer-based fuzzy control design for T-S fuzzy systems were concerned.

So far, the problem of non-fragile guaranteed cost control for fuzzy system with local bilinear model with different time-varying state and input delays has not been discussed.

In this paper, the problem of delay-dependent non-fragile guaranteed cost control is studied for the fuzzy time-varying delay systems with local bilinear model and different state and input delays. Based on the PDC scheme, new delay-dependent stabilization conditions for the closed-loop fuzzy systems are derived. No model transformation is involved in the derivation. The merit of the proposed conditions lies in its reduced conservatism, which is achieved by circumventing the utilization of some bounding inequalities for the cross-product between two vectors as in Ref. [17]. The three main contributions of this paper are the following: (1) a non-fragile guaranteed cost controller is presented for the fuzzy system with time-varying delay in both state and input; (2) some free-weighting matrices are introduced in the derivation process, where the constraint of the derivatives of time-delay, $\dot{d}(t)<1$ and $\dot{h}(t)<1$, is eliminated; and (3) the delay-dependent stability conditions for the fuzzy system are described by LMIs. Finally, simulation examples are given to illustrate the effectiveness of the obtained results.

The paper is organized as follows. Section 2 introduces the fuzzy delay system with local bilinear model, and non-fragile controller law for such system is designed based on the parallel distributed compensation approach in Section 3. Results of non-fragile guaranteed cost control are given in Section 4. Two simulation examples are used to illustrate the effectiveness of the proposed method in Section 5, which is followed by conclusions in Section 6.

Notation: Throughout this paper, the notation $P > 0 (P \geq 0)$ stands for P being real symmetric and positive definite (or positive semi-definite). In symmetric block matrices, the asterisk (*) refers to a term that is induced by symmetry, and diag{....} denotes a block-diagonal matrix. The superscript T means matrix transposition. The notion $\sum_{i,j=1}^{s}$ is an abbreviation of $\sum_{i=1}^{s}\sum_{j=1}^{s}$. Matrices, if the dimensions are not explicitly stated, are assumed to have compatible dimensions for algebraic operations.

2. SYSTEM CHARACTERIZATION AND PRESUPPOSITIONS

In this section, we introduce the T-S fuzzy time-delay system with local bilinear model. The i th rule of the fuzzy system is represented by the following form:

Plant Rule i:

$$\begin{aligned} &IF\ \vartheta_1(t)\ is\ F_{i1}\ and\ \ldots\ and\ \vartheta_v(t)\ is\ F_{iv},\ THEN \\ &\dot{x}(t) = A_i x(t) + A_{di}x(t-d(t)) + B_i u(t) + B_{hi}u(t-h(t)) + N_i x(t)u(t) + N_{di}x(t-d(t))u(t-h(t)) \\ &\quad x(t) = \phi(t), \quad t \in [-\tau_1, 0]\,, i = 1, 2, \ldots, s \end{aligned} \tag{1}$$

where F_{ij} is the fuzzy set, s is the number of fuzzy rules, $x(t) \in R^n$ is the state vector, and $u(t) \in R$ is the control input, $\theta_1(t), \theta_2(t), \ldots, \theta_v(t)$ are the premise variables. It is assumed that the premise variables do not depend on the input $u(t)$. $A_i, {}_i, N_i, N_{di} \in R^{n\times n}, Bi, B_{hi} \in R^{n\times 1}$ denote the system matrices with appropriate dimensions. $d(t)$ is a time-varying differentiable function that satisfies $0 \leq d(t) \leq \tau_1$, $0 \leq h(t) \leq \tau_2$, where τ_1, τ_2 are real positive constants as the upper bound of the time-varying delay. It is also assumed that $d'(t) \leq \sigma_1$, $h'(t) \leq \sigma_2$, and σ_1, σ_2 are known constants. The initial conditions $\varphi(t)$, $\varphi(t)$ are continuous functions of t, $t \in [-\tau, 0], \tau = \min(\tau_1, \tau_2)$.

Remark 1: The fuzzy system with time-varying state and input delays will be investigated in this paper, which is different from the system in Ref. [17]. In Ref. [17], only state time-varying delay is considered. And also, here, we assume that the derivative of time-varying delay is less than or equal to a known constant that may be greater than 1; the assumption on time-varying delay in Ref. [17] is relaxed.

By using singleton fuzzifier, product inferred and weighted defuzzifier, the fuzzy system can be expressed by the following globe model:

$$\begin{aligned} \dot{x}(t) = &\sum_{i=1}^{s} h_i(\vartheta(t))[A_i x(t) + A_{di}x(t-d(t)) + B_i u(t) + B_{hi}u(t-h(t)) + N_i x(t)u(t) \\ &+ N_{di}x(t-d(t))u(t-h(t))] \end{aligned} \tag{2}$$

where

$h_i(\vartheta(t)) = \omega_i(\vartheta(t))/\sum_{i=1}^{s}\omega_i(\vartheta(t)), \omega_i(\vartheta(t)) = \prod_{j=1}^{v}\mu_{ij}(\vartheta(t)), \mu_{ij}(\vartheta(t))$ is the grade of membership of $\theta_i(t)$ in F_{ij}. In this paper, it is assumed that $\omega_i(\vartheta(t)) \geq 0, \sum_{i=1}^{s}\omega_i(\vartheta(t)) > 0$ for all t. Then, we have the following conditions $h_i(\vartheta(t)) \geq 0, \sum_{i=1}^{s} h_i(\vartheta(t)) = 1$ for all t. In the consequent, we use abbreviation $h_i, h_{hi}, x_d(t), u_d(t), x_h(t), u_h(t),$ to replace $h_i(\vartheta(t)), h_i(\vartheta(t-h(t))), x(t-d(t)), u(t-d(t)), x(t-h(t)), u(t-h(t)),$ respectively, for convenience.

The objective of this paper is to design a state-feedback non-fragile guaranteed cost control law for the fuzzy system (2).

3. DESIGN OF A ROBUST GUARANTEED COST CONTROLLER

Extending the design concept in Ref. [17], we give the following non-fragile fuzzy control law:

$$IF\ \vartheta_1(t)\ is\ F_1^i\ and\ \ldots\ and\ \vartheta_v(t)\ is\ F_v^i$$
$$THEN\ u(t) = \frac{\rho(K_i + \Delta K_i)x(t)}{\sqrt{1 + x^T(K_i + \Delta K_i)^T(K_i + \Delta K_i)x}} = \rho \sin \theta_i = \rho \cos \theta_i (K_i + \Delta K_i)x(t) \tag{3}$$

where ρ > 0 is a scalar to be assigned, and $K_i \in R^{l\times n}$ is a local controller gain to be determined. ΔK_i represents the additive controller gain perturbations of the form $\Delta K_i = H_i F_i(t) E_{ki}$ with H_i and E_{ki} being known constant matrices, and $F_i(t)$ the uncertain parameter matrix satisfying

$$F_i^T(t)F_i(t) \leq I.\ \sin\theta_i = \frac{\bar{K}_i x(t)}{\sqrt{1+x^T\bar{K}_i^T\bar{K}_i x}},\quad \cos\theta_i = \frac{1}{\sqrt{1+x^T\bar{K}_i^T\bar{K}_i x}},\quad \theta_i \in \left[-\tfrac{\pi}{2}, \tfrac{\pi}{2}\right],\ \bar{K}_i = K_i + \Delta K_i(t) = K_i + H_i F_i(t) E_{ki}.$$

The overall fuzzy control law can be represented by

$$u(t) = \sum_{i=1}^{s} h_i \frac{\rho \bar{K}_i x(t)}{\sqrt{1 + x^T \bar{K}_i^T \bar{K}_i x}} = \sum_{i=1}^{s} h_i \rho \sin \theta_i = \sum_{i=1}^{s} h_i \rho \cos \theta_i \bar{K}_i x(t) \tag{4}$$

When there exists an input delay $h(t)$, we have that

$$u_h(t) = \sum_{l=1}^{s} h_{hl} \rho \sin \varphi_l = \sum_{l=1}^{s} h_{hl} \rho \cos \varphi_l \tilde{K}_l x_h(t) \tag{5}$$

where

$$\varphi_l = \frac{\tilde{K}_l x_h(t)}{\sqrt{1+x_h^T\tilde{K}_l^T\tilde{K}_l x_h}},\quad \cos\varphi_l = \frac{1}{\sqrt{1+x_h^T\tilde{K}_l^T\tilde{K}_l x_h}},\quad \varphi_l \in \left[-\tfrac{\pi}{2}, \tfrac{\pi}{2}\right],\quad \tilde{K}_l = K_l + \Delta K_l(t-h(t)) = K_l + H_l F_l(t-h(t)) E_{kl}.$$

So, it is natural and necessary to make an assumption that the functions h_i are well defined all $t \in [-\tau_2, 0]$, and satisfy the following properties:

$h(\vartheta(t-h(t))) \geq 0$, for i = 1,2,...,s, and $\sum_{i=1}^{s} h_i(\vartheta(t-h(t))) = 1$.

By substituting Eq. (5) into Eq. (2), the closed-loop system can be given by

$$\dot{x}(t) = \sum_{i,j,l=1}^{s} h_i h_j h_{hl}(\Lambda_{ij}x(t) + \Lambda_{dij}x_d(t) + \Lambda_{hil}x_h(t)) \tag{6}$$

where

$$\Lambda_{ij} = A_i + \rho \sin\theta_j N_i + \rho \cos\theta_j B_i \overline{K}_j,\ \Lambda_{dil} = A_{di} + \rho \sin\varphi_l N_{di},\ \Lambda_{hil} = \rho \cos\varphi_l B_{hi}\tilde{K}_l.$$

Given positive-definite symmetric matrices $S \in R^{n\times n}$ and $W \in R$, we take the cost function

$$J = \int_0^\infty [x^T(t)Sx(t) + u^T(t)Wu(t)]dt \tag{7}$$

Definition 1. The fuzzy non-fragile control law $u(t)$ is said to be non-fragile guaranteed cost if for the system (2), there exist control laws (4) and (5) and a scalar J_0 such that the closed-loop system (6) is asymptotically stable and the closed-loop value of the cost function (7) satisfies $J \le J_0$.

4. ANALYSIS OF STABILITY FOR THE CLOSED-LOOP SYSTEM

Firstly, the following lemmas are presented which will be used in the paper.

Lemma 1 [20]: *Given any matrices M and N with appropriate dimensions such that* $\varepsilon > 0$, *we have* $M^T N + N^T M \le \varepsilon M^T M + \varepsilon^{-1} N^T N$.

Lemma 2 [21]: *Given constant matrices G, E and a symmetric constant matrix S of appropriate dimensions. The inequality* $S + GFE + E^T F^T G^T < 0$ *holds, where* $F(t)$ *satisfies* $F^T(t) F(t) \le I$ *if and only if, for some* $\varepsilon > 0$, $S + \varepsilon GG^T + \varepsilon^{-1}E^T E < 0$.

The following theorem gives the sufficient conditions for the existence of the non-fragile guaranteed cost controller for system (6) with additive controller gain perturbations.

Theorem 1. *Consider system (6) associated with cost function (7). For given scalars* $\rho > 0$, $\tau_1 > 0, \tau_2 > 0, \sigma 1 > 0, \sigma_2 > 0$, *if there exist matrices* $P > 0$, $Q_1 > 0$, $Q2 > 0$, $R_1 > 0$, $R_2 > 0$, K_i, $i = 1, 2,\ldots, s$, $X_1, X_2, X_3, X_4, Y_1, Y_2, Y_3, Y_4$, *and scalar* $\varepsilon > 0$ *satisfying the inequalities (8), the system (6) is asymptotically stable and the control law (5) is a fuzzy non-fragile guaranteed cost control law, moreover,*

$$J \le x^T(0)Px(0) + \int_{-d(0)}^{0} x^T(s)Q_1x(s)ds + \int_{-\tau_1}^{0}\int_{\theta}^{0} \dot{x}^T(s)R_1\dot{x}(s)dsd\theta$$
$$+ \int_{-h(0)}^{0} x^T(s)Q_2x(s)ds + \int_{-\tau_2}^{0}\int_{\theta}^{0} \dot{x}^T(s)R_2\dot{x}(s)dsd\theta = J_0$$

$$\begin{bmatrix} T_{ijl} & * & * \\ \tau_1 X^{\mathrm{T}} & -\tau_1 R_1 & * \\ \tau_2 Z^{\mathrm{T}} & 0 & -\tau_2 R_2 \end{bmatrix} < 0, \quad i, j, l = 1, 2, \ldots, s \tag{8}$$

where

$$T_{ijl} = \begin{bmatrix} T_{11,ij} & * & * & * \\ T_{21,i} & T_{22,i} & * & * \\ T_{31,i} & T_{32,ij} & T_{33,il} & * \\ T_{41,i} & T_{42,i} & T_{43} & T_{44} \end{bmatrix},$$

$$[x^T(t)X_1 + x_d^T(t)X_2 + x_h^T(t)X_3 + \dot{x}^T(t)X_4][x(t) - x_d(t) - \int_{t-d(t)}^{t} \dot{x}(s)ds] \equiv 0,$$

$$[x^T(t)Z_1 + x_d^T(t)Z_2 + x_h^T(t)Z_3 + \dot{x}^T(t)Z_4][x(t) - x_h(t) - \int_{t-h(t)}^{t} \dot{x}(s)ds] \equiv 0,$$

$$\sum_{i,j=1}^{s} h_i h_j h_l [x^T(t)Y_1 + x_d^T(t)Y_2 + x_h^T(t)Y_4 + \dot{x}^T(t)Y_4][\Lambda_{ij}x(t) + \Lambda_{dil}x_d(t) + \Lambda_{hi}$$

Proof: Take a Lyapunov function candidate as

$$\begin{aligned} V(x(t), t) = {} & x^{\mathrm{T}}(t)Px(t) + \int_{t-d(t)}^{t} x^{\mathrm{T}}(s)Q_1 x(s)ds + \int_{-\tau_1}^{0}\int_{t+\theta}^{t} \dot{x}^{\mathrm{T}}(s)R_1\dot{x}(s)dsd\theta \\ & + \int_{t-h(t)}^{t} x^{\mathrm{T}}(s)Q_2 x(s)ds + \int_{-\tau_2}^{0}\int_{t+\theta}^{t} \dot{x}^{\mathrm{T}}(s)R_2\dot{x}(s)dsd\theta \end{aligned} \tag{9}$$

The time derivatives of $V(x(t),t)$, along the trajectory of the system (6), are given by

$$\begin{aligned} & \dot{V}(x(t), t) = 2x^{\mathrm{T}}(t)P\dot{x}(t) + x^{\mathrm{T}}(t)(Q_1 + Q_2)x(t) \\ & -(1 - \dot{d}(t))x_d^{\mathrm{T}}(t)Q_1 x_d(t) + \dot{x}^{\mathrm{T}}(t)(\tau_1 R_1 + \tau_2 R_2)\dot{x}(t) \\ & -\int_{t-\tau_1}^{t} \dot{x}^{\mathrm{T}}(s)R_1\dot{x}(s)ds - (1 - \dot{h}(t))x_h^{\mathrm{T}}(t)Q_2 x_h(t) - \int_{t-\tau_2}^{t} \dot{x}^{\mathrm{T}}(s)R_2\dot{x}(s)ds \end{aligned} \tag{10}$$

Define the free-weighting matrices as $X = [X_1^{\mathrm{T}} \ \ X_2^{\mathrm{T}} \ \ X_3^{\mathrm{T}} \ \ X_4^{\mathrm{T}}]^{\mathrm{T}}$, $Y = [Y_1^{\mathrm{T}} \ \ Y_2^{\mathrm{T}} \ \ Y_3^{\mathrm{T}} \ \ Y_4^{\mathrm{T}}]^{\mathrm{T}}$, $Z = [Z_1^{\mathrm{T}} \ \ Z_2^{\mathrm{T}} \ \ Z_3^{\mathrm{T}} \ \ Z_4^{\mathrm{T}}]^{\mathrm{T}}$, where $X_k \in R^{n\times n}$, $Y_k \in R^{n\times n}$, $Z_k \in R^{n\times n}$, $k = 1, 2, 3, 4$ will be determined later.

Using the Leibniz-Newton formula and system equation (6), we have the following identical equations:

$$
\begin{aligned}
&[x^T(t)X_1 + x_d^T(t)X_2 + x_h^T(t)X_3 + \dot{x}^T(t)X_4][x(t) - x_d(t) - \int_{t-d(t)}^{t} \dot{x}(s)ds] \equiv 0, \\
&[x^T(t)Z_1 + x_d^T(t)Z_2 + x_h^T(t)Z_3 + \dot{x}^T(t)Z_4][x(t) - x_h(t) - \int_{t-h(t)}^{t} \dot{x}(s)ds] \equiv 0, \\
&\sum_{i,j=1}^{s} h_i h_j h_l [x^T(t)Y_1 + x_d^T(t)Y_2 + x_h^T(t)Y_4 + \dot{x}^T(t)Y_4][\Lambda_{ij}x(t) + \Lambda_{dil}x_d(t) + \Lambda_{hil}x_h(t) - \dot{x}(t)] \equiv 0
\end{aligned}
\tag{11}
$$

Then, substituting Eq. (12) into Eq. (11) yields

$$
\begin{aligned}
\dot{V}(x(t),t) &= 2x^{\mathrm{T}}(t)P\dot{x}(t) + x^{\mathrm{T}}(t)(Q_1+Q_2)x(t) + \dot{x}^{\mathrm{T}}(t)(\tau_1 R_1 + \tau_2 R_2)\dot{x}(t) \\
&\quad -(1-\dot{d}(t))x_d^{\mathrm{T}}(t)Q_1 x_d(t) - (1-\dot{h}(t))x_h^{\mathrm{T}}(t)Q_2 x_h(t) \\
&\quad - \int_{t-\tau_1}^{t} \dot{x}^{\mathrm{T}}(s)R_1\dot{x}(s)ds + 2\eta^{\mathrm{T}}(t)X[x(t) - x_d(t) - \int_{t-d(t)}^{t} \dot{x}(s)ds] \\
&\quad - \int_{t-\tau_2}^{t} \dot{x}^{\mathrm{T}}(s)R_2\dot{x}(s)ds + 2\eta^{\mathrm{T}}(t)Z[x(t) - x_h(t) - \int_{t-h(t)}^{t} \dot{x}(s)ds] \\
&\quad + 2\eta^{\mathrm{T}}(t)Y\sum_{i,j,l=1}^{s} h_i h_j h_{hl}[\Lambda_{ij}x(t) + \Lambda_{dil}x_d(t) + \Lambda_{hil}x_h(t) - \dot{x}(t)] \\
&\le 2x^{\mathrm{T}}(t)P\dot{x}(t) + x^{\mathrm{T}}(t)(Q_1+Q_2)x(t) + \dot{x}^{\mathrm{T}}(t)(\tau_1 R_1 + \tau_2 R_2)\dot{x}(t) \\
&\quad -(1-\sigma_1)x_d^{\mathrm{T}}(t)Q_1 x_d(t) - (1-\sigma_2)x_h^{\mathrm{T}}(t)Q_2 x_h(t) \\
&\quad - \int_{t-d(t)}^{t} \dot{x}^{\mathrm{T}}(s)R_1(s)\dot{x}(s)ds + 2\eta^{\mathrm{T}}(t)X[x(t) - x_d(t) - \int_{t-d(t)}^{t} \dot{x}(s)ds] \\
&\quad - \int_{t-h(t)}^{t} \dot{x}^{\mathrm{T}}(s)R_2(s)\dot{x}(s)ds + 2\eta^{\mathrm{T}}(t)Z[x(t) - x_h(t) - \int_{t-h(t)}^{t} \dot{x}(s)ds] \\
&\quad + 2\eta^{\mathrm{T}}(t)Y\sum_{i,j,l=1}^{s} h_i h_j h_{hl}[\Lambda_{ij}x(t) + \Lambda_{dil}x_d(t) + \Lambda_{hil}x_h(t) - \dot{x}(t)] + x^{\mathrm{T}}(t)Sx(t) \\
&\quad + \sum_{i,j=1}^{s} h_i h_j \rho^2 x^{\mathrm{T}}(t)\overline{K}_i^{\mathrm{T}}\cos\theta_i W\overline{K}_j\cos\theta_j x(t) - [x^{\mathrm{T}}(t)Sx(t) + u^{\mathrm{T}}(t)Wu(t)]
\end{aligned}
\tag{12}
$$

where $\eta(t) = [x^{\mathrm{T}}(t),\ x_d^{\mathrm{T}}(t),\ x_h^{\mathrm{T}}(t),\ \dot{x}^{\mathrm{T}}(t)]^{\mathrm{T}}$.

Applying Lemma 1, we have the following inequalities:

$$
\begin{aligned}
&2x^{\mathrm{T}}(t)Y_1\Lambda_{ij}x(t) \le 2x^{\mathrm{T}}(t)Y_1A_ix(t) + \varepsilon\rho^2x^{\mathrm{T}}(t)Y_1{Y_1}^{\mathrm{T}}x(t) + \varepsilon^{-1}x^{\mathrm{T}}(t)(N_i^{\mathrm{T}}N_i + (B_i\overline{K}_j)^{\mathrm{T}}(B_i\overline{K}_j))x(t),\\
&2x^{\mathrm{T}}(t)Y_1\Lambda_{dil}x_d(t) \le 2x^{\mathrm{T}}(t)Y_1A_{di}x_d(t) + \varepsilon\rho^2\sin^2\phi_l x^{\mathrm{T}}(t)Y_1{Y_1}^{\mathrm{T}}x(t) + \varepsilon^{-1}x_d^{\mathrm{T}}(t)N_{di}^{\mathrm{T}}N_{di}x_d(t),\\
&2x^{\mathrm{T}}(t)Y_1\Lambda_{hil}x_h(t) \le \varepsilon\rho^2\cos^2\phi_l x^{\mathrm{T}}(t)Y_1{Y_1}^{\mathrm{T}}x(t) + \varepsilon^{-1}x_h^{\mathrm{T}}(t)(B_{hi}\tilde{K}_l)^{\mathrm{T}}(B_{hi}\tilde{K}_l)x_h(t),\\
&2x_d^{\mathrm{T}}(t)Y_2\Lambda_{ij}x(t) \le 2x_d^{\mathrm{T}}(t)Y_2A_ix(t) + \varepsilon\rho^2x_d^{\mathrm{T}}(t)Y_2{Y_2}^{\mathrm{T}}x_d(t) + \varepsilon^{-1}x^{\mathrm{T}}(t)(N_i^{\mathrm{T}}N_i + (B_i\overline{K}_j)^{\mathrm{T}}(B_i\overline{K}_j))x(t),\\
&2x^{\mathrm{T}}(t)Y_1\Lambda_{hil}x_h(t) \le \varepsilon\rho^2\cos^2\phi_l x^{\mathrm{T}}(t)Y_1{Y_1}^{\mathrm{T}}x(t) + \varepsilon^{-1}x_h^{\mathrm{T}}(t)(B_{hi}\tilde{K}_l)^{\mathrm{T}}(B_{hi}\tilde{K}_l)x_h(t),\\
&2x_d^{\mathrm{T}}(t)Y_2\Lambda_{ij}x(t) \le 2x_d^{\mathrm{T}}(t)Y_2A_ix(t) + \varepsilon\rho^2x_d^{\mathrm{T}}(t)Y_2{Y_2}^{\mathrm{T}}x_d(t) + \varepsilon^{-1}x^{\mathrm{T}}(t)(N_i^{\mathrm{T}}N_i + (B_i\overline{K}_j)^{\mathrm{T}}(B_i\overline{K}_j))x(t),\\
&2x_d^{\mathrm{T}}(t)Y_2\Lambda_{dil}x_d(t) \le 2x_d^{\mathrm{T}}(t)Y_2A_{di}x_d(t) + \varepsilon\rho^2\sin^2\phi_l x_d^{\mathrm{T}}(t)Y_2{Y_2}^{\mathrm{T}}x_d(t) + \varepsilon^{-1}x_d^{\mathrm{T}}(t)N_{di}^{\mathrm{T}}N_{di}x_d(t),\\
&2x_d^{\mathrm{T}}(t)Y_2\Lambda_{hil}x_h(t) \le \varepsilon\rho^2\cos^2\phi_l x_d^{\mathrm{T}}(t)Y_2{Y_2}^{\mathrm{T}}x_d(t) + \varepsilon^{-1}x_h^{\mathrm{T}}(t)(B_{hi}\tilde{K}_l)^{\mathrm{T}}(B_{hi}\tilde{K}_l)x_h(t),\\
&2x_h^{\mathrm{T}}(t)Y_3\Lambda_{ij}x(t) \le 2x_h^{\mathrm{T}}(t)Y_3A_ix(t) + \varepsilon\rho^2x_h^{\mathrm{T}}(t)Y_3{Y_3}^{\mathrm{T}}x_h(t) + \varepsilon^{-1}x^{\mathrm{T}}(t)(N_i^{\mathrm{T}}N_i + (B_i\overline{K}_j)^{\mathrm{T}}(B_i\overline{K}_j))x(t),\\
&2x_h^{T}(t)Y_3\Lambda_{dil}x_d(t) \le 2x_h^{T}(t)Y_3A_{di}x_d(t) + \varepsilon\rho^2\sin^2\phi_l x_h^{T}(t)Y_3{Y_3}^{\mathrm{T}}x_h(t) + \varepsilon^{-1}x_d^{T}(t)N_{di}^{T}N_{di}x_d(t),\\
&2x_h^{\mathrm{T}}(t)Y_3\Lambda_{hil}x_h(t) \le \varepsilon\rho^2\cos^2\phi_l x_h^{\mathrm{T}}(t)Y_3{Y_3}^{\mathrm{T}}x_h(t) + \varepsilon^{-1}x_h^{\mathrm{T}}(t)(B_{hi}\tilde{K}_l)^{\mathrm{T}}(B_{hi}\tilde{K}_l)x_h(t),\\
&2\dot{x}^{\mathrm{T}}(t)Y_4\Lambda_{ij}x(t) \le 2\dot{x}^{\mathrm{T}}(t)Y_4A_ix(t) + \varepsilon\rho^2\dot{x}^{\mathrm{T}}(t)Y_4{Y_4}^{\mathrm{T}}\dot{x}(t) + \varepsilon^{-1}x^{\mathrm{T}}(t)(N_i^{\mathrm{T}}N_i + (B_i\overline{K}_j)^{\mathrm{T}}(B_i\overline{K}_j))x(t),\\
&2\dot{x}^{\mathrm{T}}(t)Y_4\Lambda_{dil}x_d(t) \le 2\dot{x}^{\mathrm{T}}(t)Y_4A_{di}x_d(t) + \varepsilon\rho^2\dot{x}^{\mathrm{T}}(t)Y_4{Y_4}^{\mathrm{T}}\dot{x}(t) + \varepsilon^{-1}x_d^{\mathrm{T}}(t)N_{di}^{\mathrm{T}}N_{di}x_d(t),\\
&2\dot{x}^{\mathrm{T}}(t)Y_4\Lambda_{hil}x_h(t) \le \varepsilon\rho^2\cos^2\phi_l\dot{x}^{\mathrm{T}}(t)Y_4{Y_4}^{\mathrm{T}}\dot{x}(t) + \varepsilon^{-1}x_h^{\mathrm{T}}(t)(B_{hi}\tilde{K}_l)^{\mathrm{T}}(B_{hi}\tilde{K}_l)x_h(t)
\end{aligned}
\tag{13}
$$

Substituting Eq. (13) into Eq. (12) results in

$$
\begin{aligned}
\dot{V}(x(t),t) \le{}& \sum_{i,j,l=1}^{s} h_ih_jh_{hl}\eta^{\mathrm{T}}(t)T_{ijl}\eta(t) - \int_{t-d(t)}^{t}\dot{x}^{\mathrm{T}}(s)R_1\dot{x}(s)ds - \int_{t-h(t)}^{t}\dot{x}^{\mathrm{T}}(s)R_2\dot{x}(s)ds\\
&-2\eta^{\mathrm{T}}(t)X\int_{t-d(t)}^{t}\dot{x}(s)ds - 2\eta^{\mathrm{T}}(t)Z\int_{t-h(t)}^{t}\dot{x}(s)ds - [x^{\mathrm{T}}(t)Sx(t) + u^{\mathrm{T}}(t)Wu(t)]\\
\le{}& \sum_{i,j,l=1}^{s} h_ih_jh_{hl}\eta^{\mathrm{T}}(t)(T_{ijl} + \tau_1XR_1^{-1}X^{\mathrm{T}} + \tau_2ZR_2^{-1}Z^{\mathrm{T}})\eta(t)\\
&-\int_{t-d(t)}^{t}\left(\eta^{\mathrm{T}}(t)X + \dot{x}^{\mathrm{T}}(s)R_1\right)R_1^{-1}\left(\eta^{\mathrm{T}}(t)X + \dot{x}^{\mathrm{T}}(s)R_1\right)^{\mathrm{T}}ds\\
&-\int_{t-h(t)}^{t}\left(\eta^{\mathrm{T}}(t)Z + \dot{x}^{\mathrm{T}}(s)R_2\right)R_2^{-1}\left(\eta^{\mathrm{T}}(t)X + \dot{x}^{\mathrm{T}}(s)R_2\right)^{\mathrm{T}}ds - [x^{\mathrm{T}}(t)Sx(t) + u^{\mathrm{T}}(t)Wu(t)]\\
\le{}& \sum_{i,j,l=1}^{s} h_ih_jh_{hl}\eta^{\mathrm{T}}(t)(\tilde{T}_{ijl} + \tau_1XR_1^{-1}X^{\mathrm{T}} + \tau_2ZR_2^{-1}Z^{\mathrm{T}})\eta(t) - [x^{\mathrm{T}}(t)Sx(t) + u^{\mathrm{T}}(t)Wu(t)]
\end{aligned}
\tag{14}
$$

where

$$\tilde{T}_{ijl} = \begin{bmatrix} \tilde{T}_{11,ij} & * & * & * \\ T_{21,i} & T_{22,i} & * & * \\ T_{31,i} & T_{32,ij} & T_{33,il} & * \\ T_{41,i} & T_{42,i} & T_{43} & T_{44} \end{bmatrix}, \quad \tilde{T}_{11,ij} = T_{11,ij} + \rho^2 \overline{K}_i^{\mathrm{T}} \cos\theta_i W \overline{K}_j \cos\theta_j - \rho^2 \overline{K}_i^{\mathrm{T}} W \overline{K}_j.$$

In light of the inequality $\overline{K}_i^{\mathrm{T}} W \overline{K}_j + \overline{K}_j^{\mathrm{T}} W \overline{K}_i \leq \overline{K}_i^{\mathrm{T}} W \overline{K}_i + \overline{K}_j^{\mathrm{T}} W \overline{K}_j$, we have

$$\dot{V}(x(t),t) \leq \sum_{i,j,l=1}^{s} h_i h_j h_{hl} \eta^{\mathrm{T}}(t)(T_{ijl} + \tau_1 X R_1^{-1} X^{\mathrm{T}} + \tau_2 Z R_2^{-1} Z^{\mathrm{T}})\eta(t) - [x^{\mathrm{T}}(t) S x(t) + u^{\mathrm{T}}(t) W u(t)] \quad (15)$$

Applying the Schur complement to Eq. (8) yields

$$T_{ii} + +\tau_1 X R_1^{-1} X^{\mathrm{T}} + \tau_2 Z R_2^{-1} Z^{\mathrm{T}} < 0, T_{ij} + T_{ji} + 2\tau_1 X R_1^{-1} X^{\mathrm{T}} + 2\tau_2 Z R_2^{-1} Z^{\mathrm{T}} < 0.$$

Therefore, it follows from Eq. (15) that

$$\dot{V}(x(t),t) \leq -[x^{\mathrm{T}}(t) S x(t) + u^{\mathrm{T}}(t) W u(t)] < 0 \quad (16)$$

which implies that the system (6) is asymptotically stable.

Integrating Eq. (16) from 0 to T produces

$$\int_0^T [x^{\mathrm{T}}(t) S x(t) + u^{\mathrm{T}}(t) W u(t)] dt \leq -V(x(T),T) + V(x(0),0) < V(x(0),0)$$

Because of $V(x(t),t) \geq 0$ and $V^{\cdot}(x(t),t) < 0$, thus

$$\lim_{T \to \infty} V(x(T),T) = c,$$

where c is a nonnegative constant. Therefore, the following inequality can be obtained:

$$J \leq x^{\mathrm{T}}(0) P x(0) + \int_{-d(0)}^{0} x^{\mathrm{T}}(s) Q_1 x(s) ds + \int_{-\tau_1}^{0} \int_{\theta}^{0} \dot{x}^{\mathrm{T}}(s) R_1 \dot{x}(s) ds d\theta + \int_{-h(0)}^{0} x^{\mathrm{T}}(s) Q_2 x(s) ds \\ + \int_{-\tau_2}^{0} \int_{\theta}^{0} \dot{x}^{\mathrm{T}}(s) R_2 \dot{x}(s) ds d\theta = J_0 \quad (17)$$

This completes the proof.

Remark 2: In the derivation of Theorem 1, the free-weighting matrices $X_k \in R^{n\times n}$, $Y_k \in R^{n\times n}$, $k = 1, 2, 3, 4$ are introduced, the purpose of which is to reduce conservatism in the existing delay-dependent stabilization conditions, see Ref. [17].

In the following section, we shall turn the conditions given in Theorem 1 into linear matrix inequalities (LMIs). Under the assumptions that Y_1, Y_2, Y_3, Y_4 are non-singular, we can define the matrix $Yi^{-T}=\lambda Z$, $i = 1, 2, 3, 4$, $Z = P^{-1}$, $\lambda > 0$.

Pre- and post-multiply (8) and (9) with $\Theta = diag\{Y_1^{-1}, Y_2^{-1}, Y_3^{-1}, Y_4^{-1}, Y_4^{-1}, Y_4^{-1}\}$ and $\Theta^T = diag\{Y_1^{-T}, Y_2^{-T}, Y_3^{-T}, Y_4^{-T}, Y_4^{-T}, Y_4^{-T}\}$, respectively, and letting $\overline{Q}_1 = Y_1{}^{-1}Q_1Y_1{}^{-T}$, $\overline{Q}_2 = Y_1{}^{-1}Q_2Y_1{}^{-T}$, $\overline{R}_k = Y_4{}^{-1}R_kY_4{}^{-T}, k = 1, 2, \overline{X}_i = Y_i{}^{-1}X_iY_i{}^{-T}, \overline{Z}_i = Y_i{}^{-1}Z_iY_i{}^{-T}$, i = 1, 2, 3, 4, we obtain the following inequality (18), which is equivalent to (8):

$$\begin{bmatrix} \overline{T}_{11,ij} & * & * & * & * & * \\ \overline{T}_{21,i} & \overline{T}_{22,i} & * & * & * & * \\ \overline{T}_{31,i} & \overline{T}_{32,i} & \overline{T}_{33,il} & * & * & * \\ \overline{T}_{41,i} & \overline{T}_{42,i} & \overline{T}_{43} & \overline{T}_{44} & * & * \\ \tau_1 X_1 & \tau_1 X_2 & \tau_1 X_3 & \tau_1 X_4 & -\tau_1 \overline{R}_1 & * \\ \tau_2 \overline{Z}_1 & \tau_2 \overline{Z}_2 & \tau_2 \overline{Z}_3 & \tau_2 \overline{Z}_4 & 0 & -\tau_2 \overline{R}_2 \end{bmatrix} < 0, \quad i, j, l = 1, 2, \ldots, s \tag{18}$$

where

$$\overline{T}_{11,ij} = \overline{Q}_1 + \overline{Q}_2 + \overline{X}_1 + \overline{X}_1^{T} + \lambda A_i Z + \lambda Z A_i^{T} + \lambda^2 ZSZ + 2\varepsilon\rho^2 I + 4\varepsilon^{-1}\lambda^2 Z N_i^{T} N_i Z$$
$$+ Z_1 + Z_1^{T} + 4\varepsilon^{-1}\lambda^2 (B_i \overline{K}_j Z)^{T} (B_i \overline{K}_j Z) + \rho^2 \lambda^2 Z \overline{K}_i^{T} W \overline{K}_i Z,$$

$$\overline{T}_{21,i} = -\overline{X}_1^{T} + \overline{X}_2 + \overline{Z}_2 + \lambda A_i Z + \lambda Z A_{di}^{T}, \quad \overline{T}_{31,i} = \overline{Z}_3 - \overline{Z}_1 + \overline{X}_3 + \lambda A_i Z,$$

$$\overline{T}_{41,i} = \lambda^2 Z + \lambda A_i Z - \lambda Z + \overline{X}_4 + \overline{Z}_4,$$

$$\overline{T}_{22,i} = -(1-\sigma_1)\overline{Q}_1 - \overline{X}_2 - \overline{X}_2^{T} + \lambda A_{di} Z + \lambda Z A_{di}^{T} + 2\varepsilon\rho^2 I + 4\varepsilon^{-1}\lambda^2 Z N_{di}^{T} N_{di} Z,$$

$$\overline{T}_{32,i} = -\overline{X}_3 - \overline{Z}_2 + \lambda A_{di} Z - \lambda Z A_{di}^{T}, \ \overline{T}_{42,i} = -\overline{X}_4 + \lambda A_i Z - \lambda Z,$$

$$\overline{T}_{33,il} = -(1-\sigma_2)\overline{Q}_2 - \overline{Z}_3 - \overline{Z}_3^{T} + 4\varepsilon^{-1}\lambda^2 (B_{hi}\tilde{K}_l Z)^{T} B_{hi}\tilde{K}_l Z + 2\varepsilon\rho^2 I,$$

$$\overline{T}_{43} = -\overline{Z}_4 - \lambda Z, \ \overline{T}_{44} = \tau_1 R_1 + \tau_2 R_2 - \lambda Z - \lambda Z^{T} + 2\varepsilon\rho^2 I.$$

Applying the Schur complement to Eq. (18) results in

$$\Gamma_{ijl} = \begin{bmatrix} \Phi_{11,i} & * & * \\ \Phi_{21,ij} & \Phi_{22} & * \\ \Phi_{31,il} & 0 & \Phi_{33} \end{bmatrix} < 0, \quad i, j, l = 1, 2, \ldots, s \tag{19}$$

where

$$\Phi_{11,i} = \begin{bmatrix} \overline{\overline{T}}_{11,i} & * & * & * & * & * \\ \overline{T}_{21,i} & \overline{\overline{T}}_{22,i} & * & * & * & * \\ \overline{T}_{31,i} & \overline{T}_{32,i} & \overline{\overline{T}}_{33,il} & * & * & * \\ \overline{T}_{41,i} & \overline{T}_{42,i} & \overline{T}_{43} & \overline{T}_{44} & * & * \\ \tau_1 X_1 & \tau_1 X_2 & \tau_1 X_3 & \tau_1 X_4 & -\tau_1 \overline{R}_1 & * \\ \tau_2 \overline{Z}_1 & \tau_2 \overline{Z}_2 & \tau_2 \overline{Z}_3 & \tau_2 \overline{Z}_4 & 0 & -\tau_2 \overline{R}_2 \end{bmatrix}$$

With

$$\overline{\overline{T}}_{11,i} = \overline{Q}_1 + \overline{Q}_2 + \overline{X}_1 + \overline{X}_1^{\mathrm{T}} + \lambda A_i Z + \lambda Z A_i^{\mathrm{T}} + Z_1 + Z_1^{\mathrm{T}} + 2\varepsilon\rho^2 I,$$
$$\overline{\overline{T}}_{22,i} = -(1-\sigma_1)\overline{Q}_1 - \overline{X}_2 - \overline{X}_2^{\mathrm{T}} + \lambda A_{di} Z + \lambda Z A_{di}^{\mathrm{T}} + 2\varepsilon\rho^2 I,$$
$$\overline{\overline{T}}_{33} = -(1-\sigma_2)\overline{Q}_2 - \overline{Z}_3 - \overline{Z}_3^{\mathrm{T}} + 2\varepsilon\rho^2 I.$$

$$\Phi_{21,ij} = \begin{bmatrix} \lambda Z & 0 & 0 & 0 \\ \lambda N_i Z & 0 & 0 & 0 \\ \lambda B_i \overline{K}_j Z & 0 & 0 & 0 \\ \rho\lambda \overline{K}_j Z & 0 & 0 & 0 \end{bmatrix}$$

$$\Phi_{31,il} = \begin{bmatrix} 0 & \lambda B_{hi} \tilde{K}_l Z & 0 & 0 \\ 0 & 0 & \lambda N_{di} Z & 0 \end{bmatrix}$$

$$\Phi_{22} = \begin{bmatrix} -S^{-1} & 0 & 0 & 0 \\ 0 & -\frac{\varepsilon}{4} I & 0 & 0 \\ 0 & 0 & -\frac{\varepsilon}{4} I & 0 \\ 0 & 0 & 0 & -W^{-1} \end{bmatrix}$$

$$\Phi_{33} = \begin{bmatrix} -\frac{\varepsilon}{4} I & 0 \\ 0 & -\frac{\varepsilon}{4} I \end{bmatrix}$$

Obviously, the closed-loop fuzzy system (6) is asymptotically stable, if for some scalars $\lambda > 0$, there exist matrices $Z > 0, \overline{Q} > 0, \overline{R} > 0$ and $\overline{X}_1, \overline{X}_2, \overline{X}_3, \overline{K}_i, i = 1, 2, .., s$ satisfying the inequality (19).

Theorem 2. *Consider the system (6) associated with cost function (7). For given scalars $\rho > 0$, $\tau_1 > 0, \tau_2 > 0, \sigma_1 > 0, \sigma_2 > 0$ and $\lambda > 0, \delta > 0$, if there exist matrices $Z > 0, \overline{Q}_1 > 0, \overline{R}_1 > 0, \overline{Q}_2 > 0, \overline{R}_2 > 0$ and $\overline{X}_1, \overline{X}_2, \overline{X}_3, \overline{X}_4, M_i, i = 1,2,\ldots,s$ and scalar $\varepsilon > 0$ satisfying the following LMI (20), the system (6) is asymptotically stable and the control law (5) is a fuzzy non-fragile guaranteed cost control law*

$$\begin{bmatrix} \Theta_{1,ijl} & * \\ \Theta_{2,ijl} & \Theta_3 \end{bmatrix} < 0, \quad i, j, l = 1, 2, \ldots, s \tag{20}$$

Moreover, the feedback gains are given by

$$K_i = M_i Z^{-1}, i = 1, 2, \ldots, s$$

and

$$J \leq x^{\mathrm{T}}(0)Px(0) + \int_{-d(0)}^{0} x^{\mathrm{T}}(s)Q_1 x(s)ds + \int_{-\tau_1}^{0}\int_{\theta}^{0} \dot{x}^{\mathrm{T}}(s)R_1\dot{x}(s)dsd\theta + \int_{-h(0)}^{0} x^{\mathrm{T}}(s)Q_2 x(s)ds$$
$$+ \int_{-\tau_2}^{0}\int_{\theta}^{0} \dot{x}^{\mathrm{T}}(s)R_2\dot{x}(s)dsd\theta = J_0$$

where

$$\Theta_{2,ijl} = \begin{bmatrix} \lambda E_{kj}Z & 0 & 0 & 0 & 0 & 0 & 0 & 0 & 0 & 0 & 0 & 0 \\ \lambda E_{ki}Z & 0 & 0 & 0 & 0 & 0 & 0 & 0 & 0 & 0 & 0 & 0 \\ 0 & 0 & \lambda E_{kl}Z & 0 & 0 & 0 & 0 & 0 & 0 & 0 & 0 & 0 \\ 0 & 0 & 0 & 0 & 0 & 0 & 0 & 0 & (B_iH_j)^{\mathrm{T}} & 0 & 0 & 0 \\ 0 & 0 & 0 & 0 & 0 & 0 & 0 & 0 & 0 & \rho H_i^{\mathrm{T}} & 0 & 0 \\ 0 & 0 & 0 & 0 & 0 & 0 & 0 & 0 & 0 & 0 & 0 & (B_{hi}H_l)^{\mathrm{T}} \end{bmatrix},$$

$$\Theta_3 = \mathrm{diag}\{-\delta I, -\delta I, -\delta I, -\delta^{-1}I, -\delta^{-1}I, -\delta^{-1}I\},$$

$$\Theta_{1,ijl} = \begin{bmatrix} \bar{\bar{T}}_{11,i} & * & * & * & * & * & * & * & * & * & * & * \\ \bar{T}_{21,i} & \bar{\bar{T}}_{22,i} & * & * & * & * & * & * & * & * & * & * \\ \bar{T}_{31,i} & \bar{T}_{32,i} & \bar{T}_{33} & * & * & * & * & * & * & * & * & * \\ \bar{T}_{41,i} & \bar{T}_{42,i} & \bar{T}_{43,i} & \bar{T}_{44} & * & * & * & * & * & * & * & * \\ \tau_1\bar{X}_1 & \tau_1\bar{X}_2 & \tau_1\bar{X}_3 & \tau_1\bar{X}_4 & -\tau_1\bar{R}_1 & * & * & * & * & * & * & * \\ \tau_2\bar{Z}_1 & \tau_2\bar{Z}_2 & \tau_2\bar{Z}_3 & \tau_2\bar{Z}_4 & 0 & -\tau_2\bar{R}_{21} & * & * & * & * & * & * \\ \lambda Z & 0 & 0 & 0 & 0 & 0 & -S^{-1} & * & * & * & * & * \\ \lambda N_i Z & 0 & 0 & 0 & 0 & 0 & 0 & -\frac{1}{4}I & * & * & * & * \\ \lambda B_i M_j & 0 & 0 & 0 & 0 & 0 & 0 & 0 & -\frac{1}{4}I & * & * & * \\ \rho\lambda M_i & 0 & 0 & 0 & 0 & 0 & 0 & 0 & 0 & -W^{-1} & * & * \\ 0 & \lambda N_{di}Z & 0 & 0 & 0 & 0 & 0 & 0 & 0 & 0 & -\frac{1}{4}I & * \\ 0 & 0 & \lambda B_{hi}M_l & 0 & 0 & 0 & 0 & 0 & 0 & 0 & 0 & -\frac{1}{4}I \end{bmatrix}$$

Proof: At first, we prove that the inequality (20) implies the inequality (19). Applying the Schur complement to Eq. (20) results in

$$\Phi_{1,ijl} + \delta \begin{bmatrix} 0 & 0 & 0 \\ 0 & 0 & 0 \\ 0 & 0 & 0 \\ 0 & 0 & 0 \\ 0 & 0 & 0 \\ 0 & 0 & 0 \\ 0 & 0 & 0 \\ 0 & 0 & 0 \\ B_iH_j & 0 & 0 \\ 0 & \rho H_i & 0 \\ 0 & 0 & 0 \\ 0 & 0 & B_{hi}H_l \end{bmatrix} \begin{bmatrix} 0 & 0 & 0 & 0 & 0 & 0 & 0 & 0 & (B_iH_j)^{\mathrm{T}} & 0 & 0 & 0 \\ 0 & 0 & 0 & 0 & 0 & 0 & 0 & 0 & 0 & \rho H_i^{\mathrm{T}} & 0 & 0 \\ 0 & 0 & 0 & 0 & 0 & 0 & 0 & 0 & 0 & 0 & 0 & (B_{hi}H_l)^{\mathrm{T}} \end{bmatrix}$$

$$
+\delta^{-1}\begin{bmatrix} (\lambda E_{kj}Z)^{\mathrm{T}} & (\lambda E_{ki}Z)^{\mathrm{T}} & 0 \\ 0 & 0 & 0 \\ 0 & 0 & (\lambda E_{kl}Z)^{\mathrm{T}} \\ 0 & 0 & 0 \\ 0 & 0 & 0 \\ 0 & 0 & 0 \\ 0 & 0 & 0 \\ 0 & 0 & 0 \\ 0 & 0 & 0 \\ 0 & 0 & 0 \\ 0 & 0 & 0 \\ 0 & 0 & 0 \end{bmatrix}\begin{bmatrix} \lambda E_{kj}Z & 0 & 0 & 0 & 0 & 0 & 0 & 0 & 0 & 0 & 0 & 0 \\ \lambda E_{ki}Z & 0 & 0 & 0 & 0 & 0 & 0 & 0 & 0 & 0 & 0 & 0 \\ 0 & 0 & \lambda E_{kl}Z & 0 & 0 & 0 & 0 & 0 & 0 & 0 & 0 & 0 \end{bmatrix} < 0 \tag{21}
$$

Using Lemma 2 and noting $M_i = K_iZ$, by the condition (21), the following inequality holds:

$$
\Phi_{1,ijl}+\begin{bmatrix} 0 & * & * & * & * & * & * & * & * & * & * & * \\ 0 & 0 & * & * & * & * & * & * & * & * & * & * \\ 0 & 0 & 0 & * & * & * & * & * & * & * & * & * \\ 0 & 0 & 0 & 0 & * & * & * & * & * & * & * & * \\ 0 & 0 & 0 & 0 & 0 & * & * & * & * & * & * & * \\ 0 & 0 & 0 & 0 & 0 & 0 & * & * & * & * & * & * \\ 0 & 0 & 0 & 0 & 0 & 0 & 0 & * & * & * & * & * \\ 0 & 0 & 0 & 0 & 0 & 0 & 0 & 0 & * & * & * & * \\ \lambda B_i\Delta\bar{K}_jZ & 0 & 0 & 0 & 0 & 0 & 0 & 0 & 0 & * & * & * \\ \rho\lambda\Delta\bar{K}_iZ & 0 & 0 & 0 & 0 & 0 & 0 & 0 & 0 & 0 & * & * \\ 0 & 0 & 0 & 0 & 0 & 0 & 0 & 0 & 0 & 0 & 0 & * \\ 0 & 0 & \lambda B_{hi}\Delta\tilde{K}_lZ & 0 & 0 & 0 & 0 & 0 & 0 & 0 & 0 & 0 \end{bmatrix} < 0 \tag{22}
$$

where $\Delta\tilde{K}_i = \Delta K_i(t-d(t))$.

Therefore, it follows from Theorem 1 that the system (6) is asymptotically stable and the control law (5) is a fuzzy non-fragile guaranteed cost control law. Thus, we complete the proof.

Now consider the cost bound of

$$
J \le x^{\mathrm{T}}(0)Px(0)+\int_{-d(0)}^{0} x^{\mathrm{T}}(s)Q_1x(s)ds+\int_{-\tau_1}^{0}\int_{\theta}^{0}\dot{x}^{\mathrm{T}}(s)R_1\dot{x}(s)dsd\theta+\int_{-h(0)}^{0}x^{\mathrm{T}}(s)Q_2x(s)ds
$$

$$
+\int_{-\tau_2}^{0}\int_{\theta}^{0}\dot{x}^{\mathrm{T}}(s)R_2\dot{x}(s)dsd\theta = J_0
$$

Similar to Ref. [23], we supposed that there exist positive scalars $\alpha_1, \alpha_2, \alpha_3, \alpha_4, \alpha_5$, such that $Z^{-1} \le \alpha_1 I,\ \frac{1}{\lambda^2}P\overline{Q}_1P \le \alpha_2 I,\ \frac{1}{\lambda^2}P\overline{Q}_2P \le \alpha_3 I,\ \frac{1}{\lambda^2}P\overline{R}_1P \le \alpha_4 I,\ \frac{1}{\lambda^2}P\overline{R}_2P \le \alpha_5 I$. Then, define $S_{Q1}=\overline{Q}_1^{\,-1}, S_{Q2}=\overline{Q}_2^{\,-1}, S_{R1}=\overline{R}_1^{\,-1}, S_{R2}=\overline{R}_2^{\,-1}$ by Schur complement lemma, we have the following inequalities:

$$\begin{bmatrix} -\alpha_1 I & I \\ I & -Z \end{bmatrix} \leq 0, \begin{bmatrix} -\alpha_2 I & \frac{1}{\lambda}P \\ \frac{1}{\lambda}P & S_{Q1} \end{bmatrix} \leq 0, \begin{bmatrix} -\alpha_3 I & \frac{1}{\lambda}P \\ \frac{1}{\lambda}P & S_{Q2} \end{bmatrix} \leq 0, \begin{bmatrix} -\alpha_4 I & \frac{1}{\lambda}P \\ \frac{1}{\lambda}P & S_{R1} \end{bmatrix} \leq 0,$$
$$\begin{bmatrix} -\alpha_5 I & \frac{1}{\lambda}P \\ \frac{1}{\lambda}P & S_{R2} \end{bmatrix} \leq 0, \begin{bmatrix} Z & I \\ I & Z \end{bmatrix} \geq 0, \begin{bmatrix} S_{Q1} & I \\ I & \overline{Q}_1 \end{bmatrix} \geq 0, \begin{bmatrix} S_{Q2} & I \\ I & \overline{Q}_2 \end{bmatrix} \geq 0, \tag{23}$$
$$\begin{bmatrix} S_{R1} & I \\ I & \overline{R}_1 \end{bmatrix} \geq 0, \begin{bmatrix} S_{R2} & I \\ I & \overline{R}_2 \end{bmatrix} \geq 0,$$

Using the idea of the cone complement linear algorithm in Ref. [24], we can obtain the solution of the minimization problem of upper bound of the value of the cost function as follows:

$$\text{minimize}\{trace(PZ + S_{Q1}\overline{Q}_1 + S_{Q2}\overline{Q}_2 + S_{R1}\overline{R}_1 + S_{R2}\overline{R}_2 + \alpha_1 x^{\mathrm{T}}(0)x(0) + \alpha_2 \int_{-d(0)}^{0} x^{\mathrm{T}}(s)x(s)ds$$
$$+\alpha_4 \int_{-\tau_1}^{0} \int_{\theta}^{0} \dot{x}^{\mathrm{T}}(s)\dot{x}(s)dsd\theta + \alpha_3 \int_{-h(0)}^{0} x^{\mathrm{T}}(s)x(s)ds + \alpha_5 \int_{-\tau_2}^{0} \int_{\theta}^{0} \dot{x}^{\mathrm{T}}(s)\dot{x}(s)dsd\theta\} \tag{24}$$

subject to (20), (23), $\varepsilon > 0, \overline{Q}_1 > 0, \overline{Q}_2 > 0, \overline{R}_1 > 0, \overline{R}_2 > 0, Z > 0, \alpha_i > 0, i = 1, \ldots, 5$

Using the following cone complement linearization (CCL) algorithm [24] can iteratively solve the minimization problem (24).

5. SIMULATION EXAMPLES

In this section, the proposed approach is applied to the Van de Vusse system to verify its effectiveness.

Example: Consider the dynamics of an isothermal continuous stirred tank reactor for the Van de Vusse

$$\begin{aligned} \dot{x}_1 &= -50x_1 - 10x_1^3 + u(10 - x_1) + u(t-h) + u(t-h)(0.5x_1(t-d) + 0.2x_2(t-d)) + 5x_2(t-d) \\ \dot{x}_2 &= 50x_1 - 100x_2 - u(t-h) + u(t-h)(0.3x_1(t-d) - 0.2x_2(t-d)) + 10x_2(t-d) - 5x_1(t-d) \end{aligned} \tag{25}$$

From the system equation (25), some equilibrium points are tabulated in Table 1. According to these equilibrium points, $[x_e\ u_e]$, which are also chosen as the desired operating points, $[x'_e\ u'_e]$, we can use the similar modeling method that is described in Ref. [16].

Table 1. Data for equilibrium points

x^T_e	x^T_{de}	u_e	u_{de}
[2.0422 1.2178]	[2.0422 1.2178]	20.3077	20.3077
[3.6626 2.5443]	[3.6626 2.5443]	77.7272	77.7272
[5.9543 5.5403]	[5.9543 5.5403]	296.2414	296.2414

Thus, the system (25) can be represented by

R^1: if x_1 is about 2.0422 *then*

$$\dot{x}_\delta(t) = A_1 x_\delta(t) + A_{d1} x_{d\delta}(t) + B_1 u_\delta(t) + B_{h1} u_{d\delta}(t) + N_1 x_\delta(t) u_\delta(t) + N_{d1} x_{d\delta}(t) u_{h\delta}(t)$$

R^2: if x_1 is about 3.6626, then

$$\dot{x}_\delta(t) = A_2 x_\delta(t) + A_{d2} x_{d\delta}(t) + B_2 u_\delta(t) + B_{d2} u_{h\delta}(t) + N_2 x_\delta(t) u_\delta(t) + N_{d2} x_{d\delta}(t) u_{h\delta}(t) \quad (26)$$

R^3: *if* x_1 *is about* 5.9543,*hen*

$$\dot{x}_\delta(t) = A_3 x_\delta(t) + A_{d3} x_{d\delta}(t) + B_3 u_\delta(t) + B_{d3} u_{h\delta}(t) + N_3 x_\delta(t) u_\delta(t) + N_{d3} x_{d\delta}(t) u_{h\delta}(t)$$

where

$$A_1 = \begin{bmatrix} -75.2383 & 7.7946 \\ 50 & -100 \end{bmatrix}, A_2 = \begin{bmatrix} -98.3005 & 11.7315 \\ 50 & -100 \end{bmatrix}, A_3 = \begin{bmatrix} -122.1228 & 8.8577 \\ 50 & -100 \end{bmatrix},$$

$$N_1 = N_2 = N_3 = \begin{bmatrix} -1 & 0 \\ 0 & -1 \end{bmatrix}; B_1 = B_2 = B_3 = \begin{bmatrix} 10 \\ 0 \end{bmatrix}; A_{d1} = A_{d2} = A_{d3} = \begin{bmatrix} 0 & 5 \\ 10 & -5 \end{bmatrix},$$

$$N_{d1} = N_{d2} = N_{d3} = \begin{bmatrix} 0.5 & 0.2 \\ 0.3 & -0.2 \end{bmatrix}, B_{h1} = B_{h2} = B_{h3} = \begin{bmatrix} 1 \\ 0 \end{bmatrix}, x_\delta = x(t) - x'_e,$$

$$u_\delta = u(t) - u'_e, x_{d\delta} = x(t-d) - x'_{de}, u_{h\delta} = u(t-d) - u'_{he}.$$

The cost function associated with this system is given with $S = \begin{bmatrix} 1 & 0 \\ 0 & 1 \end{bmatrix}, W = 1.$ The controller gain perturbation ΔK of the additive form is give with $H_1 = H_2 = H_3 = 0.1$, $E_{k1} = [0.05 \quad -0.01]$, $E_{k2} = [0.02 \quad 0.01]$, $E_{k3} = [-0.01 \quad 0]$.

The membership functions of state x_1 are shown in Figure 1.

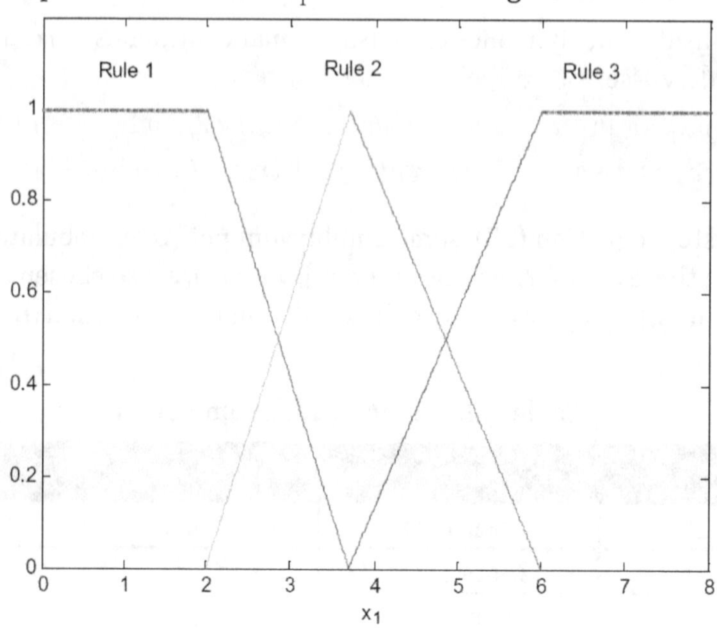

Figure 1. Membership functions.

Then, solving LMIs (23) and (24) for ρ = 0.45, λ = 1.02 and $\delta=0.11,{}_1=\tau_2=2,\sigma_1=0,\sigma_2=0$ gives the following feasible solution:

$$P=\begin{bmatrix}4.2727 & -1.3007\\ -1.3007 & 6.4906\end{bmatrix}, Q_1=\begin{bmatrix}14.1872 & -1.9381\\ -1.9381 & 13.0104\end{bmatrix}, Q_2=\begin{bmatrix}3.1029 & 1.2838\\ 1.2838 & 2.0181\end{bmatrix},$$

$$R_1=\begin{bmatrix}8.3691 & -1.3053\\ -1.3053 & 7.0523\end{bmatrix}, R_1=\begin{bmatrix}5.2020 & 2.2730\\ 2.2730 & 1.0238\end{bmatrix}, \varepsilon=1.8043,$$

$$K_1=[-0.4233\ \ -0.5031], K_2=[-0.5961\ \ -0.7049], K_1=[-0.4593\ \ -0.3874].$$

Figures 2–4 illustrate the simulation results of applying the non-fragile fuzzy controller to the system (25) with x'_e=[3.66262.5443]^T and u'_e=77.7272 under initial condition $\varphi(t)$ = [1.2 -1.8]^T, $t \in$ [-2 0]. It can be seen that with the fuzzy control law, the closed-loop system is asymptotically stable and an upper bound of the guaranteed cost is J_0 = 292.0399. The simulation results show that the fuzzy non-fragile guaranteed controller proposed in this paper is effective.

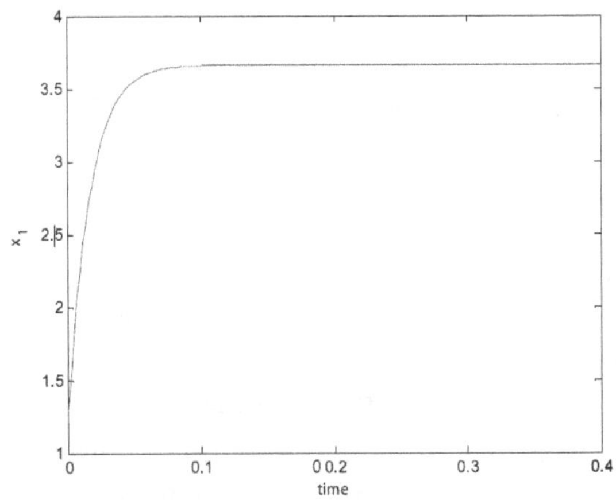

Figure 2. State responses of $x_1(t)$.

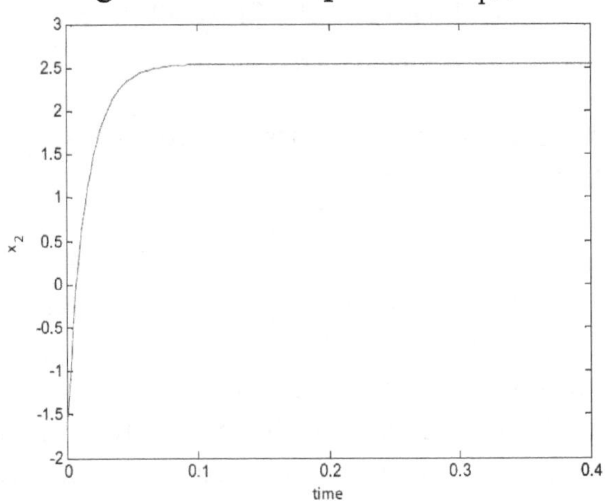

Figure 3. State responses of $x_2(t)$.

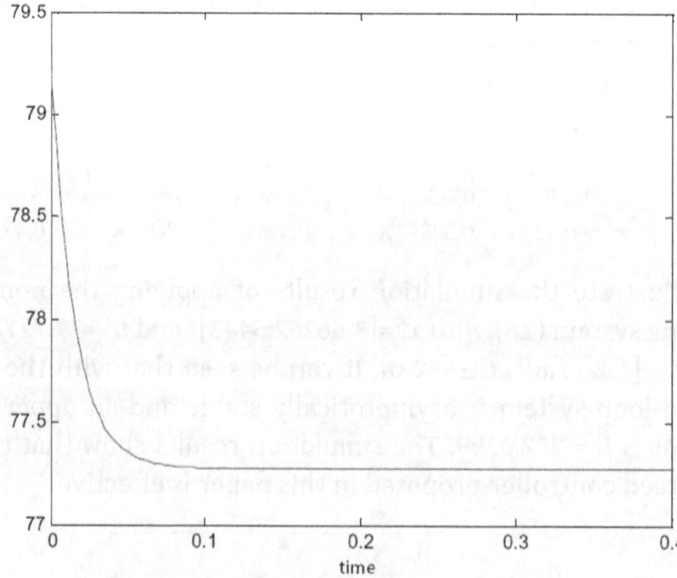

Figure 4. Control trajectory of system.

6. CONCISE SUMMARY

This chapter examines the issue of non-fragile assured cost control for a category of fuzzy time-varying delay systems characterized by local bilinear models. Utilizing Lyapunov stability theory and LMI techniques, sufficient conditions for the delay-dependent asymptotic stability of the closed-loop T-S fuzzy local bilinear system have been derived. Furthermore, the developed fuzzy controller has ensured adherence to the cost function-bound requirement. The efficacy of the devised approach has been validated by the simulation example. Future study will further investigate the resilient non-fragile guaranteed cost control and resilient non-fragile H-infinite control based on the fuzzy bilinear model.

AT A GLANCE

This chapter addresses the non-fragile assured cost control issue for a category of Takagi-Sugeno (T-S) fuzzy time-varying delay systems characterized by local bilinear models and variable state and input delays. A non-fragile guaranteed cost state-feedback controller is formulated to ensure that the closed-loop T-S fuzzy local bilinear control system achieves delay-dependent asymptotic stability, while also constraining the performance of the closed-loop fuzzy system to a specified upper limit in the presence of additive controller gain perturbations. The linear matrix inequality (LMI) technique is utilized to establish sufficient conditions for the presence of desired non-fragile assured cost controllers. The simulation examples demonstrate that the proposed method is both effective and practicable.

REFERENCES

1. Pang CT, Lur YY. On the stability of Takagi-Sugeno fuzzy systems with time-varying uncertainties. IEEE Transactions on Fuzzy Systems. 2008;16:162–170
2. Zhou SS, Lam J, Zheng WX. Control design for fuzzy systems based on relaxed non-quadratic stability and H∞ performance conditions. IEEE Transactions on Fuzzy Systems. 2007;15:188–198
3. Zhou SS, Li T. Robust stabilization for delayed discrete-time fuzzy systems via basis-dependent Lyapunov-Krasovskii function. Fuzzy Sets and Systems. 2005;151:139–153
4. Gao HJ, Liu X, Lam J. Stability analysis and stabilization for discrete-time fuzzy systems with time-varying delay. IEEE Transactions on Systems, Man, Cybernetics, Part B. 2009;39:306–316
5. Wu HN, Li HX. New approach to delay-dependent stability analysis and stabilization for continuous-time fuzzy systems with time-varying delay. IEEE Transactions on Fuzzy Systems. 2007;15:482–493
6. Chen B, Liu XP. Delay-dependent robust H∞ control for T-S fuzzy systems with time delay. IEEE Transactions on Fuzzy Systems. 2005;13:238–249
7. Chen M, Feng G, Ma H, Chen G. Delay-dependent H∞ filter design for discrete-time fuzzy systems with time-varying delays. IEEE Transactions on Fuzzy Systems. 2009;17:604–616
8. Zhang J, Xia Y, Tao R. New results on H∞ filtering for fuzzy time-delay systems. IEEE Transactions on Fuzzy Systems. 2009;17:128–137
9. Keel LH, Bhattacharryya SP. Robust, fragile, or optimal?. IEEE Transactions on Automatic Control. 1997;42:1098–1105
10. Yang GH, Wang JL, Lin C. H∞ control for linear systems with additive controller gain variations. International Journal of Control. 2000;73:1500–1506
11. Yang GH, Wang JL. Non-fragile H∞ control for linear systems with multiplicative controller gain variations. Automatica. 2001;37:727–737
12. Zhang BY, Zhou SS, Li T. A new approach to robust and non-fragile H∞ control for uncertain fuzzy systems. Information Sciences. 2007;177:5118–5133
13. Yee JS, Yang GH, Wang JL. Non-fragile guaranteed cost control for discrete-time uncertain linear systems. International Journal of Systems Science. 2001;32:845–853
14. Mohler RR. Bilinear Control Processes. New York, NY: Academic; 1973

15. Elliott DL. Bilinear Systems in Encyclopedia of Electrical Engineering. New York, NY: Wiley; 1999
16. Li THS, Tsai SH. T-S fuzzy bilinear model and fuzzy controller design for a class of nonlinear systems. IEEE Transactions on Fuzzy Systems. 2007;15:494–505
17. Tsai SH, Li THS. Robust fuzzy control of a class of fuzzy bilinear systems with time-delay. Chaos, Solitons and Fractals. 2009;39:2028–2040
18. Li THS, Tsai SH, et al. Robust H∞ fuzzy control for a class of uncertain discrete fuzzy bilinear systems. IEEE Transactions on Systems, Man, Cybernetics, Part B. 2008;38:510–526
19. Dong J, Wang Y, Yang G. Control synthesis of continuous-time T-S fuzzy systems with local nonlinear models. IEEE Transactions on Systems, Man, Cybernetics, Part B. 2009;39:1245–1258
20. Zhang G, Li JM. Non-fragile guaranteed cost control of discrete-time fuzzy bilinear system. Journal of Systems Engineering and Electronics. 2010;21:629–634
21. Ho DWC, Niu Y. Robust fuzzy design for nonlinear uncertain stochastic systems via sliding-mode control. IEEE Transactions on Fuzzy Systems. 2007;15:350–358
22. Yang DD, Cai KY. Reliable guaranteed cost sampling control for nonlinear time-delay systems. Mathematics and Computers in Simulation. 2010;80:2005–2018
23. Chen WH, Guan ZH, Lu XM. Delay-dependent output feedback guaranteed cost control for uncertain time-delay systems. Automatica. 2004;44:1263–1268
24. Chen B, Lin C, Liu XP, Tong SC. Guaranteed cost control of T–S fuzzy systems with input delay. International Journal of Robust Nonlinear Control. 2008;18:1230–1256
25. Chen B, Liu XP, Tong SC, Lin C. Observer-based stabilization of T–S fuzzy systems with input delay. IEEE Transactions on Fuzzy Systems. 2008;16:652–663
26. Chen B, Liu X, Tong S, Lin C. guaranteed cost control of T-S fuzzy systems with state and input delay. Fuzzy Sets and Systems. 2007;158:2251–2267
27. Du BZ, Lam J, Shu Z. Stabilization for state/input delay systems via static and integral output feedback. Automatica. 2010;46:2000–2007
28. Kim JH. Delay-dependent robust and non-fragile guaranteed cost control for uncertain singular systems with time-varying state and input delays. International Journal of Control, Automation, and Systems. 2009;7:357–364
29. Li L, Liu XD. New approach on robust stability for uncertain T–S fuzzy systems with state and input delays. Chaos, Solitons and Fractals. 2009;40:2329–2339

30. Yu KW, Lien CH. Robust H-infinite control for uncertain T–S fuzzy systems with state and input delays. Chaos, Solitons and Fractals. 2008;37:150–156

31. Yue D, Lam J. Non-fragile guaranteed cost control for uncertain descriptor systems with time-varying state and input delays. Optimal Control Applications and Methods. 2005;26:85–105

32. Zhang G, Li JM. Non-fragile guaranteed cost control of discrete-time fuzzy bilinear system with time-delay. Journal of Dynamic Systems, Measurement and Control. 2014;136:044502–044504

33. Yue HY, Li JM. Output-feedback adaptive fuzzy control for a class of nonlinear systems with input delay and unknown control directions. Journal of the Franklin Institute. 2013;350:129–154

34. Yue HY, Li JM. Adaptive fuzzy tracking control for a class of perturbed nonlinear time-varying delays systems with unknown control direction. International Journal of Uncertainty, Fuzziness and Knowledge-based Systems. 2013;21:497–531

35. Wang RJ, Lin WW, Wang WJ. Stabilizability of linear quadratic state feedback for uncertain fuzzy time-delay systems. IEEE Transactions on Systems, Man, Cybernetics, Part B. 2004;34:1288–1292

36. Xia ZL, Li JM, Li JR. Delay-dependent fuzzy static output feedback control for discrete-time fuzzy stochastic systems with distributed time-varying delays. ISA Transaction. 2012;51:702–712

37. Xia ZL, Li JM. Switching fuzzy filtering for nonlinear stochastic delay systems using piecewise Lyapunov-Krasovskii function. International Journal of Fuzzy Systems. 2012;14:530–539

38. Li JR, Li JM, Xia ZL. Delay-dependent generalized H2 fuzzy static-output-feedback control for discrete T-S fuzzy bilinear stochastic systems with mixed delays. Journal of Intelligent and Fuzzy Systems. 2013;25:863–880

39. Xia ZL, Li JM, Li JR. Passivity-based resilient adaptive control for fuzzy stochastic delay systems with Markovian switching. Journal of the Franklin Institute-Engineering and Applied Mathematics. 2014;351:3818–3836

40. Li JM, Li YT. Robust stability and stabilization of fractional order systems based on uncertain T-S fuzzy model with the fractional order. Journal of Computational and Nonlinear Dynamics. 2013;8:041005

41. Li YT, Li JM. Stability analysis of fractional order systems based on T-S fuzzy model with the fractional order α: $0<\alpha<1$. Nonlinear Dynamics. 2014;78:2909–2919

42. Li YT, Li JM. Decentralized stabilization of fractional order T-S fuzzy

interconnected systems with multiple time delays. Journal of Intelligent and Fuzzy Systems. 2016;30:319–331

43. Li JM, Yue HY. Adaptive fuzzy tracking control for stochastic nonlinear systems with unknown time-varying delays. Applied Mathematics and Computation. 2015;256:514–528

44. Yue HY, Yu SQ. Adaptive fuzzy tracking control for a class of stochastic nonlinearly parameterized systems with distributed input delay. Journal of the Franklin Institute-Engineering and Applied Mathematics. 2016;353:713–734

45. Duan RR, Li JM, Zhang YN, Yang Y, Chen GP. Stability analysis and H-inf control of discrete T-S fuzzy hyperbolic systems. International Journal of Applied Mathematics and Computer Science. 2016;26:133–145

46. Wang JX, Li JM. Stability analysis and feedback control of T-S fuzzy hyperbolic delay model for a class of nonlinear systems with time-varying delay. Iranian Journal of Fuzzy Systems. 2016;13:111–134

47. Li JR, Li JM, Xia ZL. Observer-based fuzzy control design for discrete time T-S fuzzy bilinear systems. International Journal of Uncertainty, Fuzziness and Knowledge-Based Systems. 2013;21:435–454

3 Chapter FUZZY INTERPOLATION SYSTEMS AND THEIR APPLICATIONS

1. INTRODUCTION: AN OVERVIEW

Since their inception in the 1960s, fuzzy logic and fuzzy sets have effectively served as instruments for addressing uncertainty and ambiguity, finding applications across numerous domains, including [1–6]. The predominant fuzzy systems are fuzzy rule-based inference systems, consisting of a rule base and an inference engine. Various inference engines were developed to accommodate distinct scenarios, such the Mamdani inference engine [7] and the TSK inference engine [8]. Rule bases are typically derived from expert knowledge or acquired through data analysis. The TSK model generates precise outputs owing to its polynomial rule implications in TSK-style rule bases, whereas the Mamdani model is more advantageous for managing conclusions derived from human natural language due to its fuzzy rule implications. Despite their extensive uses, conventional fuzzy inference processes are only effective with dense rule bases that comprehensively encompass the whole input domain. Fuzzy interpolation systems (FISs) were introduced to resolve the aforementioned issue [9], and they also facilitate complexity reduction for fuzzy models with excessively intricate (dense) rule bases. Conventional fuzzy inference methods are inapplicable when only a spare rule base exists and a given input does not intersect with any rule antecedent. Nonetheless, FISs can still derive a conclusion by fuzzy interpolation in these scenarios, hence augmenting the utility of traditional fuzzy inference systems. FISs can enhance the efficiency of intricate fuzzy inference systems by eliminating rules that can be precisely interpolated or extrapolated through other rules within a complicated rule base. Numerous fuzzy interpolation techniques utilizing Mamdani-style rule bases have been documented in the literature, including Refs. [9–20], demonstrating effective applications in decision-making assistance, prediction, and control, among other areas. Fuzzy inference systems (FISs) have been created to enhance TSK-style sparse fuzzy rule bases by augmenting the conventional TSK fuzzy inference system [21]. This methodology was formulated using an adapted similarity degree metric that facilitates the comprehensive application of all rules during the inference phase to produce a holistic outcome.

The adjusted similarity metric ensures that the degree of similarity between any specified input and any rule antecedent remains above zero, even in the absence of overlap. Consequently, all rules in the rule base may be activated to varying degrees, so contributing to the final outcome. As a result, a conclusion can still be derived even when no rule antecedent coincides with the provided observation. The extended TSK fuzzy model combines the benefits of the TSK model and fuzzy interpolation, enabling it to derive precise inference outcomes from TSK-style fuzzy rule bases that are sparse, dense, or unevenly distributed. Fuzzy Inference Systems (FISs) have been effectively utilized in practical applications. In certain real-world situations, neither comprehensive expert knowledge nor a complete dataset is accessible or easily obtained to create uniformly distributed dense rule bases. Consequently, FISs have been utilized in these circumstances. For example, a FIS has been utilized in building assessment in the research of Molnárka et al. [22] to assist real estate firms in making decisions on residential building maintenance when essential pertinent data has been compromised. Reference [23] demonstrates the successful application of a FIS system to diminish complexity and enhance the efficiency of a fuzzy home heating control system. Bai et al. [24] utilized a FIS to calibrate parallel machine tools for industrial applications. A behavior-based fuzzy control system is shown in Ref. [25], which utilizes a Fuzzy Inference System (FIS) to make decisions in the presence of an imperfect knowledge base. Recently, FISs have been employed to enhance network quality of service [26] and facilitate network intrusion detection [27].

2. FUZZY INFERENCE SYSTEMS

The fuzzy inference process is fundamentally an iterative computational paradigm grounded in fuzzy set theory, fuzzy if-then logic, and fuzzy reasoning. Each iteration accepts an input that may be an observation or a previously deduced result, whether precise or ambiguous. Subsequently, these inputs are utilized to activate the rules inside a specified rule base, and the output constitutes the aggregation of the inferred results from all activated rules. Typically, there are two principal methods for constructing a rule basis for a specific situation. The initial method involves the direct translation of expert information into rules, resulting in fuzzy inference systems sometimes referred to as fuzzy expert systems or fuzzy controllers [28]. In this instance, rules serve as ambiguous representations of expert knowledge, and the resulting rule base provides a high semantic level and strong generalization potential. The challenge of constructing rule bases for intricate situations has led to the emergence of a data-driven approach to rule base development, wherein fuzzy rules are derived from data using machine learning techniques instead of relying on expert knowledge [29, 30]. Conversely, the rule bases constructed in this manner exhibit a deficiency in comprehensibility and transparency. Rule bases can be classified into two categories based on the formulation of the implications of the fuzzy rules that constitute the rule base. Mamdani-style fuzzy rules incorporate fuzzy terms or linguistic values in the consequences, whereas TSK-style fuzzy rules express the consequences as polynomial functions of precise inputs.

2.1. Inference with Mamdani-style rule bases

There are a number of fuzzy inference mechanisms that can be utilised to derive a consequence from a given observation using a Mamdani rule base. The two most significant modes are the compositional rule of inference (CRI) [31] and analogy-based reasoning [24, 33], which are introduced below.

2.1.1. Compositional rule of inference

The introduction of CRI marks the era of fuzzy inference [31]. Given a rule 'IF x is A, THEN y is B'' and an observation 'x is A^*', the conclusion B^* can be generated through CRI as:

$$\mu_{B^*}(v) = sup_{u \in U_x} T\Big(\mu_{A^*}(u), \mu_R(u, v)\Big), \tag{1}$$

where T is a triangular norm, *sup* represents supremum, and R is the relationship between variables x and y. Essentially, CRI is a fuzzy extension of classical modus ponens which can be viewed from two perspectives. Firstly, classical modus ponens only supports predicates concerning singleton elements, but CRI is able to deal with predicates which concern a set of elements in the variable domain. This is achieved by representing a fuzzy rule as a fuzzy relation over the Cartesian product of the domains of the antecedent and consequent variables. Various fuzzy implication relations have been proposed [7, 32–34], each of which may have its own properties and therefore is suitable for a certain group of applications. Secondly, classical modus ponens only supports Boolean logic, but CRI supports multi-value logic. That is, CRI is able to deal with predicates with partial truth values, which are implemented by a compositional operator sup T, where T represents a t-norm [35].

A number of existing fuzzy reasoning methods based on CRI have been developed [36, 37], including the first successful practical approach, that is, the Mamdani inference [28]. This approach is also the most commonly seen fuzzy methodology in physical control systems thus far. It was originally proposed as an attempt to control a steam engine and boiler combination by synthesising a set of linguistic control rules obtained from experienced human operators. Mamdani inference implements CRI using *minimum* as the t-norm operator due to its simplicity. In particular, the inferred result from each fired rule is a fuzzy set which is transformed from the rule consequence by restricting the membership of those elements whose memberships are greater than the firing strength. The firing strength is also sometimes termed the satisfaction degree, which is the supremum within the variable domain of the minimum of the rule antecedent and the given observation. A defuzzification process is needed when crisp outputs are required.

2.1.2. Analogy-based fuzzy inference

Despite the success of CRI in various fuzzy system applications, it suffers various criticisms including its complexity and vague underlying semantics [34, 38]. This has led to another group of fuzzy reasoning approaches which are based on similarity degree, usually called analogy-based fuzzy reasoning [38–41]. Similarity considerations play a major role in human cognitive processes [42], so do they in approximate reasoning. It is intuitive that if a given observation is similar to the antecedent of a rule, the conclusion from the observation should also be similar to the consequence of the rule. Different to CRI-based fuzzy reasoning, analogy-based fuzzy reasoning does not require the construction of a fuzzy relation. Instead, it is based on the degree of similarity (given a certain similarity metric) between the given observation and the antecedent of a rule. Utilising the computed similarity degree, the consequence of the fired rule can be modified to the consequence of the given observation.

Approximate analogical reasoning schema is a typical analogy-based fuzzy inference approach [34, 38]. In this method, rules are fired according to the similarity degrees between a given observation and the antecedents of rules. If the degree of similarity between the given observation and the antecedent of a rule is greater than a predefined threshold value, the rule will be fired and the consequence of the observation is deduced from the rule consequence by a given modification procedure. Another analogy-based fuzzy inference approach was proposed in Refs. [39, 40], which particularly targets medical diagnostic problems. This approach is based on the cosine angle between the two vectors that represent the actual and the user's specified values of the antecedent variable. Several modification procedures can be found in Refs. [43, 44]. Particularly, a fuzzy reasoning method which employs similarity measures based on the degree of subsethood between the propositions in the antecedent and a given observation is proposed in Ref. [45]. This method has also been extended to consider the weights of the propositions in the antecedent [46]. Analogy-based fuzzy inference approaches usually arrive at solutions with more natural appeal than those introduced in the last section.

2.2. Inference with TSK-style rule bases

The TSK fuzzy inference system was proposed for the direct generation of crisp outputs [8]. In difference with the Mamdani-style fuzzy rule bases, TSK-style rule bases are usually generated from data using a clustering algorithm such as K-Means and an algorithm to determine the number of clusters such as Ref. [47]. Also, the consequence of a TSK fuzzy rule is a polynomial function rather than a fuzzy set. A typical TSK fuzzy rule can be defined as:

$$\text{IF } x_1 \text{ is } A_1 \wedge \ldots \wedge x_m \text{ is } A_m \text{ THEN } z = f(x_1, \ldots, x_m), \tag{2}$$

where $A_1,...A_m$ are fuzzy values with regard to antecedent variables x_1, ..., x_m respectively, and $f(x_1,...,x_m)$ is a crisp polynomial function of crisp inputs determining the crisp output value. The rule consequent polynomial functions $f(x_{1,...,}x_m)$ are usually zero order or first order. For simplicity, suppose that a TSK-style rule base is formed by two-antecedent rules as follows:

$$\begin{aligned} R_i &: \text{IF } x \text{ is } A_i \wedge y \text{ is } B_i \text{ THEN } z = f_i(x, y) \\ R_j &: \text{IF } x \text{ is } A_j \wedge y \text{ is } B_j \text{ THEN } z = f_j(x, y). \end{aligned} \tag{3}$$

Suppose that (x_0,y_0) is the crisp input pair, then the inference process can be shown in Figure 1. As the input values overlap with both rule antecedents, both rules are fired. Using rules R_i and R_j, the given input then leads to system outputs $f_i(x_0,x_0)$ and $f_j(x_0,x_0)$, respectively. The consequences from both rules are then integrated using weighted average function, where the values of weights represent the matching degrees between the given input and the rule antecedents (often referred to as firing strengths). Assume that $\mu(x0)$ and $\mu_{Bi}(y0)$ are the matching degree between inputs (x_0 and y_0) and rule antecedents (A_i and B_i), respectively. The firing strength of rule R_i, denoted as α_i, is calculated as:

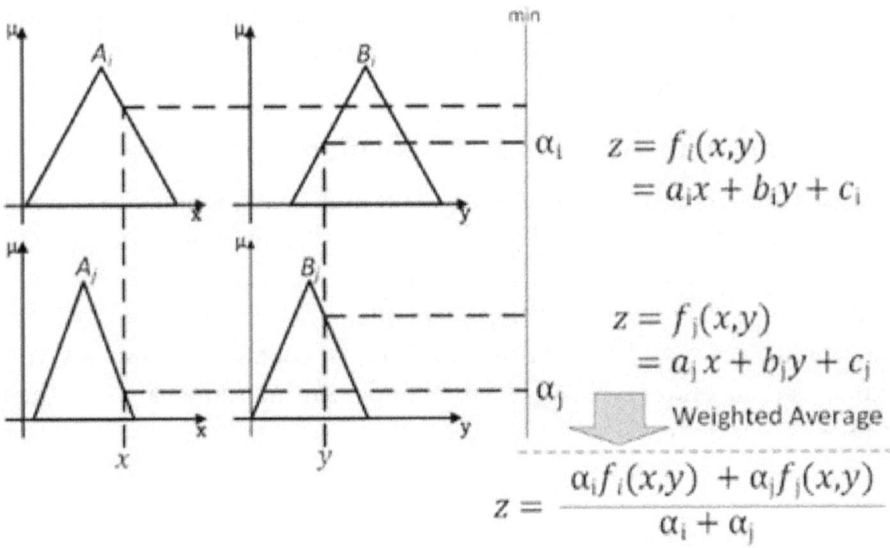

Figure 1. TSK fuzzy inference [21].

$$\alpha_i = \mu_{A_i}(x_0) \wedge \mu_{B_i}(y_0), \tag{4}$$

where $\wedge$ stands for a t-norm operator. Different implementations can be used for the t-norm operator, with the minimum operator being used most widely. Of course, if another system input (x_1, y_1) is presented and it is not covered by the rule base, the matching degrees between this new input and rule antecedents of R_i and R_j are equal to 0. In this case, no rule will be fired, and thus traditional TSK is not applicable.

3. FUZZY INTERPOLATION WITH SPARSE MAMDANI-STYLE RULE BASES

FISs based on Mamdani-style rule bases can be categorised into two classes. One group of approaches were developed based on the decomposition and resolution principle, termed as 'resolution principle-base interpolation'. In particular, the approach represents each fuzzy set as a series of α-cuts ($\alpha \in (0,1]$), and the α-cut of the conclusion is computed from the α-cuts of the observation and the α-cuts of rules. The final fuzzy set is assembled from all the α-cut consequences using the resolution principle [48–50]. The other group of fuzzy interpolation approaches were developed using the analogy reasoning system, thus termed as 'analogy-based fuzzy interpolation'. This group of approaches firstly generates an intermediate rule whose antecedent maximally overlaps with the given observation, then the system output is produced from the observation using the intermediate rule. Two representative approaches of the two classes, the KH approach [10] and the scale and move transformation-based approach [9, 51, 52], are discussed in this section based on simple rule bases with two antecedent rules. Despite of the simple examples used herein, both of these approaches have been extended to work with multiple multi-antecedent rules.

3.1. Resolution principle-based interpolation

Single step interpolation approaches are computationally efficient, such as the KH approach proposed in Refs. [9, 10, 53]. Following these approaches, all variables involved in the reasoning process must satisfy a partial ordering, denoted as $\prec$ [31]. According to the decomposition principle, a normal and convex fuzzy set A can be represented by a series of α-cut intervals, each denoted as $A\alpha$, $\alpha \in (0,1)$. Given fuzzy sets A_i and A_j which are associated with the same variable, the partial ordering $A_i \prec A_j$ is defined as:

$$\inf\{A_{i\alpha}\} < \inf\{A_{j\alpha}\} \text{ and } \sup\{A_{i\alpha}\} < \sup\{A_{j\alpha}\}, \quad \forall\alpha \in (0,1], \tag{5}$$

where $\inf\{A_{i\alpha}\}$ and $\sup\{A_{i\alpha}\}$ denote the infimum and supremum of $A_{i\alpha}$, respectively.

Take the KH approach as an example here. For simplicity, suppose there are two fuzzy rules: If x is A_i, then y is B_i, and If x is A_j then y is B_j, shorten as $A_i \Rightarrow B_i$ and $A_j \Rightarrow B_j$, respectively. Also, suppose that these two rules are adjacent, in other words, there is no rule $A \Rightarrow B$ existing such that $A_i \prec A \prec A_j$ or $A_j \prec A \prec A_i$. Given an observation A^* which satisfies $A_i \prec A^* \prec A_j$ or $A_j \prec A^* \prec A_i$, a conclusion B^* can be computed as:

$$\frac{D(A_{i\alpha}, A^*_\alpha)}{D(A^*_\alpha, A_{j\alpha})} = \frac{D(B_{i\alpha}, B^*_\alpha)}{D(B^*_\alpha, B_{j\alpha})}, \tag{6}$$

where given any $0 < \alpha \leq 1$, the distance $D(A_{i\alpha}, A_{j\alpha})$ between the α-cuts $A_i\alpha$ and $A_j\alpha$ is defined by the interval $[D^L(Ai\alpha,Aj\alpha),D^U(Ai\alpha,Aj\alpha)]$ with:

$$D^L(A_{i\alpha}, A_{j\alpha}) = inf\{A_{j\alpha}\} - inf\{A_{i\alpha}\}, D^U(A_{i\alpha}, A_{j\alpha}) = sup\{A_{j\alpha}\} - sup\{A_{i\alpha}\}. \tag{7}$$

Following Eqs. (4) and (5), the following is resulted:

$$\begin{cases} min\{B^*_\alpha\} = \dfrac{\dfrac{inf(B_{i\alpha})}{D^L(A_{i\alpha}, A^*_\alpha)} + \dfrac{inf(B_{j\alpha})}{D^L(A^*_\alpha, A_{j\alpha})}}{\dfrac{1}{D^L(A_{i\alpha}, A^*_\alpha)} + \dfrac{1}{D^L(A^*_\alpha, A_{j\alpha})}} \\ max\{B^*_\alpha\} = \dfrac{\dfrac{sup(B_{i\alpha})}{D^U(A_{i\alpha}, A^*_\alpha)} + \dfrac{sup(B_{j\alpha})}{D^U(A^*_\alpha, A_{j\alpha})}}{\dfrac{1}{D^U(A_{i\alpha}, A^*_\alpha)} + \dfrac{1}{D^U(A^*_\alpha, A_{j\alpha})}} \end{cases} \tag{8}$$

For simplicity, let

$$\begin{cases} \Lambda^L_\alpha = \dfrac{\inf\{A^*_\alpha\} - \inf\{A_{i\alpha}\}}{\inf\{A_{j\alpha}\} - \inf\{A_{i\alpha}\}} \\ \Lambda^U_\alpha = \dfrac{\sup\{A^*_\alpha\} - \sup\{A_{i\alpha}\}}{\sup\{A_{j\alpha}\} - \sup\{A_{i\alpha}\}} \end{cases} \tag{9}$$

Also, denote $\Lambda = [\Lambda^L_\alpha, \Lambda^U_\alpha]$ hereafter. From this, Eq. (8) can be re-written as:

$$\begin{cases} \min\{B^*_\alpha\} = (1 - \Lambda^L_\alpha)\inf\{B_{i\alpha}\} + \Lambda^L_\alpha \inf\{B_{j\alpha}\} \\ \max\{B^*_\alpha\} = (1 - \Lambda^U_\alpha)\sup\{B_{i\alpha}\} + \Lambda^U_\alpha \sup\{B_{j\alpha}\} \end{cases} \tag{10}$$

This means $B^*_\alpha = [\min\{B^*_\alpha\}, \max\{B^*_\alpha\}]$ is generated. The final consequence B^* is then reassembled as:

$$B^* = U_{\alpha \in (0,1]} \alpha B^*_\alpha. \tag{11}$$

The KH approach may generate invalid interpolated results [54], which is usually called 'the abnormal problem'. To eliminate this deficiency, a number of modifications or improvements have been proposed, including Refs. [9, 10, 13, 14, 18, 53, 55–60]. Approaches such as Refs. [15, 16, 61–63] also belong to this group.

3.2. Analogy-based interpolation

The scale and move transformation-based fuzzy interpolation [51, 52, 64] is a representative approach in the analogy-based interpolation group. For simplicity, following the same assumption of a simple rule base containing two rules with two antecedents, the transformation-based approach is shown in Figure 2 and outlined as follows.

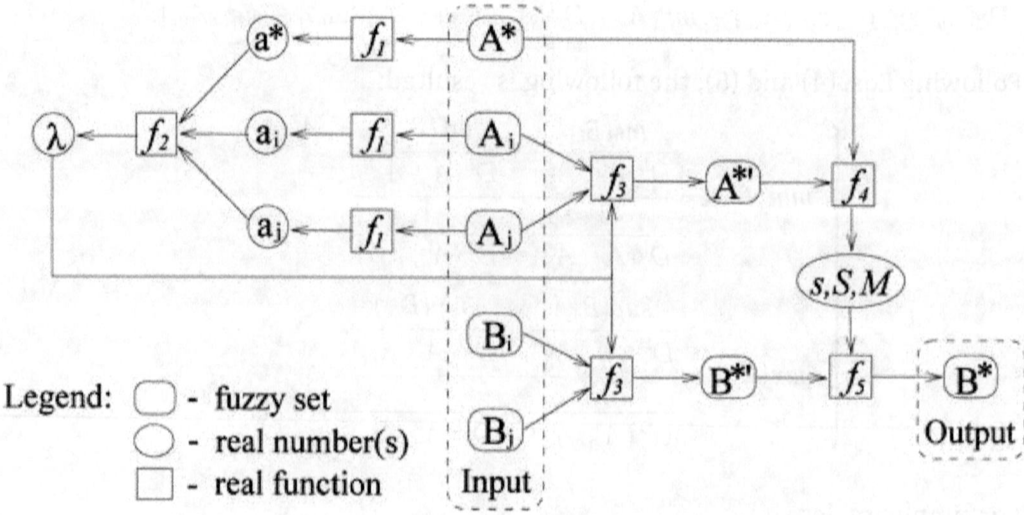

Figure 2. Transformation-based fuzzy interpolation [12].

Given neighbouring rules If x is A_i, then y is B_i, and If x is A_j then y is B_j and observation A^*, this method first maps fuzzy sets A_i, A_j and A^* to real numbers a_i, a_j and a^* (named as *representative values*) respectively, using real function f_1. Then, the location relationship between A^* and rule antecedents (A_i and A_j) is computed. This is achieved by another mapping function f_2, which results in the *relative placement factor* λ. In contrast to the resolution-based interpolation approaches, the generated *relative placement factor* in analogy-based fuzzy interpolation approach is a crisp real number. Finally, linear interpolation is implemented using mapping function f_3 of λ, which leads to the intermediate rule $'A^{*\prime} \Rightarrow B^{*\prime}$.

Note that the *representative value* of intermediate rule antecedent A∗'$A^{*\prime}$ equals to that of A^* (the given observation), although A∗'$A^{*\prime}$ and A^* are not identical for most of the situations. In the scale and move transformation-based fuzzy interpolation approach, the similarity degree between two fuzzy sets A^* and $A^{*\prime}$ with the same representative value is expressed as the scale rate s, scale ratio SS and move rate MM, which is obtained by real function f_4. From this, the consequence B^* is calculated from B∗'$B^{*\prime}$ using a transformation function f_5 which imposes the similarity degree between A^* and $A^{*\prime}$. Different approaches have been developed for intermediate rule generation and final conclusion production from the intermediate rule [17, 55, 63, 65].

3.3. Adaptive fuzzy interpolation

Fuzzy interpolation strengthens the power of fuzzy inference by enhancing the robustness of fuzzy systems and reducing the systems' complexity. Common to both classes of fuzzy interpolation approaches discussed above is the fact that interpolation is carried out in a linear manner. This may conflict with the nature of some realistic problems and consequently this may lead to inconsistencies during rule interpolation processes. Adaptive fuzzy interpolation was proposed to address this [12, 66–68]. It was developed upon FIS approaches, which detects inconsistencies, locates possible fault candidates and modifies the candidates in order to remove all the inconsistencies.

Each pair of neighbouring rules is defined as a *fuzzy reasoning component* in adaptive fuzzy interpolation. Each fuzzy reasoning component takes a fuzzy value as input and produces another as output. The process of adaptive interpolation is summarised in Figure 3. Firstly, the interpolator carries out interpolation and passes the interpolated results to the truth maintenance system (ATMS) [69, 70], which records the dependencies between an interpolated value (including any contradiction) and its proceeding interpolation components. Then, the ATMS relays any β_0-contradictions (i.e. inconsistency between two different values for a common variable at least to the degree of a given threshold β_0 ($0 \le \beta_0 \le 1$)) as well as their dependent fuzzy reasoning components to the general diagnostic engine (GDE) [71] which diagnoses the problem and generates all possible component candidates. After that, a modification process takes place to correct a certain candidate to restore consistency by modifying the original linear interpolation to become first-order piecewise linear.

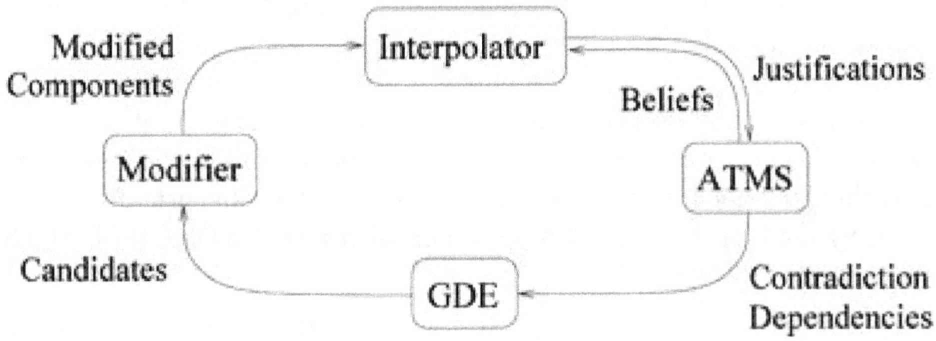

Figure 3. Adaptive fuzzy interpolation [12].

The adaptive approach has been further generalised [11, 72, 73], which allows the identification and modification of observations and rules, in addition to that of interpolation procedures that were addressed in Ref. [12]. This is supported by introducing extra information of certainty degrees associated with such basic elements of FIS. The work also allows for all candidates for modification to be prioritised, based on the extent to which a candidate is likely to lead to all the detected contradictions, by extending the classic ATMS and GDE. This study has significantly improved the efficiency of the work in Ref. [12] by exploiting more information during both the diagnosis and modification processes. Another alternative implementation of the adaptive approach has also been reported in Ref. [74].

3.4. Sparse rule base generation

A Mamdani-style fuzzy rule base is usually implemented through either a data-driven approach [75] or a knowledge-driven approach [76]. The data-driven approach using artificial intelligence approach extracts rules from data sets, while the knowledge-driven approach generates rules by human expert. Due to the limited availability of expert knowledge, data-driven approaches have been increasingly widely applied. However, the application of such approaches usually requires a large amount of training data, and it often leads to dense rule bases to support conventional fuzzy inference systems, despite of the availability of rule simplification approaches such as Refs. [77, 78].

A recent development or rule base generation has been reported with compact sparse rule bases targeted [79]. This approach firstly partitions the problem domain into a number of sub-regions and each sub-region is expressed as a fuzzy rule. Then, the importance of each sub-region is analysed using curvature value by artificially treating the problem space as a geography object (and high-dimensional problem space is represented as a collection of sub-three-dimensional spaces). Briefly, the profile curvature of a surface expresses the extent to which the geometric object deviates from being 'flat' or 'straight', the curvature values of the sub-regions are then calculated to represent how important they are in terms of linear interpolation. Given a predefined threshold, important sub-regions can be identified, and their corresponding rules are selected to generate a raw sparse rule base. The generated raw rule base can then be optimised by fine-tuning the membership functions using an optimisation algorithm. Generic algorithm has been widely used for various optimisation problems, such as Ref. [80], which has also been used in the work of Ref. [79].

Compared to most of the existing rule base generation approaches, the above approach differs in its utilisation of the curvature value in rule selection. Mathematically, curvature is the second derivate of a surface or the slope of slope. The profile curvature [81] is traditionally used in geography to represent the rate at which a surface slope changes whilst moving in the direction, which represents the steepest downward gradient for the given direction. Given a sub-region $f(x, y)$ and a certain direction, the curvature value is calculated as the directional derivative which refers to the rate at which any given scalar field is changing. The overall linearity of a sub-region can thus be accurately represented as the maximum profile curvature value on all directions. From this, those rules corresponding to sub-regions with higher profile curvature values (with respect to a given threshold) are selected, which jointly form the sparse rule base to support fuzzy rule interpolation.

FISs relax the requirement of complete expert knowledge or large data sets covering the entire input domain from the conventional fuzzy inference systems. However, it is still difficult for some real-world applications to obtain sufficient data or expert knowledge for rule base generation to support FISs. In addition, the generated rule resulted from most of the existing rule base generation approaches are fixed and cannot support changing situations. An experience-based rule base generation and adaptation approach for FISs has therefore been proposed for control problems [82]. Briefly, the approach initialises the rule base with very limited rules first. Then, the initialised rule base is revised by adding accurate interpolated rules and removing out-of-date rules guided by the performance index from a feedback mechanism and the performance experiences of rules.

4. FUZZY INTERPOLATION WITH SPARSE TSK-STYLE RULE BASE

The traditional TSK inference system has been extended to work with sparse TSK fuzzy rule base [21]. This approach, in the same time, also enjoys the benefit from its original version, which directly generates crisp outputs. The extended TSK inference approach is built upon a modified similarity measure which always generates greater than zero similarity degrees between observations and rule antecedents even when they do not overlap at all. Thanks to this property, a global consequence can always be generated by integrating the results from all rules in the rule base.

4.1. Rule firing strength

The modified similarity measure is developed from the work described in Ref. [83]. Suppose there are two fuzzy sets A and A' in a normalised variable domain. Without loss generality, a fuzzy set with any membership can be approximated by a polygonal fuzzy membership function with n odd points. Therefore, A and A' can be represented as A=(a_1,a_2,...a_n) and A'=(a'_1,a'_2,...a'_n'), as shown in Figure 4. The similarity degree $S(A,A')$ between A and A' is computed as:

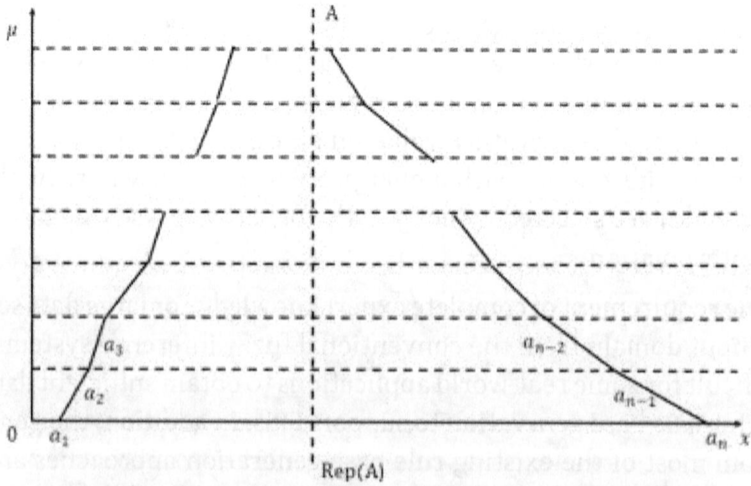

Figure 4. An arbitrary fuzzy set with n odd points.

$$S(A,A') = \left(1 - \frac{\sum_{i=1}^{n} |a_i - a'_i|}{n}\right)(DF)^{\tilde{B}(supp_A, supp_{A'})} \frac{\min\left(\mu(C_A), \mu(C_{A'})\right)}{\max\left(\mu(C_A), \mu(C_{A'})\right)}, \quad (12)$$

where c_A is the centre of gravity of fuzzy sets A, and $\mu(c_A)$ is the membership of the centre of gravity of fuzzy set A; DF represents a distance factor which is a function of the distance between two concerned fuzzy sets, and $B(supp_A, supp_A')$ is defined as follows:

$$B(supp_A, supp_{A'}) = \begin{cases} 1, & if supp_A + supp_{A'} \neq 0, \\ 0, & if supp_A + supp_{A'} = 0, \end{cases} \quad (13)$$

where $supp_A$ and $supp_A'$ are the supports of A and A', respectively.

In Eq. (13), $(supp_A, upp_A')$ is used to determine whether distance factor is considered. That is, if both A and A' are of crisp values, the distance factor DF will not take into consideration during the calculation of the similarity degree; otherwise, DF will be considered. The centre of gravity of a fuzzy set is commonly approximated as the average of its odd points. That is:

$$c_A = \frac{a_1 + a_2 + \ldots + a_n}{n}, \quad (14)$$

$$\mu(c_A) = \frac{\mu(a_1) + \mu(a_2) + \ldots + \mu(a_n)}{n}. \quad (15)$$

The distance factor DF is represented as:

$$DF = 1 - \frac{1}{1 + e^{-hd+5}} \quad (16)$$

where d is the distance between the two fuzzy sets, and $h(h > 0)$ is a sensitivity factor. The smaller the value of h is, the more sensitive the similarity degree to their distance is. The value of h is usually within the range of (20, 60), but the exact value is problem specific.

4.2. Fuzzy interpolation

Using the modified similarity measure as traduced above, the similarity between any given observation and a rule antecedent is always greater than zero. This means that all the rules in the rule base are fired for inference. Therefore, if only a sparse rule base is available and a given observation is not covered by the sparse rule base, a consequence still can be generated by firing all the rules in the rule base. The inference process is summarised as below:

Calculate the matching degree $(A^*,_i)$ and $S(B^*,B_i)$ between each pair of rule antecedent (A_i, B_i) and the input values (A^*, B^*) based on Eq. (12).

Determine the firing strength of each rule by integrating the matching degrees between the input items and rule antecedents as calculated in Step 1:

$$\alpha_i = S(A^*, A_i) \wedge S(B^*, B_i). \tag{17}$$

Compute the consequence of each rule in line with the given input and the polynomial function in rule consequent:

$$f_i(A^*, A_i) = \alpha_i \cdot c_{A^*} + b_i \cdot c_{B^*} + c_i. \tag{18}$$

Obtain the final result z by integrating the sub-consequences from all m rules in the rule base:

$$z = \frac{\sum_{i=1}^{n} \alpha_i f_i(A^*, B^*)}{\sum_{i=1}^{n} \alpha_i}. \tag{19}$$

5. APPLICATIONS OF FUZZY INTERPOLATION

Fuzzy interpolation systems have been successfully applied to a number of real-world problems including Refs. [23, 22, 25, 52, 57], two of which are reviewed in the section below.

5.1. Truck backer-upper control

Backing a trailer truck to a loading dock is a challenging task for all yet the most skilled truck drivers. Due to the difficulties, this challenge has been used as a control benchmark problem with various solutions proposed [75, 84, 85]. For instance, an artificial neural network has been applied to this problem, but a large amount of training data is required [84]. An adaptive fuzzy control system was also proposed for this problem, but the generation of the rule base is computationally expensive. Another solution combines empirical knowledge and data [85]. That is, a combined fuzzy rule base is generated by joining the previously generated rules (data-driven) and linguistic rules (expert knowledge-driven). More recently, a supervisory control system was proposed with fewer number of state variables required due to its capability to the decomposition of the control task, thus relieving the curse of dimensionality [86].

Fuzzy interpolation system has also been applied to the trailer truck backer-upper problem [52] to further reduce the system complexity. The problem can be formally formulated as $\theta=(x,y,\emptyset)$. Variables x and y represent the coordinate values corresponding to horizontal and vertical axes; Ø refers to the azimuth angle between the truck's onward direction and the horizontal axis; and θ is the steering angle of the truck. Given that enough clearance is present between the truck and loading lock in most cases, variable y can be safely omitted and hence results in a simplified formula to $\theta=f(x,\emptyset)$. By evenly partitioning each variable domain into three fuzzy sets, nine (i.e. 3*3) fuzzy rules were generated using FISMAT [87] and each of which is denoted as IF x is A AND Ø is B THEN θ is CIF x is A AND Ø is B THEN θ is C, where A, B and C are three linguistic values. Noting that domain partitions appear to be symmetrical in some sense, the three rules which are flanked by other rule pairs were removed from the rule base resulting a more compact rule base with only six fuzzy rules.

If the traditional fuzzy inference system were applied, the sparse rule base would cause a sudden break of the truck for some situations as no rule would be fired when the truck is in the position that can be represented by the omitted rules. In this case, fuzzy interpolation is naturally applied and the sudden break problem can be avoided. In addition, thanks to the great generalisation ability of the fuzzy interpolation systems, smooth performance is also demonstrated compared to the conventional fuzzy inference approaches. This study clearly demonstrates that fuzzy interpolation systems are able to simplify rule bases and support inferences with sparse rule bases.

5.2. Heating system control

Residential energy waste significantly adds to CO2 emissions in the UK, with approximately 60% of residential energy utilized for space heating. Numerous heating controllers have been designed to minimize energy loss in unoccupied houses, typically featuring programmability and utilizing multiple sensors. These systems can effectively deactivate heating systems when a property is empty [88–92], but they lack the capability to intelligently preheat the property prior to the users' return without manual intervention or unnecessarily prolonging the operation of the heating systems. A sophisticated home heating controller has been created with a Fuzzy Inference System (FIS), enabling efficient heating by precisely forecasting users' home arrival times based on historical and current location data from portable devices [23]. Figure 5 illustrates the comprehensive flow chart of the smart home heating system. The controller initially retrieves the resident's position and movement data. Four categories of residents' location and mobility information must be taken into account: At Home, Way Back Home, Leaving Home, and Static (i.e., at Special Location). The user's present location and movement status are accurately determined utilizing the GPS data from the user's portable devices. If the resident's current state is At Home, the algorithm concludes; if the resident's state is Leaving Home, indicating they are departing, the boiler is deactivated, and the system will reassess the resident's location and movement data after a specified interval. Additionally, the time to arrive home (TAH) is estimated, and the time required to preheat the home to a pleasant temperature (TPH) is computed, based on the resident's present circ umstances and the surrounding environment.

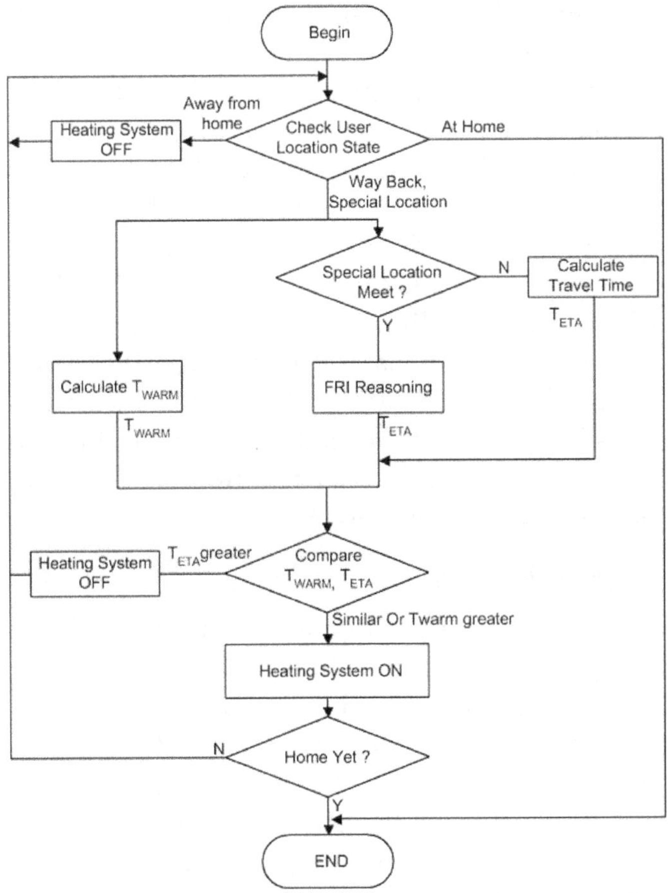

Figure 5. The flow chart of the heating controller [23].

The user's present modes of transportation (i.e., driving, walking, or bicycle) can be identified through the application of a naïve Bayes classifier utilizing GPS data. The journey distance and duration from the current location to home can be assessed with the Google Distance Matrix API. Be aware that the duration spent at various sites might change markedly, and distinct residents typically allocate varying amounts of time at the same specific location due to their individual lifestyles. The duration of residents' stay at the current place is thus calculated by fuzzy interpolation algorithms due to the problem's complexity. The fuzzy interpolation engine specifically accepts five fuzzy inputs and generates a single fuzzy output, which estimates the time required to return home. The five inputs include the present location, the day of the week, the time of day, the duration spent at the current location, and the projected travel time from the current location to home. If each input domain is fuzzy partitioned into 5 to 13 categories, it will yield tens of thousands of rules, necessitating substantial resources during inference. The suggested system has identified the 72 most critical rules, creating a sparse rule base that enhances fuzzy rule interpolation and substantially improves system performance. After calculating the home time, the residence can be precisely preheated according to a heating gain table tailored to the specific circumstances and environment of the property in question [91]. This work has been applied to a four-bedroom detached house with a total hearing space of 100 $m^2(floor\ area)\times 2.4\ m(hight)$. The house is heated by a 15 kW heating boiler. The research demonstrates that the fuzzy inference-based controller has effectively diminished the boiler's heating duration and enhanced the precision of house preheating. Notwithstanding the success of the aforementioned applications, there exists potential for FISs to be utilized in more extensive and complex real-world challenges, particularly in the domain of system control. Robotics has become central in the control domain, executing jobs ranging from fundamental robotic calligraphy systems [94] to intricate operations necessitating hand-eye (camera) synchronization [95]. FISs may potentially be utilized in sophisticated domains of robotics, necessitating additional research.

6. CONCISE SUMMARY

This chapter examined fuzzy interpolation systems and their applications in control theory. Two primary categories of fuzzy interpolation methods exist, utilizing the two predominant forms of fuzzy rule bases: Mamdani-style and TSK-style rule bases, which complement the two generally employed fuzzy inference techniques: Mamdani inference and TSK inference. The chapter also discusses the uses of fuzzy interpolation systems, illustrating the efficacy of these approaches. FISs can be enhanced further despite their promising performance. Type-2 fuzzy inference systems (FISs) have been previously proposed in the literature; nevertheless, the use of type-2 FISs in real-world scenarios necessitates additional exploration. Additionally, further theoretical research of FISs is required to quantitatively substantiate the convergence properties of the methodologies. Moreover, the majority of current fuzzy interpolation methods are designed to complement existing fuzzy inference models. It is intriguing to explore the creation of a cohesive platform that amalgamates existing fuzzy models and fuzzy inference methods, enabling the new system to leverage the advantages of both methodologies.

AT A GLANCE

Fuzzy inference methods offer a straightforward and efficient resolution to intricate non-linear issues, successfully implemented in various real-world situations. Conventional fuzzy inference systems may experience issues with sparse, overly complex, or imbalanced rule bases due to the unequal distribution of data within the problem space, irrespective of its volume. Fuzzy interpolation resolves this issue. It facilitates fuzzy inferences using sparse rule bases when the rule base fails to encompass a specific input, and it streamlines excessively dense rule bases by approximating particular rules with adjacent ones. This chapter comprehensively examines various fuzzy interpolation methods and their modifications, focusing on the interpolation mechanism (inference engine) and the development of sparse rule bases. This chapter revisits notable uses of fuzzy interpolation in control, validating its methodologies and showcasing its effectiveness and potential for broader applications.

REFERENCES

1. Fu X, Zeng XJ, Wang D, Xu D, Yang L. Fuzzy system approaches to negotiation pricing decision support. Journal of Intelligent & Fuzzy Systems. 2015;29(2):685–699
2. Yang L, Neagu D. Integration strategies for toxicity data from an empirical perspective. In: 2014 14th UK Workshop on Computational Intelligence (UKCI). 2014. pp. 1–8
3. Yang L, Neagu D. Toxicity risk assessment from heterogeneous uncertain data with possibility-probability distribution. In: 2013 IEEE International Conference on Fuzzy Systems (FUZZ-IEEE). 2013. pp. 1–8
4. Yang, Neagu D, Cronin MTD, Hewitt M, Enoch SJ, Madden JC, Przybylak K. Towards a fuzzy expert system on toxicological data quality assessment. Molecular Informatics. 2012;32(1):65–78
5. Yang L, Neagu D. Towards the integration of heterogeneous uncertain data. In: 2012 IEEE 13th International Conference on Information Reuse & Integration (IRI). 2012. pp. 295–302

6. Guo Q, Qu Y, Deng A, Yang L. A new fuzzy-rough feature selection algorithm for mammographic risk analysis. In: 2016 12th International Conference on Natural Computation, Fuzzy Systems and Knowledge Discovery (ICNC-FSKD). 2016. pp. 934–939
7. Mamdani EH. Application of fuzzy logic to approximate reasoning using linguistic synthesis. IEEE Transactions on Computers. 1976;26(12):1182–1191
8. Takagi T, Sugeno M. Fuzzy identification of systems and its applications to modeling and control. IEEE Transactions on Systems, Man, and Cybernetics. 1985;1:116–132
9. Kóczy L, Hirota K. Approximate reasoning by linear rule interpolation and general approximation. International Journal of Approximate Reasoning. 1993;9(3): 197–225
10. Kóczy L, Hirota K. Interpolative reasoning with insufficient evidence in sparse fuzzy rule bases. Information Sciences. 1993;71(1–2):169–201
11. Yang L, Chao F, Shen Q. Generalised adaptive fuzzy rule interpolation. IEEE Transactions on Fuzzy Systems. 2016. DOI: 10.1109/TFUZZ.2016.2582526
12. Yang L, Shen Q. Adaptive Fuzzy Interpolation. IEEE Transactions on Fuzzy Systems. 2011;19(6):1107–1126
13. Yang L, Shen Q. Closed form fuzzy interpolation. Fuzzy Sets and Systems. 2013;225:1–22
14. Tikk D, Baranyi P. Comprehensive analysis of a new fuzzy rule interpolation method. IEEE Transactions on Fuzzy Systems. 2000;8(3):281–296
15. Kovács S. Extending the fuzzy rule interpolation "FIVE" by fuzzy observation. In: Computational Intelligence, Theory and Applications. Springer; 2006. pp. 485–497
16. Johanyák ZC, Kovács S. Fuzzy rule interpolation based on polar cuts. In: Computational Intelligence, Theory and Applications. Springer; 2006. pp. 499–511
17. Baranyi P, Kóczy LT, Gedeon TD. A generalized concept for fuzzy rule interpolation. IEEE Transactions on Fuzzy Systems. 2004;12(6):820–837
18. Dubois D, Prade H. On fuzzy interpolation. International Journal of General System. 1999;28(2–3):103–114
19. Jin S, Diao R, Quek C, Shen Q. Backward fuzzy rule interpolation. IEEE Transactions on Fuzzy Systems. 2014;22(6):1682–1698
20. Chen C, Parthaláin NM, Li Y, Price C, Quek C, Shen Q. Rough-fuzzy rule interpolation. Information Sciences. 2016;351:1–17

21. Li J, Qu Y, Shum HP, Yang L. TSK inference with sparse rule bases. In: Advances in Computational Intelligence Systems. 2017. pp. 107–123
22. Molnárka GI, Kovács S, Kóczy LT. Fuzzy rule interpolation based fuzzy signature structure in building condition evaluation. In: 2014 IEEE International Conference on Fuzzy Systems (FUZZ-IEEE). 2014. pp. 2214–2221
23. Li J, Yang L, Shum HP, Sexton G, Tan Y. Intelligent home heating controller using fuzzy rule interpolation. In: 2015 15th UK Workshop on Computational Intelligence (UKCI). 2015
24. Bai Y, Zhuang H, Wang D. Apply fuzzy interpolation method to calibrate parallel machine tools. The International Journal of Advanced Manufacturing Technology. 2012;60(5–8):553–560
25. Kovács S, Koczy LT. Application of interpolation-based fuzzy logic reasoning in behaviour-based control structures. In: 2004 IEEE International Conference on Fuzzy Systems. 2004
26. Li J, Yang L, Fu X, Chao F, Qu Y. Dynamic QoS solution for enterprise networks using TSK fuzzy interpolation. In: 2017 IEEE International Conference on Fuzzy Systems (FUZZ-IEEE). 2017
27. Yang L, Li J, Fehringer G, Barraclough P, Sexton G, Cao Y. Intrusion detection system by fuzzy interpolation. In: 2017 IEEE International Conference on Fuzzy Systems (FUZZ-IEEE). 2017
28. Mamdani EH, Assilian S. An experiment in linguistic synthesis with a fuzzy logic controller. International Journal of Man-Machine Studies. 1975;7(1):1–13
29. Castro JL, Zurita JM. An inductive learning algorithm in fuzzy systems. Fuzzy Sets and Systems. 1997;89(2):193–203
30. Chen SM, Lee SH, and Lee CH. A new method for generating fuzzy rules from numerical data for handling classification problems. Applied Artificial Intelligence. 2001;15(7):645–664
31. Zedeh LA. Outline of a new approach to the analysis of complex systems and decision processes. IEEE Transactions on Systems, Man, and Cybernetic. 1973;3:28–44
32. Bandler W, Kohout L. Fuzzy power sets and fuzzy implication operators. Fuzzy Sets and Systems. 1980;4(1):13–30
33. Turksen IB. Four methods of approximate reasoning with interval-valued fuzzy sets. International Journal of Approximate Reasoning. 1989;3(2):121–142
34. Turksen IB, Zhong Z. An approximate analogical reasoning approach based on similarity measures. IEEE Transactions on Systems, Man, and Cybernetics. 1988;18(6):1049–1056

35. Klement EP, Mesiar R, Pap E. Triangular norms. Position paper I: Basic analytical and algebraic properties. Fuzzy Sets and Systems. 2004;143(1):5–26

36. Mizumoto M, Zimmermann HJ. Comparison of fuzzy reasoning methods. Fuzzy Sets and Systems. 1982;8(3):253–283

37. Nakanishi H, Turksen IB, Sugeno M. A review and comparison of six reasoning methods. Fuzzy Sets and Systems. 1993;57(3):257–294

38. Turksen IB, Zhong Z. An approximate analogical reasoning schema based on similarity measures and interval-valued fuzzy sets. Fuzzy Sets and Systems. 1990;34(3): 323–346

39. Chen SM. A new approach to handling fuzzy decision-making problems. IEEE Transactions on Systems, Man, and Cybernetics. 1988;18(6):1012–1016

40. Chen SM. A Weighted fuzzy reasoning algorithm for medical diagnosis. Decision Support Systems. 1994;11(1):37–43

41. Yeung DS, Tsang ECC. A comparative study on similarity-based fuzzy reasoning methods. IEEE Transactions on Systems, Man, and Cybernetics, Part B (Cybernetics). 1997;27(2):216–227

42. Tversky A. Features of similarity. Psychological Review. 1977;84(4):327

43. Yeung DS, Tsang ECC. Fuzzy knowledge representation and reasoning using petri nets. Expert Systems with Applications. 1994;7(2):281–289

44. Yeung DS, Tsang ECC. Improved Fuzzy Knowledge Representation and Rule Evaluation Using Fuzzy Petri Nets and Degree of Subsethood. International Journal of Intelligent Systems. 1994;9(12):1083–1100

45. Kosko B. Neural Networks and Fuzzy Systems: A Dynamical Systems Approach to Machine Intelligence. New Jersey: Prentice-Hall; 1992

46. Yeung DS, Ysang ECC. A multilevel weighted fuzzy reasoning algorithm for expert systems. IEEE Transactions on Systems, Man, and Cybernetics-Part A: Systems and Humans. 1998;28(2):149–158

47. Thorndike RL. Who belongs in the family? Psychometrika. 1953;18(4):267–276

48. Klir GJ, Yuan B. Fuzzy Sets and Fuzzy Logic: Theory and Applications. Prentice Hall PTR; 1995

49. Robinson JA. A machine-oriented logic based on the resolution principle. Journal of the ACM (JACM). 1965;12(1):23–41

50. Zadeh LA. Quantitative fuzzy semantics. Information Sciences. 1971;3(2):159–176

51. Huang Z, Shen Q. Fuzzy interpolative reasoning via scale and move transformations. IEEE Transactions on Fuzzy Systems. 2006;14(2):340–359

52. Huang Z, Shen Q. Fuzzy interpolation and extrapolation: A practical approach. IEEE Transactions on Fuzzy Systems. 2008;16(1):13–28

53. Kóczy LT, Hirota K. Size reduction by interpolation in fuzzy rule bases. IEEE Transactions on Systems, Man, and Cybernetics, Part B (Cybernetics). 1997;27(1):14–25

54. Yan S, Mizumoto M, Qiao WZ. Reasoning conditions on Koczy's interpolative reasoning method in sparse fuzzy rule bases. Fuzzy Sets and Systems. 1995;75(1):63–71

55. Chen SM, Ko YK. Fuzzy interpolative reasoning for sparse fuzzy rule-based systems based on ?-cuts and transformations techniques. IEEE Transactions on Fuzzy Systems. 2008;16(6):1626–1648

56. Hsiao WH, Chen SM, Lee CH. A new interpolative reasoning method in sparse rule-based systems. Fuzzy Sets and Systems. 1998;93(1):17–22

57. Wong KW, Tikk D, Dedeon TD, Kóczy LT. Fuzzy rule interpolation for multidimensional input spaces with applications: A case study. IEEE Transactions on Fuzzy Systems. 2005;13(6):809–819

58. Yam Y, Kóczy LT. Representing membership functions as points in high-dimensional spaces for fuzzy interpolation and extrapolation. IEEE Transactions on Fuzzy Systems. 2000;8(6):761–772

59. Yeung Y, Wong ML, Baranyi P. Interpolation with function space representation of membership functions. IEEE Transactions on Fuzzy Systems. 2006;14(3):398–411

60. Yang L, Chen C, Jin N, Fu X, Shen Q. Closed form fuzzy interpolation with interval type-2 fuzzy sets. In: 2014 IEEE International Conference on Fuzzy Systems (FUZZ-IEEE). 2014. pp. 2184–2191

61. Chang YC, Chen SM, Liau CJ. Fuzzy interpolative reasoning for sparse fuzzy-rule-based systems based on the areas of fuzzy sets. IEEE Transactions on Fuzzy Systems. 2008;16(5):1285–1301

62. Tikk D, Joó I, Kóczy L, Várlaki P, Moser B, Gedeon TD. Stability of interpolative fuzzy KH controllers. Fuzzy Sets and Systems. 2002;125(1):105–119

63. Ughetto L, Dubois D, Prade H. Fuzzy interpolation by convex completion of sparse rule bases. In: 2000 IEEE International Conference on Fuzzy Systems (FUZZ-IEEE). 2000. pp. 465–470

64. Shen Q, Yang L. Generalisation of scale and move transformation-based fuzzy interpolation. Journal of Advanced Computational Intelligence and Intelligent Informatics. 2011;15(3):288–298

65. Jenei S, Klement EP, Konzel R. Interpolation and extrapolation of fuzzy quantities-The multiple-dimensional case. Soft Computing. 2002;6(3–4):258–270

66. Yang L, Shen Q. Towards adaptive interpolative reasoning. In: 2009 IEEE International Conference on Fuzzy Systems (FUZZ-IEEE). 2009. pp. 542–549

67. Yang L, Shen Q. Extending adaptive interpolation: From triangular to trapezoidal. In: 2009 9th UK Workshop on Computational Intelligence (UKCI). 2009. pp. 25–30

68. Yang L, Shen Q. Adaptive fuzzy interpolation and extrapolation with multiple-antecedent rules. In: 2010 IEEE International Conference on Fuzzy Systems (FUZZ-IEEE). 2010. pp. 1–8

69. Kleer JD. An assumption-based TMS. Artificial Intelligence. 1986;28(2):127–162

70. Kleer JD. Extending the ATMS. Artificial Intelligence. 1986;28(2):163–196

71. Kleer JD, Williams BC. Diagnosing multiple faults. Artificial Intelligence. 1987;32(1):97–130

72. Yang L, Shen Q. Adaptive fuzzy interpolation with prioritized component candidates. In: 2011 IEEE International Conference on Fuzzy Systems (FUZZ-IEEE). 2011. pp. 428–435

73. Yang L, Shen Q. Adaptive fuzzy interpolation with uncertain observations and rule base. In 2011 IEEE International Conference on Fuzzy Systems (FUZZ-IEEE). 2011. pp. 471–478

74. Cheng SH, Chen SM, Chen CL. Adaptive fuzzy interpolation based on ranking values of polygonal fuzzy sets and similarity measures between polygonal fuzzy sets. Information Sciences. 2016;342:176–190

75. Wang LX, Mendel JM. Generating fuzzy rules by learning from examples. IEEE Transactions on Systems, Man, and Cybernetics. 1992;22(6):1414–1427

76. Johanyák ZC, Kovács S. Sparse fuzzy system generation by rule base extension. In: 2007 11th IEEE International Conference on Intelligent Engineering Systems. 2007. pp. 99–104

77. Bellaaj H, Ketata R, Chtourou M. A new method for fuzzy rule base reduction. Journal of Intelligent & Fuzzy Systems. 2013;25(3):605–613

78. Tao CW. A reduction approach for fuzzy rule bases of fuzzy controllers. IEEE Transactions on Systems, Man, and Cybernetics, Part B (Cybernetics). 2002;32(5):668–675

79. Tan Y, Li J, Wonders M, Chao F, Shum HP, Yang L. Towards sparse rule base generation for fuzzy rule interpolation. In: 2016 IEEE International Conference on Fuzzy Systems (FUZZ-IEEE). 2016. pp. 110–117

80. Cowton J, Yang L. A smart calendar system using multiple search techniques. In: 2015 15th UK Workshop on Computational Intelligence (UKCI). 2015

81. Peckham SD. Profile, plan and streamline curvature: A simple derivation and applications. In: Proceedings of Geomorphometry. 2011. pp. 27–30

82. Li J, Shum PH, Fu X, Sexton G, Yang L. Experience-based rule base generation and adaptation for fuzzy interpolation. In: 2016 IEEE International Conference on Fuzzy Systems (FUZZ-IEEE). 2016. pp. 102–109

83. Chen SJ, Chen SM. Fuzzy risk analysis based on similarity measures of generalized fuzzy numbers. IEEE Transactions on Fuzzy Systems. 2003;11(1):45–56

84. Nguyen D, Widrow B. The truck backer-upper: An example of self-learning in neural networks. In: Proceedings of the International Joint Conference on Neural Networks (IJCNN). 1989;2:357–363

85. Kong SG, Kosko B. Adaptive fuzzy systems for backing up a truck-and-trailer. IEEE Transactions on Neural Networks. 1992;3(2):211–223

86. Riid A, Rustern E. Fuzzy logic in control: Truck backer-upper problem revisited. In: 2001 IEEE International Conference on Fuzzy Systems (FUZZ-IEEE). 2001;1:513–516

87. Lotfi A. Fuzzy Inference Systems Toolbox for Matlab (FISMAT). 2000

88. Haissig C. Adaptive fuzzy temperature control for hydronic heating systems. IEEE Control Systems. 2000;20(2):39–48

89. Lu J, Sookoor T, Srinivasan V, Gao G, Holben B, Stankovic J, Field E, Whitehouse K. The Smart Thermostat: Using Occupancy Sensors to Save Energy in Homes. In: Proceedings of the 8th ACM Conference on Embedded Networked Sensor Systems. 2010. pp. 211–224

90. Nevius M, Pigg S. Programmable thermostats that go berserk: Taking a social perspective on space heating in Wisconsin. In: Proceedings of the 2000 ACEEE Summer Study on Energy Efficiency in Buildings. 2000;8:e44

91. Scott, J, Bernheim Brush AJ, Krumm J, Meyers B, Hazas M, Hodges S, Villar N. PreHeat: Controlling home heating using occupancy prediction. In: Proceedings of the 13th International conference on Ubiquitous Computing. 2011 pp. 281–290

92. Von Altrock, C, Arend HO, Krausse B, Steffens C. Customer-adaptive fuzzy logic control of home heating system. In: 1994 IEEE International Conference on Fuzzy Systems (FUZZ-IEEE). 1994;3:1713–1718

93. Friedman N, Geiger D, Goldszmidt M. Bayesian network classifiers. Machine Learning. 1997;29(2–3):131–163

94. Chao F, Huang Y, Zhang X, Shang C, Yang L, Zhou C, Hu H, Lin CM. A robot calligraphy system: From simple to complex writing by human gestures. Engineering Applications of Artificial Intelligence. 2017;59:1–14

95. Chao F, Zhu Z, Lin CM, Hu H, Yang L, Shang C, Zhou C. Enhanced robotic hand-eye coordination inspired from human-like behavioral patterns. IEEE Transactions on Cognitive and Developmental Systems. 2017. DOI: 10.1109/TCDS.2016.2620156

4 Chapter

EXECUTING SOPHISTICATED FUZZY ANALYSIS FOR BUSINESS PLANNING SYSTEMS

1. INTRODUCTION: AN OVERVIEW

In the initial section of the chapter, the author analyzes the difficulties in converting subject descriptions from normal language into a formal model. Facilitating collaboration among experts in many fields (technology, information, business, etc.) is sometimes a complex challenge. In collaborative efforts, there is a significant necessity to assess model clarity, efficacy, and efficiency for experts across many fields of study. We define efficacy as the accurate interpretation of a subject in the real world by a model. Efficiency is determined by the proportion of the service states of the model. In the subsequent section, the author demonstrates the application of the methodology in virtual training design. Formulating targeted training methodologies for adult lifelong learners is a highly intricate endeavor. We employed fuzzy-based research to identify the optimal learning methods for various student demographics and course categories. This section focuses on the adaptation and implementation of virtual training for individuals with hearing impairments.

2. EXPANSION FOR THE FUZZY MODEL DEFINITION DOMAIN

2.1. Introduction

Information systems are widely spread in our daily life: offices, industry, and home. Each system represents or controls an extra object, such as household appliances or industrial control devices. Efficiency of those information and control systems depends on a wide range of factors. Wide-known methods for control systems' synthesis demand the formal model represented in analytical format [1]. The main problem of the discrete automate models is the strong necessity of formalization at the beginning of the process. It is a rather difficult task for complicated technological and information processes. Moreover, it is more appropriate to set a task of the adaptive control system with parametric adaptation features synthesis.

Let us consider the control system for an informational object as a "black box" with two sets of inputs and outputs [3]. Two main inputs are a math (or analytical) model, and data. One of the important features is a number and percentage of internal states of an information system/control device. Internal states are always necessary to control stability of work, prevent user's mistakes, etc. These internal states and their properties require additional computing resources. The author examines an approach aimed at balancing advantages and resource usage of internal states [2].

Meanwhile these both inputs are (or can be) cross-dependent. Creating an analytical model as the first step of the system synthesis is based on intrinsic and semantic analysis of input data. Establishment of collaboration between specialists in different knowledge areas (technology, information, business, or others) is usually one of the main problems and challenges almost in all practice-based projects [3]. Factor analysis or lingual models may provide partial decision of the problem. Lingual models combine a system description in both formal and descriptive terms. A restriction of linguistic models is in their insufficient formalization and a high level of dependence on subjective expert appraisals.

Fuzzy models implementation ensures an analysis of processes in technical and information-based systems with nonlinear and/or multifactor-based behavior [1, 5]. The author presents the approach that combines model analysis based on natural language with strict formal systems by implementing fuzzy logic approaches.

2.2. General approach based on lingual models

We consider a process of analytical model synthesis as a first phase of a control model synthesis. The first important step at this phase is transition of a description based on natural language to a formal model. We should consider an interaction between specialists in different knowledge areas while creating such models as they may use different terms. Another important task is establishing back coupling between input data and information/control system. It is quite necessary for creating information and control systems with adaptive features. We implement fuzzy logic to solve the task of information system synthesis for complicated technological objects.

Implementing fuzzy logic ensures the opportunity of a semantic analysis of the object description made on natural-based language due to partial entrance of an element in one or several sets. The conceptual structure of the research area is represented as a set of abstract entities relying on concepts and terms of both a natural language and fuzzy sets. Next, we need to extend the existing fuzzy sets' model as an analytical model contains features, appropriate to elements of analyzable object, and own properties of the model, that provide its integrity.

Second step should be to deal with input data analysis to adapt the model and control system to environment changes. We do not need to change our model but make it adaptive. We need to establish clear borders for research object and its analytical model. To resolve this task, we need also to extend current fuzzy sets and models' methods and features. In this work, we present an extension range of definition of fuzzy sets and its elements' logical division to two sets: corresponding to object properties and internal properties of a system itself.

2.3. Model analysis in a partially defined environment based on fuzzy sets

Any analytical model has several internal features that ensure its internal integrity and reliability of a control system to controlled object. One may implement internal model states for internal data integrity control, exchange, additional logic control, etc.

Let's name a set of objects to be synthesized as *S* = {*a*[*i*]}. Each noun in a lingual model compares to an object *a*[*i*] with value (weight) *a*(*i*) and functionality *H*(*a*[*i*]). Lets' define an expansion of a definition range values for fuzzy attributes to imaginary area:

$$\begin{cases} j*a(i), & \text{if an object exists only in a formal model} \\ a(i)=a(i) & \text{if an object compares to a modelling subject} \end{cases} \tag{1}$$

Based on the above, it is obvious that model S always consists of objects with both rational and imaginary features and functionality weights. *H*+ and *H*-[4, 6, 7]. In this case, total number of object *a*[*i*] features is h(n).

1. Number of features compared to lingual model is h+.
2. Total number of object features in fuzzy model– S.
3. Weights of each feature is *s*(*i*).
4. A functional as a summary (vector) weight of features divided into rational and imaginary parts *a*[*i*] -> *H*.

A model functional parameter may be represented as the following complex number:

$$H = H + +jH - \tag{2}$$

2.4. Extending fuzzy logic procedures for analysis and synthesis of information systems

Below, we show the extension is correct and does not break the fuzzy logic postulates. The definition area would be the following:

$$H\ (A|B) \in [0;\infty[+\mathrm{j}([0,\infty[) \tag{3}$$

Lemmas:

If A is the element of a real object and B is the internal element:

$$H\ (A|B) \in [0;\infty[\tag{4}$$

If both A, B are internal elements:

$$H\ (A|B) \in j([0;\infty[) \tag{5}$$

$$H(U) = \infty \tag{6}$$

U is a universal set.

We provided a transformation of the traditional fuzzy sets axioms and features to a specific fuzzy model for using complex-based definition area to specify features of models for synthesis of control automation systems.

2.4.1. Fuzzy model synthesis based on lingual model analysis for technological and information objects

First, we need to transform a lingual model based on subject area terms and definitions to a metalanguage-based model. This model still may include some terms from its predecessor—the certain lingual model. Let us mark sets of keywords, and consider them as fuzzy model objects. At the first step, we may consider all nouns as fuzzy model objects [1–3].

Next let's define a transition between fuzzy objects:

$$\left\{ a[i] \overset{\text{fi}}{\rightarrow} a[i+1], \text{fi is an } a[i] \text{ object's method} \right. \tag{7}$$

If weight *a(i)* is an imaginary one, then the object *a*[*i*] does not match any object in a lingual model, and only meta model contains it.

Then, based on δ-operation, as it is defined in common fuzzy logic, we provide the following transformation between logically tied objects:

$$\begin{cases} a[i]\ \delta i a[i+1] \ \Rightarrow a[i+1] \\ a[i]\ \delta i+1\ [i+1] \ \Rightarrow a[i] \overset{\text{fi}+1}{\longrightarrow} a[i+2] \end{cases} \tag{8}$$

So, we can formalize an object's method based on a 2-step δ-operation.

Let's introduce a weight of a fuzzy object *H* (*M*)=*A*. We consider Σ*H*(*a*[*i*])>=*H*(*P*)), as the model contains both objects and their methods *f*(*i*), that define a consequence of transitions between objects in a model. *f*(*i*) methods are described by γ and δ operations. δ -operation defines a consequence of operations, events, and nodes of a model, and γ -operation defines weights (for example, possibility, availability, resilience, etc.) for nodes, events, operations. If we take into consideration both types and an optimization factor $H(M)=A\rightarrow \max$, then the following equality is true:

$$H(P) = H(a[1]\gamma H(a[2]\gamma\ \gamma H(a[i]\gamma\ \gamma H(a[n] \tag{9}$$

The following set of conditions provides model and based on it control system integrity:

$$\begin{cases} \mathrm{H(M)} \quad \rightarrow \max; \\ \mathrm{H(M)} = \Gamma a[i]\ \Delta a[i]; \\ \nabla a[i], \mathrm{H}(a[i] \neq 0; \\ \nabla a[i], \mathrm{H}(a[i]) \rightarrow \max; \\ \gamma \mathrm{i} \in a[i],\ \delta \mathrm{i} \in\ a[i]; \\ \mathrm{f(i)} = \gamma \mathrm{i} \cup \delta \mathrm{i} \end{cases} \tag{10}$$

The following operations provide structuration of fuzzy sets in A:

1. Establish relations between sets:

1.1. Relation of entrance Ai ∈ Aj

1.2. Relation of inheritance. Ai inherits Aj if a(i) ∈ Ai has the fixed set of values, and methods no less than a(j) ∈ A(j).

2. Let's introduce meta-set M, which describes a set of sets:

$$\mathrm{Ai}:\ \mathrm{M} = \cup(\mathrm{n(i)}, \cup \mathrm{h(i,j)}), \tag{11}$$

And a system of their internal relationships, where *n(i)*, Ai's unique identifier (usually string name); *h(i,j)*, a weight of relation between Ai and Aj sets.

An optimization of model parameters provides by consequent iterations. A rate, which weights reliability between fuzzy meta-model and preceding lingual model, is one of the most important optimization criteria. It is based on a feature $\nabla[i] \in M$, that indicates a belongingness of a set element to a real or imaginary areas.

2.4.2. Fuzzy model synthesis algorithm based on a preceding lingual model

1. Allocate described essences based on the linguistic analysis. Create a set S of subessences.
2. Allocate a set of described attributes of essences. Based on it create a set T of fuzzy model attributes as mentioned in [2, 3].

3. Create a set K of basic knowledge sources, including experts, knowledgebase, and experimental data.
4. Create set C as a united set of weight criteria of each element. Primary weight is proportional to number of its occurrences into subsets *C*(*i*).
5. Create subsets *S*(*i*) as based on the S set. *S*(*i*) elements are united into a certain subset according to *C*(*i*) criteria.
6. Create sets H(j) containing experts', knowledgebase articles and other data sources' weights: $C_i \rightarrow H_j(Ci)$.
7. Create sets *A*(*i*) including attributes *a*(*i*), as Ai ∇ S,t(i)A1, An, where

$$n >= 1, \nabla h(j)(Ai) = f(h(j), A(i), Hj, [Ai] \setminus Ci).$$

8. Establish fuzzy model M (Ai, H[a(j)].

Synthesis of a control system for technical and/or information object is based on the defined fuzzy set M.

2.5. Scope definition for fuzzy sets usage in control and simulation systems for technical and information systems

While designing formal model (including fuzzy one), it is necessary to estimate the following items:

- Possibility of implementing a certain formalization method to a certain object;
- Degree of model compliance to an object. Consider fuzzy set A, containing *a*[*i*] elements. Each *a*[*i*] is compliant with an object feature or internal feature of a model. We consider a method of compliance between model and object based on research of mutual consistency of elements. The method is based on a ∪-operation, and indicates subsets, which can describe an object behavior if used in common [2, 3]:

$$M = A \cup B \cup C \cup \, \cup N \qquad (12)$$

2.5.1. Internal consistency criteria

1. Redundancy property:

$$\forall H(a[i]\beta a[j] > 0$$

2. Compliance property:

$$\forall H(a[i]|a[j] \neq \infty$$

3. Efficiency and resilience balance:

$$\begin{cases} \sum a[i](\mathrm{h}+) > 0 \\ \sum a[i](\mathrm{h}-) > 0 \\ \sum a[i](\mathrm{h}) > 0 \end{cases} \tag{13}$$

2.5.2. Compliance to modeling object criteria

(1) Compliance between M - model transformations results, and an object fact features: implementing of any possible track of operations of M model cannot lead an object to a prohibited state.

$$\forall \mathrm{H}(Xi(Ai)) > 0, \tag{14}$$

or

Implementing of any possible track in M, model can transform the controlled system or object R into a possible state or common null state

$$\forall H(Xi(Ai)) >= 0 \tag{15}$$

Thus, we consider a fuzzy model is applicable in case that its real functional is positive and there is at least one set of allowed methods that transform a control system from its initial to final state.

2.6. Conclusion

Finally, we found that the synthesis of control systems based on descriptive models of natural language may be adequately implemented based on fuzzy sets. Logical separation of elements of fuzzy sets, in which the real domain includes the attributes and functional elements that describe the state of an object, but to the imaginary one—own internal state of the model and the management system that are required to make it operational. Based on this logical separation, we may estimate effectiveness and resilience of control system.

Finally, the authors resume that automated systems' synthesis is appropriately presented and formalized by fuzzy sets' models. Fuzzy logic definition area has been extended to an imaginary area. We established the logical division of model components to real and imaginary areas per their role. Internal objects of a model are presented in the imaginary area, and objects that describe the modeling system to the real area. We introduced necessary functional extensions for fuzzy logic to operate with logical extension.

Transformation algorithm is developed, and we recommend the certain implementation area for it.

3. DEFINING APPROPRIATE TRAINING METHODS FOR LIFELONG LEARNING ORGANIZATION

3.1. Introduction

The author has developed a method and algorithms of fuzzy analysis for lingual models with complex digits' implementation. The author used the approach that differentiates native data, attributes of an object and internal model data, and attributes [2, 3]. Dividing these classes into real and imaginary leads to the decrease of dimensions in a model, and in this way to the decrease of computing capacity. This fact allows to decrease the risk of incorrect interpretation of results, and it provides also an opportunity to estimate costs of efficiency.

This article is devoted to implementing fuzzy analysis to define and implement various virtual training methods in a lifelong learning educational organization and reaches the highest possible satisfaction level by different categories of adult students as defined in Ref. [4].

3.2. Big challenges in lifelong learning

The lifelong professional learning training center offers short-term trainings and postdiploma programs to upgrade professional skills or gain a new specialization for adult professionals. A lot of students take multiple courses as bundles or periodically in accordance to new versions of software, technological equipment, or professional standards. It's of great importance for the training organization to analyze big data interdependencies to find out trends, develop new courses, make targeted offers for students, and create specific training methods for certain client groups. Since 2009, the author has been deeply involved into developing and implementing various virtual—online, and blended—training methods. During this work, the author carried out a regular analysis of data from different sources to determine customer requirements, demands for courses, and ways of their representation, technical, and mythological opportunities [4, 5]. The goal of those continual research efforts was the development of strongly targeted training options for certain student groups and courses. The fuzzy-based modeling is considered as the most appropriate approach to the task, because students', customers', trainers', and other staff's feedback, requirements, as well as demand estimations, are mostly represented as a nonformal or mixed way. For example, rating A in a feedback means "more than I can expect." It is obvious that the level of expectations differs among students, and customer representatives.

3.3. Opportunities and threats in lifelong learning

Based on M_o_R™ and Total Risk Management® concepts fundamental characteristics of any risk define organization behavior for it, for example: tolerance level, impact, mitigation and contingency strategies, management level, as well as level of financial reserves. While examining risk nature we often consider that a single risk belongs to different characteristic sets. For example, a risk of incorrect professional behavior can belong to human and organizational, and technical sets simultaneously. Therefore, we can create a fuzzy description of a risk:

$$r(i) \in O, r(i) \in T, \mathrm{r(i)}\ H \tag{16}$$

where **O, T, H,** fuzzy sets (organizational, technical, human features).

Implementing risk analysis in fuzzy terms ensures complex analysis for risk source, impact, mitigation, and contingency. The author examined complex risk analysis for portfolio (both projects and operational activities) of virtual learning methods in a lifelong adult training center. As a service-based and private user (a student) oriented business, its success depends dramatically on a subjective personalized opinion of students and partially of corporate HR managers. Their feedbacks are represented both in partially formalized manner, and comments in a natural language.

Another challenge concerns representing risk dependencies, or so-called domino effect. As it's a rather complicated task to formalize risks interdependencies, we can implement an approach, starting with an informal description in a natural language with further formalizing it by means of fuzzy-based algorithms.

The fuzzy analysis is the very appropriate tool to transfer statements in a natural language into a formal model, and explore threats and opportunities. The fuzzy analysis is implemented as described by the author in Ref. [5]. Identifying and analyzing risks, and their interdependencies, we include both negative (threat) and positive (opportunity) parts of risk analysis with the primary aim for finding new opportunities for development and quality improvement.

Main threats for an adult professional training organization are in customer dissatisfaction, and on the opposite main opportunities are based on reaching continuous education of students personally, and corporate customers. Let's examine a simple example about modern technology-based virtual learning implementation, and consider online learning process. There are several main opportunities of online learning for an educational organization, which are:

1. Improve organization innovative brand.
2. Attracting more students from distant regions.
3. More students in a class in a certain group.

On the other side, there are threats:

1. Changing in teaching methods.
2. Decrease of teaching quality.
3. Student's and corporate customers' rejection of new training method.
4. Technical issues.

Those risks—both positive and negative—are well-known when we talk about them in a natural language, but training organization's decision-making process requires qualitative and quantitative estimates. As shown in reference [1] we can implement fuzzy analysis to transform natural language to a weak formalized fuzzy model, by placing model-internal risks into an imaginary area, and objective risks into a real area of the model.

3.4. Building current data analysis with fuzzy logic

We investigated our students, trainers, corporate clients, and internal administrative staff feedbacks to discover additional training opportunities.

In 2009 we started online webinar trainings, which are held as simultaneous trainings in groups consisting of online (webinar), and class-based students (named as "webinar-in-class™"). The example of feedbacks is given in a Table 1. Total number of feedbacks: 10,000+ student feedbacks, 700+ by trainers, 1000+ by training center administrative staff, and 500+ by corporate customers' representatives.

Table 1. An excerpt from webinar-N-class studies.

Rate	Parameter (%)			
	Trainer	Technical facilities	Course	Willing for further training
A	32	27	45	42
B	48	51	42	47
C	12	14	10	7
D	5	6	2	2
E	3	2	1	1

We compared and analyzed feedbacks of webinars with the excerpt feedbacks of traditional class-based trainings. As the "specialist computer training center (CCT)" has been operated since 1991, we extracted feedbacks for class trainings for previous 5 years, e.g., we included 25,000 students', 5000 trainers', 5000 administrative staff's, and 2000 customer representatives' feedbacks into comparative analysis against "webinar-in-class" feedbacks, which are presented in Table 2.

Table 2. An excerpt from traditional class-based studies.

Rate	Parameter (%)			
	Trainer	Technical facilities	Course	Willing for further training
A	39	42	41	42
B	46	41	44	47
C	11	13	13	7
D	3	4	2	2
E	1	2	1	1

The "trapeze" form of a fuzzy interpretation, as shown at Figure 1, is used to represent fuzzy component, because rating values are subjective and personal oriented. For example, we use ratings from "1" or "E" (minimum value, means that a client is completely dissatisfied) to "5" or "A" (maximum value, indicates that a service exceeds customer's expectations). Rate "3" or "C" indicates customer's general satisfaction. These estimations are personally based and depend on a lot of factors, such as professional specialization, job function and rank, individual specialties. Only ratings "E" and "A" strongly indicate satisfactory level. For example, rating "3" or "C" at design or HR trainings is mainly considered as more dissatisfied than satisfied. On the other side a "C" rating at business or project management trainings is mainly considered as satisfied' and rating A" is very rare, because business managers are mostly not as emotional.

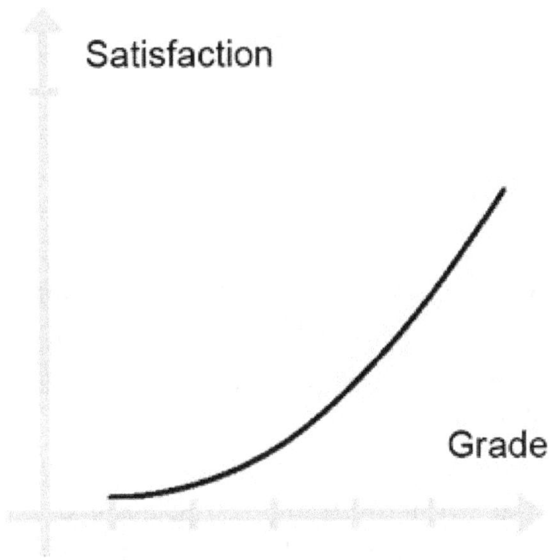

Figure 1. Trapeze interpretation of satisfaction level.

In fact, the total number of analyzed attributes is more than 100, and it changes regularly to follow customer, market demands, technical, and methodological facilities.

Below is a partial list of main attributes in the information system, which contain basic data and we consider them as a real area attributes in an analytical model:

- Student, Client/Customer, Learning format, Country/region, Course, Year/season/month, Vendor, Product Trainer, Trainer rating, Course rating, Training method rating, Number of courses taken by a student afterwards

The total number of real area attributes is more than 50.

Next I show an excerpt from a list of more than 25 additional (information model based) attributes. We consider these attributes in an imaginary area of our fuzzy model:

- Total rating of a training method, comparative rating to class-based training, comparative rating of webinars against class and self-learning combined method, views and filters across client types, regions, time, season, etc.

To build an integral customer satisfaction rating we use multidimensional fuzzy analysis of different partial (single-parameter) rankings as shown at Figure 2. Also, different filters and constraints are implemented to localize problems, challenges and find grow-points.

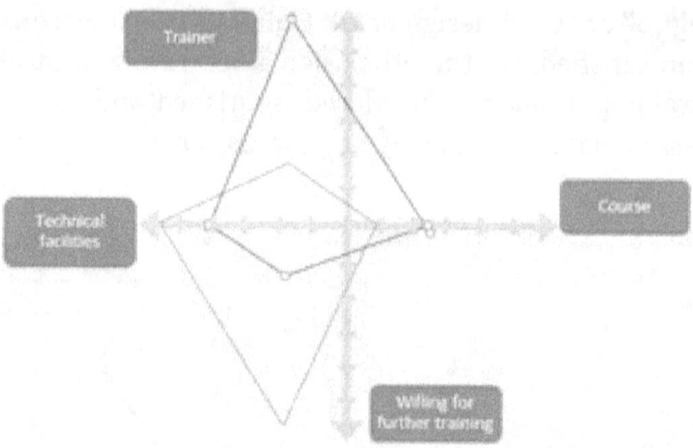

Figure 2. Determining the satisfaction area on an example of four parameter analysis.

3.5. Investigating students' satisfaction against educational organization efficiency

We can investigate an integral satisfaction/dissatisfaction level of a training group based on the following attributes: trainer ratings, course ratings, willingness to continue training at a next course, and ratings of technical facilities. Each set is fuzzy and contains ranking values (ratings) by each student.

In each case (for a student/group/trainer/course, etc.) we form a fuzzy area, in which we may consider that a course/study/trainer/class, etc. are satisfactory. According to company goals and statistical data we can also implement additional weights for attributes. For example, a trainer rank has coefficient "1, 5" and class ranking—"0, 8". Attribute "willing for further training" will have the maximum weight coefficient "2", as it's the most important factor for commercial adult learning company. This model shows a subjective satisfaction level as a set (family) of polygons. Each polygon draws by connecting points, which reflect partial ratings. For example, an area of subjective satisfaction is defined by the following partial ratings:

1. Trainer rating = 2 AND
2. Course rating = 4 AND
3. Technical facilities = 3 AND
4. Willing to continue education = 5.

In this case, we define that a group is partially satisfied/partially dissatisfied with a trainer, but if most of students of a certain group are ready to continue education at next courses, we can mark this group as "satisfied."

For complex estimation of job effectiveness of an adult life-long training organization we developed more complicated approaches, which include statistical data based on more than 70 attributes (both in real and imaginary areas) and collected them in multidimensional databases—OLAP cubes. This cube has the appropriate number (more than 70) of dimensions, and we need to build a set of fuzzy models, and optimize them for daily calculations and analysis. A real part of a model includes basic facts, and on the other side an imaginary part of a model includes filters, views, and additional states of a model or database. Due to this approach, we decrease number of dimensions to 50 in total, which leads to decrease in computing capacity requirements. Practical result: we have an opportunity to process analytical reports in a real-time mode, and postpone few complicated reports for nonworking hours (night time and Sundays).

Let us examine a comparatively simple set of fuzzy sets, which describes an integral satisfactory factor and training organization efficiency:

- By course, a certain trainer and/or trainer group, a company—customer, learning location, a certain period, a training method, a training branch (e.g., Management, ITSM, software development, HR, etc.).

The developed model contains both real area attributes, which reflect basic states, and imaginary area attributes, which reflect temporary, service states, filter conditions, identifiers, etc.

Below, I show an excerpt from a model. Attributes named in a lingual model terms to simplify understanding

$$\begin{cases} M1 = (CN + St + Tr + TM) + j(V + TP + ST), \\ M2 = (CN + Tr + CCF + TCF + WFT) + j(TP(i) + ACR(i) + ATR(i) + V), \\ M3 = (CN + TL + C + TM) + j(TP(i) + ACR(i) + AMR(i) + V), \end{cases} \quad (17)$$

where CN, Course name; St, Student identifier; Tr, trainer identifier; V, view name; TP(i), selected time period; St, threshold level of students' satisfaction for the certain model; TM, training method; CCF, course cash flow; TCF, cash flow on courses by a certain trainer; WFT, student's willing for continuous education; ACR, average course ranking for selected period; ATR, average trainer ranking for selected period; TL, training location; C, corporate customer name; and AMR, average ranking of a certain training method for selected period.

Set M1 reflects mean level of students' satisfaction for a certain course and a certain trainer for selected time period.

Set M2 reflects current level of economic efficiency of a certain trainer, based on dynamic trend of students' satisfaction across a number of time periods (for example, month to month or quarter to quarter).

Set M3 reflects dynamics of corporate customers' satisfaction for a certain training location, course, and a training method. This set gives a control how a certain training location provides quality for a certain course and a training method, for example, webinar, or blended, or self-paced, etc.

3.6. Defining training methods and models

Adult learning training organization should offer various training opportunities for its students, such as long- and short-term trainings, class-based, virtual, blended, synchronous, and asynchronous, etc. Based on the analysis model at our Specialist CCT we develop balanced cost-effective vs. "student satisfactory" training methods for precisely defined customer audience and course bundles.

While analyzing results of modeling, we find several maximums. Each of the maximums is characterized by a certain set of parameters, as shown in Figure 3.

For example, the maximum satisfaction level which is found in an analytical cube is defined by the following attributes:

- Student total expenses (minimal);
- High quality skills received;
- High qualified trainers;

- Number of students in a class (maximal);
- Collaboration facilities in the group (maximal);
- Technical expenses are minimal.

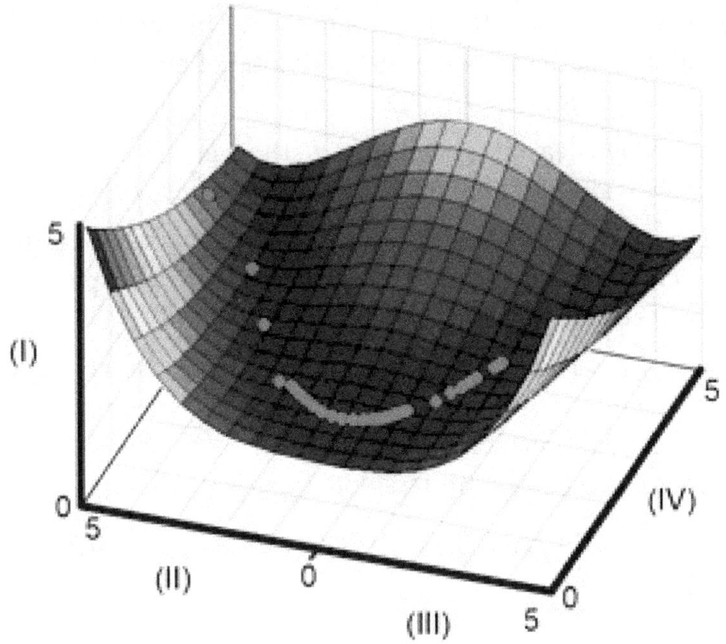

Figure 3. Multidimensional estimation, where I – Individual approach level; II – Trainer level; III – Technical facilities; IV – Training center economical effectiveness.

-

The “webinar-in-class” training option is the best for mentioned attributes, because a webinar-based student has an opportunity to study anywhere, and has no travel or accommodation expenses. Simultaneously, he/she has an access to the best trainers, and can collaborate with classmates in a class and other webinar students, using a very simple software. The webinar students have an access to the same technical facilities such as labs. Thus, the webinar-in-class training method becomes very popular solution for students studying technical (Microsoft, Oracle, CISCO, etc.) courses, as well as project, and IT service management courses.

We worked further to analyze maximums, and another point is based on the following attributes:

- Appropriate (frequent) course start date;
- Individual approach;
- Introvert students;

- Students with high level of self-organization capabilities;
- Full classes;
- Low expenses on training organization by a training center.

In this example, we see more attributes, and it was a bit more difficult to create an appropriate training method. The result was a kind of blended learning, which we named an "open learning."

- Blended "open learning" training method is the best solution for very specialized trainings, and self-organized students;
- Unlimited webinar subscription is for students, who are unemployed or wish to change profession. Both categories need cheap training at a large number of courses.

To resume I want to stress that the estimating process is everlasting, as well as optimization of the research model. While preparing this article, one more training method was developed, which proves the efficiency of the described approach. The described methods won several professional awards [6, 7].

4. IDENTIFYING SPECIAL TOOLS FOR VIRTUAL TRAINING OF HARD OF HEARING PEOPLE

4.1. Introduction

More than 10% of people on the Earth suffers from different hearing impairs as the World Health Organization data shows. Many of them are young people, or employees, which are involved into lifelong learning. They have difficulties with taking both class and virtual trainings, if they do not have or use special hearing aids. Meanwhile, many young people do not use special devices due to medical recommendations or having scruples.

Based on our research of students' with hearing impairs demands our research team deliver a special computer-based technology—named as Petralex©. It is implemented in mobile Apps and Windows driver, which works as a personal hearing assistant [6]. A student passes an "*in cito*" hearing test and the application creates a personal hearing profile for each place or environment (for example, public transport, café, car, room at home, classroom, and workplace). Mobile App acts as a hearing aid in a smartphone, so a student can easily attend classes. The Windows-based driver creates a Virtual audio device (VAD), which adapts streaming audio for both online and asynchronous virtual learning to an appropriate user's hearing profile.

Different virtual training methods—synchronous, asynchronous, blended—which are defined in Ref. [4], include online and/or off-line listening in videos, podcasts, as well as online training delivery, including real-time discussion with a trainer and classmates. So, students with hearing disabilities should have opportunities to be involved into the entire training process.

4.2. Synchronous learning methods

A student with partial hearing losses can feel uncomfortable while studying in class, or at online webinar. If a student studies in a class, he or she can implement the Petralex® mobile app as a hearing assistant, and involve in-depth into a learning process. In a synchronous learning training content is delivered in an online mode as shown in Figure 4. A student accesses it using the special audio driver, which ensures audio stream adjustment to personal hearing profile. As a result, a student can attend studies for a long time—up to 8 hours per day—due to improved hearing tolerance, reducing fatigue for long listening sessions, and attenuation of excessive sound pressure [6, 7].

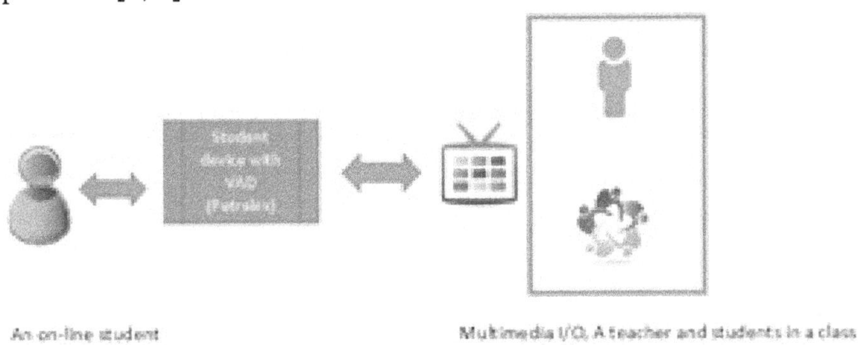

Figure 4. The learning schema for synchronous virtual learning.

4.2.1. Typical learning cases for online trainings

1. A trainer explains learning materials: a student studies at home. S/he connects to an online tool (for example, Skype®, Citrix GoTo®, WebEx, Adobe® Connect®, or any other), activates "My room" profile. Next, our driver transforms audio stream in a real-time mode with only 10–50 ms delay, so a student can hear a teacher clearly, have concentration on studying, and ask or answer questions; present his/her work, and discuss with other students in a real-time mode.
2. A business game: business games and other forms of group studies are very popular according to our model (Eq. (16)). An online student usually plays a role of a virtual team member or help desk agent. Implementing audio-driver provides both parties an opportunity to collaborate in a clear mode without delays.

4.3. Asynchronous learning methods

In this section let us consider different scenarios for asynchronous learning of students with partial hearing losses. The most popular tools for asynchronous learning are: learning management systems (LMS), stream and offline video, and audio services.

As shown in Figure 5, an external audio signal from a learning tool goes through the virtual audio driver, which transforms it according to an activated user hearing profile. Thanks to it, a student can study anywhere. At our training center, and with our partners, the following scenarios, as shown in Figure 5, were tested:

In a special class,

At home,

At a workplace, and

On a beach.

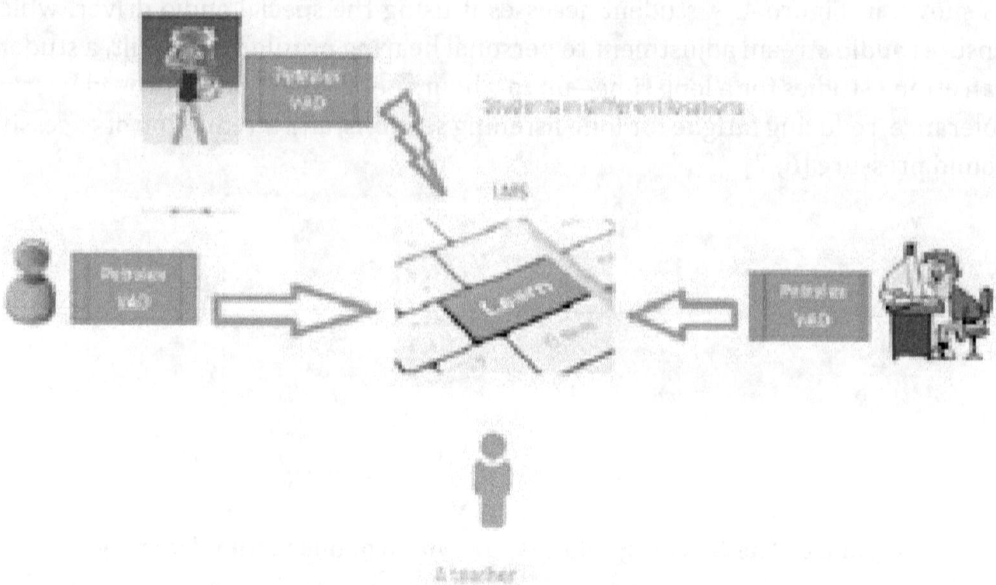

Figure 5. Asynchronous learning scenarios.

Our training approaches have demonstrated the effectiveness of short-term learning for adults with busy schedules. In our model, we included a special category for lifelong business students. Lifelong learners frequently study on their holidays or weekends. They are usually very driven, so they can effortlessly balance their leisure and academic activities. As an example, a student establishes a "beach" profile and can listen to learning records in a way that is suitable for his or her hearing.

5. RESUME

Implementing extended definition area for fuzzy set analysis provides vast opportunities for representation of control objects by information systems, their analysis and optimization. Based on implementation of fuzzy analysis the author succeeded in creating and launching various virtual training models for lifelong learning, including people suffering with partial hearing losses.

AT A GLANCE

This chapter addresses the application of fuzzy logic for converting natural language descriptions into formal fuzzy and stochastic models, along with their subsequent optimization regarding the effectiveness and efficiency of information modeling and predictive systems. Theoretical methodologies are applied in lifelong learning enterprises for development-specific virtual training for adult learners.

REFERENCES

1. Zadeh L. A. Decision Analysis and Fuzzy logic. In: Conference on Fuzzy Sets and Soft Computing in Economics and International Finance (FSSCEF 2004). Saint-Petersburg, Russia; 2004.
2. Dintsis D. A modification of discrete-event models algebra for industrial technological systems. Devices and systems. Control, Monitoring, And Diagnostics Journal # 1, 2008.
3. Dintsis D, Simankov V. The control automation system method based on combined implementation of fuzzy logic and discrete automation systems. In: European Modelling Symposium EMS2013 - 03-C Methodologies, Tools and Operations Research; November 2013; Manchester. Print ISBN:978-1-4799-2577-3 INSPEC Accession Number:14199480. DOI: 10.1109/EMS.2013.15. IEEE. p. 87–90
4. Ferdinand D.S. Flexible Learning Environments-Theories-Trends-Issues. Monography. University of West-Indies. West Indies. St.-Augustine. DOI: 10.13140/RG.2.1.3958.2488. 2016. 32p
5. Dintsis D. Implementing Fuzzy Sets For "Big Data" Analysis Based on Large Training Centre Feedbacks. In: IEEE AEIT Annual 2015 Conference. Naples, Italy. Publisher: IEEE. October, 2015.
6. Dintsis D, Bredikhin A. Virtual Learning for People with Hearing Impairs. In: IEEE AEIT Annual 2015 Conference. Naples, Italy. Publisher: IEEE. October, 2015.
7. LERN International Award Winners 2016. https://www.flipsnack.com/LearningResourcesNetwork/2016-lern-international-award-winners.html. 2016. LERN. Chicago. [Accessed: 2017-01-11]

5 Chapter

A NOVEL METHODOLOGY FOR OPTIMIZING SCALING FACTORS OF PIDTYPE FUZZY LOGIC CONTROLLERS USING GENETIC ALGORITHM IN A DISCRETE-TIME SYSTEM

1. INTRODUCTION: AN OVERVIEW

Fuzzy logic control (FLC) has been extensively employed in numerous effective industrial applications. The inaugural FLC algorithm was executed by Mamdani in 1974. In contrast to traditional control, which relies on a mathematical model of a system, a Fuzzy Logic Controller (FLC) typically incorporates the intuition and expertise of a human operator, potentially resulting in a nonlinear connection shaped by membership functions, rules, and defuzzification processes. In this regard, FLC has demonstrated successful use in various complicated and nonlinear systems, proving to be more robust and less susceptible to parametric fluctuations compared to conventional controllers. Various forms of fuzzy logic controllers (FLCs), including proportional integral (PI), proportional derivative (PD), and proportional-integral derivative (PID), have been proposed in the literature. For instance, PI-type FLCs have been effectively employed in several physical applications, including the regulation of temperature and pressure in steam engines, as well as the control of steering and speed in automobiles. Nonetheless, the efficacy of PI-type Fuzzy Logic Controllers for higher-order and nonlinear systems may be suboptimal because to significant overshoot and excessive oscillation. PD-type FLCs are appropriate for a restricted category of systems and are not advisable in the presence of measurement noise and abrupt load disruptions. In theory, PID-type fuzzy logic controllers provide commendable performance. Nonetheless, challenges arise in the creation of an effective rule foundation and the adjustment of parameters. In the proposed PID-type FLC, the parameters are categorized into two groups: structural parameters and tuning parameters. Structural parameters encompass input/output (I/O) variables for fuzzy inference, fuzzy language sets, membership functions, fuzzy rules, inference mechanisms, and defuzzification mechanisms, typically established during offline design. Tuning parameters encompass I/O scaling factors (SF) and membership function (MF) parameters, which can be computed during online controller adjustments to improve process performance [1]. The judicious selection of input and output scaling factors is crucial as they substantially influence the dynamics of the fuzzy controller. This prompts academics to investigate the most effective approach for determining optimal PID-type FLC parameters. Diverse procedures or methods have been employed to date. In Reference [2], Qiao and Mizumoto introduced a method for adjusting PID-type Fuzzy Logic Controller parameters based on a peak observer mechanism. This self-tuning technique progressively reduces the equivalent integral control component of the fuzzy controller in accordance with the system reaction time. Additionally, Woo et al. [3] devised a methodology grounded in two empirical functions that grew with the system's error data. In Ref. [1], Guzelkaya et al. introduced a method that modifies the scaling factors associated with the derivative and integral components through a fuzzy inference mechanism. Nonetheless, the primary drawbacks of all these PID-type FLC tuning methods are the challenging selection of their respective parameters and procedures. A differential search algorithm (DSA) meta-heuristic technique is proposed to systematically optimize the scaling factors of the PID-type fuzzy logic controller (FLC) to address these challenges, as referenced in [4]. The fuzzy control design is established as a restricted optimization problem, which is effectively resolved using an enhanced DSA. This suggested technique incorporates many optimization criteria, including integral square error (ISE) and maximum overshoot, to ensure enhanced resilience and performance control objectives. This chapter implements a genetic algorithm (GA)-based heuristic optimization technique to achieve superior performance relative to the Simulink design optimization (SDO) technique. GA aims to minimize the discrepancy between the system output and the intended trajectory, commencing from a flat output variable produced using the flatness property. Diverse performance metrics may be utilized. This study use the integral of squared error (ISE) index to reduce the discrepancy between the output and the desired flat trajectory. The methods are implemented inside a discrete-time framework, utilizing an electronic throttle valve as a case study.

2. PID-TYPE FUZZY LOGIC CONTROLLER DESCRIPTION

In this study, we will deal with fuzzy PID-type controllers formed using one PD-type FLC with an integrator at the output.

The PID-type fuzzy logic controller structure is shown in Figure 1 [5], where K_e and K_d (K_e,$\in R^+$) are the input scaling factors are α and β ($\alpha, \beta \in R^+$) the output scaling factors.

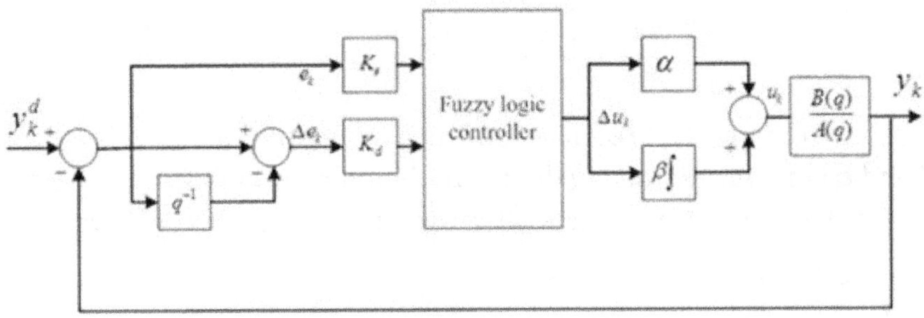

Figure 1. PID-type FLC.

The inputs variables, well known as the error ek between the desired trajectory y^d_k and the measure y_k, as well as the error variation Δe_k given by Eqs. (1) and (2) where Te is the sampling period.

$$e_k = y_k^d - y_k \tag{1}$$

$$\Delta e_k = \frac{e_k - e_{k-1}}{T_e} \tag{2}$$

The output variable Δu_k of such a controller is the variation of the control signal u_k which can be defined as Eq. (3).

$$\Delta u_k = \frac{u_k - u_{k-1}}{T_e} \tag{3}$$

The output of the PID-type fuzzy is given by Eq. (4) (5)

$$\begin{aligned} u_k &= \alpha \Delta u_k + \beta \int \Delta u_k dt \\ &= \alpha(A + PK_e e_k + DK_d \Delta e_k) \\ &+ \beta \int (A + PK_e e_k + DK_d \Delta e_k) dt \\ &= \alpha A + \beta A t + (\alpha K_e P + \beta K_d D) e_k \\ &+ \beta K_e P \int e_k dt + \alpha K_d D \Delta e_k \end{aligned} \tag{4}$$

Thus, the equivalent control components of the PID-type FLC are such that

Proportional gain: $\alpha K_e P$+$\beta K_d D$

Integral gain: $\beta K_e P$

Derivative gain: $\alpha K_d D$

where the terms P and D are given by Eqs. (5) and (6) [2].

$$P = \frac{\Delta u_{(i+1)j} - \Delta u_{ij}}{e_{i+1} - e_i} \tag{5}$$

$$D = \frac{\Delta u_{i(j+1)} - \Delta u_{ij}}{\Delta e_{j+1} - \Delta e_j} \tag{6}$$

The fuzzy controllers with a product-sum inference method, centroid defuzzification method and triangular uniformly distributed membership functions for the inputs and a crisp output proposed in Refs. [2, 6] are used in our case of study.

Table 1 gives the linguistic levels, assigned to the variables e_k, Δe_k and Δu_k, as follows: *NL*: negative large; *N*: negative; *ZR*: zero; *P*: positive; *PL*: positive large.

Table 1. Fuzzy rules-base.

$e_k \backslash \Delta e_k$	N	ZR	P
N	NL	N	ZR
ZR	N	ZR	P
P	ZR	P	PL

3. SCALING FACTORS TUNING USING GENETIC ALGORITHM

In this work, a new method is proposed for tuning the coefficients of PID-type FLCs. This method adjusts the input scaling factor corresponding to the derivative coefficient and the output scaling factor corresponding to the integral coefficient of the PID-type FLC using genetic algorithm, as shown in Figure 2. The integral of squared error (ISE) index has been used in order to minimize the error between the output and the desired flat trajectory.

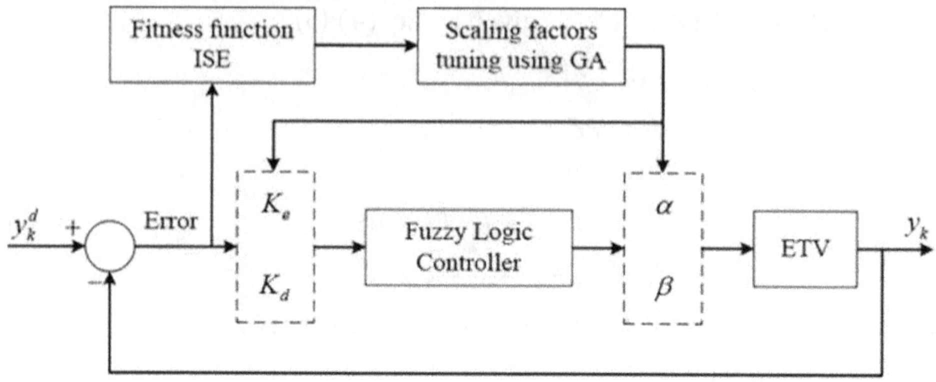

Figure 2. PID-type FLC scaling factors tuning.

3.1. Genetic algorithm

Genetic algorithm was first proposed by Holland [7]. It is a heuristic optimization technique inspired by the mechanism of natural selection. It is used in order to solve highly complex problems. GA starts with an initial population containing a number of parameters, where each one is regarded as the genes of a chromosome and can be structured by a string of concatenated values. Each chromosome represents a solution of the problem and its performance is evaluated based on fitness function.

In the beginning, an initial chromosome population is randomly generated. The chromosomes are candidate solutions to the problem. Then, the fitness values of all chromosomes are evaluated by calculating the objective function. So, a group of the best chromosomes is selected based on the fitness of each individual. In this 'surviving' population, the genetic operators of crossover and mutation are applied in order to create the next population solution. The above steps are repeated until a specific termination criterion is found.

- **Reproduction:** Create a part of the new population by simply copying without changing the selected individuals from the present population. Also, new population has the possibility of selection by already developed solutions [8].
- **Crossover:** Create new individuals as offspring of two parents. It is a recombination operator that combines selected subparts called crossover points of two parent chromosomes. The individuals resulting in this way are the offspring [8].
- **Mutation:** Create a new individual for the new population by randomly mutating a selected individual. The modifications can consist of changing one or more values in the representation or adding/deleting parts of the representation [8].

To compute the fitness of each chromosome, the objective functions are used. Many authors use integral of time multiplied by absolute error (ITAE), mean of the squared error (MSE), integral of absolute error (IAE) and integral of the squared error (ISE) as performance index [9, 10].

In this chapter, the method of tuning PID-type FLC parameters using GA consists in finding the optimal I/O scaling factors, which minimize the defined objective function, chosen as the ISE in order to specify more performance in terms of tracking a desired trajectory.

If (t) is the desired trajectory and y is the output trajectory, then error $e(t)$ is

$$e(t) = y^d(t) - y(t) \tag{7}$$

and the ISE can be defined by

$$ISE = \int_0^t e(t)^2\, dt \tag{8}$$

Fitness function is taken as inverse of error, i.e., performance index.

$$Fitness\ value = \frac{1}{Performance\ index} \tag{9}$$

3.2. Tuning procedure

The overall flowchart for optimization using GA is shown in Figure 3. Initially, a number of populations N have been generated for the scaling factors K_e, α and β. Each individual of these N sets in the current population is evaluated using the objective function ISE. Based on the values of the objective function, out of these N possible solutions, the good solutions are retained and the others are eliminated. A new population is formed by applying the genetic operators (reproduction, crossover and mutation) to these selected individuals. This process of production of a new generation and its evaluation is performed repetitively. The algorithm continues until the population converges to the stop criterion.

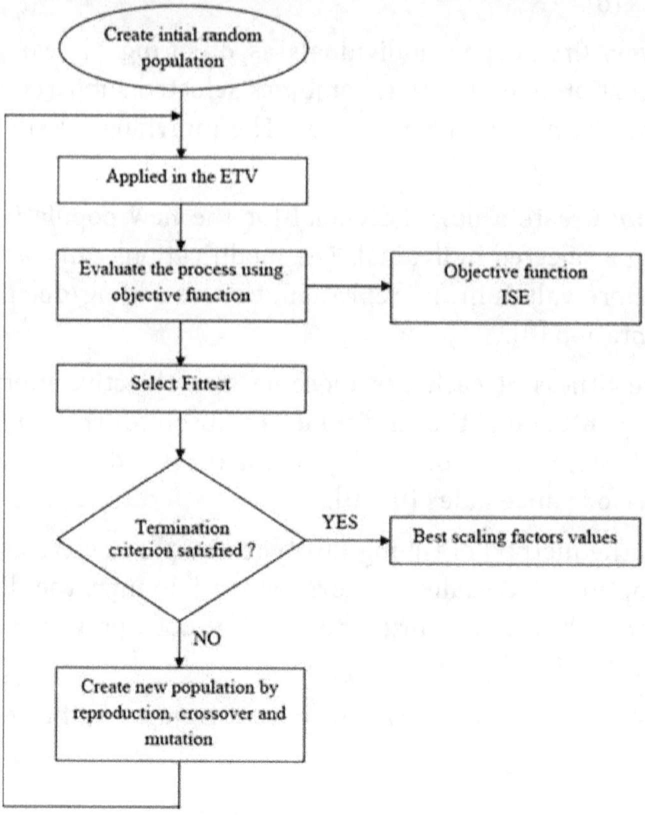

Figure 3. Flowchart of the GA optimization algorithm.

4. FLATNESS AND TRAJECTORY PLANNING

The flat property has been introduced in Ref. [11] for continuous-time nonlinear systems. It can be stated in a discrete-time version which leads for the design of a control which ensures a tracking of a desired trajectory. One major property of differential flatness is that the state and the input variables can be directly expressed, without integrating any differential equation, in terms of the flat output and a finite number of its derivatives. The flatness approach will be used in this chapter in a discrete-time framework.

The studied dynamic linear discrete system is described by Eq. (10).

$$A(q)y_k = B(q)u_k \tag{10}$$

where q is the forward operator, u_k and y_k are the input and the output, respectively, and $A(q)$ and $B(q)$ are polynomials defined by

$$A(q) = q^n + a_{n-1}q^{n-1} + \ldots + a_1 q + a_0 \tag{11}$$

$$B(q) = b_{n-1}q^{n-1} + \ldots + b_1 q + b_0 \tag{12}$$

where the parameters a_i and b_i are constants, $i=0,1,\ldots,n-1$. The partial state of such a dynamic system can be considered as a discrete flat output z_k which can be expressed as a function of input and output signals as following

$$A(q)z_k = u_k \tag{13}$$

$$B(q)z_k = y_k \tag{14}$$

Often, the real output signal y_k to be controlled is not a flat output. Then, it is necessary to plan a desired trajectory for the flat output [11] and to consider thereafter the relation (14).

The open loop control law can be determined by the following relations [12].

$$u^d(t) = f(z^d(t), \ldots, z^{d(r+1)}(t)) \tag{15}$$

$$y^d(t) = g(z^d(t), \ldots, z^{d(r)}(t)) \tag{16}$$

where f and g are the vectorial functions. Then, it is sufficient to find a desired continuous flat trajectory $t \mapsto (t)$ that must to be differentiable at the $(r+1)$ order.

The polynomial interpolation technique is used in order to plan the desired flat trajectory (t). Let consider the state vector $z^d(t)=(z^d(t)\ \dot{z}^d(t) \ldots z^{d(r+1)}(t))^T$ containing the desired continuous flat output and its successive derivatives. The expression of (t) can be given in Eq. (17); t_0 and t_f are the two moments known in advance.

$$Z^d(t) = M_1(t-t_0)c_1(t_0) + M_2(t-t_0)c_2(t_0,t_f) \quad (17)$$

where M_1 and M_2 are such as [12]

$$M_1 = \begin{pmatrix} 1 & t & \cdots & \frac{t^{n-1}}{(n-1)!} \\ 0 & 1 & \cdots & \frac{t^{n-2}}{(n-2)!} \\ \vdots & \ddots & \ddots & \vdots \\ 0 & \cdots & 0 & 1 \end{pmatrix} \quad (18)$$

$$M_2 = \begin{pmatrix} \frac{t^n}{n!} & \frac{t^{n+1}}{(n+1)!} & \cdots & \frac{t^{2n-1}}{(2n-1)!} \\ \frac{t^{n-1}}{(n-1)!} & \frac{t^n}{n!} & \cdots & \frac{t^{(n-2)}}{(n-2)!} \\ \vdots & \ddots & \ddots & \vdots \\ t & \cdots & \frac{t^{n-1}}{(n-1)!} & \frac{t^n}{n!} \end{pmatrix} \quad (19)$$

and the vectors c_1 and c_2 defined by

$$c_1 = Z^d(t_0) \quad (20)$$

$$c_2 = M_2^{-1}(t_f - t_0)(Z^d(t_f) - M_1(t_f - t_0)Z^d(t_0)) \quad (21)$$

Then, the output desired trajectory y^d_k is defined. In the discrete-time framework, the real output y_k has asymptotically to track this such as Eq. (22).

$$y_k^d = B(q)z_k^d \quad (22)$$

In the following section of this chapter, the efficiency of the proposed methodology for tuning PID-type FLC scaling factors has been validated on a discrete-time system: an electronic throttle valve for a defined desired trajectory generated using the flatness property and compared to the Simulink design optimization (SDO) technique.

5. CASE OF STUDY: ELECTRONIC THROTTLE VALVE (ETV)

Throttle valve is one of the most important devices in the engine management system. In conventional engine, the amount of airflow into the combustion system has been adjusted by the throttle valve, which is connected mechanically to an accelerator pedal [13]. The electronic throttle body (ETB) regulates air inflow into the car engine. Compared to the mechanical throttle, a well-controlled ETB can reduce fuel consumption.

5.1. System modelling

The case of the ETV is described in Figure 4.

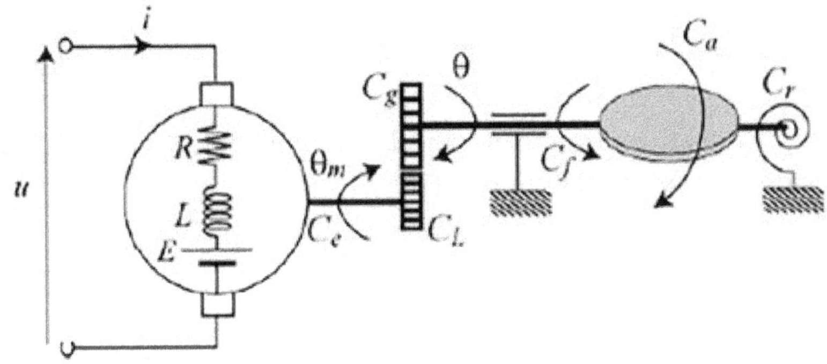

Figure 4. Electronic throttle valve system.

The electrical part is modelled by Eq. (23)

$$u(t) = L\frac{d}{dt}i(t) + Ri(t) + k_v\omega_m(t) \tag{23}$$

where L is the inductance R is the resistance $u(t)$ and $i(t)$ are the voltage and the armature current, respectively, k_v is an electromotive force constant and $\omega_m(t)$ is the motor rotational speed.

The mechanical part of the throttle is modelled by a gear reducer characterized by its reduction ratio γ such as Eq. (24)

$$\gamma = \frac{C_g}{C_L} \tag{24}$$

where C_L is the load torque and C_g is the gear torque. The mechanical part is modelled according to Eq. (25), such that [14, 15].

$$J\frac{d}{dt}\omega_m(t) = C_e - C_f - C_r - C_a \tag{25}$$

and

$$\frac{d}{dt}\theta(t) = (180/\pi/\gamma)\omega_m(t) \tag{26}$$

where $\theta(t)$ is the throttle plate angle, J is the overall moment of inertia, C_e=$k_e i(t)$ is the electrical torque where K_e is a constant, C_f is the torque caused by mechanical friction, C_r is the spring resistive torque and C_a is the torque generated by the airflow. The electronic throttle valve involves two complex nonlinearities due to the nonlinear spring torque C_r and the friction torque C_f. They are given by their static characteristics [16]. The static characteristic of the nonlinear spring torque C_r is defined by

$$C_r = \frac{k_r}{\gamma}(\theta - \theta_0) + D\,\mathrm{sgn}(\theta - \theta_0) \tag{27}$$

For $\theta_{min} \leq \theta \leq \theta_{max}$, k_r, is the spring constant, D is a constant, θ_0 is the default position and sgn(.) is the following signum function

$$\mathrm{sgn}(\theta - \theta_0) = \begin{cases} 1, \text{ if } \theta \geq \theta_0 \\ -1, \text{ else} \end{cases} \tag{28}$$

The friction torque function C_f of the angular velocity of the throttle plate can be expressed as

$$C_f = f_v \omega + f_c\,\mathrm{sgn}(\omega) \tag{29}$$

where f_v and f_c are two constants. By substituting in Eq. (25) the expressions of C_g, C_f and C_r and by neglecting the torque generated by the airflow C_a, the two nonlinearities sgn(θ-θ_0) and sgn(ω) and the two constants $\frac{k_r}{\gamma}\theta_0$ and f_v the transfer function of the linear model becomes (30) [15].

$$H(q) = \frac{(180/\pi/\gamma)k_e}{JLq^3 + JRq^2 + (k_e k_v + Lk_s)q + Rk_s} \tag{30}$$

with $k_s = (180/\pi/\gamma^2)k_r$ and q as the Laplace operators.

5.2. Simulation results

The identified parameters of (q) are given in Table 2 at 25°C temperature [15].

Table 2. Model's parameters.

Parameters	Values
R(Ω)	2.8
L(H)	0.0011
ke (N.m/A)	0.0183
kv (v/rad/s)	0.0183
J (kg.m²)	4×10^{-6}
γ	16.95

The corresponding discrete-time transfer function is given by Eq. (31) for the sample time $T_e = 0.002s$.

$$H(q^{-1}) = \frac{0.007833q^{-1} + 0.01396q^{-2} + 0.0007724q^{-3}}{1 - 1.948q^{-1} + 0.954q^{-2} - 0.006152q^{-3}} \quad (31)$$

The desired continuous time flat trajectory (t) can be computed according to the following polynomial form

$$z^d(k) = \begin{cases} \frac{cst1}{B(1)}, & \text{if } 0 \leq k \leq k_0 \\ Poly_1(k), & \text{if } k_0 \leq k \leq k_1 \\ \frac{cst2}{B(1)}, & \text{if } k_1 \leq k \leq k_2 \\ Poly_2(k), & \text{if } k_2 \leq k \leq k_3 \\ \frac{cst1}{B(1)}, & \text{if } k \geq k_3 \end{cases} \quad (32)$$

where *cst*1 and *cst*2 are constants, k0=3s, k1=6s, k2=10s and k3-15s are the instants of transitions, $B(1)$ is the static gain between the flat output z_k and the output signal y_k for each operating mode and poly1(k) and poly2(k) are polynomials calculated using the technique of polynomial interpolation.

The desired trajectory is then given in Figure 5.

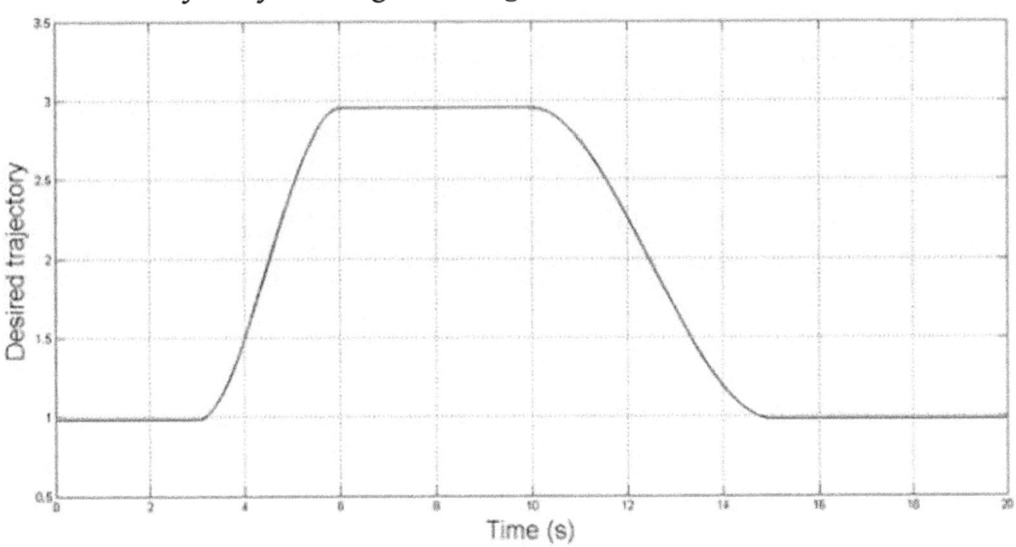

Figure 5. Desired trajectory.

The obtained optimal I/O scaling factors (Ke, Kd, α, β) for Simulink design optimization technique and GA are summarized in Table 3.

Table 3. PID-type fuzzy scaling factors values.

SF\Method	SDO	GA
Ke	0.1144	0.0086
Kd	1.5997	0.4612
α	2.6239	0.1108
β	0.0001	0.1934

The obtained results are given in Figures 6–9.

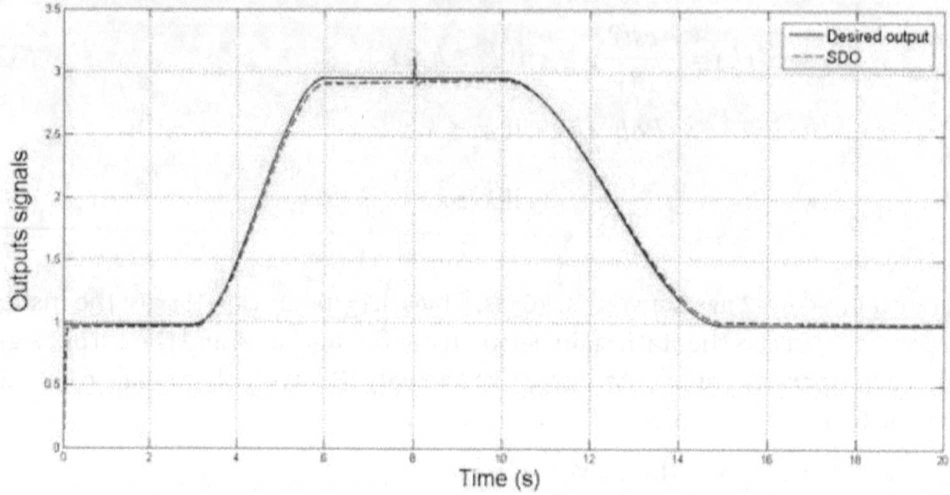

Figure 6. System outputs using Simulink design optimization.

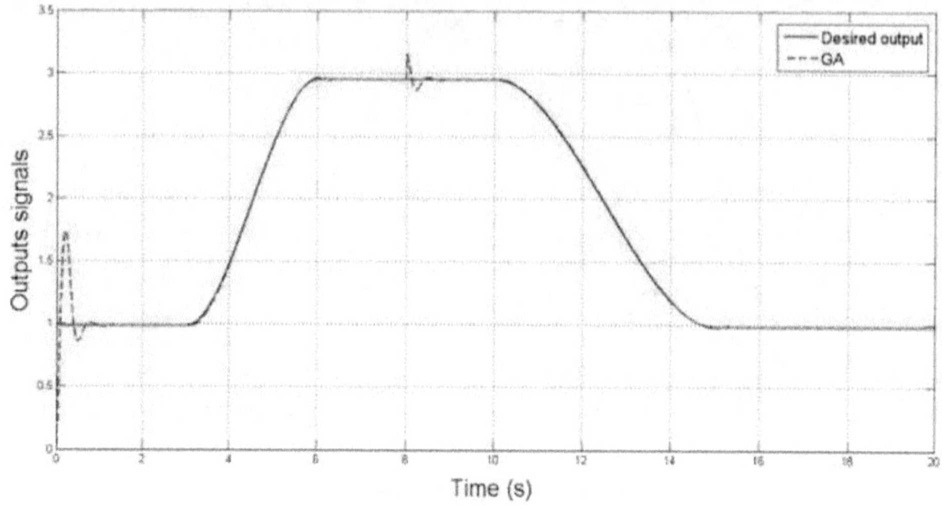

Figure 7. System outputs using genetic algorithm.

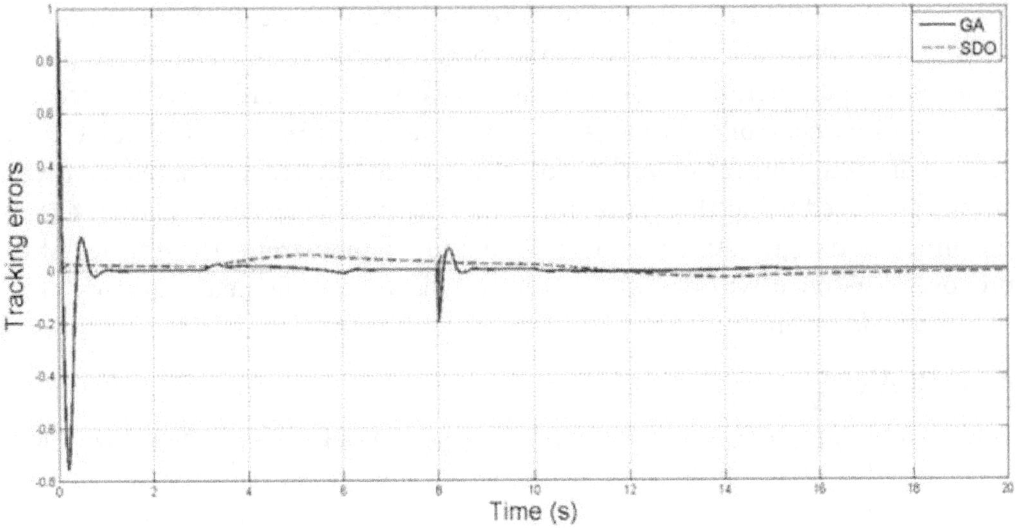

Figure 8. Tracking errors.

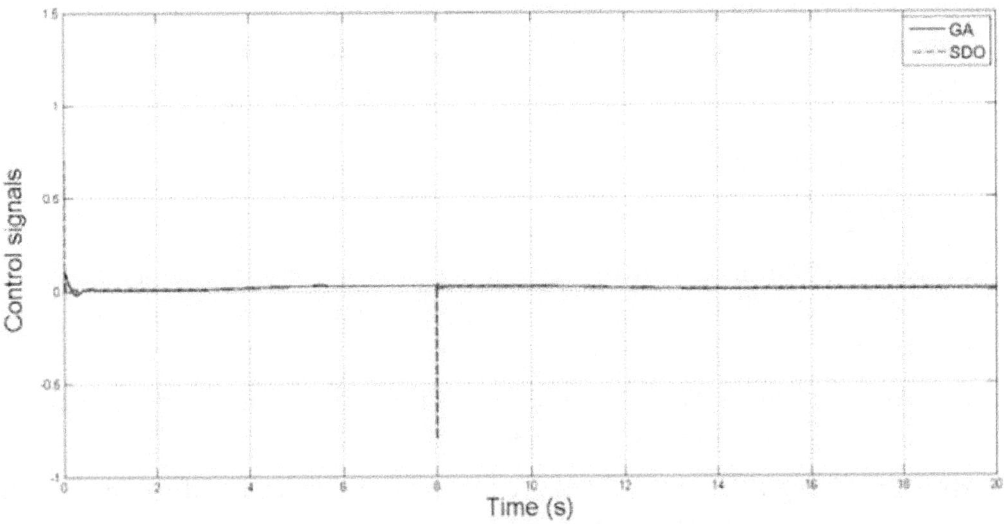

Figure 9. Control signals.

Figures 8 and 9 illustrate the solutions obtained by the Simulink design optimization technique and genetic algorithm tuning utilizing the integral of squared error criterion. A comparative examination revealed superior outcomes using the GA tuning strategy. All findings for the generated scaling factor values are satisfactory and demonstrate the efficacy of the suggested GA tuning approach in tracking the desired trajectory while rejecting disturbances, in contrast to the SDO technique.

6. CONCISE SUMMARY

This chapter offered an optimization technique for tuning the settings of a PID-type fuzzy logic controller (FLC). The objective is to employ a genetic algorithm

(GA)-based heuristic optimization method to address very complicated issues. To enhance the robustness and performance of the proposed GA-tuned PID-type FLC, optimization parameters such as integral square error (ISE) are taken into account. The suggested controller is implemented on an electronic throttle valve (ETV) inside a discrete-time framework to follow a desired trajectory initiated from a flat output produced using the flatness feature. The performance evaluation against the Simulink design optimization (SDO) method demonstrates the efficacy of the proposed GA-tuned strategy in accurately following a planned trajectory while rejecting disruptions.

AT A GLANCE

This chapter proposes a proportional-integral-derivative (PID)-type fuzzy logic controller (FLC) for a discrete-time system to follow a desired trajectory derived from the flatness characteristic. To enhance the performance of the proposed controller, a genetic algorithm (GA) is employed to minimize the integral of the squared error (ISE) for the online tuning of the input and output PID-type FLC scaling factors. The proposed controller is implemented in an electronic throttle valve (ETV). Genetic approach (GA) tuning demonstrates superior and more robust performance relative to the Simulink Design Optimization (SDO) approach regarding trajectory tracking and disturbance rejection.

REFERENCES

1. Guzelkaya M, Eksin I, Yesil E. Self-tuning of PID-type fuzzy logic controller coefficients via relative rate observer. Engineering Applications of Artificial Intelligence. 2003;16(3):227–236
2. Qiao WZ, Mizumoto M. PID type fuzzy controller and parameters adaptive method. Fuzzy Sets and Systems. 1996;78:23–35
3. Woo ZW, Chung HY, Lin JJ. A PID type fuzzy controller with self-tuning scaling factors. Fuzzy Sets and Systems. 2000;115:321–326
4. Toumi F, Bouallègue S, Haggège J, Siarry P. Differential search algorithm-based approach for PID-type fuzzy controller tuning. In: International Conference on Control, Engineering & Information Technology (CEIT'14); 2014. pp. 329–334
5. Gritli W, Gharsallaoui H, Benrejeb M. PID-type fuzzy scaling factors tuning using genetic algorithm and Simulink design optimization for electronic throttle valve. In: International Conference on Control, Decision and Information Technologies (CoDIT'16); 6-8 April; Malta. 2016. pp. 216–221
6. Galichet S, Foulloy L. Fuzzy controllers: Synthesis and equivalences. IEEE Transactions on Fuzzy Systems. 1995;3:140–148

7. Holland JJ. Adaptation in Natural and Artificial Systems, University of Michigan Press; I975
8. Kim JS, Kim JH, Park JM, Park SM, Choe WY, Heo H. Auto tuning PID controller based on improved genetic algorithm for reverse osmosis plant. International Journal of Computer, Electrical, Automation, Control and Information Engineering. 2008;2(11):3707–3712
9. Gaing ZL. Particle swarm optimization approach for optimum design of PID controller in AVR system. IEEE Transactions on Energy Conversion. 2004;19:384–391
10. Mahony TO, Downing CJ, Fatla K. Genetic algorithm for PID parameter optimization: minimizing error criteria. In: Process Control and Instrumentation; 26–28 July; 2000. pp. 148–153
11. Fliess M, Levine J, Martin P, Rouchon P. On differentially flat non linear systems. In: Proc IFAC Symposium on Nonlinear Control Systems Design (NOLCOS), IFAC, Laxenburg, Austria; 1992, pp. 408–412
12. Gharsallaoui H, Ayadi M, Benrejeb M, Borne P. Robust flatness-based multi-controllers approach. Studies in Informatics and Control. 2010;19(4):357–368
13. Costin M, Schaller R, Maiorana M, Purcell J. An architecture for electronic throttle control systems. In SAE Technical Paper, SAE International; 2003.
14. Lebbal Ml, Chafouk H, Hoblos G, Lefebvre D. Modelling and identification of non-linear systems by a multimodel approach: Application to a throttle valve. International Journal Information and Systems Science. 2007;3:67–87
15. Yang C. Model-based analysis and tuning of electronic throttle controllers. In: Visteon Corporation, SAE Paper; March 8-11; Detroit, Michigan. 2004.
16. Aidi I, Ayadi M, Benrejeb M, Borne P. Flatness-based control of throttle valve using neural observer. International Journal of Research and Surveys. 2012;12:333–344

6 Chapter

UTILIZATION OF FUZZY LOGIC IN ENERGY CONVERSION SYSTEMS FOR UNMANNED AERIAL VEHICLES

1. INTRODUCTION: AN OVERVIEW

Type UAV (Unmanned Aerial Vehicle) platforms have recently undergone significant development owing to their extensive range of applications. UAVs are capable of executing many missions such as military reconnaissance, combat operations, agricultural monitoring, and surveillance of disaster zones, all while demonstrating superior economic and energy efficiency compared to traditional aviation. Unmanned Aerial Vehicles (UAVs) provide numerous advantages, including diminished manufacturing and operational costs, low energy consumption, safe access to hazardous areas unsuitable for manned aircraft, the capacity to infiltrate more restricted spaces inaccessible to conventional aircraft, and, owing to recent technological advancements, the capability to conduct continuous flight missions lasting from several days to several weeks. The advancement of UAVs is anticipated to significantly contribute to High Altitude Pseudo-Satellites (HAPS) with a continuous flight endurance of approximately five years. The attainment of extended flight time was facilitated by the utilization of hybrid power sources that incorporate a minimum of two energy kinds. Currently, the predominant energy sources are high-capacity batteries or supercapacitors recharged by renewable energy, particularly advanced solar cells produced recently. Fuel cells have excellent efficiency, making them a viable choice for electrically powered manned aircraft. This approach is thoroughly examined and intended for use in the imminent future. Airbus has created an electric aircraft capable of two trips, each lasting around one hour. At present, efforts are on to create a four-person aircraft utilizing a hybrid power system, comprising a high-capacity battery and a two-stroke combustion engine functioning as an electric generator. The flight duration is anticipated to increase to two hours. The research department of Airbus is contemplating the substitution of the thermal engine with a fuel cell to extend flight time. Titan Aerospace conducted tests in August 2013 for a conceptual vehicle intended for future high-altitude SOLAR 50 and 60 UAVs. SUN SOLAR 50 was slated for flight in 2014. sun 50 was designed as an unmanned aerial vehicle (UAV) that utilizes high-capacity batteries and 3000 sun cells for its power supply. The anticipated maximum flight period of 50 SOLAR is projected to be 5 years ultimately. It is intended to be outfitted with telecommunications and recognition systems, atmospheric sensors, and other components. SUN 60 is designed for a payload capacity of 125 kg [1–3]. The High Altitude Long Endurance Boeing (HALE) UAS 2012 focuses on the development of high-altitude unmanned aerial systems for missions that facilitate communications regarding the decommissioning of current infrastructure and extended surveillance operations in areas of interest. The Phantom Eye Project employs a hydrogen internal combustion engine utilizing cryogenically stored hydrogen. The Solar Eagle Project, initiated by Boeing, anticipates the utilization of high-efficiency solar cells with solid oxide fuel cells (SOFC). The Wingspan aircraft is anticipated to have a wingspan of 120 meters and will be propelled by six electric motors utilizing permanent magnets [1–3]. Project Zephyr 7 High Altitude Pseudo-satellites by Airbus Defence & Space is intended for observation, communication, and monitoring services over areas spanning tens of thousands of square kilometers. This UAV possesses the record for the longest flight duration—336 hours. Solar energy has been utilized for propulsion and storage in lithium-ion batteries with capacities of 3 kWh for nighttime operation [1–3]. Given that the parameters of power sources, such as solar cells or fuel cells, fluctuate significantly based on operating conditions, and that onboard UAV loads (cameras, sensors, communication systems, etc.) typically necessitate stabilized voltage, the power buses of UAVs must be regulated to maintain a constant voltage. This can be accomplished by employing power converters to modify output sources to meet the specifications of bus power sources. The majority of onboard UAV power sources are direct current (DC), and the loads are predominantly of DC types, with the exception of certain propulsion motors. The following are the requisite types of DC-to-DC converters that exhibit enhanced efficiency and minimal weight for use in UAVs. The most convenient converters in this context are those that have not undergone transformation, including buck, boost, and buck-boost converters. They are appropriate for UAV applications as they adjust the power source voltage to match the power bus voltage by modulating the duty cycle of the command pulses. Numerous strategies for converter control exist in the literature, varying in sophistication, including fuzzy techniques. These strategies provide stable control and possess the advantage of not necessitating a comprehensive mathematical model of the system, allowing for relatively straightforward tuning using numerical simulations. This chapter discusses the construction of models for electrical systems in UAVs, focusing on three types of DC-DC converters utilized on board, and the creation of fuzzy controllers to stabilize the output voltage. The mathematical models created for these converters account for the average parameters (voltage and current) while disregarding their fluctuations. While all models are averaged, the methodology for acquiring these models involves introducing circuit voltage or current sources to account for the DC parameters of switches, coils, or diodes included in converters. We find this procedure to be more logically plausible than deriving equations segmented by two operational periods of the converter and subsequently averaging them, as demonstrated in Ref. [4].

2. BOOST CONVERTER

The primary configuration of the boost converter is illustrated in Figure 1a, while the corresponding schematic for continuous mode is provided in Figure 1b. The operation of this converter comprises two stages. Initially, when the switch is in position 1, the inductor current increases, allowing the inductor to store energy in its electromagnetic field. In the second step, when the switch is in position 2, the inductor is in series with the input source and transfers its energy to the capacitor. As the inductor current diminishes, the inductor functions as a voltage source in series with the input source, resulting in the capacitor being charged to a voltage above the input voltage. Consequently, the boost converter elevates the input voltage [5, 6]

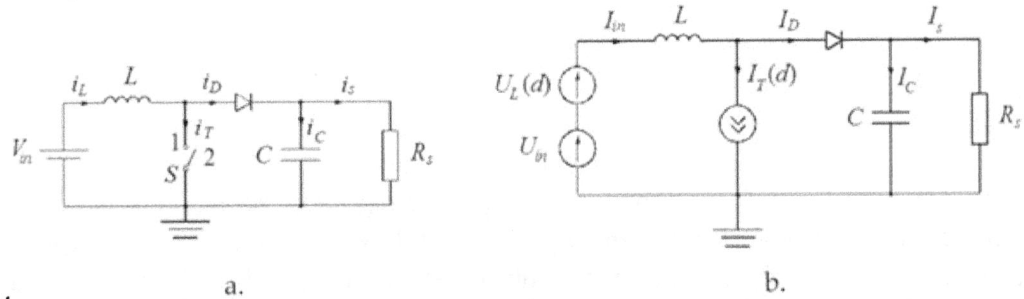

Figure 1. Boost converter. (a) Principal scheme, (b) Equivalent scheme for continuous mode.

Diagram in Figure 1a is equivalent in terms of DC components with the diagram in Figure 1b. Given that, coil behaviour was modelled in terms of averaged voltage with a voltage source dependent on the duty cycle of the control pulses $U_L(d)$, and the current through the switch was modelled by a current source dependent on the duty cycle of control pulses $I_T(d)$.

Coil voltage was found to be $U_L = \frac{d \cdot U_{in}}{1-d}$ so as to respect the converter transfer characteristic in steady state. Regarding the currents, it was considered that $I_T = I_D$ in terms of the average values. In stationary regime, coil *L* behaves like a voltage source $U_L(d)$. At a duty cycle variation, in transitory regime. In these conditions, one can write relations:

$$\begin{cases} \frac{dI_{in}}{dt} = -\frac{1}{L}U_C + \frac{U_{in}}{L(1-d)} \\ \frac{dU_C}{dt} = \frac{1}{2C}I_{in} - \frac{U_C}{CR_S} \end{cases} \quad (1)$$

For linearization, notations, Iin=Iin0+ΔIin, UC=UC0+ΔUC , Uin=Uin0+ΔUin, , d=d0+Δd have been used. Relation obtained after linearization is (2)

$$\begin{cases} \begin{bmatrix} \Delta I_{in} \\ \Delta U_C \end{bmatrix} = \begin{bmatrix} 0 & -\frac{1}{L} \\ \frac{1}{2C} & -\frac{1}{CR_S} \end{bmatrix} \begin{bmatrix} \Delta I_{in} \\ \Delta U_C \end{bmatrix} + \begin{bmatrix} \frac{1}{L(1-d)} & \frac{U_{in}}{L(1-d)^2} \\ 0 & 0 \end{bmatrix} \begin{bmatrix} \Delta U_{in} \\ \Delta d \end{bmatrix} \\ [\Delta U_{out}] = [0 \quad 1] \begin{bmatrix} \Delta I_{in} \\ \Delta U_C \end{bmatrix} + [0 \quad 0] \begin{bmatrix} \Delta I_{in} \\ \Delta d \end{bmatrix} \end{cases} \quad (2)$$

Using this system, one can deduce the transfer functions given in Eqs. (3) and (4)

$$H_1(s) = \frac{\Delta U_{out}(s)}{\Delta U_{in}(s)} = \frac{\frac{1}{2LC(1-d)}}{s^2 + s\frac{1}{CR_S} + \frac{1}{2LC}} = \frac{k_1}{(s-s_1)(s-s_2)} \quad (3)$$

$$H_2(s) = \frac{\Delta U_{out}(s)}{\Delta d(s)} = \frac{\frac{U_{in}}{2LC(1-d)^2}}{s^2 + s\frac{1}{CR_S} + \frac{1}{2LC}} = \frac{k_2}{(s-s_1)(s-s_2)} \quad (4)$$

In Figure 2, simulation scheme of boost converter in open loop is shown. They were used ΔU_{in} and Δ_d as input, the output voltage was considered as ΔU_{out}. Simulated boost converter parameters were as follows ,=47 μF, ,*C*=2.2 mF, R_S=1 Ω, ,*d*=0.5, U_{in}=24 V. In the simulations using transfer functions, variations of output voltage about output voltage reference, when there are some variations in input voltage and duty cycle compared to reference values, are obtained. To capture the absolute values, some sum blocks are introduced in Figure 2 whose variations overlap reference values achieved through the transfer or the absolute values of the input parameters. Simulation results are presented in Figure 3a–c.

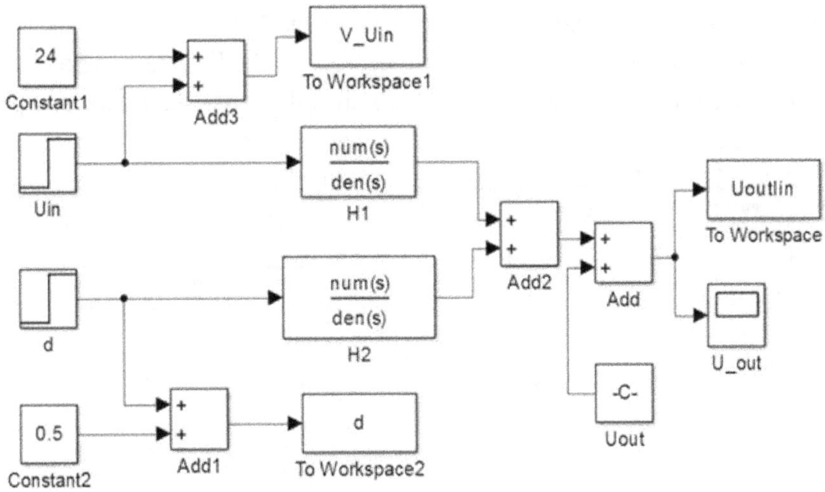

Figure 2. Open loop simulation scheme of boost converter.

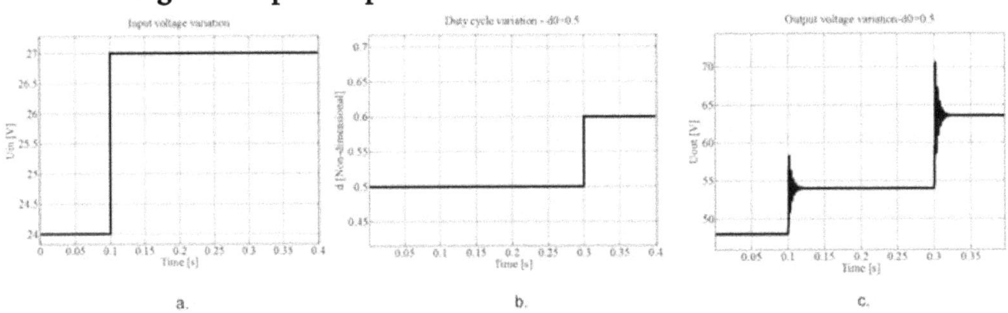

Figure 3. Boost converter response: (a) input voltage variation; (b) duty cycle variation; (c) output voltage variation.

Fuzzy control technique for DC-DC converters is applied in some studies in literature [7–10]. Other control techniques for these converters are studied in Refs. [11–17].

Closed loop scheme of the boost converter, implemented in MATLAB/Simulink is shown in Figure 4. One used a fuzzy controller with two inputs, output voltage error with respect to the nominal output voltage and its integral, so one can say that this is a PI fuzzy controller. The advantage of the fuzzy controllers is that one can modify conveniently the control characteristics slopes in order to obtain the desired system behaviour. Usually one wishes a higher slope for points far from the origin and smaller slope for points near the origin. In this way, when the system is far from the nominal functioning point, one can accelerate its return to the nominal point. When it comes near the nominal point, parameters variations are sluggish in order to obtain a good stability in the nominal point. Fuzzy system in Figure 4 has a particularity for an error and its integral are defined by two separate fuzzy controllers with four membership functions on input and output. In this way, one can simplify the controller tuning. There are necessarily eight inference rules (four rules for each controller) in contrast with the classical fuzzy controller with two inputs and one output having 16 inference rules. This difference increases once the membership function for each input increases. In Figure 5, the membership functions for fuzzy proportional controller P and its control surface are presented.

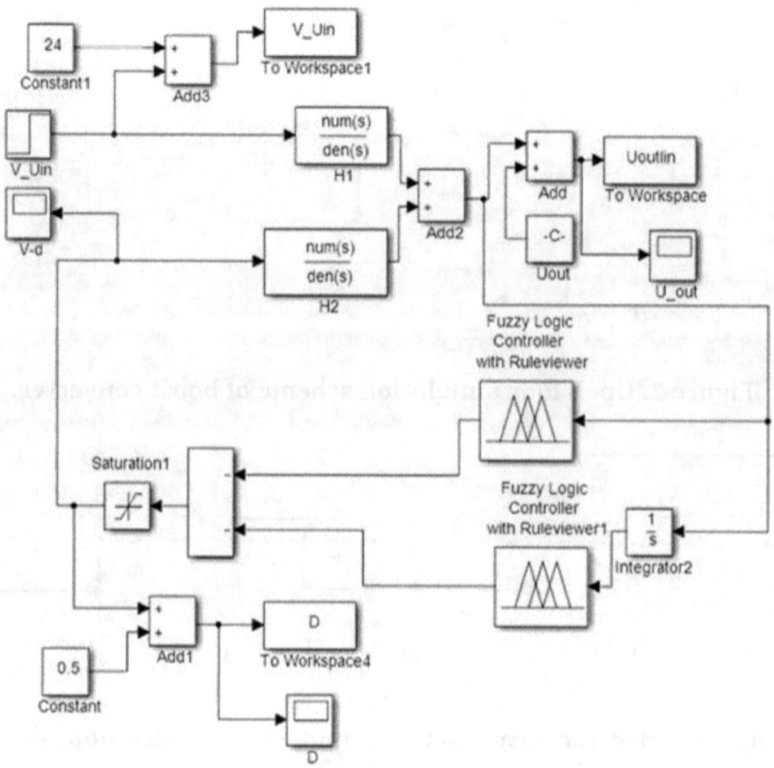

Figure 4. Closed loop simulation scheme for boost converter.

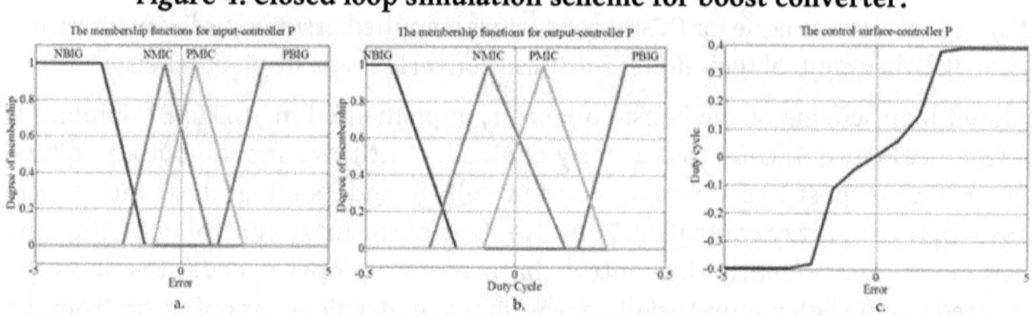

Figure 5. P-fuzzy controller. (a) Input membership functions; (b) Output membership functions; (c) Control surface.

In Figure 6, the membership functions for fuzzy integrator controller I and its control surface are presented. Converter operation with these fuzzy controllers has been tested to a step type signal and ramp type signal. In hybrid power systems of UAVs containing fuel cell or solar cells, parameters variations are not very sudden, so it is exciting to study the converters behaviour at sluggish signals, such as the ramp type which is taken into consideration here. In Figure 7, the boost converter behaviour at step signal is shown and in Figure 8, its behaviour to a ramp signal with a slope of 3 V/s is shown.

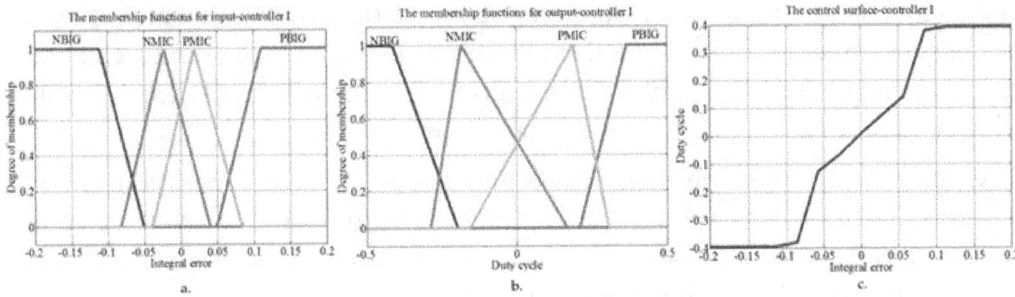

Figure 6. I-fuzzy controller. (a) Input membership functions; (b) Output membership functions; (c). Control surface.

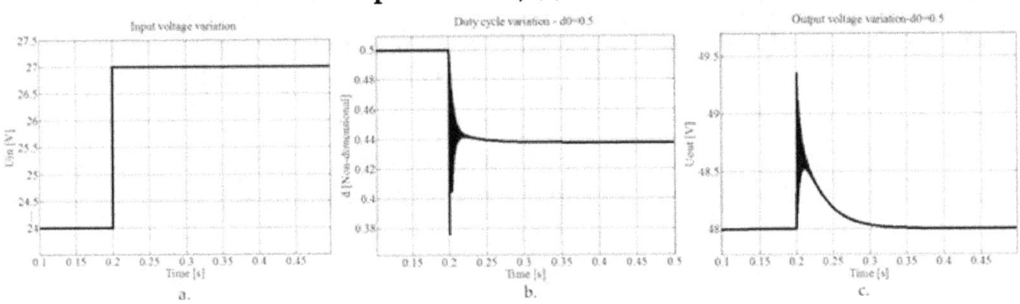

Figure 7. Closed loop behaviour of boost converter: (a) input variation; (b) duty cycle variation; (c) output voltage variation.

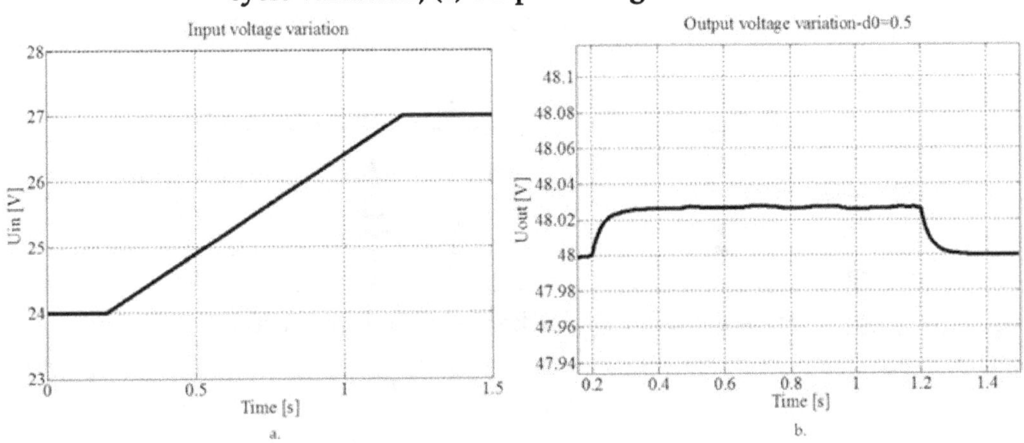

Figure 8. Boost converter behaviour at ramp signal: (a) input voltage variation; (b) output voltage variation.

Linguistic terms for both the controllers and for both inputs and outputs are NBIG (large negative), NMC (small negative), PMC (small positive) and PBIG (big positive). For both controllers, inference rules are: If input is NBIG, then output is NBIG; If input is NMC, then output is NMC; If input is PMC, then output is PMC; If input is PBIG, then output is PBIG.

For defining the controllers in this case, it is intended to simplify them as much so as to reduce the computation time required. In this purpose, one eliminated the membership function corresponding to linquistic term ZERO, which is used in the usual manner in fuzzy controllers. By conveniently change of the membership function one obtained control surfaces in concordance with the followed strategy - lower slope around the origin and greater slope farther origin.

The behaviour of the converter closed loop is one adequate operation of power systems on UAVs. Peak voltages at step input are attenuated and voltage returns quickly enough to the prescribed value. For the ramp input, deviation from set point is only 0.025 V, which is a very good performance.

3. BUCK CONVERTER

Principal scheme of the buck converter is shown in Figure 9. This converter has also two steps in its functioning. In the second step when the switch is in position 1, the capacitor loads through the inductor and inductor current increases, so the inductor behaves as a voltage source with inverse polarity with respect to the input voltage. In this manner on the capacitor, one can obtain a voltage smaller than the input voltage. In the second step when the switch is in position 2, energy accumulated in the inductor is pushed on the capacitor and thus Inductor current decreases, so it behaves as a voltage source with the same polarity as the input voltage but with a smaller value. So, the buck converter decreases the input voltage. In Figure 10, equivalent scheme for continuous mode is presented.

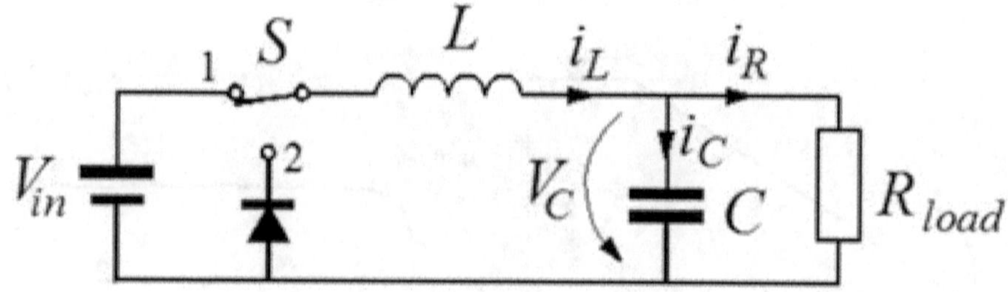

Figure 9. Principal scheme of buck converter.

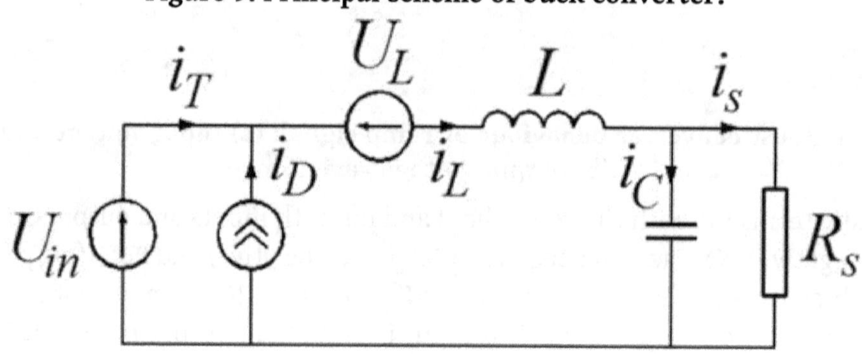

Figure 10. Equivalent scheme for continuous mode of buck converter.

Following Figure 10, one can write the following relations: ,U_{in}-U_L=U_{out}=d·U_{in}, ,i_T+i_D=i_L=i_S, U_{in}·i_T=U_{out}·i_S. After several transformations, following system of equations give:

$$\begin{cases} U_{\text{out}} = U_{\text{in}}d - L\dfrac{d}{dt}\left(\dfrac{I_T}{d}\right) \\ \dfrac{U_{\text{out}}}{R_S} = I_T\dfrac{1}{d} - C\dfrac{dU_{\text{out}}}{dt} \end{cases} \tag{5}$$

By linearization, like the case of the boost converter, resulted linearized model is:

$$\begin{cases} \Delta \dot{I}_T = -\dfrac{d_0}{L}\Delta U_{\text{out}} + \dfrac{d_0^2}{L}\Delta U_{\text{in}} + \dfrac{V_{\text{in}0}\cdot d_0}{L}\Delta d + \dfrac{I_{T0}}{d_0}\Delta \dot{d} \\ \Delta \dot{U}_{\text{out}} = \dfrac{1}{C\cdot d_0}\Delta I_T - \dfrac{1}{R_S\cdot C}\Delta U_{\text{out}} - \dfrac{I_{T0}}{C\cdot d_0^2}\Delta d \end{cases} \tag{6}$$

Applying Laplace in system zero initial conditions, Eq. (6) results in transfer functions of the form

$$\frac{\Delta U_{\text{out}}}{\Delta U_{\text{in}}} = \frac{R_S\cdot d_0}{s^2(R_S\cdot L\cdot C) + L\cdot s + R_S} \tag{7}$$

$$\frac{\Delta U_{\text{out}}}{\Delta d} = \frac{R_S\cdot U_{\text{in}0}}{s^2(R_S\cdot L\cdot C) + L\cdot s + R_S} \tag{8}$$

Buck converter parameters considered in this chapter are , L=56.5 µF,C=166.7 µF, R_S=2 Ω,d=0.5, $U_{in}0$=24 V. $U_{out}0$=12 V. Simulation scheme for the buck converter with transfer functions, open loop is shown in Figure 11. Simulation results are presented in Figure 12. Utilizing a closed loop scheme and the same fuzzy controllers, like in the case of boost converter shown in Figure 4, closed-loop simulation results are obtained shown in Figure 13. For the buck converter control, a classical PI fuzzy controller has been designed with two inputs, the error between the output voltage and the voltage prescribed and integral of this error. Simulation scheme in this case is the one in Figure 14. The membership functions and control surface of this controller are presented in Figure 15. The inference rules of fuzzy PI controller here are defined in table of this Figure 16. Buck converter behaviour with this control is shown in Figure 17. Note that the peak voltage obtained from step input is reduced to one third of the peak voltage obtained with the controller in the previous case (fuzzy controllers P and I are put in parallel), and the time to restore the prescribed value is also diminished.

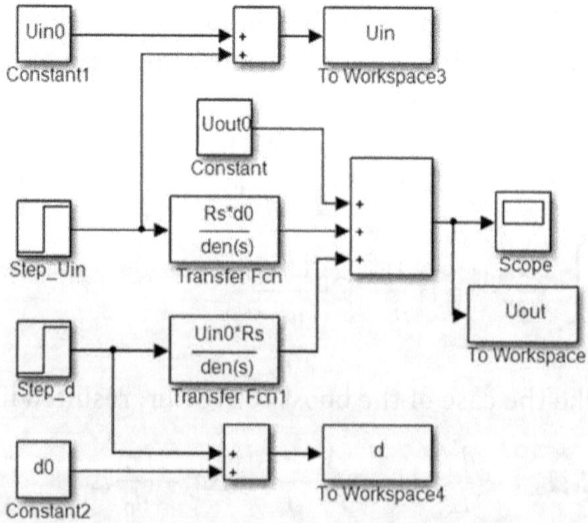

Figure 11. Open loop simulation scheme for buck converter.

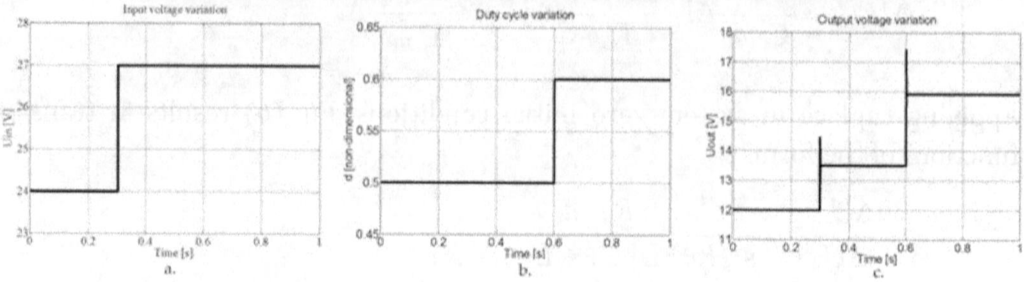

Figure 12. Open loop behaviour of buck converter: (a) input voltage variation; (b) duty cycle variation; (c) output voltage variation.

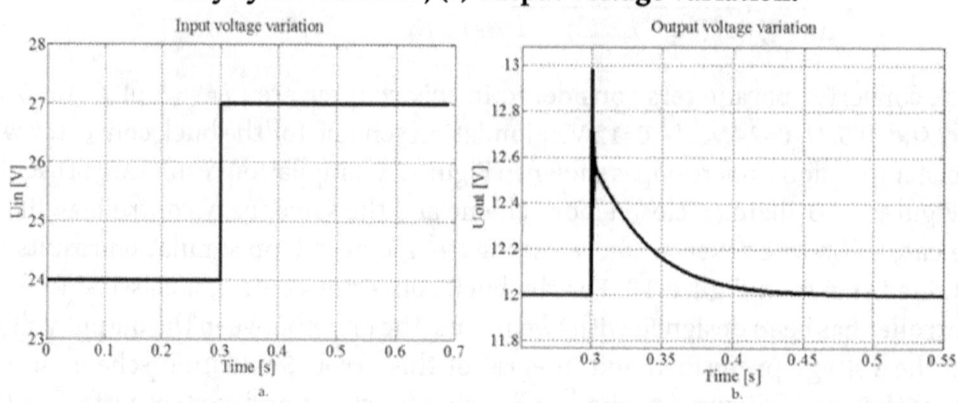

Figure 13. Closed loop behaviour of buck converter with two fuzzy controllers in parallel P+I: (a) input voltage variation; (b). output voltage variation.

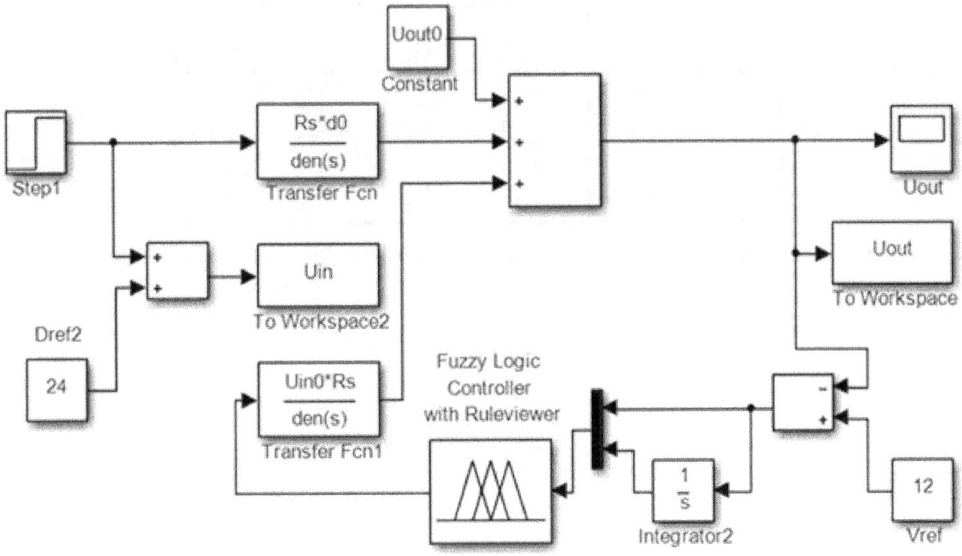

Figure 14. Closed loop simulation scheme for buck converter with one PI fuzzy controller with two inputs.

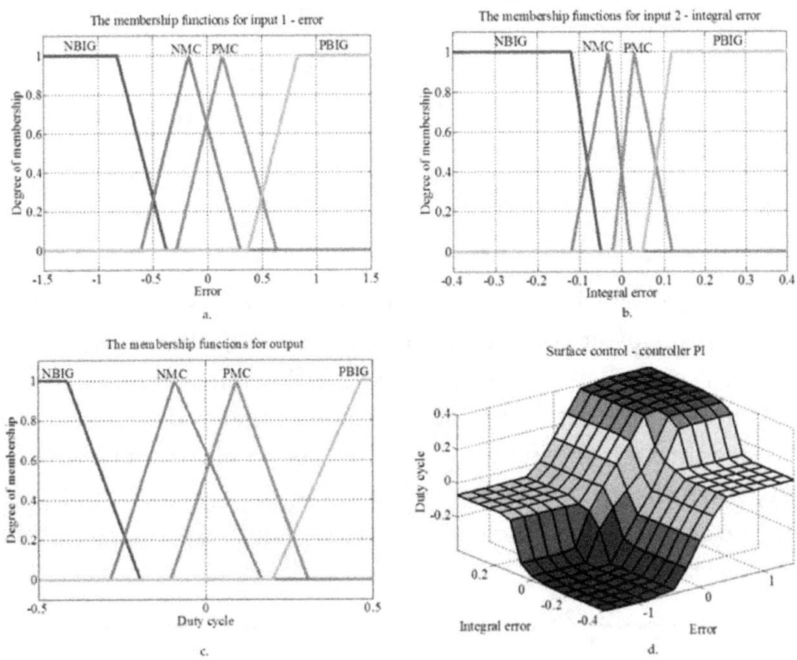

Figure 15. PI fuzzy controller with two inputs: (a) membership functions for error input; (b) membership functions for integral error input; (c) membership functions for output; (d) control surface.

		ERROR			
		NBIG	NMC	PMC	PBIG
ERROR INTEGRAL	NBIG	NBIG	NBIG	PMC	PMC
	NMC	NBIG	NBIG	PMC	PBIG
	PMC	NBIG	NMC	PBIG	PBIG
	PBIG	NMC	NMC	PBIG	PBIG

Figure 16. Inference rules for PI fuzzy controller with two inputs.

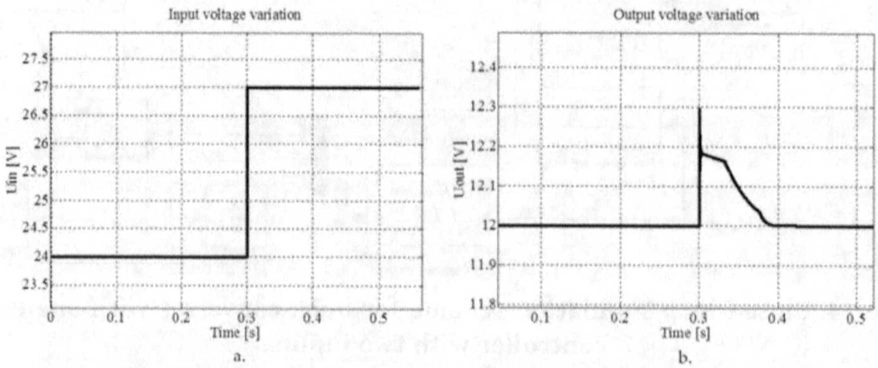

Figure 17. Closed loop behaviour of buck converter with PI fuzzy controller with two inputs: (a) input voltage variation; (b) output voltage variation.

So, we can say that this second fuzzy PI controller with two inputs provides better behaviour, but at the expense of greater computation time, so microcontroller that is implemented should have better performance.

4. BUCK-BOOST CONVERTER

Principle scheme of the buck-boost converter is in Figure 18. This converter outputs a higher or a lower voltage than the input voltage but with inverse polarity, so it is known as controller-inverter. It also functions in two steps. In step 1, when the switch is in position 1, the diode is inversely polarized and the current flows from the source through the inductor and charges its electromagnetic field. In the second step, when the switch is in position 2, the energy stored in the inductor is released on the capacitor and the output load with inverse polarity of the voltage.

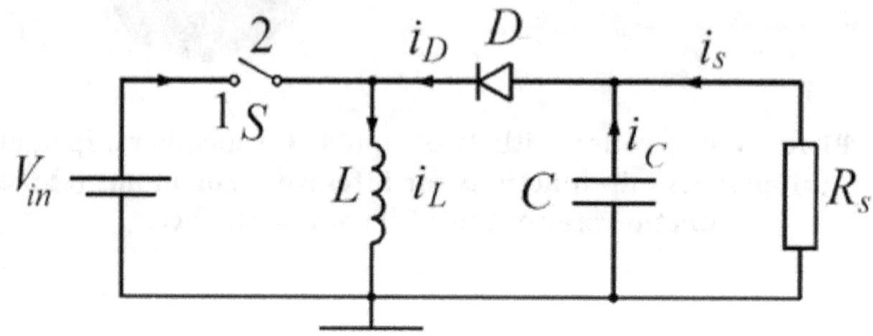

Figure 18. Principle scheme of buck-boost converter.

From the point of view of the average values, it was considered a voltage drop occurs on the switch. Also one considered the coil average voltage is equal with continuous output voltage. Following Figure 19, one can write the following relations: i_S+i_C= , +i_T=i_L, UL= $U_{\text{in}}\cdot\frac{d}{d-1}$, $i_T(d-1) = i_D{\cdot}d$.

It has considered among others stationary input-output characteristics of buck-boost converter. In addition, one can write the following equations:

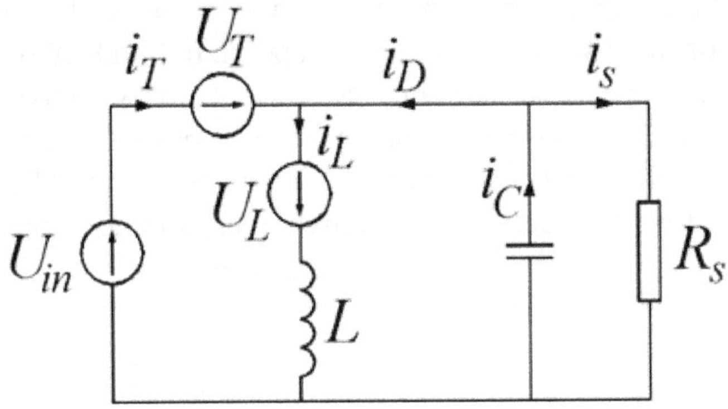

Figure 19. Equivalent scheme for continuous mode of buck-boost converter.

$$U_{\text{in}} + U_T - U_L = -L\frac{di_L}{dt} \tag{9}$$

$$U_{\text{in}} = -L\frac{di_L}{dt} + R_S i_S \tag{10}$$

$$C\frac{dU_{\text{out}}}{dt} + \frac{U_{\text{out}}}{R_S} + i_T = i_L \tag{11}$$

After several transformations, we obtain:

$$U_{\text{out}} - U_{\text{in}}\frac{d}{d-1} = -L\cdot C\frac{d^2U_{\text{out}}}{dt^2}\cdot\frac{1}{d-1} + \frac{L}{R_S}\frac{dU_{\text{out}}}{dt}\cdot\frac{1}{d-1} \tag{12}$$

By linearization and applying Laplace transform in null initial conditions, transfer functions result as follows:

$$\frac{\Delta U_{\text{out}}(s)}{\Delta U_{\text{in}}(s)} = \frac{\frac{d_0}{d_0-1}}{-L\cdot C\frac{d_0}{d_0-1}s^2 - \frac{L}{R_S\cdot}\cdot\frac{d_0}{d_0-1}s + 1} \tag{13}$$

$$\frac{\Delta U_{\text{out}}(s)}{\Delta d(s)} = \frac{-U_{\text{in}0}\frac{1}{(d_0-1)^2}}{-L\cdot C\frac{d_0}{d_0-1}s^2 - \frac{L}{R_S\cdot}\cdot\frac{d_0}{d_0-1}s + 1} \tag{14}$$

Parameters used in the simulation buck-boost converter are: L=56.5 μF, C=166.7 μF, R_S=2 Ω, d=0.5, U_{in0}=15 V, U_{out0}=-15 V. Scheme simulation for buck-boost converter in open circuit is similar to that for buck converter in Figure 11. Responses to step signals applied to both voltage input, and on the duty cycle are shown in Figure 20. Although for this converter two versions of fuzzy controllers were made, one version with two fuzzy controllers disposed in parallel, one for the proportional component and one for the integrative component, and a second version with a single fuzzy controller with two inputs (proportional and integrative) and one output (duty cycle variation). For the first variant, fuzzy controllers are shown in Figures 21 and 22. Buck-boost converter case has chosen five membership functions, one of which corresponds to the term linguistic ZERO (ZE in Figures 21 and 22), the other membership functions corresponding to the same linguistic terms as with previous fuzzy regulators. The inference rules in this case were the same for both controllers: If the input is NBIG, then output is NBIG; If the input is NMIC, then output is NMIC; If the input is ZE, then output is ZE; If the input is PMIC, then output is PMIC; If the input is PBIG, then output is PBIG.

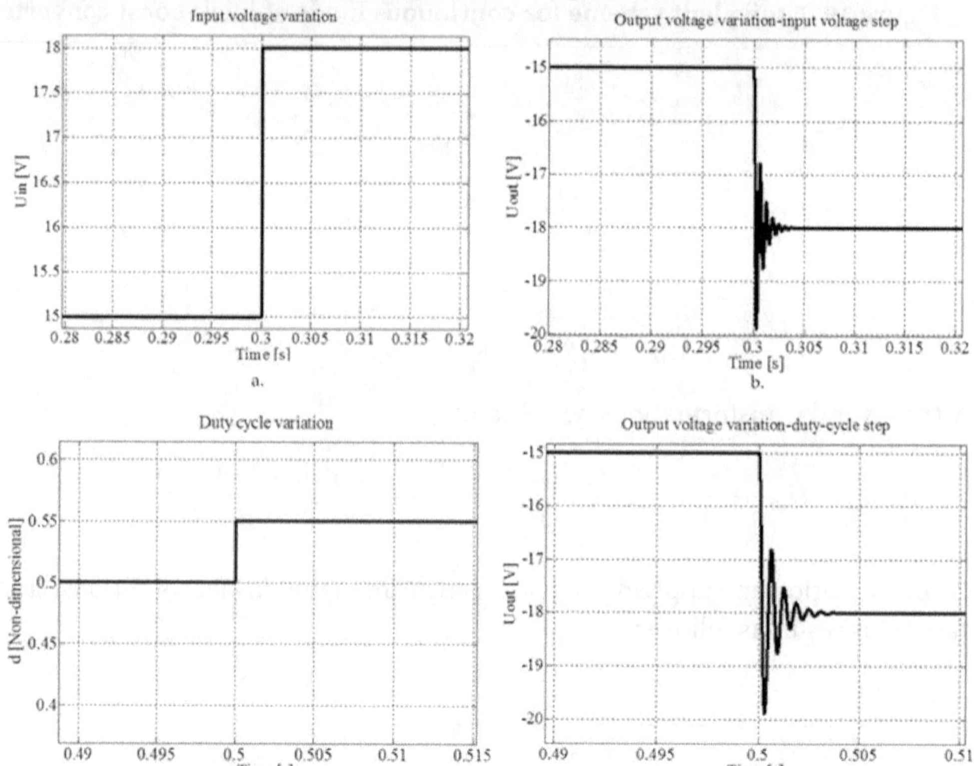

Figure 20. Buck-boost converter behaviour in open loop: (a) input voltage step; (b) output voltage at input voltage step; (c) duty cycle step; (d) output voltage at duty cycle step.

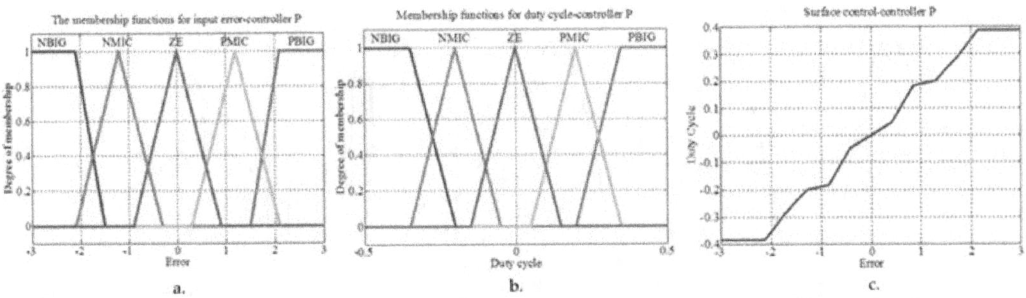

Figure 21. P fuzzy controller for buck-boost converter: (a) membership functions for input; (b) membership functions for output; (c) control surface.

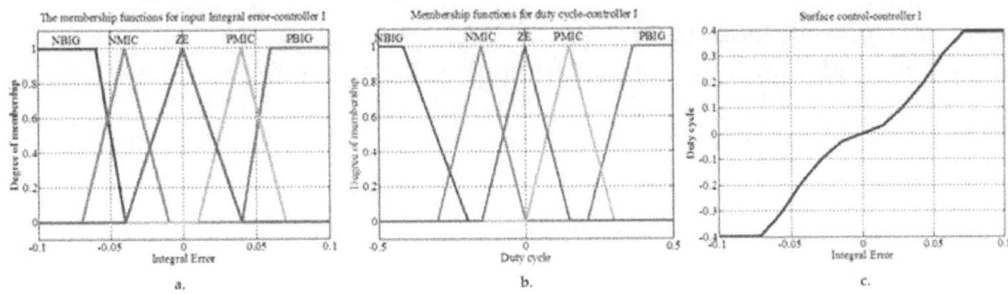

Figure 22. I fuzzy controller for buck-boost converter: (a) membership functions for input; (b) membership functions for output; (c) control surface.

Closed-loop simulation scheme for this case is similar to that for boost converter in Figure 4 but with transfer functions and controllers shown in buck-boost converter. Buck-boost converter's response in a closed loop and step input are shown in Figure 23.

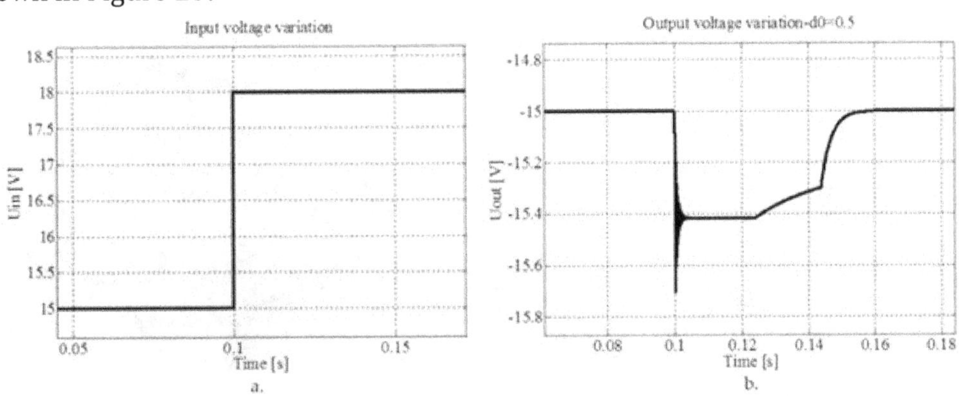

Figure 23. Closed loop behaviour of buck-boost converter with two fuzzy controllers in parallel (P+I): (a) input voltage variation; (b) output voltage variation.

The behaviour of the converter for sinusoidal input voltage variation has been studied in this case which is shown in Figure 24.

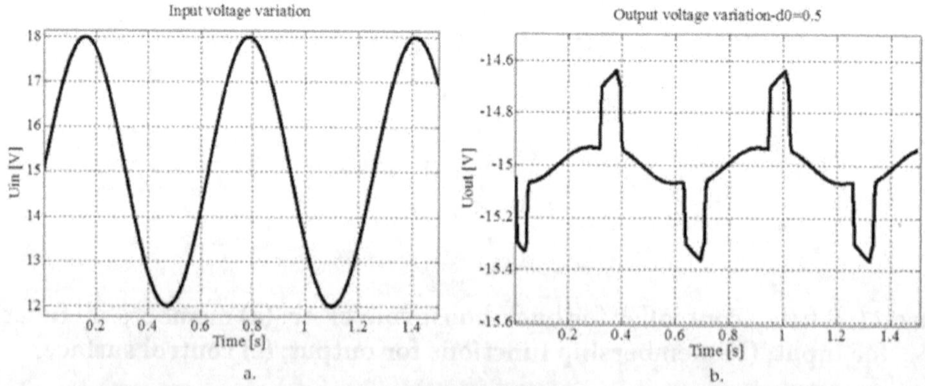

Figure 24. Closed loop behaviour of buck-boost converter with two fuzzy controllers in at sinusoidal input: (a) input voltage variation; (b) output voltage variation.

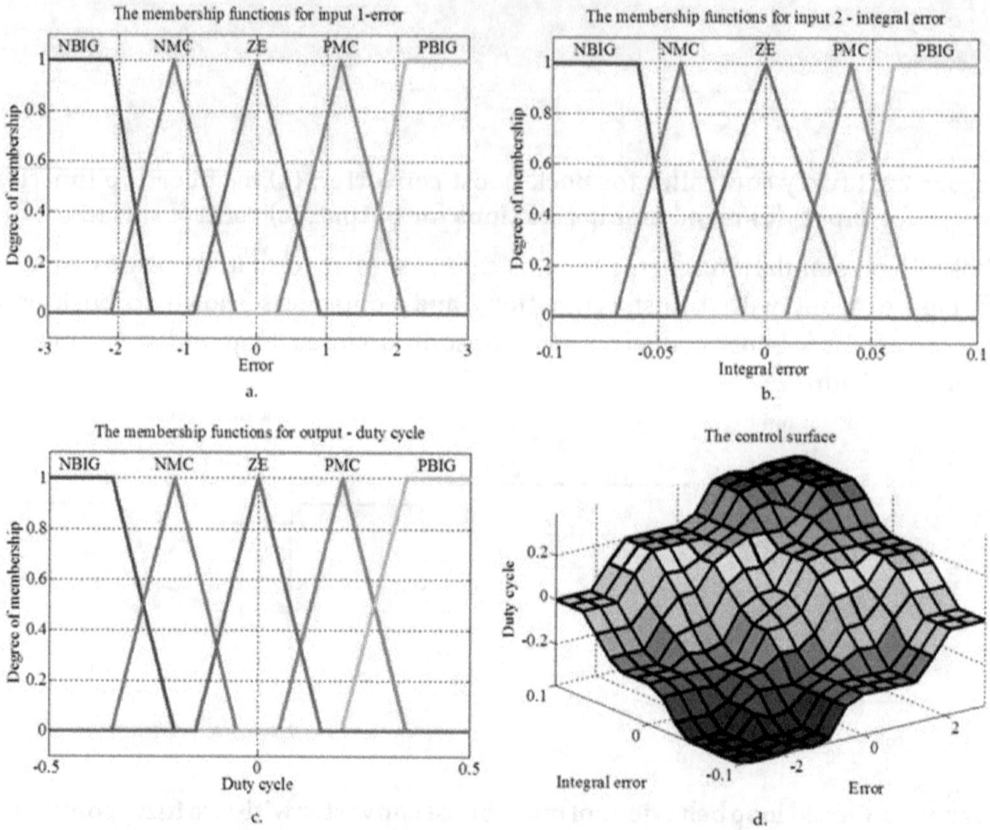

Figure 25. PI fuzzy controller with two inputs for buck-boost converter: (a) membership functions for proportional input; (b) membership functions for integral input; (c) membership functions for output; (d) control surface.

We notice here a better behaviour of fuzzy controller at step signal, the restore time to prescribed value is lower than the other controller presented converters but notice a worsened behaviour for sinusoidal signal response. At zero-crossing of the input signal, one obtained jumps in output voltage. The amplitude is not large, about 0.3 V, but these jumps means higher harmonics induced on power bus and thus worsening power quality on board of the UAVs. A second version of fuzzy controller designed for buck-boost converter is that of a controller with two inputs and one output, with five membership functions on each input and the output. This fuzzy controller is shown in Figure 25. Inference rules for this fuzzy controller are presented in table in Figure 26.

		ERROR				
		NBIG	NMC	ZE	PMC	PBIG
ERROR INTEGRAL	NBIG	NBIG	NBIG	NMC	NMC	ZE
	NMC	NBIG	NMC	NMC	ZE	PMC
	ZE	NMC	NMC	ZE	PMC	PMC
	PMC	NMC	ZE	PMC	PMC	PBIG
	PBIG	ZE	PMC	PMC	PBIG	PBIG

Figure 26. Inference rules for PI fuzzy controller.

In this instance, a favorable response is noted with step input; nevertheless, with sinusoidal input, voltage peaks reemerge at the zero crossing of the input. This indicates a further decline in energy quality on board.

5. CONCISE SUMMARY

This chapter delineates several applications of fuzzy control techniques to DC-DC converters in standard topologies (boost, buck, buck-boost). Initially, averaged mathematical models for each converter are presented, derived from analogous methods for continuous operation. Transfer functions for each converter are derived from these averaged models. These transfer functions are relevant for transient regimes under the small perturbation hypothesis. It is essential to consider that the averaged models and transfer functions are accessible for the continuous conduction mode of each converter. To achieve the continuous conduction mode, each of the analyzed converters requires a sufficiently high switching frequency. This frequency is not expressly present in the transfer functions but is dictated by the circuit components. The fuzzy controllers created for the analyzed converters are relatively uncommon. Initially, fuzzy controllers with four membership functions—two for each input and output—were employed for the boost and buck converters. To minimize the computational volume for each operational step, triangular membership functions were employed at the midpoint of the interval, while trapezoidal membership functions were utilized at the boundaries. Nonetheless, one adhered to the objective of diminishing the quantity of membership functions. To enhance the stability of the converters, efforts were made to diminish the control surface slope at the origin while maintaining a steeper slope away from the origin, so facilitating a quicker return of the converter to the intended operating position. Through a judicious selection of membership functions, this objective can be achieved with a limited number of functions, as illustrated in Figures 5c, 6c, and to a lesser extent in Figure 15d. To further decrease the volume of calculus at each operational step and to streamline the development of inference rules, the feasibility of decomposing a PI fuzzy controller with two inputs into two smaller fuzzy controllers, each with a single input, was examined. The newly developed fuzzy controllers operate in parallel, with one receiving the error between the actual output voltage and the target output voltage, while the other receives the integral of this error. This approach allows for the reduction of inference rules from $n \times n$ to $2 \times n$, where n represents the quantity of membership functions. Fuzzy controllers designed for the boost converter were effectively utilized to regulate the buck converter. For the buck converter, a traditional PI fuzzy controller was constructed with two inputs and one output, using four membership functions—two for each input and one for the output (refer to Figure 15). The performance achieved with this controller has been enhanced. The output voltage peaks are diminished, and the time required to revert to the specified voltage is also decreased (refer to Figure 17). Figure 17 illustrates alterations in response slope as the controller transitions between control intervals and the return acceleration upon the integrative component reaching the threshold of the membership function PMC. One notes a further reduction in slope response as the error diminishes, accompanied by a gradual convergence of the actual output voltage to the specified value. For the buck-boost converter, a traditional fuzzy converter was tested with an odd number of membership functions for each input, one of which corresponded to the linguistic phrase ZERO. The objective was to achieve a little slope of the control surface near the origin and a greater slope at a distance from it for this controller. The designed controllers are illustrated in Figures 21, 22, and 25. The performance of the buck-boost converter is optimal when a step input is applied, as illustrated in Figures 23 and 27. However, issues arise when sinusoidal input is employed. Figures 24 and 27 illustrate the output voltage peaks occurring as the input crosses zero. This indicates a decline in energy quality on board.

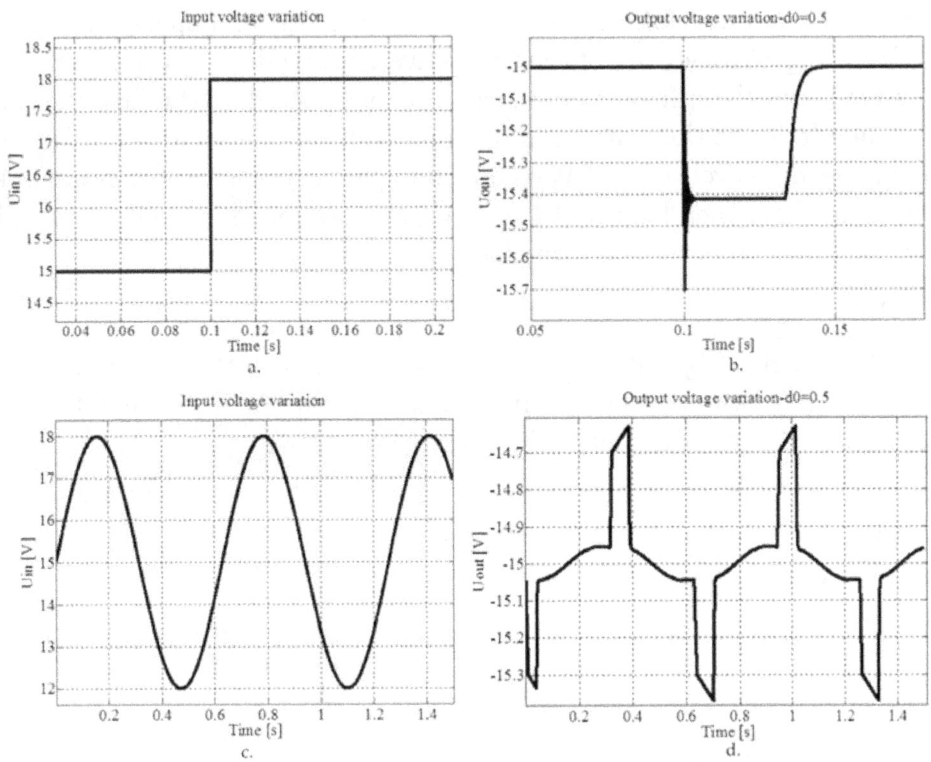

Figure 27. Buck-boost converter behaviour with PI fuzzy controller: (a) voltage step input; (b) response to step input; (c) sinusoidal voltage input; (d) response to sinusoidal input.

As an important conclusion, one can say that control with fuzzy technique offers a special flexibility to the controller design. One can reach very good performances even for simplified configurations of the controller. Possibility to obtain a convenient control surface for each application is an advantage that cannot be reached with other control techniques. Although, in some cases, design of a good fuzzy controller needs some experience and testing of many variants, in order to select the better one. Simplistic definition of the membership functions and inference rules can lead either to inconvenient control surfaces or to linear ones. In the second case, the fuzzy controller could be replaced very well with a classical P or PI controller, with same performances.

AT A GLANCE

This chapter aims to demonstrate several applications of fuzzy controllers in the automatic regulation of boost, buck, and buck-boost DC-to-DC converters. For the mathematical modeling of these converters, averaged models were employed, considering that the controlled parameter is the average output voltage. Only the continuous conduction mode was addressed in the mathematical model. For each converter, a brief description of the primary schematic and operational mode is provided, along with the presentation of the averaged model. The transfer functions are derived, and then, fuzzy controllers are constructed to stabilize the output voltage in relation to fluctuations in the input voltage. The control is achieved by adjusting the duty cycle of the PWM command pulses. Implemented systems, both in closed loop and open loop configurations, are developed in MATLAB/Simulink, and simulation results are also provided.

REFERENCES

1. P.O' Neil. Boeing High Altitude Long Endurance (HALE) UAS, available at https://higherlogicdownload.s3.amazonaws.com/AUVSI/656942e4-4448-41c3-877d-0c5f3ea40e63/UploadedImages/KnowledgeVault/HALE.pdf
2. N. Owano. Titan Aerospace readies solar-powered, long-endurance UAVs; 22 August 2013. www.boeing.com/defense/phantom-eye [Internet].
3. H. Ross. www.solarimpulse.com [Internet]; May 2009.
4. F. Misoc. A comparative study of dc–dc converters' effects on the output characteristics of direct ethanol fuel cells and NiCd batteries [dissertation]. Thesis of Dissertation Kansas State University, Manhattan, Kansas; 2007.
5. M. H. Rashid. Power Electronics Circuits, Devices, and Applications. 3rd edition. University of West Florida, Pearson Prentice Hall; 2003.
6. R. W. Erickson, D. Maksimovic. Fundamentals of Power Electronics. 2nd edition. 912 p. Kluwer Academic Publishers 2001, Sixth Printing 2004, DOI: ISBN 0-7923-7270-0.
7. G. Feng, W. Zhang, Y.-F. Liu. An Adaptive Current Mode Fuzzy Logic Controller for Dc-to-Dc Converters. Eighteenth Annual IEEE Applied Power Electronics Conference and Exposition; Miami Beach, Florida, 2003; pp. 983–989.
8. P. Mattavelli, L. Rossetto, G. Spiazzi, P. Tenti. General-Purpose Fuzzy Controller for Dc/Dc Converters. In: Applied Power Electronics Conference, Dallas, Texas 1995.
9. A. G. Perry, G. Feng, Y.-F. Liu, P.C. Sen. A New Design Method for PI-like Fuzzy Logic Controllers for DC-to-DC Converters. In: 35th Annual IEEE Power Electronics Specialists Conference; 2004; pp. 3751–3757.
10. A. Gad, M. Farooq. Application of Fuzzy Logic in Engineering Problems. Proceedings of the 27th annual conference of the IEEE Industrial Electronics Society (IECON 01), Denver, CO, 29 November-2 December 2001; p. 2044–2049.
11. A. Johansson. DC-DC converters-dynamic model design and experimental verification [thesis]. Lund University: Doctoral Dissertation in Industrial

Automation Department of Industrial Electrical Engineering and Automation; 2004.

12. M. Salimi, A. Zakipour. Direct voltage regulation of DC-DC buck converter in a wide range of operation using adaptive input-output linearization. In: IEEE Transactions on Electrical and Electronic Engineering, January 2015; pp. 85?91. DOI: ISSN 1931–4973.
13. J. Chen, D. Maksimovic, R. Erickson. Buck-boost PWM converters having two independently controlled switches. In: Proc. IEEE Power Electronics Specialists Conference; Vancouver Canada, June 17–21, 2001; pp. 736–741.
14. L. Guo. Design and implementation of digital controllers for buck and boost converters using linear and nonlinear control methods [dissertation]. Auburn University, Alabama; 2006.
15. H. Fadali. Fuel cell distributed generation, power conditioning, control and energy management [dissertation]. Master of Applied Science in Electrical and Computer Engineering, Ontario, Canada; 2008.
16. P. Hemachander, A. D. VimalRaj, M. Sudhakaran. Analysis design and implementation f soft single switched boost converter. In: International Journal of Computer Applications; 26(11), July 2011; pp. 38–46.
17. P. S. Priambodo, D. Sukoco, W. Purnomo, H. Sudibyo, D. Hartanto., Electric Energy Management and Engineering in Solar Cell System. Chapter 12 in book "Solar cells ? Research and Application Perspectives" editor Arturo Morales-Acevedo, published by INTECH on 6th March 2013; pp. 327–351.

7 Chapter

FUZZY LOGIC FOR MANAGING ENERGY IN A HOME POWER SYSTEM THAT USES RENEWABLE ENERGY SOURCES

1. INTRODUCTION: AN OVERVIEW

In recent decades, renewable energy sources have experienced significant annual growth rates, particularly solar energy, which provides competitive, environmentally friendly, and cost-effective options available at a mass production scale. Renewable energy generation units situated near energy loads are beneficial as transportation energy losses are virtually eradicated. The energy requirements for an average residential residence are quite modest and may primarily be met through renewable sources. The efficient collection of energy from the two primary renewable sources, wind and solar, is a fundamental objective of the energy sector. Despite the seeming abundance and no negative impact of these energy, the sustainability of the system is determined only by the methods of extraction and utilization employed. Consequently, it can be asserted that a system is effectively harnessing energy from these sources sustainably if its total efficiency is elevated and the generated energy is controlled to minimize adverse impacts on the power grid. In other words, utilizing or storing locally generated energy from renewable sources is more efficient than injecting it into the grid. The electricity grid remains unaffected by a tiny production facility exhibiting high dynamic behavior. Furthermore, if the system is engineered for overall cost efficiency, the quantity of photovoltaic panels, wind turbine output, and battery storage capacity can be reduced to achieve an optimum solution where the power grid functions solely as an energy buffer source. Future renewable systems must combine all energy-dependent applications into a centrally controlled system to achieve optimal cost-performance efficiency. Consequently, an intriguing approach is to integrate electric vehicles with residential electrical networks. Recent concepts such as car-to-grid are being introduced, necessitating the integration of electric vehicle energy storage into residential electrical systems to optimize overall system efficiency. A summary of solar heat pump systems is provided in Reference [2]. Garcia et al. [3] propose an optimal energy management system for a freestanding hybrid system comprising wind turbine, solar, hydrogen, and battery technologies, utilizing fuzzy logic. Ben Salah et al. [4] and Athari and Ardehali [5] employ fuzzy logic to enhance the integration of renewable energy into residential or decentralized small power grid applications. A number of studies authored by the same research group demonstrate optimal control of various combinations of power sources that supply a heat pump and additional loads. Reference [6] employs a wind/PV and power grid combination, whereas References [7] and [8] utilize a PV/diesel/battery and power grid combination and a fuel cell/wind/PV and power grid combination, respectively. These three studies demonstrate cost reductions achieved through optimal control strategies. The subsequent sections delineate the power system under examination, the exposition of fuzzy logic energy management, the simulation outcomes, and a conclusion.

2. POWER SYSTEM UNDER STUDY

This chapter introduces a model wherein two power systems engage through two replaceable tiny batteries: a lightweight electric car and a household partially powered by renewable energy sources. Figure 1 illustrates the comprehensive schematic of the analyzed system. The power generation system consists of renewable energy conversion devices: photovoltaic panels and a wind turbine, with both components linked at the outputs on a DC-Link via MPPT-based converters. The DC-Link serves as the primary power conduit of the system, to which all DC house loads are also linked. This power system is interconnected with the grid via a bidirectional DC-AC converter. The storage component is a battery from a lightweight electric vehicle. This low-capacity battery is interchangeable with a secondary battery placed in the car. The heat pump load is a crucial component of the proposed system, primarily as it serves as the principal electric load for balancing energy from renewable sources. A crucial aspect of the proposed system is the minimization of battery storage capacity; the two batteries are utilized alternately by the vehicle and are mostly refilled from renewable sources. The battery is operable by the home electrical system just in the absence of the power grid. It is essential to note that, for a viable solution, the battery must be lightweight to facilitate easy replacement.

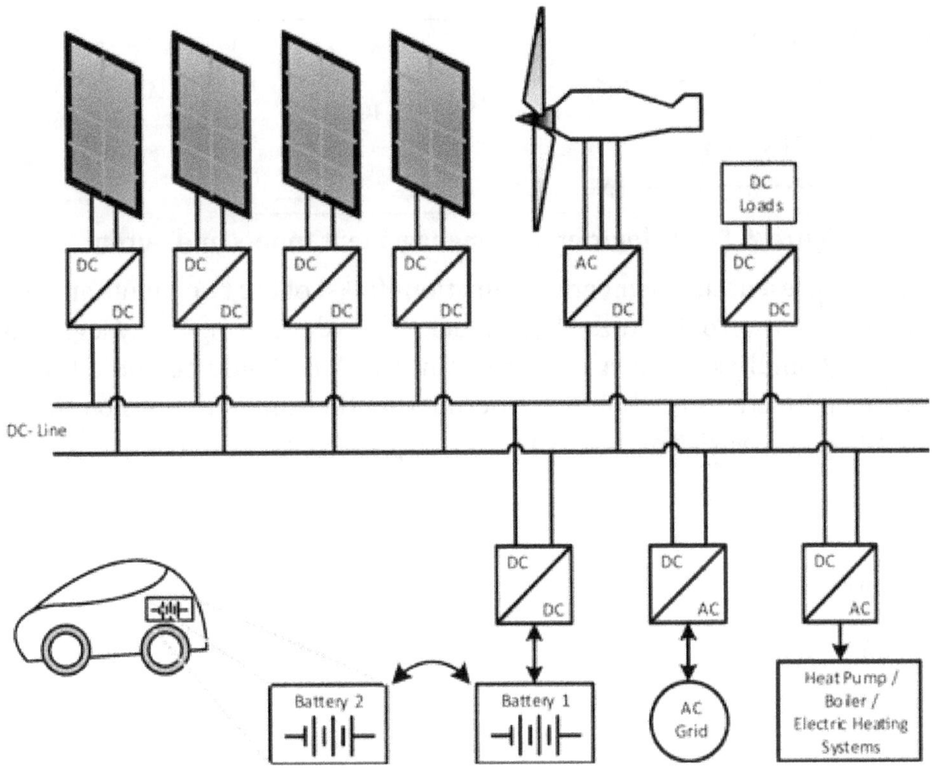

Figure 1. Schematic of the power system.

The primary aim of this project is to create an energy management control system utilizing fuzzy logic, designed to maximize the utilization of energy from renewable sources while minimizing or eliminating energy injection into the power grid and maintaining a minimal energy storage capacity. Electrical energy can also be sourced from the power grid. This approach achieves the optimal cost-performance balance with minimal storage capacity and maximal exploitation of renewable energy.

3. FUZZY LOGIC ENERGY MANAGEMENT

The fuzzy logic energy management is developed with the aim of splitting the renewable energy power to be either stored into the battery or transformed into thermal energy, with a multiplication factor by a heat pump. The fuzzy algorithm is using as inputs the battery state of charge (SoC) and the required heat energy (hot water and heating during cold periods). The output is the battery power (P_{batt}), the heat pump power (P_{hp}) being calculated with the simple Eq. (1):

$$P_{hp} = 1 - P_{batt} \tag{1}$$

Figure 2 presents the two inputs and one output of the fuzzy logic supervisor.

Figure 2. Fuzzy logic energy management inputs and output.

Figures 3–5 present the membership functions (MFs) of the two inputs and output. The implementation of the fuzzy logic supervisor is done using Fuzzy logic toolbox and Matlab/Simulink® software from Mathworks®. It should be noted that those membership functions are built considering that the sum of them is on the entire interval 1 and the variation on both axes is expressed in per unit.

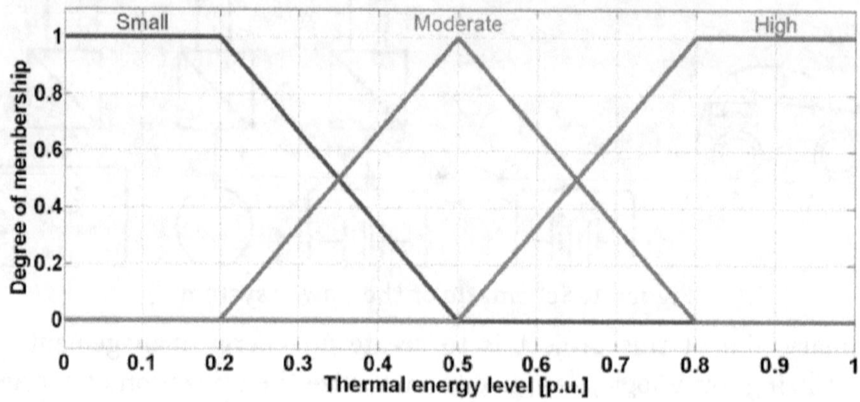

Figure 3. MFs for the first input of the fuzzy logic management system.

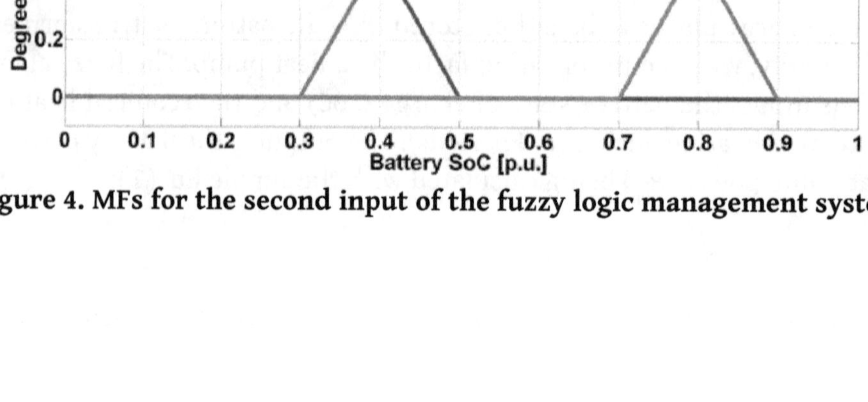

Figure 4. MFs for the second input of the fuzzy logic management system.

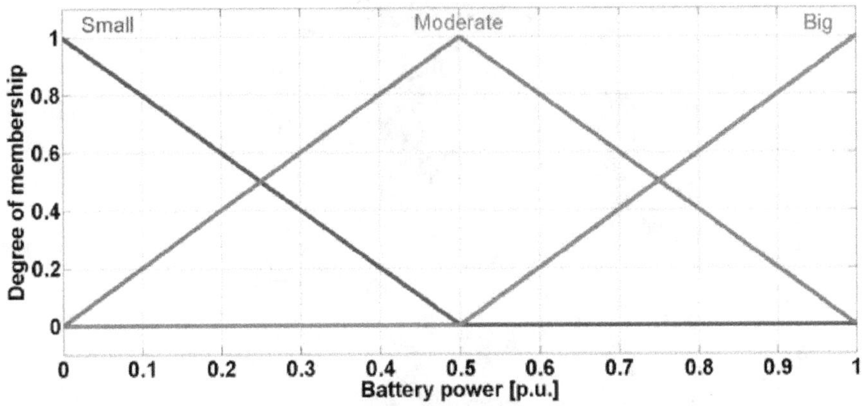

Figure 5. The output of the fuzzy logic management system.

The rule base of the fuzzy algorithm is presented in Table 1. It contains nine rules according to the fact that each input has three MFs, all of them being considered in the rule-editing phase. These rules were established considering the next principle: the thermal energy level and the battery SoC should increase similarly if they have a similar level (in per unit), but if one is lower than the other, it will be favored in order to reach a similar level (in per unit).

Table 1. Rule base of the fuzzy logic power management.

Battery power		Battery SoC Small	Battery SoC Moderate	Battery SoC High
Thermal energy level	Small	Moderate	Small	Small
	Moderate	Big	Moderate	Small
	High	Big	Big	Small

The defuzzification method used is centroid, which returns the center of area under the response surface. Figure 6 presents the response surface generated for this fuzzy logic supervisor. As it can be seen from this figure, the output variation does not reach the full range between 0 and 1 (its range is between 0.163 and 0.837). Thus, in order to extend the output range, the output result is subtracted with 0.163 and then multiplied with 1.485.

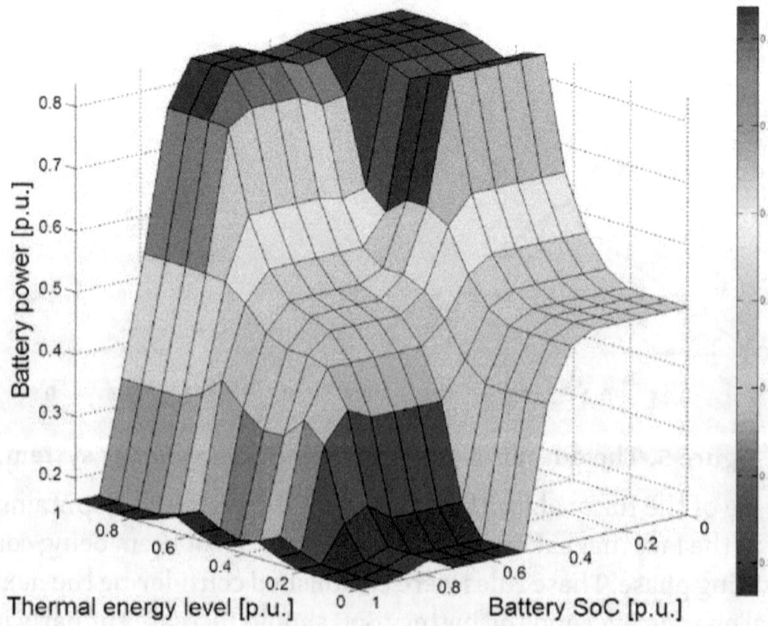

Figure 6. Response surface for the fuzzy logic supervisor.

4. SIMULATION RESULTS

The following figures present the simulation results for a power system with the following assumptions:

- Wind turbine and PV panels have an installed power of 1 kW each.
- The electric vehicle battery has a capacity of 1 kWh with an SoC of 40% when the charge process starts.
- It is considered that the vehicle is an ultralight version having two interchangeable batteries: one that equips the vehicle and the other left at home for charging.
- The heat pump has a coefficient of performance (COP) of 3, thus it consumes one-third electrical energy and generates three times more heat energy.
- The necessary heat energy considered is 30 kWh, thus a 10 kWh of electrical energy is required for the heat pump.
- The heat pump has the possibility to adjust power according to renewable energy production.
- Simulations are made for 1 day (24 h).

Figures 7 and 8 present the power and energy produced by the wind/PV hybrid system. The PV power curve has similar variations with some classic production curves for a sunny day, and the wind power has approximate variations according to a wind measurement from Brasov area (Romania).

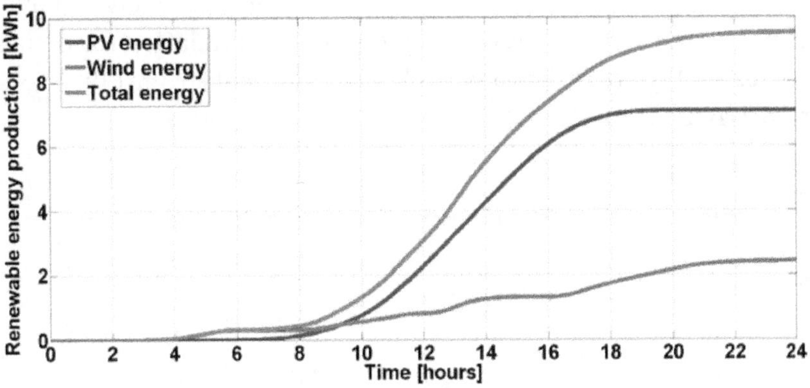

Figure 7. Renewable sources power curves.

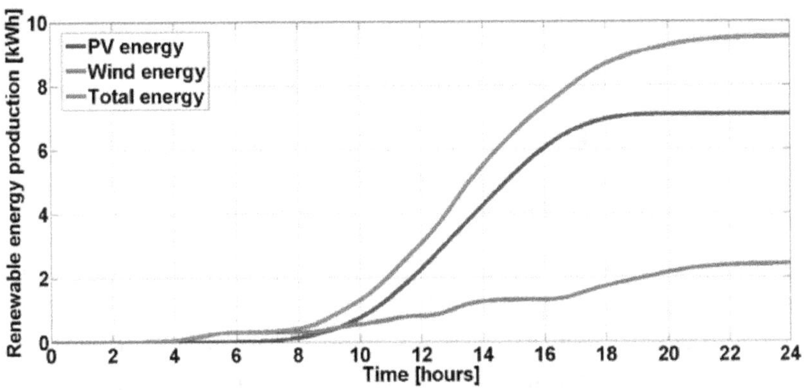

Figure 8. Renewable energy production.

The necessary electrical and thermal powers required are presented in Figure 9. The electrical power consumption is taken from a figure presented in Ref. [9] and the thermal power is estimated considering 30 kWh of needed energy (for a cold period) and no energy consumption during workhours.

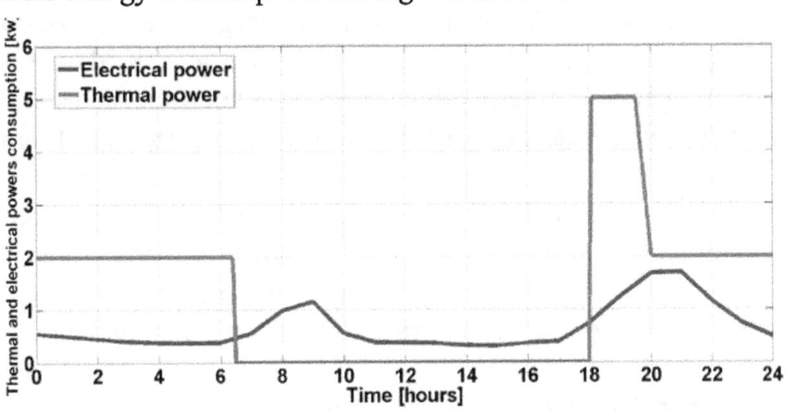

Figure 9. Estimated consumption of thermal and electrical powers.

Figure 10 presents the power grid failures. Even if it is unlikely to have two grid voltage drops during 1 day, it was considered in the simulation for demonstration purposes. Each voltage drop has a period of 15 min, first, from 14h30 to 14h45 and second from 21h30 to 21h45.

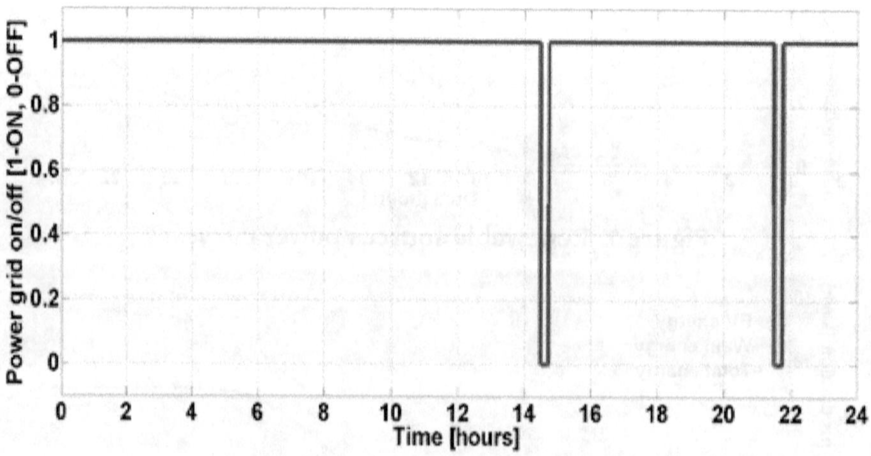

Figure 10. Power grid failure.

Figures 11 and 12 present the thermal energy level and the battery SoC, respectively, of the heat pump and battery powers, for the considered application. Both the thermal energy level and the SoC have an initial level. Due to the fact that at 18h00, the thermal power consumption restarts, and even if the heat pump is still working, the thermal energy level decreases (Figure 11). At the same time (18h00), the batteries are switched, the charged one is placed on the electric vehicle and the discharged one is put to charge. In order to use mainly the renewable energy to charge the battery and considering that the thermal energy demand is high, during the remaining hours of the day (18h00–24h00), the charging of the battery is stopped. By the end of the day, the thermal energy level and the battery SoC are lower compared with the situation at the beginning of the day. It is clear that not all required energy can be obtained from the renewable sources, thus, some has to come from the power grid to arrive to the necessary level. From Figure 12 one can observe that the first power grid drop is easily covered from renewable energy sources, just a sudden drop in the heat pump power can be observed. But the second one takes a lot of energy from the battery, which arrives to an SoC of about 5%. If the power grid loss would be longer, the feeding of all the loads would not be possible, and if considered appropriate, only the mandatory loads (pumps, lighting, etc.) should be fed. Or, if charging the discharged one is started at 18h00, the SoC would not decrease that much (Figure 13). Of course, in this case, the thermal energy level will decrease. For this situation, the heat pump and battery working powers are changed (Figure 14).

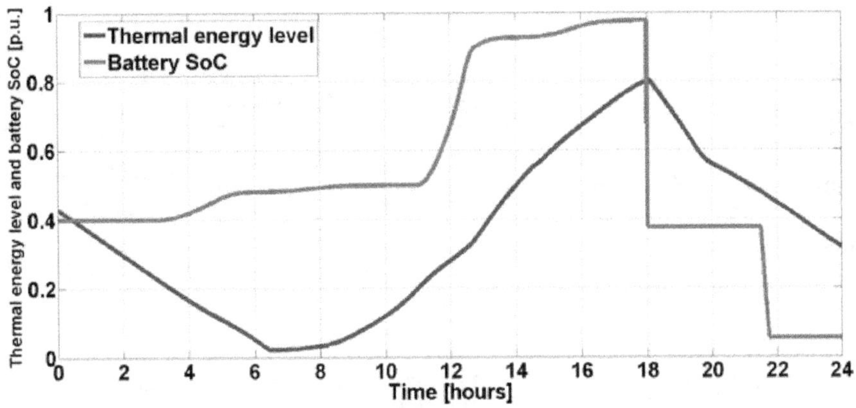

Figure 11. Thermal energy level and the battery SoC.

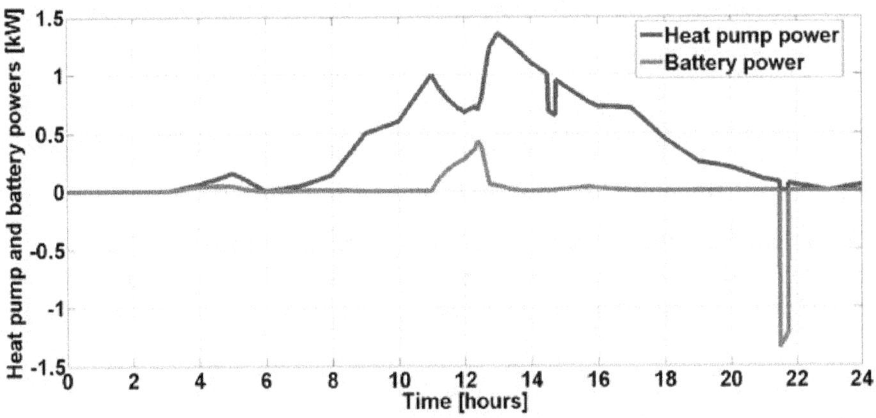

Figure 12. Heat pump power and battery power.

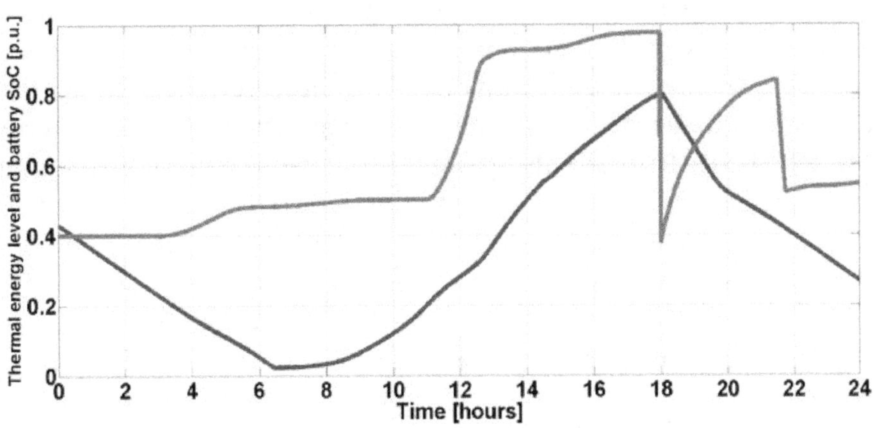

Figure 13. Thermal energy level and the battery SoC for the immediate charging of the discharged battery.

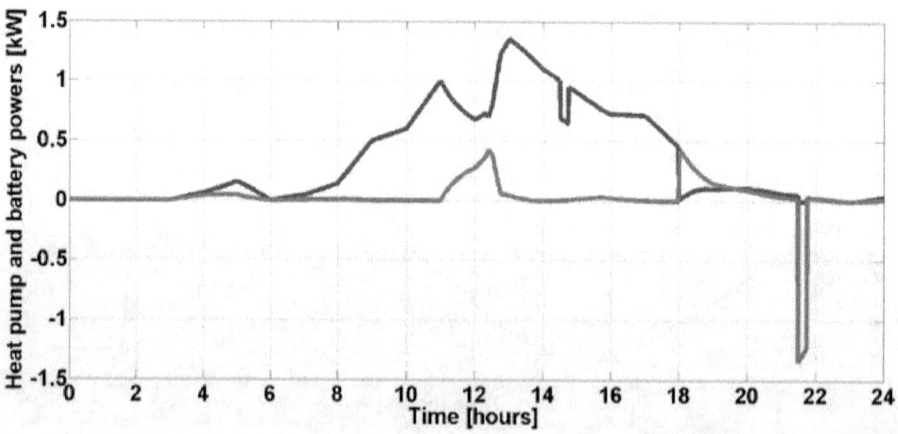

Figure 14. Heat pump and battery powers for the immediate charging of the discharged battery.

5. CONCISE SUMMARY

An energy management method utilizing fuzzy logic has been suggested and proven through simulations for residential applications. This algorithm facilitates the allocation of renewable energy for battery charging and for powering a heat pump that generates thermal energy. The findings indicate that the battery attains approximately 97% charge, while the thermal energy level from renewable sources is about 88% in the first scenario (discharged battery remains uncharged during the evening) and approximately 83% in the second scenario (discharged battery is charged during the evening). The remaining energy requirements should be supplied by the power grid.

AT A GLANCE

A fuzzy logic energy management algorithm is suggested for a hybrid wind/photovoltaic (PV) power generation unit, a battery for an electric car, and a heat pump for use in homes. The idea involves two separate power systems: a light electric car and a home. These systems work together using light batteries that can be switched out. They are also powered by a renewable energy system that includes PV panels, a wind generator, and the right MPPT-based converters. The main parts of the idea are the heat pump load, which turns heat into electricity and is the system's major electric load, and the storage element, which the vehicle uses and can be charged from renewable sources. It is suggested that fuzzy tools be used with the given algorithm to create a good energy management control system that makes the best use of renewable energy. It was found that most of the energy needed to charge the battery and power the heat pump can come from clean, natural sources.

REFERENCES

1. Guille C, Gross G. A conceptual framework for the vehicle-to-grid (V2G) implementation. Energy Policy. 2009;37(11):4379-4390. ISSN: 0301-4215

2. Sparber W, Vajen K, Herkel S, Ruschenburg J, Thür A, Fedrizzi R, D'Antoni M. Overview on solar heat pump systems and review of monitoring results. In: ISES World Solar Congress, 28 August -2 September 2011, Kessel, Germany; Published by International Solar Energy Society, Freiburg Germany:1-12

3. Garcia P, Torreglosa JP, Fernandez LM, Jurado F. Optimal energy management system for standalone wind turbine/photovoltaic/hydrogen/battery hybrid system with supervisory control based on fuzzy logic. International Journal of Hydrogen Energy. 2013;23:14146-14158

4. Ben Salah C, Chaabene M, Ben Ammar M. Multi-criteria fuzzy algorithm for energy management of a domestic photovoltaic panel. Renewable Energy. 2008;33(5):993-1001. ISSN: 0960-1481

5. Athari MH, Ardehali MM. Operational performance of energy storage as function of electricity prices for on-grid hybrid renewable energy system by optimized fuzzy logic controller. Renewable Energy. 2016;85:890-902

6. Sichilalu S, Mathaba T, Xia X. Optimal control of a wind–PV-hybrid powered heat pump water heater. Applied Energy. 2017;185:1173-1184

7. Sichilalu S, Xia X. Optimal energy control of grid tied PV–diesel–battery hybrid system powering heat pump water heater. Solar Energy. 2015;115:243-254

8. Sichilalu S, Tazvinga H, Xia X. Optimal control of a fuel cell/wind/PV/grid hybrid system with thermal heat pump load. Solar Energy. 2016;135:59-69

9. Arif MT, Oo AMT, Ali ABMS. Estimation of energy storage and its feasibility analysis. In: Zobaa AF, editor. Energy Storage—Technologies and Applications. InTech, Rijeka, Croatia; 2013: 41-78. ISBN: 978-953-51-0951-8

Chapter 8

RESILIENT ADAPTIVE FUZZY CONTROL FOR A CATEGORY OF SWITCHING POWER CONVERTERS

1. INTRODUCTION: AN OVERVIEW

Switching power converters are progressively supplanting traditional linear power converters owing to their compactness, lightweight design, high efficiency, and broader input voltage range. As personal computers, mobile communication devices, and automobile electronics rapidly improve and gain popularity, the demand for the stability and efficiency of converters is increasing. The phase-shifted pulse-width modulated (PSPWM) full-bridge soft switched power converter [1, 2] and its subsequent modifications [3–8] have emerged as a prevalent circuit topology among switching power converters, attributed to numerous advantageous features, such as diminished switching losses and stresses, along with the obviation of primary snubbers. The circuit can operate at high switching frequencies while enhancing power density and conversion efficiency. Feedback control has been integrated into switching power converters to enhance both the stability and performance robustness of the output voltage. Despite its beneficial attributes, feedback control for soft switched PSPWM full-bridge converters is limited to basic linear time-invariant designs, such as proportional-integral (PI) or lead-lag compensators derived from a linearized model [9–12]. As indicated by [13, 14], the dynamics of a PSPWM full-bridge converter are considerably more complex than those of a basic buck converter, owing to the heightened number of topological stages and the influence of input voltage, output voltage, and load current on the PWM duty cycle. A trade-off must be determined between establishing a simple model (e.g., linearized model) and a complex model (e.g., switched model) for control design purposes. The initial step in model-based control design often involves developing a mathematical model of suitable complexity that accurately represents the desired dynamics. Such models can typically be derived by simplifying a complex model, known as model reduction, or by linearizing a nonlinear model around a designated operating point. The transfer function is a well-established modeling technique for the design of linear controllers. A state variable model is typically a requirement for more sophisticated control design. Numerous modeling methodologies for switching power converters with intricate topologies have been presented in the literature [15–20]. The majority have achieved success in modeling the «local» behavior (i.e., tiny signal model) or analyzing the fundamental properties. Nevertheless, a limited number of the outcomes can be applied to the subsequent phase of control design. A range of essential information for control design cannot be obtained from those modeling approaches that are not control-centric. The critical information encompasses the stability, controllability, and observability of the open-loop system. This chapter presents a control-centric modeling and analysis methodology, accompanied by a robust adaptive fuzzy control design for switching power converters with complicated topologies, utilizing PSPWM full-bridge power converters as a design exemplar. The chapter›s outline is as follows: Sections 2 and 3 illustrate the methodology for developing a control-centric mathematical model for a PSPWM full-bridge soft-switched power converter system. The differential equations and the associated state variable model are formulated for each operational interval. The subsystem models for all intervals are consolidated to create a comprehensive model. The model undergoes numerical simulation and is compared to the related PSpice model to validate its accuracy. Section 4 will conduct a stability analysis of the system. Stability analysis is conducted for each interval subsystem of the developed model to ascertain its internal/BIBO stability. Section 5 will perform a study of controllability and observability for the system. The controllability and observability of the subsystems are examined to identify which signals or variables can be influenced by control efforts and which can be inferred through output feedback control mechanisms. The developed comprehensive model is utilized for sophisticated control design. For instance, by eliminating uncontrollable and unobservable variables and dynamics, an LPV gain-scheduling control architecture can be implemented as demonstrated in Ref. [21]. Sections 6 and 7 present model reduction and robust adaptive fuzzy control design. Section 8 presents the conclusion and future work.

2. CONTROL-CENTRIC MATHEMATICAL MODEL

This section will provide a concise description of the operation of a PSPWM full-bridge DC-DC power converter. Be advised that there are eight operational intervals. The operation of adjacent intervals is discontinuous due to switching. This indicates that the settings and initial conditions vary as the converter transitions. A thorough control-oriented state variable model for each operational interval will be constructed for subsequent analysis and numerical simulation. The circuit diagram of the converter is shown in Figure 1. Figure 2 is the waveform timing for various signals in the converter, where i_{LlK} is the primary current, v_{ab} is the voltage between a and b, i_L is the secondary current, v_s is the secondary voltage, Q_A,Q_B,Q_C, and Q_D are the four switches, Δ_D is the duty cycle loss, and ZVS delay is the dead time.

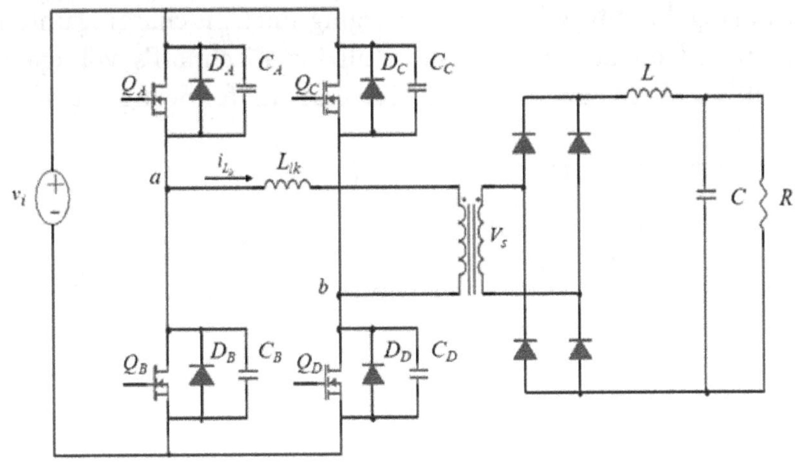

Figure 1.Circuit topology of a PSPWM full-bridge converter.

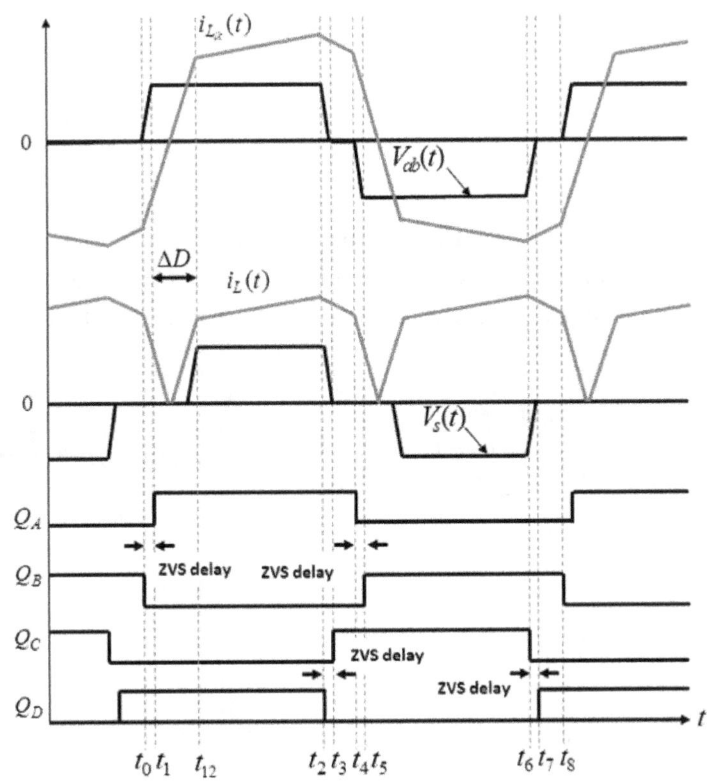

Figure 2.Waveform timing for a PSPWM full-bridge converter.

2.1. Positive half cycle: trailing-leg (passive-to-active) transition (t_0~t_1)

During this operation interval, only Q_D is conducting. Figure 3 shows the equivalent circuit of state 1. Initial conditions are $v_{CA}(t_0)=v_i$, $v_{CB}(t_0)=0$, $v_{CC}(t_0)=v_i$, and $v_{CD}(t_0)=0$. During this interval, C_A is discharging and C_B is charging until v_{CA} equals zero. Q_A is turned on at zero voltage. Utilizing Kirchhoff's voltage law (KVL) and Kirchhoff's current law (KCL), we arrive at the following set of differential equations:

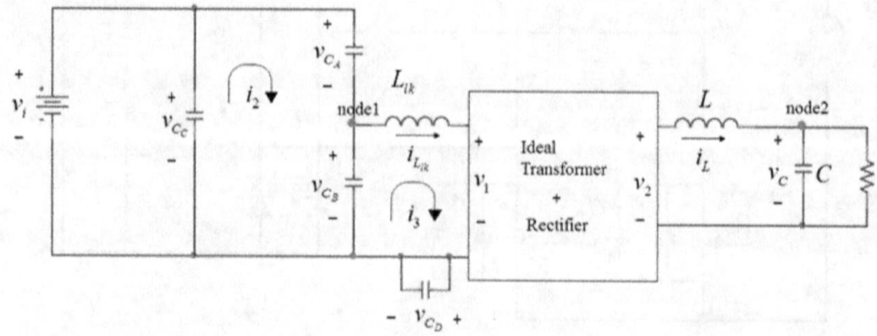

Figure 3.The equivalent circuit of operation interval 1 (t_0~t_1).

$$\frac{dv_{C_A}(t)}{dt}=\frac{i_{L_{lk}}}{C_A+C_B},\ \frac{dv_{C_B}(t)}{dt}=-\frac{i_{L_{lk}}}{C_A+C_B},\ \frac{dv_{C_C}(t)}{dt}=0,\ \frac{dv_{C_D}(t)}{dt}=0 \tag{1}$$

$$-\frac{di_{L_{lk}}}{dt}=\frac{di_L}{dt}=\frac{-1}{n^2L_{lk}+L}v_C+\frac{n}{n^2L_{lk}+L}v_{C_A}+\frac{-n}{n^2L_{lk}+L}v_i,\ \frac{dv_C(t)}{dt}=\frac{1}{C}i_L-\frac{1}{RC}v_C \tag{2}$$

Define $\vec{x}(t)=[i_{L_{lK}}(t)\ \ i_L(t)\ \ v_C(t)\ \ v_{C_A}(t)\ \ v_{C_B}(t)\ \ v_{C_C}(t)\ \ v_{C_D}(t)]^T$, where i_{LlK} is leakage inductance current, i_L is inductance current, v_C is output voltage, v_{CA} is the voltage across C_A, v_{CB} is the voltage across C_B, v_{CC} is the voltage across C_C, and v_{CD} is the voltage across C_D. Therefore, a state variable model can be obtained as follows:

$$\dot{\vec{x}}(t)=\begin{bmatrix} 0 & 0 & \frac{n}{n^2L_{lk}+L} & \frac{-n^2}{n^2L_{lk}+L} & 0 & 0 & 0 \\ 0 & 0 & \frac{-1}{n^2L_{lk}+L} & \frac{n}{n^2L_{lk}+L} & 0 & 0 & 0 \\ 0 & \frac{1}{C} & -\frac{1}{RC} & 0 & 0 & 0 & 0 \\ \frac{1}{C_A+C_B} & 0 & 0 & 0 & 0 & 0 & 0 \\ -\frac{1}{C_A+C_B} & 0 & 0 & 0 & 0 & 0 & 0 \\ 0 & 0 & 0 & 0 & 0 & 0 & 0 \\ 0 & 0 & 0 & 0 & 0 & 0 & 0 \end{bmatrix}\vec{x}(t)+\begin{bmatrix} \frac{n^2}{n^2L_{lk}+L} \\ \frac{-n}{n^2L_{lk}+L} \\ 0 \\ 0 \\ 0 \\ 0 \\ 0 \end{bmatrix}v_i(t) \tag{3}$$

where v_i is input voltage, $N_2/N_1=n$ is the transformer turns ratio, L is inductance, C is capacitance, and R is resistance.

2.2. Positive half cycle: active region (t_1~t_2)

During this operation interval, Q_A and Q_D are conducting. Initially, due to leakage inductance, the secondary side will experience a short period of "no energy" passing through, called duty cycle loss. Figure 4 shows the equivalent circuit for this period. Initial conditions are $v_{CA}(t1)=0$, $v_{CB}(t_1)=vi$, $v_{CC}(t_1)=vi$, and $v_{CD}(t_1)=0$. Similarly, we may derive a set of differential equations and the corresponding state variable model for this short period is

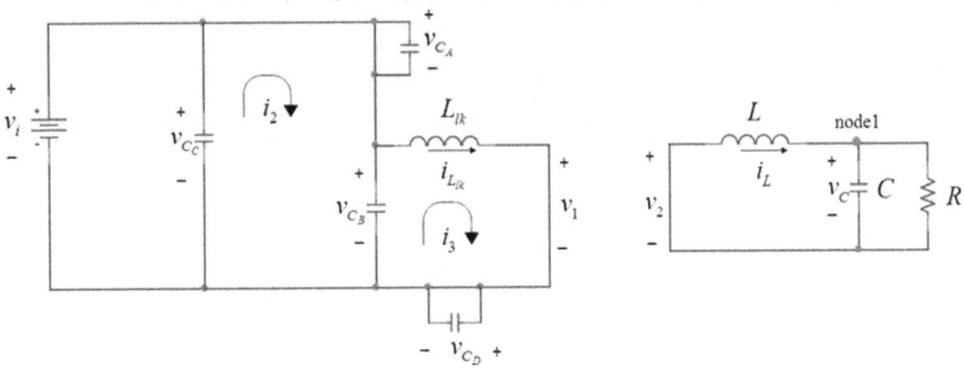

Figure 4.The equivalent circuit of the period of duty cycle loss (t_1~t_2).

$$\dot{\vec{x}}(t)=\begin{bmatrix}0&0&0&0&0&0&0\\0&0&-\frac{1}{L}&0&0&0&0\\0&\frac{1}{C}&-\frac{1}{RC}&0&0&0&0\\0&0&0&0&0&0&0\\0&0&0&0&0&0&0\\0&0&0&0&0&0&0\\0&0&0&0&0&0&0\end{bmatrix}\vec{x}(t)+\begin{bmatrix}\frac{1}{L_{lk}}\\0\\0\\0\\0\\0\\0\end{bmatrix}v_i(t) \tag{4}$$

After the short period of duty cycle loss, energy passes through the transformer again. Figure 5 shows the equivalent circuit. Initial conditions are $v_{CA}(t_{12})=0$, $v_{CB}(t_{12})=v_i$, $v_{CC}(t_{12})=v_i$, and $v_{CD}(t_{12})=0$. We may derive a set of differential equations and the corresponding set of differential equations and state variable model are

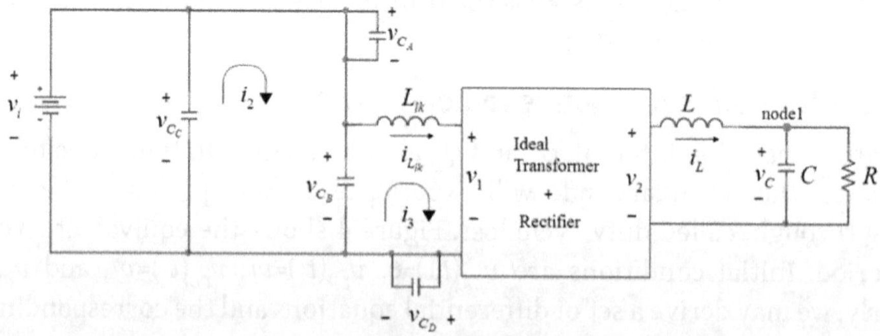

Figure 5.The equivalent circuit of operation interval 2 (t_1~t_2).

$$\dot{\vec{x}}(t)=\begin{bmatrix} 0 & 0 & -\frac{n}{n^2L_{lk}+L} & 0 & 0 & 0 & 0 \\ 0 & 0 & -\frac{1}{n^2L_{lk}+L} & 0 & 0 & 0 & 0 \\ 0 & \frac{1}{C} & -\frac{1}{RC} & 0 & 0 & 0 & 0 \\ 0 & 0 & 0 & 0 & 0 & 0 & 0 \\ 0 & 0 & 0 & 0 & 0 & 0 & 0 \\ 0 & 0 & 0 & 0 & 0 & 0 & 0 \\ 0 & 0 & 0 & 0 & 0 & 0 & 0 \end{bmatrix}\vec{x}(t)+\begin{bmatrix} \frac{n^2}{n^2L_{lk}+L} \\ \frac{n}{n^2L_{lk}+L} \\ 0 \\ 0 \\ 0 \\ 0 \\ 0 \end{bmatrix}v_i(t) \quad (5)$$

The duration of duty cycle loss may be derived based on falling range of i_L is equal to rising range of i_L, i.e.,

$$\frac{t_{passive}}{n^2L_{lk}+L}v_C+\frac{\Delta D}{L}v_C=\left(\frac{-1}{n^2L_{lk}+L}v_C+\frac{n}{n^2L_{lk}+L}v_i\right)\times(t_{active}-\Delta D)\phi=\frac{t_{passive}}{t_{passive}+t_{active}}\times 180^\circ \quad (6)$$

where ΔD is duty cycle loss time, $t_{passive}$ is the time of passive region, t_{active} is the time of active region, and φ is the difference of phase between Q_A and Q_C.

2.3. Positive half cycle: leading-leg (active-to-passive) transition (t_2~t_3)

During this operation interval, only Q_A is conducting. Initial conditions are $v_{CA}(t_2)=0$, $v_{CB}(t_2)=v_i$, $v_{CC}(t_2)=v_i$, and $v_{CD}(t_2)=0$. In this interval, C_C is discharging and C_D is charging until v_{CC} equals zero. Q_C is turned on at zero voltage. Applying KVL/KCL in a similar way, we obtain

$$\dot{\vec{x}}(t)=\begin{bmatrix}0 & 0 & -\frac{n}{n^2L_{lk}+L} & 0 & 0 & 0 & -\frac{n^2}{n^2L_{lk}+L}\\ 0 & 0 & -\frac{1}{n^2L_{lk}+L} & 0 & 0 & 0 & -\frac{n}{n^2L_{lk}+L}\\ 0 & \frac{1}{C} & -\frac{1}{RC} & 0 & 0 & 0 & 0\\ 0 & 0 & 0 & 0 & 0 & 0 & 0\\ 0 & 0 & 0 & 0 & 0 & 0 & 0\\ -\frac{1}{C_C+C_D} & 0 & 0 & 0 & 0 & 0 & 0\\ \frac{1}{C_C+C_D} & 0 & 0 & 0 & 0 & 0 & 0\end{bmatrix}\vec{x}(t)+\begin{bmatrix}\frac{n^2}{n^2L_{lk}+L}\\ \frac{n}{n^2L_{lk}+L}\\ 0\\ 0\\ 0\\ 0\\ 0\end{bmatrix}v_i(t) \quad (7)$$

2.4. Positive half cycle: Passive region (t_3~t_4)

During this interval, Q_A and Q_C are conducting. Initial conditions are $v_{CA}(t_3)$=0, $v_{CB}(t_3)$=v_i, $v_{CC}(t_3)$=0, and $v_{CD}(t_3)$=v_i. Applying KVL/KCL in a similar way, we obtain

$$\dot{\vec{x}}(t)=\begin{bmatrix}0 & 0 & -\frac{n}{n^2L_{lk}+L} & 0 & 0 & 0 & 0\\ 0 & 0 & -\frac{1}{n^2L_{lk}+L} & 0 & 0 & 0 & 0\\ 0 & \frac{1}{C} & -\frac{1}{RC} & 0 & 0 & 0 & 0\\ 0 & 0 & 0 & 0 & 0 & 0 & 0\\ 0 & 0 & 0 & 0 & 0 & 0 & 0\\ 0 & 0 & 0 & 0 & 0 & 0 & 0\\ 0 & 0 & 0 & 0 & 0 & 0 & 0\end{bmatrix}\vec{x}(t) \quad (8)$$

2.5. Negative half cycle

The subsequent four operation intervals basically "mirror" those in positive cycle. Therefore, the derivations are omitted for brevity.

3. SOLUTION AND NUMERICAL SIMULATION

Numerical simulation based on a typical PSPWM full-bridge power converter circuit (with parameters: transformer turns ratio n = 0.5, V_i=160 volt, V_o=50 volt, R = 6 Ω, C = 940 μF, L = 300μH, L_{lk}=20μH, C_A=C_B=C_C=C_D=5nF, f_s=50 kHz) in our laboratory is performed. A "realistic" model of the circuit is built using PSpice, and the developed mathematical model is realized using MATLAB/Simulink. Comparison of the simulation results validates the correctness and effectiveness of the established model.

4. STABILITY ANALYSIS

4.1. Zero-state response

A SISO system (A,B,C) with proper rational transfer function $G(s)=C(sI-A)^{-1}B$ is BIBO stable if and only if every pole of $G(s)$ has a negative real part or, equivalently, lies inside the left-half s-plane. For both positive and negative half cycles, we can obtain the following transfer function for each operation interval:

Trailing-leg (passive-to-active) transition:

$$G_1 = \frac{-nR(C_A + C_B)s}{RC(C_A + C_B)(n^2L_{lk} + L)s^3 + (C_A + C_B)(n^2L_{lk} + L)s^2 + (R(C_A + C_B) + n^2RC)s + n^2} \quad (9)$$

pole=p11, p12, p13 (in complicated form)

Active region (duty cycle loss): $G_{2loss}=0$

Active region:

$$G_2 = \frac{nR}{RC(n^2L_{lk} + L)s^2 + (n^2L_{lk} + L)s + R}$$

$$pole = \frac{-(n^2L_{lk} + L) \pm \sqrt{(n^2L_{lk} + L)^2 - 4R^2C(n^2L_{lk} + L)}}{2RC(n^2L_{lk} + L)} \quad (10)$$

Leading-leg (active-to-passive) transition:

$$G_3 = \frac{nR(C_C + C_D)s}{RC(C_C + C_D)(n^2L_{lk} + L)s^3 + (C_C + C_D)(n^2L_{lk} + L)s^2 + (R(C_C + C_D) + n^2RC)s + n^2} \quad (11)$$

pole=p31,p32,p33(in complicated form)

Passive region: no input:

Due to “mirroring” operation, corresponding intervals in positive and negative half cycles will have the same transfer function. The pole location for the operation interval of trailing-leg and leading-leg transition depends further on the values of the circuit elements. The poles for the operation interval of active region have negative real parts due to

$$(n^2L_{lk} + L) > \sqrt{(n^2L_{lk} + L)^2 - 4R^2C(n^2L_{lk} + L)} \quad (12)$$

Hence, the system is BIBO stable within this interval. Note that there is pole/zero cancelation for all intervals, which implies that each interval subsystem is either uncontrollable or unobservable.

4.2. Zero-input response

The equation $\dot{x}=Ax$ is marginally stable if and only if all eigenvalues of A have zero or negative real parts and those with zero real parts are simple root of the minimal polynomial of A. The equation $\dot{x}=Ax$ is asymptotically stable if and only if all eigenvalues of A have negative real parts. For both positive and negative half cycles, we can obtain the following set of eigenvalues for each operation interval:

Trailing-leg (passive-to-active) transitions:

$$A_1 = \begin{bmatrix} 0 & 0 & \frac{n}{n^2L_{lk}+L} & \frac{-n^2}{n^2L_{lk}+L} & 0 & 0 & 0 \\ 0 & 0 & \frac{-1}{n^2L_{lk}+L} & \frac{n}{n^2L_{lk}+L} & 0 & 0 & 0 \\ 0 & \frac{1}{C} & \frac{-1}{RC} & 0 & 0 & 0 & 0 \\ \frac{1}{C_A+C_B} & 0 & 0 & 0 & 0 & 0 & 0 \\ \frac{-1}{C_A+C_B} & 0 & 0 & 0 & 0 & 0 & 0 \\ 0 & 0 & 0 & 0 & 0 & 0 & 0 \\ 0 & 0 & 0 & 0 & 0 & 0 & 0 \end{bmatrix}, \lambda = 0, 0, 0, 0, \lambda 11, \lambda 12, \lambda 13 \quad (13)$$

Active region (duty cycle loss):

$$\lambda = 0, 0, 0, 0, 0, \frac{-1 \pm \sqrt{(L-4R^2C)/L}}{2RC} \quad (14)$$

Active region:

$$\lambda = 0, 0, 0, 0, 0, \frac{-(n^2L_{lk}+L) \pm \sqrt{(n^2L_{lk}+L)^2 - 4R^2C(n^2L_{lk}+L)}}{2RC(n^2L_{lk}+L)} \quad (15)$$

Leading-leg (active-to-passive) transitions:

$$\lambda = 0, 0, 0, 0, \lambda 31, \lambda 32, \lambda 33 (\text{in complicated form}) \quad (16)$$

Passive region:

$$\lambda = 0, 0, 0, 0, 0, \frac{-(n^2L_{lk}+L) \pm \sqrt{(n^2L_{lk}+L)^2 - 4R^2C(n^2L_{lk}+L)}}{2RC(n^2L_{lk}+L)} \quad (17)$$

Since all intervals have zero eigenvalue, we need to determine whether zero is a simple root of the minimal polynomial of A. The minimal polynomial (in x) for each operation interval (positive or negative half cycle) is summarized as follows:

Trailing-leg (passive-to-active) transitions:

$$x^4 + \frac{1}{RC}x^3 + \frac{n^2C + C_A + C_B}{C(C_A+C_B)(n^2L_{lk}+L)}x^2 + \frac{n^2}{RC(C_A+C_B)(n^2L_{lk}+L)}x \quad (18)$$

Active region (duty cycle loss):

$$x^3 + \frac{1}{RC}x^2 + \frac{1}{LC}x \tag{19}$$

Active region:

$$x^3 + \frac{1}{RC}x^2 + \frac{1}{C(n^2L_{lk} + L)}x \tag{20}$$

Leading-leg (active-to-passive) transitions:

$$x^4 + \frac{1}{RC}x^3 + \frac{n^2C + C_C + C_D}{C(C_C + C_D)(n^2L_{lk} + L)}x^2 + \frac{n^2}{RC(C_C + C_D)(n^2L_{lk} + L)}x \tag{21}$$

Passive region:

$$x^3 + \frac{1}{RC}x^2 + \frac{1}{C(n^2L_{lk} + L)}x \tag{22}$$

Although all operation intervals have different state matrix (A), corresponding intervals in positive and negative half cycles actually possess the same set of eigenvalues. The eigenvalues for the operation interval of trailing-leg and leading-leg transition depend further on the values of the circuit elements. Both intervals of active (including duty cycle loss) and passive region are marginally stable due to

$$n^2L_{lk} + L > \sqrt{(n^2L_{lk} + L)^2 - 4R^2C(n^2L_{lk} + L)}, \quad 1 > \sqrt{\frac{L - 4R^2C}{L}} \tag{23}$$

and zero is the simple root of the minimal polynomial.

5. CONTROLLABILITY/OBSERVABILITY ANALYSIS

The stability analysis indicates that all interval subsystems have uncontrollable or unobservable modes. We may decompose the state variable model of each subsystem into controllable and uncontrollable parts, and follow by decomposing each part into observable and unobservable parts to obtain

$$\begin{bmatrix} \dot{\bar{x}}_{co} \\ \dot{\bar{x}}_{c\bar{o}} \\ \dot{\bar{x}}_{\bar{c}o} \\ \dot{\bar{x}}_{\bar{c}\bar{o}} \end{bmatrix} = \left[\begin{array}{c|ccc} \bar{A}_{co} & 0 & \bar{A}_{13} & 0 \\ \hline \bar{A}_{21} & \bar{A}_{c\bar{o}} & \bar{A}_{23} & \bar{A}_{24} \\ 0 & 0 & \bar{A}_{\bar{c}o} & 0 \\ 0 & 0 & \bar{A}_{43} & \bar{A}_{\bar{c}\bar{o}} \end{array}\right] \begin{bmatrix} \bar{x}_{co} \\ \bar{x}_{c\bar{o}} \\ \bar{x}_{\bar{c}o} \\ \bar{x}_{\bar{c}\bar{o}} \end{bmatrix} + \left[\begin{array}{c} \bar{B}_{co} \\ \hline \bar{B}_{c\bar{o}} \\ 0 \\ 0 \end{array}\right] u, \quad y = \left[\begin{array}{c|ccc} \bar{C}_{co} & 0 & \bar{C}_{\bar{c}o} & 0 \end{array}\right] \begin{bmatrix} \bar{x}_{co} \\ \bar{x}_{c\bar{o}} \\ \bar{x}_{\bar{c}o} \\ \bar{x}_{\bar{c}\bar{o}} \end{bmatrix} \tag{24}$$

The observability matrices of the controllable part for each operation interval (positive or negative half cycle) are summarized as follows:

Trailing-leg (passive-to-active) transition:

$$O_{c1} = \begin{bmatrix} 0 & \frac{-n}{C(n^2L_{lk}+L)} & \frac{n}{RC^2(n^2L_{lk}+L)} \\ \frac{-n}{C(n^2L_{lk}+L)} & \frac{n}{RC^2(n^2L_{lk}+L)} & \frac{n(n^2C+C_A+C_B)}{C^2(C_A+C_B)(n^2L_{lk}+L)^2} - \frac{n}{R^2C^3(n^2L_{lk}+L)} \\ \frac{n}{RC^2(n^2L_{lk}+L)} & oc1.1 & oc1.2 \end{bmatrix} \tag{25}$$

Active region (duty cycle loss):

$$O_{c2loss} = [0] \tag{26}$$

Active region:

$$O_{c2} = \begin{bmatrix} 0 & \frac{n}{C(n^2L_{lk}+L)} \\ \frac{n}{C(n^2L_{lk}+L)} & \frac{-n}{RC^2(n^2L_{lk}+L)} \end{bmatrix} \tag{27}$$

Leading-leg (active-to-passive) transitions:

$$O_{c3} = \begin{bmatrix} 0 & \frac{n}{C(n^2L_{lk}+L)} & \frac{-n}{RC^2(n^2L_{lk}+L)} \\ \frac{n}{C(n^2L_{lk}+L)} & \frac{-n}{RC^2(n^2L_{lk}+L)} & \frac{n}{R^2C^3(n^2L_{lk}+L)} - \frac{n(n^2C+C_C+C_D)}{C^2(C_C+C_D)(n^2L_{lk}+L)^2} \\ \frac{-n}{RC^2(n^2L_{lk}+L)} & oc3.1 & oc3.2 \end{bmatrix} \tag{28}$$

Table 1 summarizes the rank of the observability matrix. The state variables of both controllable and observable are listed in Table 2. Since equivalent transformation does not affect the eigenvalues, Eq. (24) has the same set of eigenvalues as in stability analysis. For the operation intervals of trailing/leading leg and active region, uncontrollable or unobservable states ($v_{CA'B'}$, $v_{CC'}$ and v_{CD}) are marginally stable corresponding to zero eigenvalue. Therefore, those states will stay constant within those intervals, which matches what is observed in numerical simulation. For the intervals of duty cycle loss and passive region, uncontrollable (i_L,) states are asymptotically stable, which also matches what is observed during simulation.

Table 1. Rank of the observability matrix for the controllable part.

	Interval	Rank
Positive half cycle	Trailing-leg (passive-to-active) transitions	3
	Active region (duty cycle loss)	0
	Active region	2
	Leading-leg (active-to-passive) transitions	3

	Interval	Rank
Negative half cycle	Trailing-leg (passive-to-active) transitions	3
	Active region (duty cycle loss)	0
	Active region	2
	Leading-leg (active-to-passive) transitions	3

Table 2. State variables of both controllable and observable.

	Interval	Controllable and observable
Positive half cycle	Trailing-leg (passive-to-active) transitions	i_L, v_C, v_{CB}
	Active region	i_L, v_C
	Leading-leg (active-to-passive) transitions	i_L, v_C, v_{CD}
Negative half cycle	Trailing-leg (passive-to-active) transitions	i_L, v_C, v_{CB}
	Active region	i_L, v_C
	Leading-leg (active-to-passive) transitions	i_L, v_C, v_{CD}

6. MODEL REDUCTION

The goal is to obtain a low dimensional model that encompasses the imperative response characteristics of the comprehensive model. The reduced model is then utilized for subsequent control design. For control of the "steady-state" response, we neglect the transition intervals and take only the active region and passive region into consideration. Define d is duty cycle (ON) of the converter and $d'=1-d$ (OFF). Assuming that $L_{lk} \ll L$, we may derive the following differential inclusion model

$$\begin{bmatrix} \frac{di_L(t)}{dt} \\ \frac{dv_o(t)}{dt} \end{bmatrix} = \begin{bmatrix} 0 & -\frac{1}{L} \\ \frac{1}{C} & -\frac{1}{RC} \end{bmatrix} \begin{bmatrix} i_L(t) \\ v_o(t) \end{bmatrix} + \begin{bmatrix} \frac{nd}{L} \\ 0 \end{bmatrix} v_i(t) \tag{29}$$

Let $x_1 = i_L, {}_2 = v_o, u = d, y = v_o$

$$\begin{bmatrix} \dot{x}_1 \\ \dot{x}_2 \end{bmatrix} = \begin{bmatrix} 0 & \frac{-1}{L} \\ \frac{1}{C} & \frac{-1}{RC} \end{bmatrix} \begin{bmatrix} x_1 \\ x_2 \end{bmatrix} + \begin{bmatrix} \frac{nv_i}{L} \\ 0 \end{bmatrix} u, \ddot{y} = \frac{-1}{RC^2} x_1 + \left(\frac{1}{R^2C^2} - \frac{1}{LC} \right) x_2 + \frac{nv_i}{LC} u \tag{30}$$

7. INDIRECT ADAPTIVE FUZZY CONTROL FOR UNCERTAIN SWITCHING POWER CONVERTERS SUBJECT TO EXTERNAL DISTURBANCES

In the following, we propose a robust adaptive fuzzy tracking controller for the PSPWM full-bridge soft switched power converter. Although the controller is designed based on the reduced model, its effectiveness and performance are subsequently verified with the comprehensive model.

Indirect adaptive fuzzy control with sliding model

Based on the input-output linearization concept, Eq. (30) can be represented by

$$y^{(2)} = f(x) + g(x)u, f(x) = \frac{-1}{RC^2}x_1 + \left(\frac{1}{R^2C^2} - \frac{1}{LC}\right)x_2,\ g(x) = \frac{nv_i}{LC} \quad (31)$$

The control objective is to force y to follow a given bounded reference signal ymym, under the constraint that all signals involved must be bounded. Let $e=y_m-y$, $e=(e,e^{\cdot})^T$ and $k=(k_2,k_1)^T$ be such that all roots of the polynomial $s^2+k_1s+k_2$ are in the open left half-plane. If the functions f and g are known, then the control law

$$u = \frac{1}{g(x)}[-f(x) + y_m{}^{(2)} + \boldsymbol{k}^T\boldsymbol{e}] \quad (32)$$

applied to Eq. (31) results in

$$e^{(2)} + k_1\dot{e} + k_2e = 0 \quad (33)$$

which implies that $\lim_{t\to\infty} e(t) = 0$ a main objective of control.

However, f and g are unknown. We replace f and g in Eq. (32) by the fuzzy logic systems $\hat{f}(x|\theta_f)$ and $\hat{g}(x|\theta_g)$. The resulting control law

$$u_c = \frac{1}{\hat{g}(x|\theta_g)}\left[-\hat{f}(x|\theta_f) + y_m{}^{(2)} + \boldsymbol{k}^T\boldsymbol{e}\right] \quad (34)$$

is the so-called certainty equivalent controller. We use

$$u = u_c + u_s \quad (35)$$

where the additional control term usus is called a supervisory control for stability. Substituting Eq. (35) into Eq. (31), we obtain the error equation

$$\dot{e} = \Lambda_c e + b_c\left[\hat{f}(x|\theta_f) - f(x) + (\hat{g}(x|\theta_g) - g(x))u_c - g(x)u_s\right] \quad (36)$$

where

$$\Lambda_c = \begin{bmatrix} 0 & 1 \\ -k_2 & -k_1 \end{bmatrix},\quad b_c = \begin{bmatrix} 0 \\ 1 \end{bmatrix} \quad (37)$$

Since Λ_c is a stable matrix ($|sI-\Lambda_c|=s^n+k1^{(n-1)}+\cdots+k_n$ which is stable), we know that there exists a unique positive definite symmetric $n\times n$ matrix P which satisfies the Lyapunov equation:

$$\Lambda_c{}^T P + P\Lambda_c = -Q \quad (38)$$

where Q is an arbitrary 2×2 positive definite matrix. Let $V_e = \frac{1}{2}e^TPe$, in order for

$x_i = y_m^{(i-1)} - e^{(i-1)}$ to be bounded, we require that V_e must be bounded, which means we require that $\dot{V}_e \leq 0$ when V_e is greater than a large constant $\bar{V}$. Using Eq. (35) and Eq. (38), we have

$$\begin{aligned}\dot{V}_e &= -\frac{1}{2}e^TQe + e^TPb_c[(\hat{f}(x|\theta_f) - f(x)) + (\hat{g}(x|\theta_g) - g(x))u_c - g(x)u_s] \\ &\leq -\frac{1}{2}e^TQe + |e^TPb_c|[|(\hat{f}(x|\theta_f)| + |f(x)| + |\hat{g}(x|\theta_g)u_c| + |g(x)u_c|] - e^TPb_cg(x)u_s\end{aligned} \tag{39}$$

In order to design the usus such that the right-hand side of Eq. (39) is not positive, we need to know the bounds of f and g. That is, we have to make the following assumption.

Assumption: We can determine functions $f^U(x)$, $g^U(x)$ and $g_L(x)$ such that $|f(x)| \leq f^U(x)$ and $g_L(x) \leq g(x) \leq g^U(x)$ for $x \in R^2$, where $f^U(x) < \infty$, $g^U(x) < \infty$, and $g^L(x) > 0$ for $x \in R^2$.

Based on $f^U(\underline{x})$, $g^U(\underline{x})$ and $g_L(\underline{x})$, and by observing Eq. (39), we choose the supervisory control u_s as

$$u_s = \begin{cases} \operatorname{sgn}(e^TPb_c)\dfrac{1}{g_L(x)}[|\hat{f}(x|\theta_f)| + |f^U(x)| + |\hat{g}(x|\theta_g)u_c| + |g^U(x)u_c|], & V_e \geq \bar{V} \\ 0, & V_e \leq \bar{V} \end{cases} \tag{40}$$

Substituting Eq. (40) into Eq. (39) and considering the case $V_e > \bar{V}$, we have

$$\begin{aligned}\dot{V}_e &\leq -\frac{1}{2}e^TQe + |e^TPb_c|[|\hat{f}| + |f| + |\hat{g}u_c| + |gu_c| - \frac{g}{g_L}(|\hat{f}| + f^U + |\hat{g}u_c| + |g^Uu_c|)] \\ &\leq -\frac{1}{2}e^TQe < 0\end{aligned} \tag{41}$$

In summary, using the control Eq. (35), we can guarantee that $V_e \leq \bar{V} < \infty$. Since P is positive definite, the boundedness of V_e implies the boundedness of e, which in turn implies the boundedness of x.

We employ the following fuzzy logic system:

$$\hat{f}(x|\theta_f) = \sum_{l=1}^{M} \theta_l \xi_l(x) = \theta^T \xi(x) \tag{42}$$

where $\theta = (\theta_1, \ldots, \theta_M)^T$, $\xi(x) = (\xi_1(x), \ldots, \xi_M(x))^T$, $\xi_l(x)$ is the fuzzy basis function defined by

$$\xi_l(x) = \prod_{i=1}^{2} \mu_{F_i^l}(x_i) \Big/ \sum_{l=1}^{M} \prod_{i=1}^{2} \mu_{F_i^l}(x_i) \tag{43}$$

θlθl are adjustable parameters, and μ_{Fil} are given membership functions.

We present the detailed design steps of the adaptive fuzzy controller.

Step 1 Define p_i fuzzy sets $A_i^l(l_i = 1, 2, \ldots, p_i)$ for variable $x_i(i = 1, 2)$

Step 2 There are $\prod_{i=1}^{2} p_i$ rules to construct fuzzy systems $\hat{f}(x|\theta_f)$:

$$if\ x_1\ is\ A_1^{l_1}\ \ and\ \ x_2\ is\ A_2^{l_2},\quad then\ \hat{f}\ is\ E^{l_1 \ldots l_2}$$

There are $\prod_{i=1}^{2} p_i$ rules to construct fuzzy systems $\hat{g}(x|\theta_g)$:

$$if\ x_1\ is\ B_1^{l_1}\ \ and\ \ x_2\ is\ B_2^{l_2},\quad then\ \hat{g}\ is\ H^{l_1 \ldots l_2}$$

Using product-inference rule, singleton fuzzifier, and center average defuzzifier, we obtain

$$\hat{f}(x|\theta_f) = \theta_f^T \xi_f(x),\ \hat{g}(x|\theta_g) = \theta_g^T \xi_g(x) \tag{44}$$

where

$$\xi_f^l(x) = \frac{\prod_{i=1}^{2} \mu_{A_i^l}(x_i)}{\sum_{l=1}^{p_1 \times p_2} \prod_{i=1}^{2} \mu_{A_i^l}(x_i)},\ \xi_g^l(x) = \frac{\prod_{i=1}^{2} \mu_{B_i^l}(x_i)}{\sum_{l=1}^{p_1 \times p_2} \prod_{i=1}^{2} \mu_{B_i^l}(x_i)} \tag{45}$$

Our next task is to develop an adaptive law to adjust the parameters in the fuzzy logic systems for the purpose of forcing the tracking error to converge to zero.

Define

$$\begin{aligned} \theta_f^* &= \arg\min_{\theta_f}[sup_{x \in R^2}|\hat{f}(x|\theta_f) - f(x)|] \\ \theta_g^* &= \arg\min_{\theta_g}[sup_{x \in R^2}|\hat{g}(x|\theta_g) - g(x)|] \end{aligned} \tag{46}$$

Define the minimum approximation error

$$\omega = (\hat{f}(x|\theta_f^*) - f(x)) + (\hat{g}(x|\theta_g^*) - g(x))u_c \tag{47}$$

Then the error equation can be rewritten as

$$\dot{e} = \Lambda_c e + b_c[(\hat{f}(x|\theta_f)) - \hat{f}(x|\theta_f^*)) + (\hat{g}(x|\theta_g) - \hat{g}(x|\theta_g^*))u_c + \omega] \tag{48}$$

Substituting Eq. (43) into Eq. (47), we have

$$\dot{e} = \Lambda_c e + b_c\omega + b_c[(\theta_f - \theta_f^*)^T \xi_f(x) + (\theta_g - \theta_g^*)^T \xi_g(x) u_c] \tag{49}$$

Consider the Lyapunov function candidate

$$V = \frac{1}{2}e^T P e + \frac{1}{2\gamma_1}(\theta_f - \theta_f^*)^T(\theta_f - \theta_f^*) + \frac{1}{2\gamma_2}(\theta_g - \theta_g^*)^T(\theta_g - \theta_g^*) \tag{50}$$

where γ_1 and γ_2 are positive constants. The time derivative of V along the trajectory of Eq. (48) is

$$\begin{aligned} \dot{V} = &-\frac{1}{2}e^T Q e + e^T P b_c \omega + \frac{1}{\gamma_1}(\theta_f - \theta_f^*)^T[\dot{\theta}_f + \gamma_1 e^T P b_c \xi_f(x)] \\ &+ \frac{1}{\gamma_2}(\theta_g - \theta_g^*)^T[\dot{\theta}_g + \gamma_2 e^T P b_c \xi_g(x) u_c] \end{aligned} \tag{51}$$

If we choose the adaptive law

$$\dot{\theta}_f = -\gamma_1 e^T P b_c \xi_f(x),\ \dot{\theta}_g = -\gamma_2 e^T P b_c \xi_g(x) u_c \qquad (52)$$

then from Eq. (50) we have

$$\dot{V} = -\frac{1}{2} e^T Q e + e^T P b_c \omega \qquad (53)$$

This is the best we can hope to get because the term $e^T P b_c \omega$ is of the order of the minimum approximation error. If ω=0, that is, the searching spaces for $\hat{f}$ and $\hat{g}$ are so big that f and g are included in them, then we have $V^{\cdot} \leq 0$. Eq. (51) cannot guarantee θ_f and θ_g are bounded, so we use projection algorithm. If the parameter vectors θ_f and θ_g are within the constraint sets or on the boundaries of the constraint sets but moving toward the inside of the constraint sets, then use the simple adaptive law Eq. (51). Otherwise, if the parameter vectors are on the boundaries of the constraint sets but moving toward the outside of the constraint sets, then use the projection algorithm to modify the adaptive law Eq. (51). such that the parameter vectors will remain inside the constraint sets.

$$\Omega_f = \left\{ \theta_f \in R^{\prod_{i=1}^{2} p_i} \middle| \|\theta_f\| \leq M_f \right\} \qquad (54)$$

$$\Omega_g = \left\{ \theta_g \in R^{\prod_{i=1}^{2} q_i} \middle| 0 < \varepsilon \leq \|\theta_g\| \leq M_g \right\} \qquad (55)$$

where Ω_f and Ω_g are constraint sets for θ_f and θ_g, M_f,M_g,ε are constants

$$\dot{\theta}_f = \begin{cases} -\gamma_1 e^T P b_c \xi_f(x) & \text{if } (\|\theta_f\| < M_f) \text{ or } (\|\theta_f\| = M_f \text{ and } e^T P b_c \xi_f(x) \geq 0) \\ P\{-\gamma_1 e^T P b_c \xi_f(x)\} & \text{if } (\|\theta_f\| = M_f \text{ and } e^T P b_{cc} \xi_f(x) < 0) \end{cases} \qquad (56)$$

where the projection operator $P\{*\}$ is defined as

$$P\{-\gamma_1 e^T P b_c \xi_f(x)\} = -\gamma_1 e^T P b_c \xi_f(x) + \gamma_1 e^T P b_c \frac{\theta_f \theta_f^T \xi_f(x)}{\|\theta_f\|^2} \qquad (57)$$

Whenever an element θ_{gi} of θ_g=ε, use

$$\dot{\theta}_{gi} = \begin{cases} -\gamma_2 e^T P b_c \xi_{gi}(x) u_c & \text{if } e^T P b_c \xi_{gi}(x) u_c < 0 \\ 0 & \text{if } e^T P b_c \xi_{gi}(x) u_c \geq 0 \end{cases} \qquad (58)$$

where $\xi_{gi}(x)$ is the ith component of $\xi_g(x)$.

Otherwise, use

$$\dot{\theta}_g = \begin{cases} -\gamma_2 e^T P b_c \xi_g(x) u_c & \text{if } (\|\theta_g\| < M_g) \text{ or } (\|\theta_g\| = M_g \text{ and } e^T P b_c \xi_g(x) u_c \geq 0) \\ P\{-\gamma_2 e^T P b_c \xi_g(x)\} & \text{if } (\|\theta_g\| = M_g \text{ and } e^T P b_c \xi_g(x) u_c < 0) \end{cases} \qquad (59)$$

where the projection operator $P\{*\}$ is defined as

$$P\{-\gamma_2 e^T P b_c \xi_g(x) u_c\} = -\gamma_2 e^T P b_c \xi_g(x) u_c + \gamma_2 e^T P b_c \frac{\theta_g \theta_g^T \xi_g(x) u_c}{\|\theta_g\|^2} \quad (60)$$

Theorem:

$\|\theta(t)\| \le M_f, \|\theta_g(t)\| \le M_g$, all elements of $\theta_g \ge \varepsilon$,

$$\|x(t)\| \le \|y_m\| + \left(\frac{2\overline{V}}{\lambda_{P\min}}\right)^{\frac{1}{2}} \quad (61)$$

$$|u(t)| \le \frac{1}{\varepsilon}\left(M_f + |y_m^2| + \|k\|\left(\frac{2\overline{V}}{\lambda_{P\min}}\right)^{\frac{1}{2}}\right) + \frac{1}{g_L(x)}$$
$$\left[M_f + |f^U(x)| + \frac{1}{\varepsilon}(M_g + g^U)\left(M_f + |y_m^2| + \|k\|\left(\frac{2\overline{V}}{\lambda_{P\min}}\right)^{\frac{1}{2}}\right)\right] \quad (62)$$

for all $t \ge 0$, where $\lambda_{P\min}$ is the minimum eigenvalue of P, and $y_m = (y_m, y'_m, \ldots, y_m^{(n-1)})^T$.

$$\int_0^t \|e(\tau)\|^2 d\tau \le a + b\int_0^t |\omega(\tau)|^2 d\tau \quad (63)$$

for all $t \ge 0$, where a and b are constants, and ω is the minimum approximation error defined by Eq. (46)

If □ is squared integrable, that is, $\int_0^\infty |\omega(t)|^2 dt < \infty$, then $\lim_{t\to\infty}\|e(t)\| = 0$

Design example:

The parameters of the converter are listed in Figure 6. Consider the following system:

Figure 6.Photo and parameters of a PSPWM full-bridge power converter circuit. Transformer turns ratio n = 0.5. V_i = 160 volt, V_o = 50 volt, R = 6Ω, C = 940μF L = 300μH, L_{lk} = 20μH $C_A = C_B = C_C = C_D$ = 5nF, f_s = 50 kHz.

$$y^{(2)} = f(\boldsymbol{x}) + g(\boldsymbol{x})u \quad (64)$$

$$f(\boldsymbol{x}) = \frac{-1}{RC^2}x_1 + \left(\frac{1}{R^2C^2} - \frac{1}{LC}\right)x_2,\ g(\boldsymbol{x}) = \frac{nv_i}{LC} \quad (65)$$

The design steps of the adaptive fuzzy controller are provided in the following:

$$u = u_c + u_s,\ u_c = \frac{1}{\hat{g}(\boldsymbol{x}|\boldsymbol{\theta}_g)}\left[-\hat{f}(\boldsymbol{x}|\boldsymbol{\theta}_f) + y_m{}^{(2)} + \boldsymbol{k}^T\boldsymbol{e}\right]$$
$$u_s = \begin{cases} \operatorname{sgn}(\boldsymbol{e}^TP\boldsymbol{b}_c)\frac{1}{g_L(\boldsymbol{x})}\left[|\hat{f}(\boldsymbol{x}|\boldsymbol{\theta}_f)| + |f^U(\boldsymbol{x})| + |\hat{g}(\boldsymbol{x}|\boldsymbol{\theta}_g)u_c| + |g^U(\boldsymbol{x})u_c|\right], & V_e \geq \overline{V} \\ 0, & V_e \leq \overline{V} \end{cases} \quad (66)$$

Step 1:

Let

$$\Lambda_c = \begin{bmatrix} 0 & 1 \\ -100000 & -1000 \end{bmatrix}, Q = \begin{bmatrix} 200000 & 0 \\ 0 & 1 \end{bmatrix} \quad (67)$$

Use $\Lambda_c{}^T P + P\Lambda_c = -Q$ to find P

$$P = \begin{bmatrix} 1150 & 1 \\ 1 & 0.0015 \end{bmatrix}, \lambda_{P\min} = 0.00063 \quad (68)$$

Determine the range of x

$$0 < x_1 < 20, 0 < x_2 < 60, 0 < \|\boldsymbol{x}\| < 63.24 \quad (69)$$

Determine the range of input and output

$$10 < u \leq 90, y_m = 50 \quad (70)$$

Obtain $\overline{V}$ by

$$\left(\frac{2\overline{V}}{\lambda_{P\min}}\right)^{\frac{1}{2}} \leq 13.24, \overline{V} = \frac{\lambda_{P\min}}{2}(13.24)^2 = 0.055 \quad (71)$$

Find $(x), {}^U(x), g_L(x)$ according to

$$\begin{aligned} |f(\boldsymbol{x})| &= \left|\frac{-1}{RC^2}x_1 + \left(\frac{1}{R^2C^2} - \frac{1}{LC}\right)x_2\right| \\ &= \left|\frac{-1}{6\times(940\times10^{-6})^2}x_1 + \left(\frac{1}{6^2(940\times10^{-6})^2} - \frac{1}{(300\times10^{-6})(940\times10^{-6})}\right)x_2\right| \\ &= 188,622.3027x_1 + 3,514,662.2403x_2 = f^U(\boldsymbol{x}) \end{aligned} \quad (72)$$
$$|g(\boldsymbol{x})| = \left|\frac{nv_i}{LC}\right| = \frac{0.5\times160}{(300\times10^{-6})(940\times10^{-6})} = 283,687,943.2624 = g^U(\boldsymbol{x}) = g_L(\boldsymbol{x})$$

Set the other parameters as

$$M_f = 1,000,000,000,\ M_g = 1,000,000,000,\ \varepsilon = 2,\ \gamma_1 = 10,000,000,000,\ \gamma_2 = 500,000,000 \quad (73)$$

Step 2:

Establish the following fuzzy rules

$$\begin{aligned}&\mu_{F_1^1}(x_1) = \exp\left(-\left(\tfrac{x_1-0}{2}\right)^2\right), \mu_{F_1^2}(x_1) = \exp\left(-\left(\tfrac{x_1-4}{2}\right)^2\right), \mu_{F_1^3}(x_1) = \exp\left(-\left(\tfrac{x_1-8}{2}\right)^2\right),\\ &\mu_{F_1^4}(x_1) = \exp\left(-\left(\tfrac{x_1-12}{2}\right)^2\right), \mu_{F_1^5}(x_1) = \exp\left(-\left(\tfrac{x_1-16}{2}\right)^2\right), \mu_{F_1^6}(x_1) = \exp\left(-\left(\tfrac{x_1-20}{2}\right)^2\right)\end{aligned} \quad (74)$$

$$\begin{aligned}&\mu_{F_2^1}(x_2) = \exp\left(-\left(\tfrac{x_1-0}{6}\right)^2\right), \mu_{F_2^2}(x_2) = \exp\left(-\left(\tfrac{x_1-12}{6}\right)^2\right), \mu_{F_2^3}(x_2) = \exp\left(-\left(\tfrac{x_1-24}{6}\right)^2\right),\\ &\mu_{F_2^4}(x_2) = \exp\left(-\left(\tfrac{x_1-36}{6}\right)^2\right), \mu_{F_2^5}(x_2) = \exp\left(-\left(\tfrac{x_1-48}{6}\right)^2\right), \mu_{F_2^6}(x_2) = \exp\left(-\left(\tfrac{x_1-60}{6}\right)^2\right)\end{aligned} \quad (75)$$

such that we have 36 rules

$$\hat{f}(x|\theta_f) = \theta_f^T \xi_f(x), \hat{g}(x|\theta_g) = \theta_g^T \xi_g(x) \quad (76)$$

where

$$\xi_l(x) = \prod_{i=1}^{2} \mu_{F_i^l}(x_i) \Big/ \sum_{l=1}^{36} \prod_{i=1}^{2} \mu_{F_i^l}(x_i), l = 1, 2\ldots 36 \quad (77)$$

Step 3:

Use the adaptive law as described in Eq. 51 to Eq. 59

Numerical simulation is performed by augmenting the controller and parametric adaptive law with the comprehensive open-loop model. Figure 7 is the input, output current, output voltage, and tracking error within 0–0.3s. The input is not supplied until 0.05s to allow some transient response. Note that the tracking error converges around 0.12s. Figure 8 is the zoomed input, output current, output voltage, and tracking error within 0.299–0.3s. We see that the output voltage converges to 50V and the tracking error converges to 0.

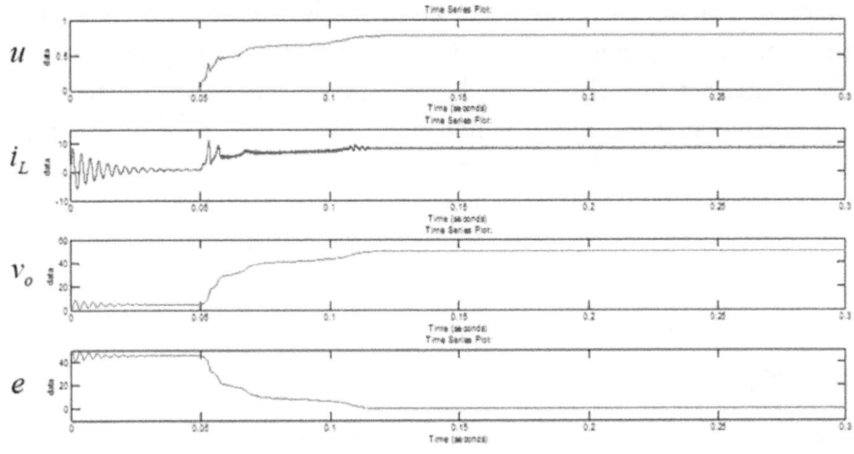

Figure 7.t = 0~0.3s.

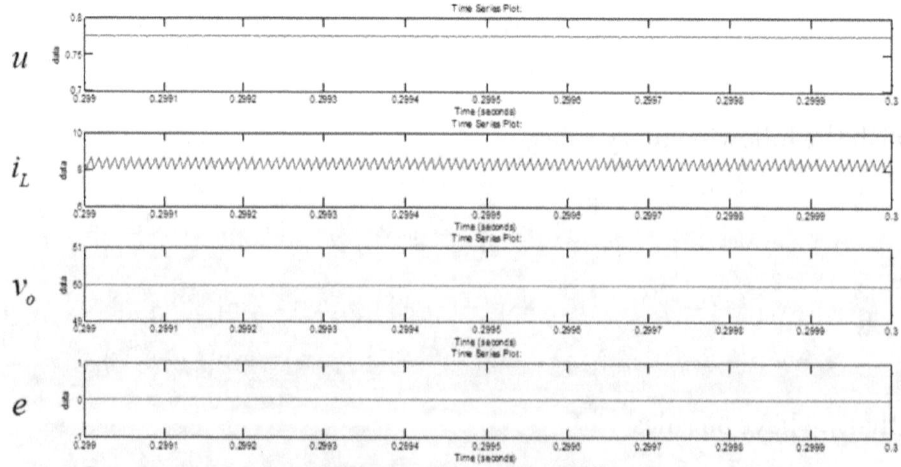

Figure 8.t = 0.299~0.3.

8. CONCISE SUMMARY

This chapter introduces a control-centric modeling and analytical methodology for a category of PWM full-bridge power converters. The findings can be applied to further categories of switching power converters featuring intricate topologies. The proposed modeling and analysis methodology offers a range of critical information for future control design, encompassing the selection of circuit element values, stability attributes of the open-loop system, and the identification of controllable and observable signals/variables, among others. Contemporary studies on feedback management of dc-dc power converters predominantly concentrate on systems featuring uncomplicated circuit topologies (buck, boost, or buck/boost). Control for soft-switched PSPWM full-bridge converters remains predominantly confined to linearized designs utilizing PI or lead-lag compensators. Traditional linearized design methodologies may disregard essential dynamics by neglecting bilinear factors. For systems characterized by nonlinearities and uncertainties that are challenging to mathematically describe accurately, fuzzy control is a viable alternative. Furthermore, this work demonstrates that desirable attributes, such as tracking and resilience, are attained by the integration of fuzzy control with parametric adaptation and sliding mode control. The experimental validation of the suggested control system is presently underway. A future objective is to construct a power factor correction (PFC) circuit to optimize the input current of off-line power sources, hence maximizing the real power derived from the mains. Another incentive to using PFC is to adhere to regulatory mandates.

AT A GLANCE

This chapter presents a control-focused modeling and analytical methodology, accompanied by a nonlinear control design for a category of switching power converters. A comprehensive model integrating the individual state variable models of the interval subsystems is constructed. The comparison with PSpice simulation validates the model's trustworthiness. This model allows for the examination of internal/BIBO stability for each interval subsystem. Furthermore, controllability and observability can be assessed to inform the subsequent control configuration. The established model is further examined for sophisticated control design, specifically robust adaptive fuzzy control.

REFERENCES

Mweene LH, Wright CA, Schlecht MF. A 1 kW 500 kHz front-end converter for a distributed power supply system. IEEE Transactions on Power Electronics. 1991;6(3):398–407.

Redl R, Sokal NO, Balogh L. A novel soft-switching full-bridge DC/DC converter: analysis, design considerations, and experimental results at 1.5 kW, 100 kHz. IEEE Transactions on Power Electronics. 1991;6(3):408–418.

Brunoro M, Vieira JLF. A high-performance ZVS full-bridge DC-DC 0-50-V/0-10-A power supply with phase-shift control. IEEE Transactions on Power Electronics. 1999;14(3):495–505.

Cho J-G, Sabate JA, Hua G, Lee FC. Zero-voltage and zero-current-switching full bridge PWM converter for high-power applications. IEEE Transactions on Power Electronics. 1996;11(4):622–628.

Hua G, Lee FC, Jovanovic MM. An improved full-bridge zero-voltage-switched PWM converter using a saturable inductor. IEEE Transactions on Power Electronics. 1993;8(4):530–534.

Jang Y, Jovanovic MM, editors. A new family of full-bridge ZVS converters. Applied Power Electronics Conference and Exposition, 2003 APEC'03 Eighteenth Annual IEEE; 2003: IEEE.

Jang Y, Jovanovic MM, Chang Y-M. A new ZVS-PWM full-bridge converter. IEEE Transactions on Power Electronics. 2003;18(5):1122–1129.

Redl R, Balogh L, Edwards DW, editors. Optimum ZVS full-bridge dc/dc converter with PWM phase-shift control: analysis, design considerations, and experimental results. Applied Power Electronics Conference and Exposition, 1994 APEC'94 Conference Proceedings 1994, Ninth Annual; 1994: IEEE Orlando, Florida (USA).

Cho J-H, Seong H-W, Jung S-M, Park J-S, Moon G-W, Youn M-J, editors. Implementation of digitally controlled phase shift full bridge converter for server power supply. Energy Conversion Congress and Exposition (ECCE), 2010 IEEE; 2010: IEEE Atlanta, Georgia (USA).

Lim J-G, Lim S-H, Chung S-K, editors. Digital control of phase-shifted full bridge PWM converter. 2007 ICPE'07 7th International Conference on Power Electronics; 2007: IEEE Daegu, Korea.

Liu Y-F, Meyer E, Liu X. Recent developments in digital control strategies for DC/DC switching power converters. IEEE Transactions on Power Electronics. 2009;24(11):2567–2577.

Tseng K-H, Chen C-L, editors. Design and hardware implementation for a full-bridge phase-shift PWM DC/DC converter system with FPGA-based PI gain-scheduling control. 2011 6th IEEE Conference on Industrial Electronics and Applications (ICIEA); 2011: IEEE Beijing, China.

Schutten MJ, Torrey DA. Improved small-signal analysis for the phase-shifted PWM power converter. IEEE Transactions on Power Electronics. 2003;18(2):659–669.

Vlatkovic V, Sabate JA, Ridley RB, Lee FC, Cho BH. Small-signal analysis of the phase-shifted PWM converter. IEEE Transactions on Power Electronics. 1992;7(1):128–135.

Chen W, Lee FC, Jovanovic M, Sabate JA, editors. A comparative study of a class of full bridge zero-voltage-switched PWM converters. Applied Power Electronics Conference and Exposition, 1995 APEC'95 Conference Proceedings 1995, Tenth Annual; 1995: IEEE Dallas, Texas (USA).

Alonge F, D'Ippolito F, Raimondi FM, Tumminaro S. Nonlinear modeling of DC/DC converters using the Hammerstein's approach. IEEE Transactions on Power Electronics. 2007;22(4):1210–1221.

Ortiz CEC. On the circuit oriented average large-signal modeling of switching power converters and its applications. Virginia Polytechnic Institute and State University; 2003 Blacksburg, Virginia (this is a doctoral dissertation).

Iannello C, Luo S, Batarseh I. Small-signal and transient analysis of a full-bridge, zero-current-switched PWM converter using an average model. IEEE Transactions on Power Electronics. 2003;18(3):793–801.

Tsai F-S. Small-signal and transient analysis of a zero-voltage-switched, phase-controlled PWM converter using averaged switch model. IEEE Transactions on Industry Applications. 1993;29(3):493–499.

Van Dijk E, Spruijt J, O'sullivan DM, Klaassens JB. PWM-switch modeling of DC-DC converters. IEEE Transactions on Power Electronics. 1995;10(6):659–665.

Chen C-C, Chen C-L, Chang J-X, Yang C-F. LPV Gain-scheduling control for a phase-shifted PWM full-bridge soft switched converter. IFAC Proceedings Volumes. 2014;47(3):6135–6140.

9 Chapter

FUZZY OPTIMIZATION CONTROL: TRANSITIONING FROM CRISP OPTIMIZATION

1. INTRODUCTION: AN OVERVIEW

This chapter introduces the section structure and abstract in section 1. Additionally, abstracts of several studies on diverse fuzzy optimization control systems are presented and analyzed. Current interests are mathematically stable and will become operational in the future, whereas Mandani-type fuzzy control is intelligible and has already been implemented in a marketing personal computer control system for a connected tank level. Section 2 presents the structure of the optimal control system created by the author and colleagues, along with recent advancements in the optimization component. The initial version of the process application comprised the optimal search component, constrained by an upper computer, standard cascade control via a decentralized digital control system, multiple input components, and an interface for these elements. The optimization search approach evolved from generic nonlinear programming to integer programming, integrating a local search with a boundary search through a straightforward pattern. Subsequently, the optimization component has been improved by employing double patterns with more assessment cells [4] and by incorporating techniques from various optimization literature [5–7], excluding border pattern search. The concept of cellular automata and the technique of Q learning were adopted and further developed at three universities and a technological institute. Zhang, Maeda, and Kawachi [10] proposed an optimization model for the allocation of river-extracted water to paddy field blocks within an irrigation system. Fuzzy linear programming is utilized to address fuzzy decision-type optimization in the model formulation, accommodating uncertainties arising from the unpredictability of hydrologic and hydraulic parameters, as well as the vagueness in management objectives. The author [11] introduces other diverse patterns and additional concepts for optimization in the literature. Recent research by Fujita, Tani, and Kawamura et al. [12] on fuzzy optimal control theory, grounded in the fuzzy maximization decision procedure for constructed structures, is presented. Section 3 proposes a single input single output (SISO) feed-forward and feedback control system utilizing a Table Base Controller with Zero (TBCZ), a variant of Table Base System (TBS), employing crisp numbers, which is anticipated to yield certain advantages. A feed-forward controller is defined as the inverse transfer function of the controlled system. A control table for three PID inputs and three types of membership functions is suggested to utilize crisp sets instead of fuzzy sets, anticipating certain advantages. Several simulation outcomes and assessments are also shown there. Section 4 presents a fuzzy decision-making method utilizing crisp numbers, incorporating both 1D-2D fuzzy evaluation and restriction membership functions, as well as a 3D representation of these functions with an overview plan. Additionally, a search method on the overview plan is proposed as a component of fuzzy optimization.

2. EXAMPLES OF CRISP AND FUZZY OPTIMIZATION CONTROL SYSTEMS

2.1. A crisp example in real plants

The optimizing control is divided into three parts, the optimization part in the upper system, the digital control part in the lower system, and the interface part where they are connected to former two parts. The method of the search for the combination pattern of the local search part and the boundary search part [4] was used as an upper system.

Figure 1 shows a system block diagram of the optimum combustion control system (MHI Operation Support System), which is a real example of crisp optimization controls [3]. The output chart of optimization part is shown in Figure 2 [3] and control logic part is shown in Figure 3 [11]. The patent [11] has more information on the optimization control.

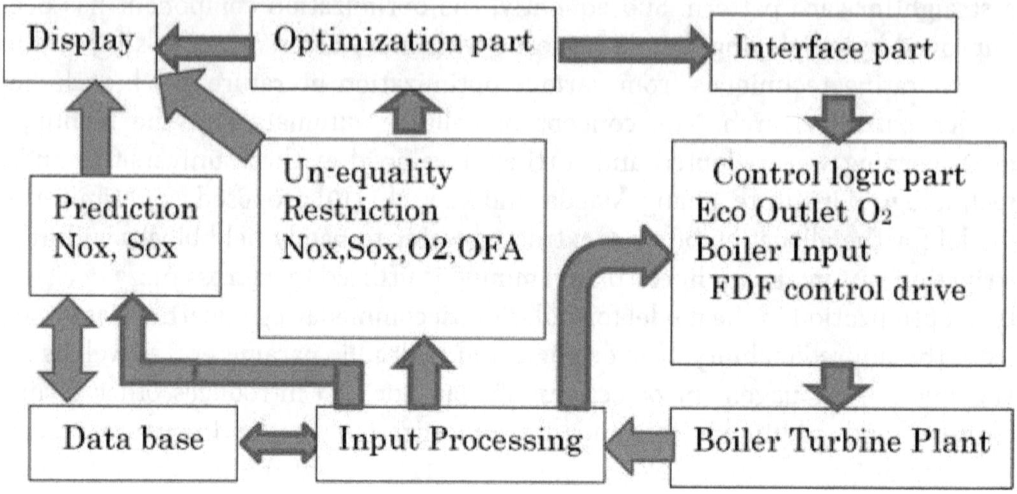

Figure 1. System block diagram of an optimization control.

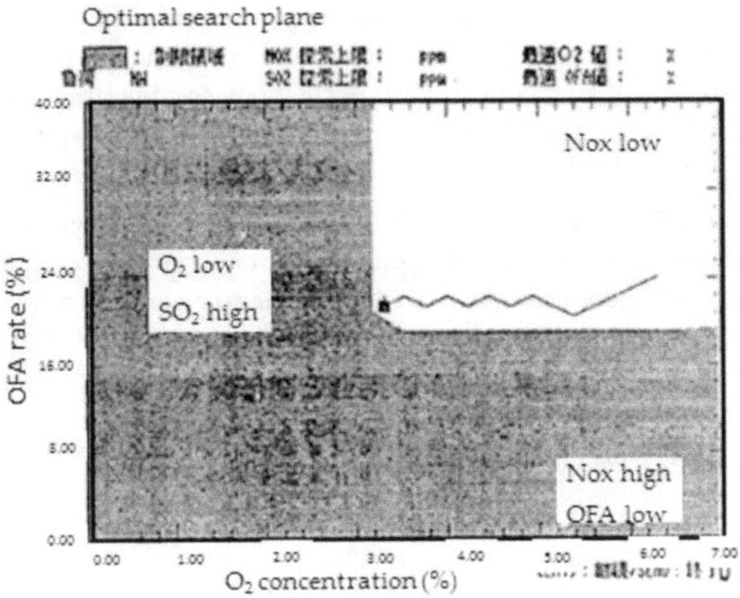

Figure 2. An example of optimization search for maximizing a boiler efficiency in OFA rate and O_2 concentration plane.

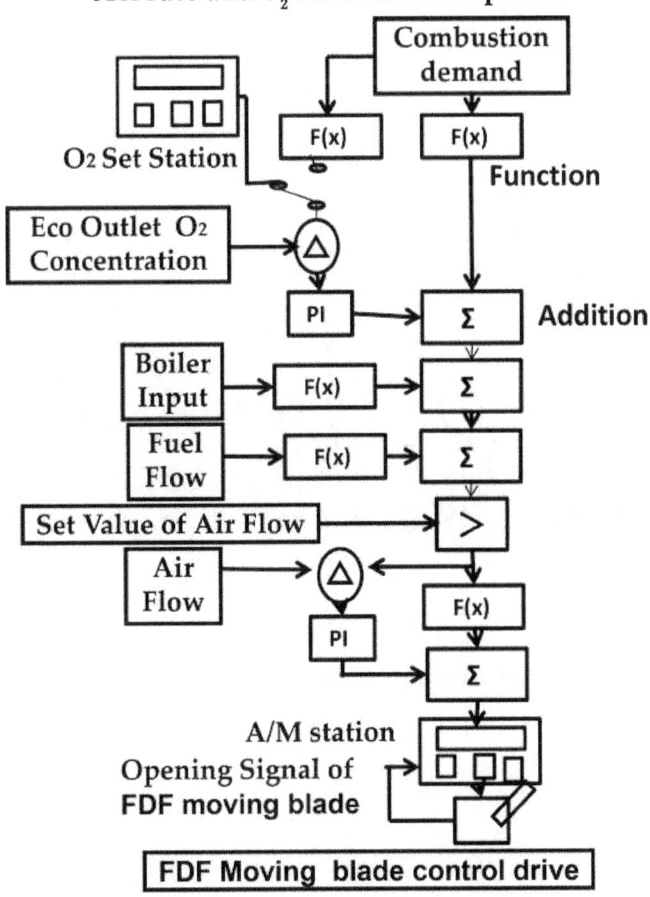

Figure 3. Eco outlet O_2 feedback control logic with optimization signal.

This example is very important, for horizontally developing the optimizing control with the three-layer structures in the future, and as an example of applying a real machine of the optimizing control in a thermal power generation process that is a multivariable system with large changes though it has not arrived at "fuzzy" optimizing control yet.

However, the initially developed artificial intelligent language and the detail search algorithm of the local search agent and the boundary search agent could not open for secret of know-how. Then, the author has developed a new evolutionary cellular automata algorithm using different patterns to low cost squares in the upper optimization control system though they are cost up. They are introduced in the next paragraph.

2.2. A crisp evolutionary algorithm for hybrid optimization control

In this paragraph, a crisp evolutionary algorithm for hybrid optimization control by multiagents is shown, because reliability and adaptability of the above hybrid optimization must be needed to increase and open.

Figures 4 and 5 show the cellular (remaining to use fuzzy number) structure of local search agents to search for the peak in the boundary line and boundary search agents to search for the peak out of the boundary line. Moreover, increasing reliability and adaptability of these search methods are shown [8].

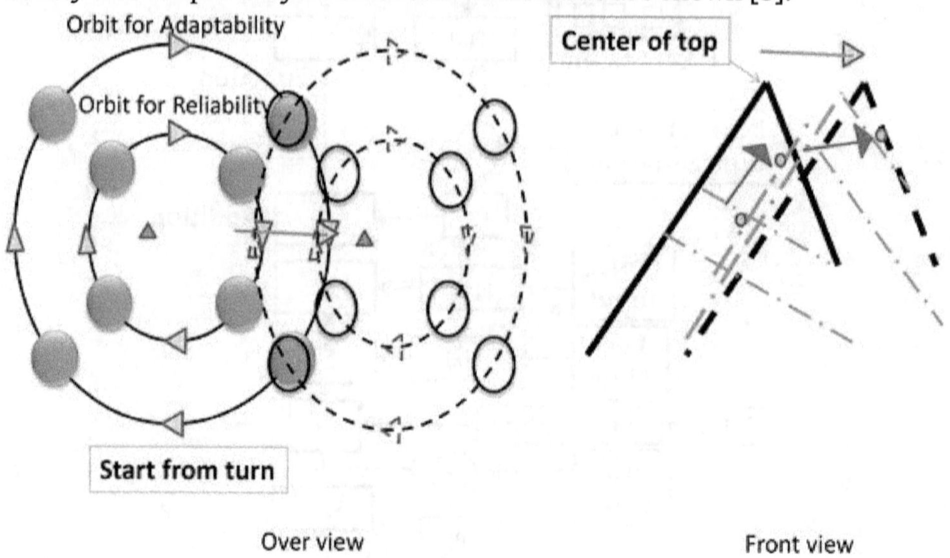

Figure 4. Cellular structure of local search agents for reliability and adaptability.

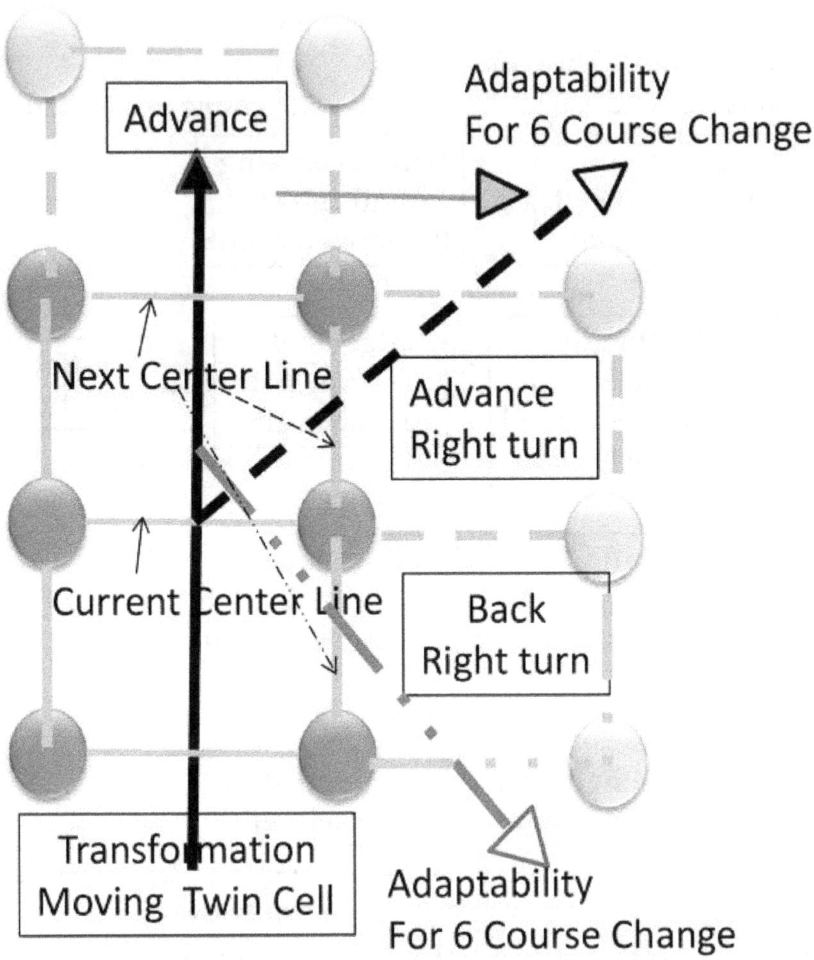

Figure 5. Cellular structure of boundary search agents for reliability and adaptability.

The following features are obtained by using eight cells group in an agent for evaluation of double circles as rotated machines as shown in Figure 4.

1. Adaptability to movement of mountains
2. Evaluation to continuous values
3. Application to multipeak

The following features are obtained by using six cells group in an agent for evaluation of double squares as moving cars with six wheels as shown in Figure 5.

1. Decrease of derailment probability
2. Evaluation to discrete values
3. Application to bifurcated boundary by using agents with different priority evaluation order

Oppositely, there are the following features in the quadrangle (square) agents adopted with an initial real system.

1. It is easy to develop because algorithm is easy.
2. There is a room to enlarge fuzzy circle.

Figure 6 shows local and boundary search algorithm using cellular automata.

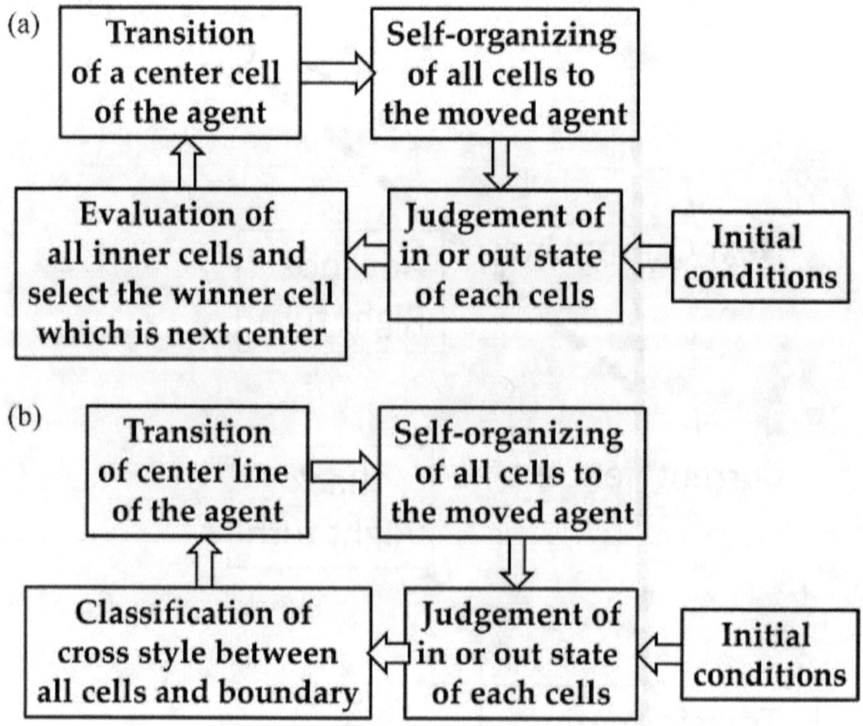

Figure 6. Cellular automata algorithm for local and boundary search algorithm. (a) Local search. (b) Boundary search.

The cellular automata algorithm has the following features.

1. Multistart points can set to any positions.
2. Four steps of center transition, self-organization of all cells to the moved agent, in or out state judgment, are repeated.

In the case of four-cell agents like four wheel cars in Figure 2, center line must be replaced to front line in automata (b) of Figure 6.

2.3. A fuzzy example of built structure

On the other hand, Fujita, Tani, and Kawamura et al. at Kobe University [12] execute fuzzy optimum control theory based on fuzzy maximization decision method (target conditions and restriction condition are expressed by fuzzy membership functions) to built structure, and it is called intelligent active control. Not only target response displacement and target control power but also structure identification values obtained from the responses of the structure and earthquake vibration forecast obtained from earthquake input measurement were used for the structural response forecast.

They decided the parameters using the fuzzy maximization decision method. Then, the effect of controls were examined by experiment and simulation about two methods for addition of control power added the feedback only of the ground vibration input acceleration power and the relative acceleration power's from base of structure.

They reported that the feedback control with optimal coefficients can imorove the rate of improvement by 30% higher or more.

This report feels the difficulty for execution of fuzzy optimization control though it can encourage the execution.

In the following section, a table base controller with zero (TBCZ) is introduced for easiness of tuning of membership functions more than conventional fuzzy control.

3. A FUZZY-LIKE TABLE BASE PID CONTROLLER WITH ZERO

The purpose of this section is to design and evaluate a Table Base Controller with Zero (TBCZ) as one of Table Base System (TBS) [13].

3.1. Table base controller with zero

3.1.1. TBCZ configuration

In this paragraph, we propose the following SISO feed-forward and feedback control system with a TBCZ in Figure 7, which is expected to provide the following advantages:

1. Fast start up
2. Small overshoot
3. Easy maintenance using the tuning knowledge of the conventional PID control

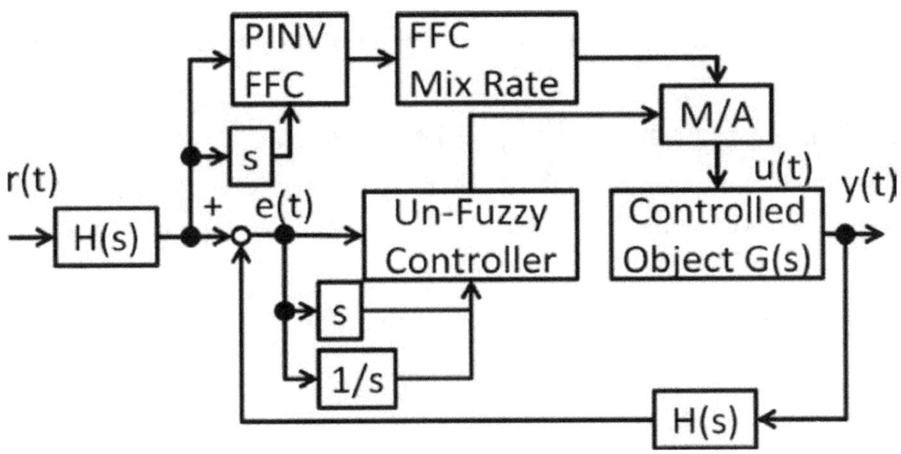

Figure 7. Configuration of a TBCZ.

Moreover, M/A means a manual/auto switch station, *s* means a differential operator, and *1/s* means an integral operator in Figure 7.

3.1.2. Table of TBC

Figure 8 shows a table of TBCZ and rectangular membership functions instead of usual triangle membership functions. Here, SUMi contains not only the proportional scaling factor SF_p and the differentiation scaling factor SF_d but also the integrator scaling factor SF_i. Finally, the scaling factor of the input SF_u must not be overlooked.

e \ Δe	N	ZERO	P
N	SUMi	SUMi	ZERO
ZERO	SUMi	ZERO	SUMi
P	ZERO	SUMi	SUMi

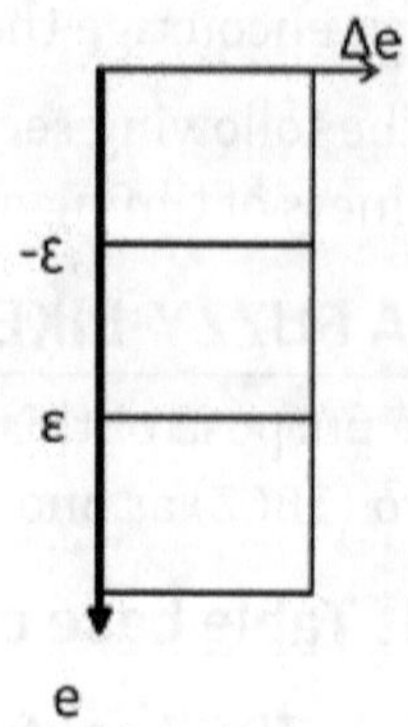

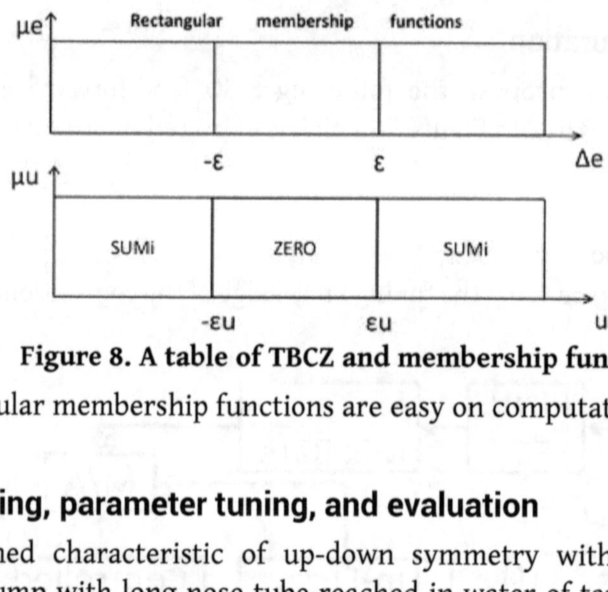

Figure 8. A table of TBCZ and membership functions.

The rectangular membership functions are easy on computation because they are crisp.

3.2. Modeling, parameter tuning, and evaluation

Here, assumed characteristic of up-down symmetry without hysteresis, and neglected pump with long nose tube reached in water of tank and PWM control dynamics, which is sufficiently fast.

Then, we modeled the control using the following simple linear transfer functions:

$$G(s) = \frac{Ke^{-Ls}}{Ts+1}; K = 30, T = 20, L = 2 \quad (1)$$

$$H(s) = \frac{K_s}{T_s s+1}; K_s = 1, T_s = 0.1 \quad (2)$$

3.2.1. Parameter tuning

3.2.1.1. Scaling factor in the TBCZ

The followings are examples of scaling factors in the TBCZ.

$$P = \frac{1}{SF_p}, \quad D = \frac{1}{SF_d}, \quad I = \frac{\frac{1}{s}}{SF_i}, \quad \varepsilon_0 = 0$$
$$\frac{1}{SF_p} = 0.15, \quad \frac{1}{SF_i} = 0.0091, \quad \frac{1}{SF_d} = 0.1 \tag{3}$$

3.2.2. Performances of 3 × 3 and 2 × 2 tables of TBCZ

An evaluation of the mean integral square error and input (MISEI) was compared for various values of the terminator ε in Figure 9.

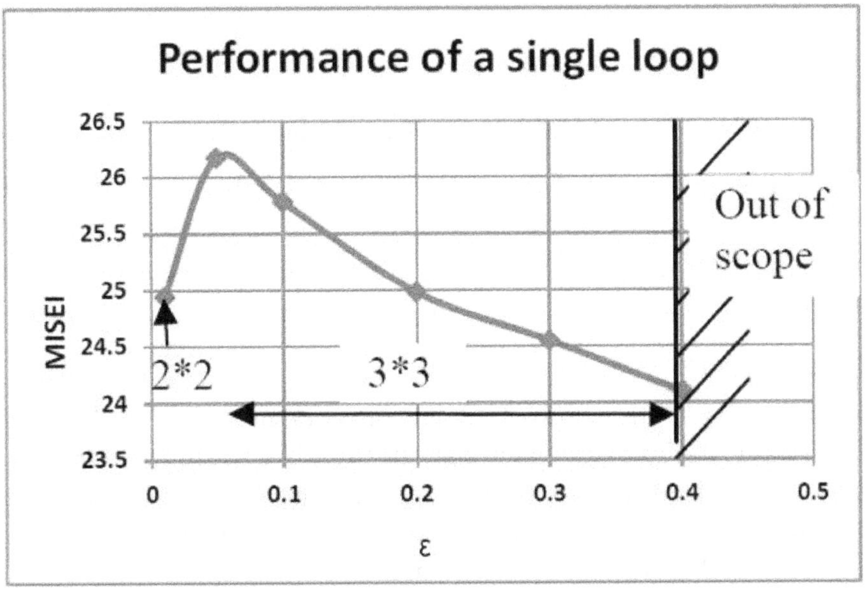

Figure 9. MISEI vs. ε (in a table of TBCZ).

The performance of ε = 0.01 is considered to be equivalent to that of the 2 × 2 table of TBCZ, it is one point except for line from 0.1 to 0.4 in Figure 9. Thus, the 3 × 3 table of TBCZ is superior than the 2 × 2 table of TBCZ since this table allows superior performance tuning. If performance is valued, then a larger table is better, although this results in high-cost and increased complexity. Performance of MISEI is superior in the case of smaller Center ZERO. There is a minimum point of MISEI on the edge of out of scope.

3.3. Other considerations to TBCZ

The robust modification of the experimental PID tuning method of Ziegler Nichols could be used here. Figure 10 shows performance MISEI of a double loop.

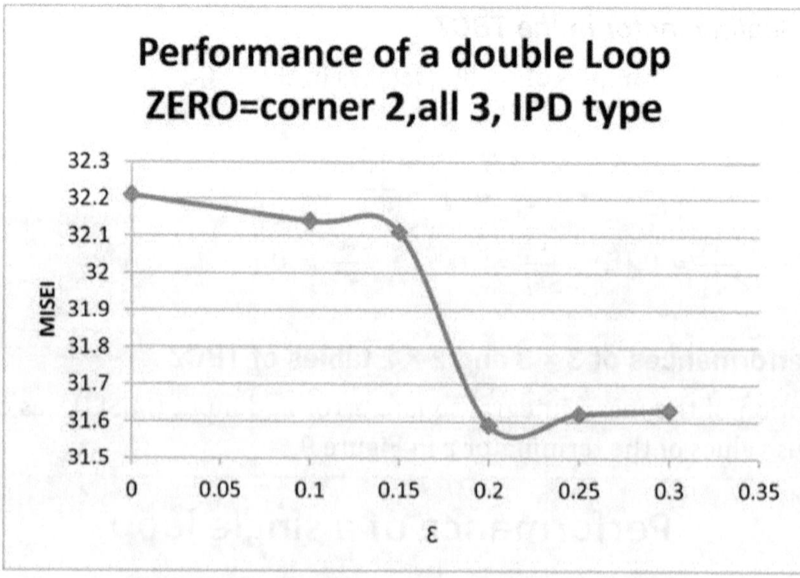

Figure 10. MISEI vs. ε (in a table of TBCZ).

The FFC mix rate in the double loop case must be less than half that in the single loop case.

The decoupling [14] study is omitted in this paragraph, then refer to in reference [13].

3.4. Subconclusion

A new concept of a TBCZ with rectangular membership functions based on crisp sets, which was featured by ZERO's in the rule table like the fuzzy-like control table as one of TBC and a feed-forward control line were proposed, and simulation results are presented for a tank level control as an example. Then, superior evaluation based on MISEI and performance was obtained.

The membership functions of the proposed TBCZ were able to easily tune only terminators, which mean the size of ZERO's through the evaluation of MISEI.

Making and tuning of the controller are easier than in conventional fuzzy control because the membership functions are rectangular without common parts in crisp sets, and the only control rules are ZERO and SUMi in a 3 × 3 control rule table.

The development of a robust compensator of integrator for no-overshoot property is a future theme. The readers can find the literature [15] on conventional system.

4. FUZZY DECISION-TYPE OPTIMIZATION

It proposes a fuzzy decision-making type optimization technique in this paragraph as an example of the problem on elevators.

Wada and Kato propose an optimization technique of fuzzy rules according to the situation by using behavior acquisition based on emotional memory that uses fuzzy sets on pleasantness and unpleasantness of a robot. And the robot is made to acknowledge pleasantness and unpleasantness by using a source of light and an experiment toward a comfortable goal evading the obstacle is done [16].

The fuzzy decision-type optimization for an elevator is proposed firstly in ordinary 2D membership description [17]. In this paragraph, it is enhanced to 3D description.

The other fuzzy optimization studies [18–22] are interesting for this study.

4.1. Maximizing decision probability methods

In this paragraph, firstly, the problem of fuzzy decision-type optimization (maximization) with subjects is defined generally.

Assuming that x_1 and x_2 are fuzzy numbers, membership functions μ_1 and μ_2 are introduced according to x_1 and x_2 and λ is a scalar for λ-cut.

$$\begin{array}{lcl} \underset{x \in X}{maximize} & & \min\{\mu_1(x_1),\ \mu_2(-x_1)\} \Leftrightarrow \\ maximize & & \lambda \\ subject & to & \alpha_1 x_1 + \beta_1 \geq \lambda,\ -\alpha_2 x_2 + \beta_2 \geq \lambda,\ x \in X \end{array} \tag{4}$$

This technique can be used for fuzzy deciding. For example, it is decided using this technique whether an "almost crowded" elevator should pass over a certain floor with "long queuing length". These two fuzzy sets "almost crowded" and "long queuing length" are described by using fuzzy numbers of passengers in the elevators and queues in the floor.

x_1 is defined to the number of passengers ride on an elevator, and x_2 is defined to the number of queuing in an elevator lobby of a floor. An example of detail equations and measurement method of person numbers can be referred to the literature [23].

The fuzzy decision method is called to maximizing (min-max) decision because it is optimize when the product (minimum) of two membership functions is max as in Figure 11.

where

$$\begin{array}{l} \lambda_{\max} = \underset{\{z_1^0,\, z_2^0\}}{\max}\ \underset{\{z_1^0,\, z_2^0\}}{\min}\ \{\mu(x_1), \mu(-x_2)\} \\ \lambda_c = \dfrac{1}{3}\lambda_{\max}, \quad \lambda_0 = 0 \end{array} \tag{5}$$

The notation of the trapezoid membership function is described by the three terms set [*left_terminator*, *center*, *right_terminator*] as same as the triangle membership function (center means the position of grade 1, and terminator means the position of grade 0) and is inserted the end point of right and left terminator by infinity mark ∞, -∞ as the following *G* and *C*. Then, the product (minimum calculation) *D* of *G* and *C* is described, if you devise it as multiplying the scalar corresponding the max value $\mu_D(x^*)$; x^* means center of *D*, then you can write as follows.

$$G = [g_1, g, \infty], \qquad C = [-\infty, c, c_2] \tag{6}$$

$$D = \mu_D(x^*)\,[g_1, x^*, c_2] \tag{7}$$

where x^*, $\mu_D(x^*)$ can be obtained easily because they are coordinate values of a cross point of two lines in the case of triangle or trapezoidal membership functions.

In multiobjective decision-making, generally it is common to narrow down to the only optimum solution by using preference function from among plural noninferior Pareto solutions, and depending on the shape of this preference function, decision maker's preference is divided into whether it is risk avoidance type, risk-oriented type, or risk neutral type.

When the preference function is convex, the case is the risk avoidance type. When it is concave, the case is the risk-oriented type.

When the case has multipurpose G_i (i = 1,...,n) and numerous restrictions C_j (j = 1,...,m), their common set *D* is defined equally as follows.

$$D = G_1 \cap G_2 \cap \cdots \cap G_n \cap C_1 \cap C_2 \cap \cdots \cap C_m \tag{8}$$

When getting a minimum of the respective membership function μ_{Gi}, μ_{Cj}, the membership function of *D* is found as μ_D.

$$\mu_D = \mu_{G_1} \wedge \mu_{G_2} \wedge \cdots \wedge \mu_{G_n} \wedge \mu_{C_1} \wedge \mu_{C_2} \wedge \cdots \wedge \mu_{C_n} \tag{9}$$

When adopting notation like Eq. (7), it can be written as follows.

$$D = \mu_D(x^*)\left[\{minx, s.t., \mu_D(x) = 0\}, x^*, \{max\, x, s.t., \mu_D(x) = 0\}\right] \tag{10}$$

Here, about (x^*, $\mu_D(x^*)$), because they are the coordinates of an intersection point of two straight lines, they can be found easily when finding the set of which *D* is composed. About both end points, the finding method is like the same.

4.2. Expendability of the proposed method

4.2.1. 3D description

Figure 11 with only a front view becomes Figure 12 with a front view, a side view and an over view when elevator passenger number x_1 and queuing line number x_2 is selected as independent logically double axis if three view description is adopted in this fuzzy optimization. This change may be increase possibility of fuzzy optimization approach.

The following Figure 12 comes next Figure 13 not making the independent double axis of regular axis and reverse axis like Figure 11 but making psychological one axis by describing variable X_3 (the underside is unpleasant and the upper side is pleasant) in addition to physical two axis by describing variables X_1 and X_2.

Here, "Almost Crowded" and "Long Queuing Length" are unequality conditions and "Passenger pleasure" is an objective function. It is the fuzzy decision-making method that "pass of an elevator to a floor" is done by maximizing decision method using min-max of three member ship functions (a blue point by minimax method). Whereas it is the crisp decision-making that the decision-making is done at the grey point of which pleasant degree is higher slightly in grey area which grade of all membership functions are one. It's a problem that the comfort level of the waiting line isn't considered against the comfort level of the passenger in this fuzzy decision-making area. Though the comfort level of the passenger is sacrificed, it can be said that the comfort level of the waiting line is considered in the crisp decision-making area.

The three side of plans (over view, front view and side view) are described like Figure 13 instead of the above plan like Figure 12 when X_2 and X_3 out of the three axes are changed like Figure 13. Usually, only front view plan is indicated as Figure 11 and others are omitted for economy and maximum decision policy. Here, overview plan is also indicated because minimum decision policy in overview plan is better than the maximum decision policy as understand if you see.

These plans permit two kinds of optimization methods, minimizing in overview plan and maximizing in front plan.

4.2.2. Search on overview plan

For searching minimum point on overview plan less than grade 1 such as Figure 14, mobile method of freely writing lines and circles like dance on the plane in order to evaluate grade values using the following equations with the rotation matrix is proposed here. This is a discrete mobile model which generates left and right double velocities v_L and v_R of interval *d*.

$$v_G(k) = v_L(k) + v_R(k); r = 0.5d \tag{11}$$

$$\omega_G(k) = \frac{1}{r}(|v_R(k)| - |v_L(k)|) \tag{12}$$

$$\Delta\theta_G(k) = h\omega_G(k); \theta_G(0) = \theta_{G0} \tag{13}$$

$$\theta_G(k+1) = \theta_G(k) + \Delta\theta_G(k) \tag{14}$$

$$v_G(k+1) = \begin{bmatrix} \cos\Delta\theta_G(k) & -\sin\Delta\theta_G(k) \\ \sin\Delta\theta_G(k) & \cos\Delta\theta_G(k) \end{bmatrix} v_G(k) \tag{15}$$

$$p_G(k+1) = p_G(k) + hv_G(k+1); p_G(0) = p_{G0} \tag{16}$$

where suffix *G* means center of the mobile object, suffix *L* means left, suffix *R* means right, *p* means position, ω means angle velocity, θ means angle from *x*-axis, and *h* means sample time.

Figure 14 shows an example pattern by a mobile agent and the reduced inner copy for design of searched course in the plane.

Addition of inverse kinematics will make easier the process of drawing figures filled in the canvas.

4.3. Action mode, effective, influence, yield analysis (AMEIYA)

A table on action mode effect, influence, yield analysis (AMEIYA) may be obtained by imitating reasoning methods from left-column action mode to right-column effect, influence and yield things.

A Q&A table used to obtain the above table on AMEIYA may also be made by readers referred to Ref. [23].

In the future, new strategic systems for fuzzy optimization control on elevators will expect to born from these tables.

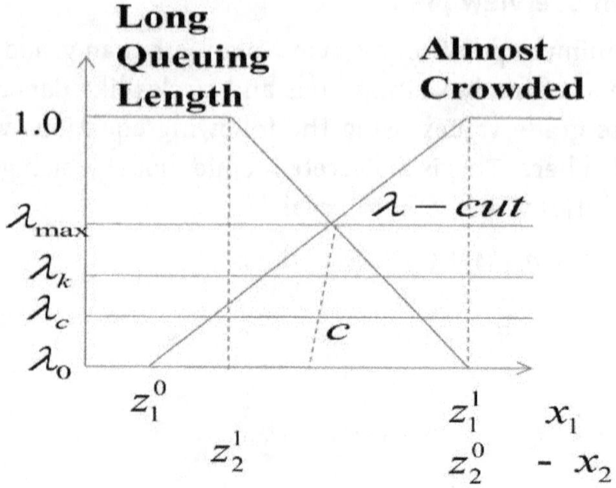

Figure 11. Membership functions of fuzzy sets "Almost crowded" and "Long Queuing Length" with a variable λ-level on fuzzy passage decision of "Maybe Pass a Waiting Floor".

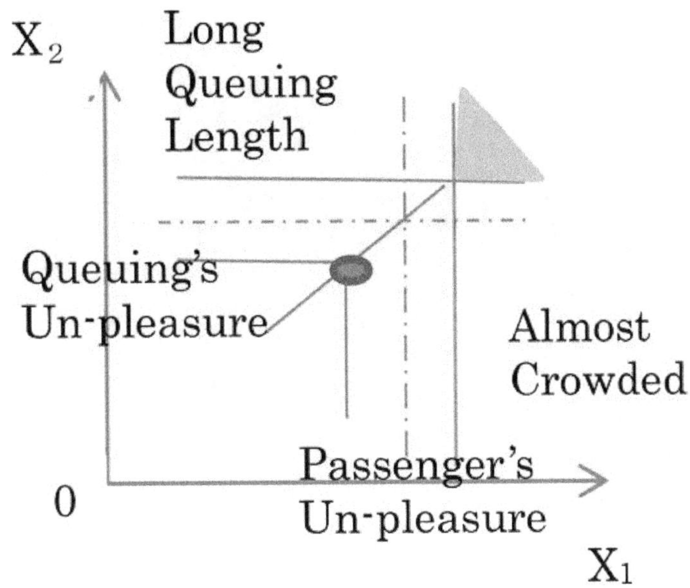

Figure 12. Over view plan on passengers? un-pleasure caused by almost crowded and queuing's unpleasure caused by long queuing length for a fuzzy optimization problem.

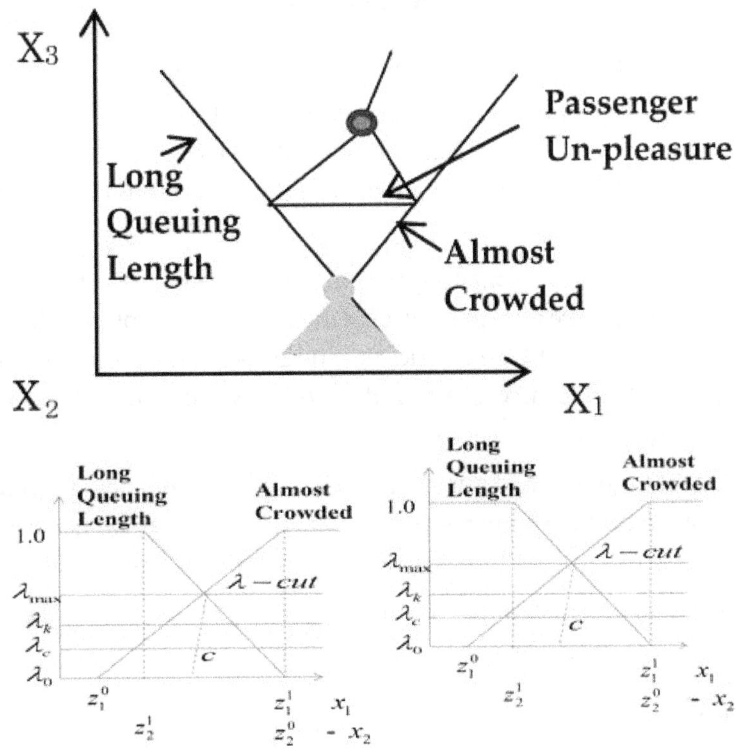

Figure 13. Three side plan in physics and psychological crossing at right angle axis of two dimensions on a fuzzy optimization problem.

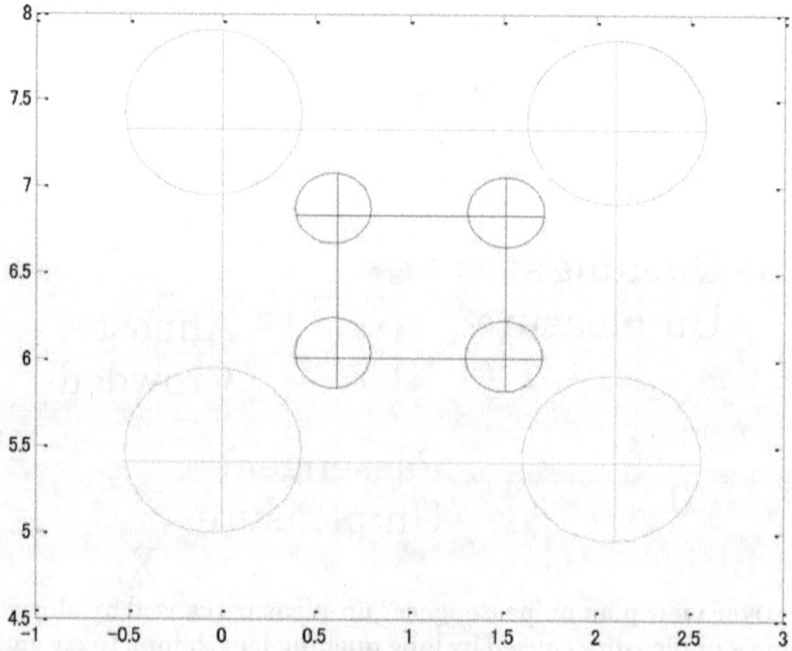

Figure 14. An example pattern of searched course for reduced repeat by a discrete mobile agent.

5. CONCLUSIONS

In this chapter, a fuzzy optimization control system by combining fuzzy decision-type optimization parts and a fuzzy-like table base PID controller parts was proposed separately instead of the conventional united crisp optimization control system.

Double future themes on fuzzy optimization, which may be used in fuzzy optimization control, were also presented. One is three side plans with an overview plan, which omitted conventional studies. The other is a new search method on a plane using a new simple discrete mobile model, which generates left and right double vector velocities.

These two separate parts are expected to be united and to be evaluated. Moreover, optimization based on fuzzy numbers and calculations may be used for absorbing uncertainty of sensors output and digitalizing of input to computer.

Moreover, some idea of strategic systems on fuzzy optimization control was provided for readers in the future.

AT A GLANCE

This part presents intriguing findings from the author's previous crisp optimization combustion control research related to actual fossil power plant boilers, focusing on the new fuzzy optimization control system design for upper and lower separation. The fuzzy decision-type optimization pertains to elevators, whereas the fuzzy table-like control with zero is designated for single-element level control of a single tank model. Furthermore, new studies by other academics regarding alternative uses are presented to ensure equity and balance.

REFERENCES

1. N. Furukawa, "Mathematical Theory of Fuzzy Optimization", Morikita Publisher, Tokyo, 1999 (in Japanese).
2. Manual of "A Coupled Tank", PID Co. Ltd, Tokyo, 2011.
3. M. Mizuno, K. Katsuki, M. Katoh, K. Mizuochi, K. Takahashi and K. Ryotaro, "Operation Assistance System for Fossil Thermal Power Plant for Sunagawa Power Station No.3 and 4 Units of The Hokkaido Electric Power Co. Inc ", MHI Technical Reports, 27, 33–36.
4. M. Katoh and K. Hasegawa, A Cooperative Heuristic Search Approach By Multi Agents, Proceedings of SMC 2000, IEEE, New York, pp. 277–279, 2000.
5. S. S. Rao. "Engineering Optimization Theory and Practice Third Edition", Wiley Interscience Publication, John Wiley & Sons, Inc, New York, Chichester, Brisbane, Toronto, Singapore, 1995.
6. Edgar, Himmelblau, Lasdon, "Optimization of Chemical Processes", Mcgrowhill, Boston, Burr Ridge, Dubuque, Madison, New York, San Francisco, St. Louice, Bangkok, Bogota, Caracas, Kuala Lumpur, Lisbon, London, Madrid, Mexico City, Milan, Montreal, New Delhi, Santiago, Seoul, Singapore, Sydney, Taipei, Toronto, 2001.
7. N. S. Rau. "Optimization Principles", Wiley Interscience, 2003.
8. M. Katoh, K. Wada and T. Nishikawa. "A Cooperative Heuristic Search Approach By Multi Agent Cellular Automata", Proceedings of JSME No.1-63, Cellular Automata Symposium, pp. 171–173, 2001.
9. M. Katoh, R. Simotani and K. Tokushige "Integrated Multiagent Course Search to Goal by ε-Greedy Learning Strategy—Dual Probability Approximation Searching-", Proceedings of SMC 2015, IEEE, New York, pp. 382–392, 2015.
10. Qin Zhang, Shigeya Maeda and Toshihiko Kawachi, "Fuzzy Optimization Model for Allocating Irrigation Water to Paddy Fields", Transactions of JSIDRE, 2007, No.249, pp.55–62, 2007.
11. M. Katoh, J P 3 7 0 8 5 8 1, B 2 2 0 0 5 . 1 0 . 1 9, (in Japanese)
12. H. Fujita, A. Tani, H. Kawamura and A. Takizawa, "Study on Activation Methods of Control Forces", Experimental Research on Intelligent Fuzzy Optimal Active Control Forces, 279/280, 2001 (in Japanese).

13. M. Katoh and T. Ueno, "Design of an Un-fuzzy Controller for a Tank Level Control", Proceedings of 2013 International Conference on Fuzzy Theory and Its Application, pp. 340–345, 2013.

14. M. Matsuzawa and M. Saeki, "Entrainment in interpersonal motor coordination of oscillating lower legs", Annual Bulletin of Showa Woman's University, Vol. 11, 9–15, 2008.

15. M. Katoh, No Overshoot Step Response of an Integral Controller with a Simple Robust Lead Inverse Compensator to Improve Safety of Vehicles, IJACE, Vol. 5, 1/11, 2016.

16. K. Wada and M. Katoh, "Behavior Acquisition of Mobile Robots with Emotional Memory System", Proceedings of SICE System Integration Conference, 69/70, 2002.

17. M. Katoh and N. Imura, "A Plan for Fuzzy Passage Decision of Elevators Using Equivalent Person Number and AMEA", IJACE, Vol. 5, Issue (1), pp. 47–52, 2012.

18. H. Suito and H. Kawarada, "An algorithm of Fuzzy Optimization", Department Bulletin Paper of Kyoto University, Kyoto. Research Information Repository. 1129, 95/107, 2000.

19. M. Sakawa and K. Katoh, "Large Scale of Multi-Objective Fuzzy Programming, Operation Research", Science of Business, 47(5), 2002, pp. 295–301 (in Japanese).

20. R.E. Bellman and Zadeh, "Decision making in a fuzzy environment", Management Science, Vol. 17, 1970, pp. 141–164.

21. Y. Gao, G. Zhang, J. Ma and J. Lu, "A λ-Cut and Goal-Programming-Based Algorithm for Fuzzy-Linear Multiple-Objective Bi-level Optimization", IEEE Transactions On Fuzzy Systems, Vol. 18, No. 1, 2010, pp. 1–13.

22. P. T. Chang and K. C. Hung, "α- Cut Fuzzy Arithmetic: Simplifying Rules and a Fuzzy Function Optimization with a Decision Variable", IEEE Transactions On Fuzzy Systems, Vol. 14, No. 4, 2006, pp. 496–510.

23. M. Katoh, T. Ozeki, and J. Sawaki, "A Strategic Question Inquiry System for Fuzzy Decision of Elevator Street", 2012 9th International Conference on Fuzzy Systems and Knowledge Discovery, 2012.

10 Chapter

AN APPROACH OF FUZZY LOGIC H_∞ FILTER IN MOBILE ROBOT NAVIGATION CONSIDERING NON-GAUSSIAN NOISE

1. INTRODUCTION: AN OVERVIEW

Employment in hazardous environments has consistently posed a threat to human safety and health. This scenario presents robotics as an alternative means to do a specific activity. To address this issue, an autonomous robot offers an effective solution requiring minimal human oversight. Since Karel Čapek invented the term 'robot' in the 1920s, robotics has undergone significant developments and evolutions. Robotics is often categorized into many types as defined by the Robotics Institute of America, including variable-sequence robots, playback robots, numerical control robots, and intelligent robots. Various robotic configurations are available, including manipulator robots, SCARA robots, spherical robots, Cartesian robots, and cylindrical robots. Recent advancements in robotic research, development, and technology have been substantial across numerous sectors and applications. The technology has proliferated and is extensively utilized in domestic products, including robotic lawn mowers, robotic vacuum cleaners, and robotic floor washers. These home robots are occasionally engineered to operate autonomously and independently using their own predefined algorithms or systems. Under these conditions, the robot must possess two primary features: navigation and autonomous capabilities. This chapter mostly addresses mobile robot navigation. Three methodologies are available for modeling mobile robot navigation [1, 2]. Initially, the mathematical modeling formulates the mathematical model of the robot and its environment by considering the robot's dimensions, settings, and environmental variables. Consequently, designers must possess a comprehensive understanding of mobile robot setups and environmental variables. Sensor data is a critical component in the information processing of mobile robot navigation. The information enhances mobile robots' trust in navigation processes. This represents the second category of navigation modeling that utilizes data obtained from sensors. A variety of sensors can be integrated into a mobile robot to facilitate navigation. This technique is less difficult than the mathematical model, as sensors yield relative measurements between the mobile robot and observable landmarks. The third category concentrates on how to endure the sounds produced during mobile robot navigation. The mobile robot employs probabilistic principles, specifically Bayes' theorem, to ascertain its position and the observed landmarks amid some uncertainty. This approach requires minimal mathematical processing and does not heavily depend on sensor information. Due to these advantages, the third type is currently widely utilized. This chapter concentrates on the probabilistic method for mobile robot navigation. Various methodologies have been suggested for the development of autonomous mobile robots, particularly employing artificial intelligence techniques, as exemplified in references [3–5]. Jaradat and Abdel-Hafez [4] have effectively utilized neural networks to analyze data acquired from sensors, IMU, and GPS for vehicle navigation. Force and angular velocity of the vehicle are measured using an IMU, while GPS is utilized for determining the vehicle's location. Two concealed layers, each comprising a minimum of six neurons, are constructed to compute and analyze vehicle movement. Their results appear to demonstrate a high accuracy in vehicle estimation. Drawing from human cognitive processes and existing literature, the mobile robot is designed to operate with minimal human oversight, utilizing artificial intelligence to monitor its surroundings. In this context, artificial intelligence is acknowledged as a potential method for delivering effective solutions. This research employs fuzzy logic control to enhance a probabilistic technique referred to as the $H\infty$ filter by examining many factors, including finite escape time, operation under non-Gaussian noise, intermittent loss of measurement data, and the impact of parameters on estimation.

2. MOBILE ROBOT NAVIGATION

The renowned Kalman filter is the most utilized method for mobile robot navigation. The method founded on minimizing mean square error employs earlier information acquired by the sensor to refine its present position. Although the Kalman filter provides dependable estimations, it remains ineffective in environments characterized by non-Gaussian noise. Noise is defined as such when it possesses a non-zero mean and exhibits characteristics of non-uniform distribution. Consequently, several novel methodologies are introduced, including the unscented Kalman filter, particle filter, and ensemble Kalman filter, which predominantly utilize numerous particles to deduce the position of the mobile robot [2]. These strategies demonstrate greater computing expense with a rise in the number of particles. They necessitate a rapid processor to compute and retain data during mobile robot observation. Therefore, a straightforward and resilient filter superior to the aforementioned strategies is desirable to address the navigation challenges. Some of the Kalman filter limitations can be listed as follows:

- The mean and correlation of the process and measurement noise need to be known at each time instant.
- The covariances of the noises must be determined beforehand, since the Kalman filter uses these covariances as design parameters.
- The Kalman filter is the minimum variance estimator if the noise is Gaussian. However, if a different cost function (such as the worst-case estimation error) should be applied to the system, then the Kalman filter may not be suitable to accomplish the objectives.

H_∞ filter is one of Kalman filter families. The filter is proven to work in non-Gaussian noise environment and it is assumed that the noises are bounded. By defining the tuning parameters known as γ, the filter can provide a better solution than Kalman filter as well as other techniques. In this chapter, the non-Gaussian noise is assumed to be bounded and acts as a random noise when mobile robot does its observations. It is also important to note that H_∞ filter is facing one issue, that is the finite escape time where the estimation can be unbounded if the system is not satisfying the designed criteria [7]. Despite these shortcomings, the H_∞ filter performance analysis is the main objective in this chapter to propose an alternative solution for mobile robot navigation. To ensure the finite escape time is not perceived during observation, fuzzy logic technique is applied for the proposed system. The system design consisting of these approaches is presented in the next section.

3. H_∞ FILTER AND FUZZY LOGIC TECHNIQUE MODELLING

One of the main parts in navigation is the simultaneous localization and mapping (SLAM) problem. It states a situation where the mobile robot builds up an environment based on its sensor readings while at the same time localizes itself on the map [8, 9]. Typically, two models that simulate the system are referred to describe how the SLAM problem is being solved. The first model is known as the kinematic model and the latter is defined as measurement model. Both models are essential to define the behaviour of the system with reference to the environmental conditions.

The kinematic model simply illustrates how the mobile robot movements are recorded. The following equation demonstrates the model.

$$X_{k+1} = F_k X_k + B_k u_k + w_k \tag{1}$$

where X is the augmented states of the mobile robot (x,y) and landmark i (x_i,y_i). F and B are the state transition matrix and the control input matrix, respectively. u and w are the control input and the associated noise during mobile robot movements.

Mobile robot measures relative distance and angle to any observed landmarks location when moving throughout the environment. The measurement model holds the following equation.

$$y_{k+1} = H_k X_k + v_k \tag{2}$$

Here, y contains the information of relative distance and angle measurements and H is showing the measurement matrix. v is the associated noise occurred during measurements.

H_∞ filter algorithm is almost similar to the Kalman filter. The only differences are notated on the state covariance and state update. For convenience, the comparison between H_∞ filter and Kalman filter is presented in Table 1.

Table 1. Kalman filter and H_∞ filter comparison [10].

	The extended Kalman filter	The H_∞ filter
The system	$X_{k+1}=f(X_k,u_k,w_k,k)$	$X_{k+1}=f(X_k,u_k,w_k,k)$
	$z_k=h(X_k,v_k,k)$	$z_k=h(X_k,v_k,k)$
		yk=DkXk$yk=DkXk$
The prediction step	$\hat{X}^-_{k+1} = f(\hat{X}_k, u_k, 0, k)$	
	$P^-_{k+1} = \nabla F_X P_k \nabla F_X^T + \nabla F_w Q_k \nabla F_w^T$	$\hat{X}_{k+1} = \nabla F_k \hat{X}_k + \nabla F_k K_k(z_k - \nabla H_k \hat{X}_k) + u_k$
The update step	$\hat{X}^+_{k+1} = \hat{X}^-_{k+1} + K_{k+1}\ \mu_{k+1}$	$P_{k+1} = \nabla F_k P_k (I - \gamma^{-2} P_k + \nabla H_k^T R_k^{-1} \nabla H_k P_k)^{-1} \tilde{N} F_k^T + Q_k$
	$P^+_{k+1}=(I_n - K_{k+1} \nabla H_i) P^-_{k+1}$	
	$\mu_{k+1} = z_{k+1} - h(\hat{X}^-_{k+1}, 0, k)$	$K_k = P_k (I - \gamma^{-2} P_k + \nabla H_k^T R_k^{-1} \nabla H_k P_k)^{-1} \nabla H_k^T R_k^{-1}$
	$S_{(\mu\mu)k+1} = \nabla H_i P^-_{k+1} \nabla H_i^T + R_{k+1} K_{k+1} = P^-_{k+1} \nabla H_i^T (S_{(\mu\mu)k+1})^{-1}$	

3.1. Fuzzy Logic-based navigation

In this chapter, the measurement innovation is referred as the main reference in designing the fuzzy logic. Different to Ref. [3] which utilizes the heading angle and relative distance range as their inputs to the system, fuzzy logic is designed in this research to process the angle and distance errors as its inputs. Our objective is to decrease those errors by configuring the fuzzy sets in producing smaller errors. By choosing the output appropriately, the effects or measurement error due to sensor inaccuracies can be minimized further.

There were also some researchers attempts to applied fuzzy Logic in navigation such as in Refs. [11–17]. Each of them demonstrated that fuzzy logic or artificial intelligence technique is capable to fuse the information obtained from the sensors for mobile robot estimation. Some of the researches are referred to evaluate the performance of this work.

Literature has stated that mobile robot has some confidence on its estimation especially when Kalman filter is applied for estimation [3]. For non-Gaussian noise characteristic, the sensors reading might be interfered and exhibits bigger error and hence results in bigger measurement noise covariance, *R*. If the gain *K* is small at all times during observations, then it is possible to have smaller measurement errors. Inspired by this fact, fuzzy logic is proposed to find the best value of measurement innovation to pursue lower error. Kobayashi et al. [16] have selected the *P*, *Q* and *R* from fuzzy logic to gain smaller uncertainties. Work by Wang et al. [18] has recognized this as one of the ways to realize smaller measurement noise even when it was being applied to the other H_∞ filter family, the Kalman filter. Mamdani technique is used for analysis purposes in determining the output of the system. The technique is proposed as it calculates the output by considering and utilizing the maximum information gained from measurement.

The general design is illustrated in Figures 1–3, that consists of the input and output and their respective fuzzy sets. The fuzzy sets are changed whenever the mobile robot moves in different motion and therefore, the values are not included. The following describes the rules of fuzzy logic that are used to define the output of the measurement innovation.

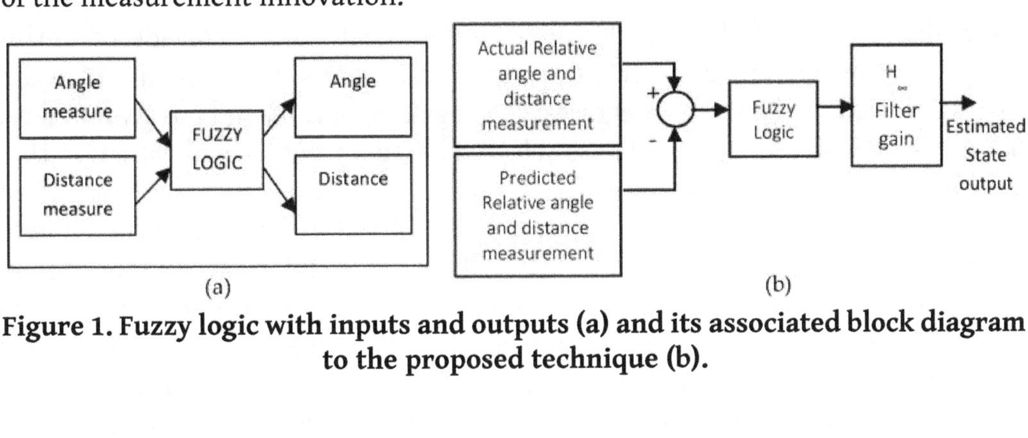

Figure 1. Fuzzy logic with inputs and outputs (a) and its associated block diagram to the proposed technique (b).

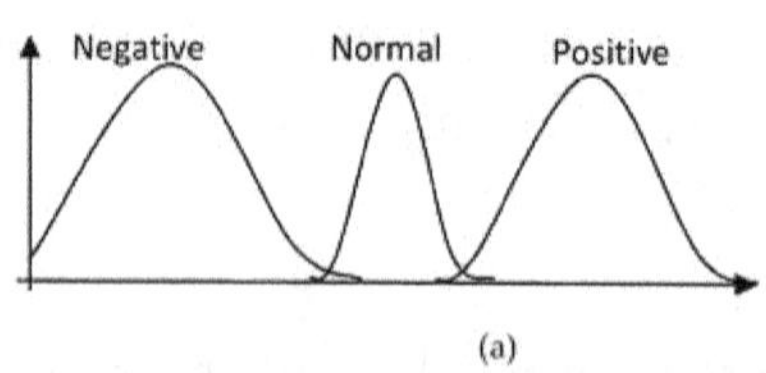

Figure 2. (a) Angle measurement and (b) distance measurement.

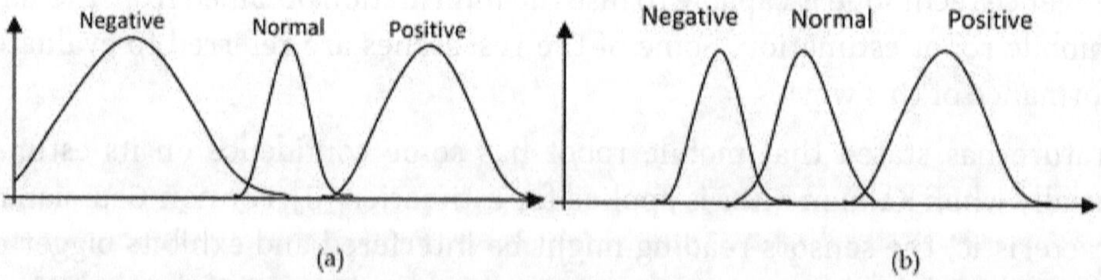

Figure 3. (a) Fuzzified angle and (b) fuzzified distance measurement.

- IF angle error is negative and distance error is negative, THEN angle is negative.
- IF angle error is negative and distance error is normal, THEN angle is normal.
- IF angle error is negative and distance error is positive, THEN angle is negative and distance is normal.
- IF angle error is positive and distance error is normal, THEN angle is negative.
- IF angle error is positive and distance error is negative, THEN distance is normal.
- IF angle error is positive and distance error is positive, THEN angle is negative and distance is normal.

Generally, Gaussian and triangular membership functions are considered in this chapter for evaluation purposes. Only three fuzzy sets are defined which are divided into three different categories: the negative, normal and positive regions. The scale of each of the fuzzy sets is selected based on the normal condition that has high errors. The value differs with each of the fuzzy sets, and it has been tuned several times to obtain the best estimation results. The tuning is taking into account the uncertainties behaviour throughout the simulation. Other than that, both angle and distance measurement characteristics are observed prior to the tuning process. Wang et al. [18] designed the membership function of the angle error to be positive at all times. However, random mobile robot's movements may also show a negative angle especially when a global coordinate system is being considered. This aspect is one of the major differences between our approach and what they have investigated.

4. RESULTS AND DISCUSSIONS

This section describes the performance of the fuzzy logic-based H_∞ filter by considering several factors such as the finite escape time problem, localization and mapping problem and the effects of noise parameters. These three issues are important to be solved concurrently during the navigation. If not, then the estimation results will not be as expected. Analysis is based on the design parameters included in Table 2.

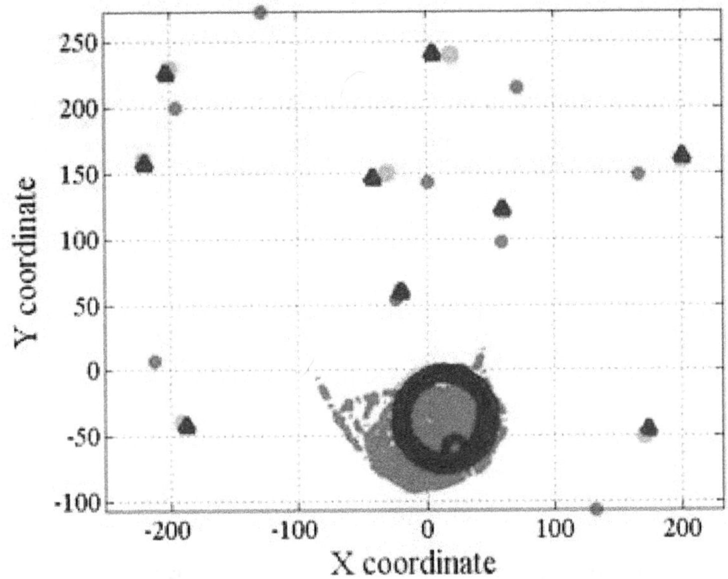

Figure 4. The mobile robot movements through the environment. Lighter colour of round shape defines the actual positions while the triangle shape and darker colour round shape presents the H_∞ Filter with Fuzzy Logic(FHF) and normal H_∞ Filter (HF) estimation performance, respectively.

Table 2. Simulation parameters.

Variables	Parameter values
Process noise:	
Q_{min}, Q_{max}	-0.002, 0.001
Measurement noise:	
$R\theta_{min}, R\theta_{max}$	-0.04, 0.01
$R_{dist\_min}, R_{dist\_max}$	-0.15, 0.3
Initial covariance:	
$P_{robot}, P_{landmark}$	0.001, 100
Simulation time	1000 (s)

Figures 4 and 5 demonstrate the estimation results when $\gamma = 0.7$, which determines that fuzzy logic-based H_∞ filter outperforms the normal H_∞ filter about the mobile robot and landmarks estimations. Erroneous estimations are perceived for the mobile robot location estimations as well as the landmarks positions. The measurement details are presented in Figure 2 for each of the estimation error. Figure 6 illustrates the state covariance update performance. It can also be seen from this figure that the proposed method attempts to avoid the finite escape time from happening.

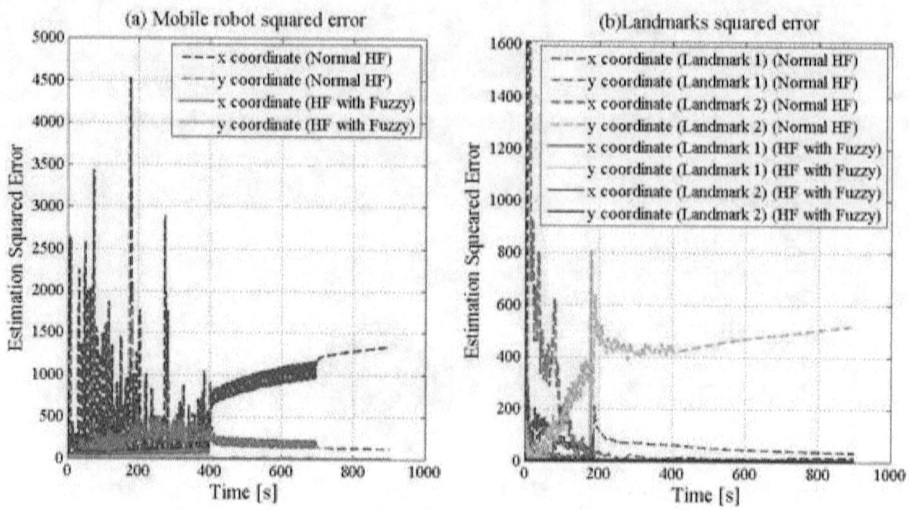

Figure 5. A performance comparison between FHF and normal HF estimations for both mobile robot (a) and landmarks (b) estimations about the errors.

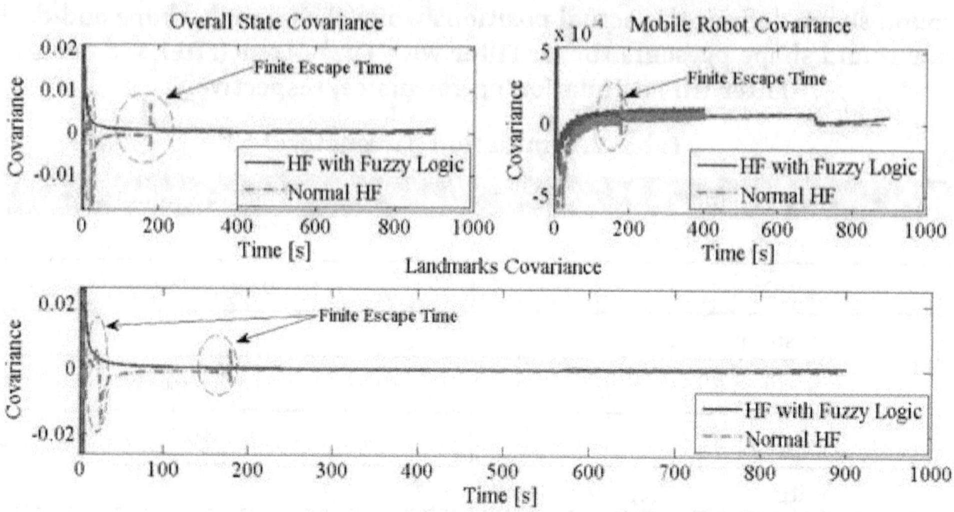

Figure 6. The state covariance conditions between H_∞ Filter (HF) and H_∞ Filter with Fuzzy Logic(FHF). Normal HF exhibits frequent Finite Escape Time(FET) compared to the H_∞ Filter with Fuzzy Logic(FHF).

4.1. Effect of changing γ

The γ effect to the proposed technique has also been analysed to ensure that our proposed technique has consistent and reliable results. Figures 7 and 8 describe the performance of the state estimations for both system of normal H_∞ filter and fuzzy logic-based H_∞ filter, respectively, when $\gamma = 0.23$, which is smaller than the previous case. It is plotted clearly that normal H_∞ filter exhibits erroneous results compared to our proposed technique.

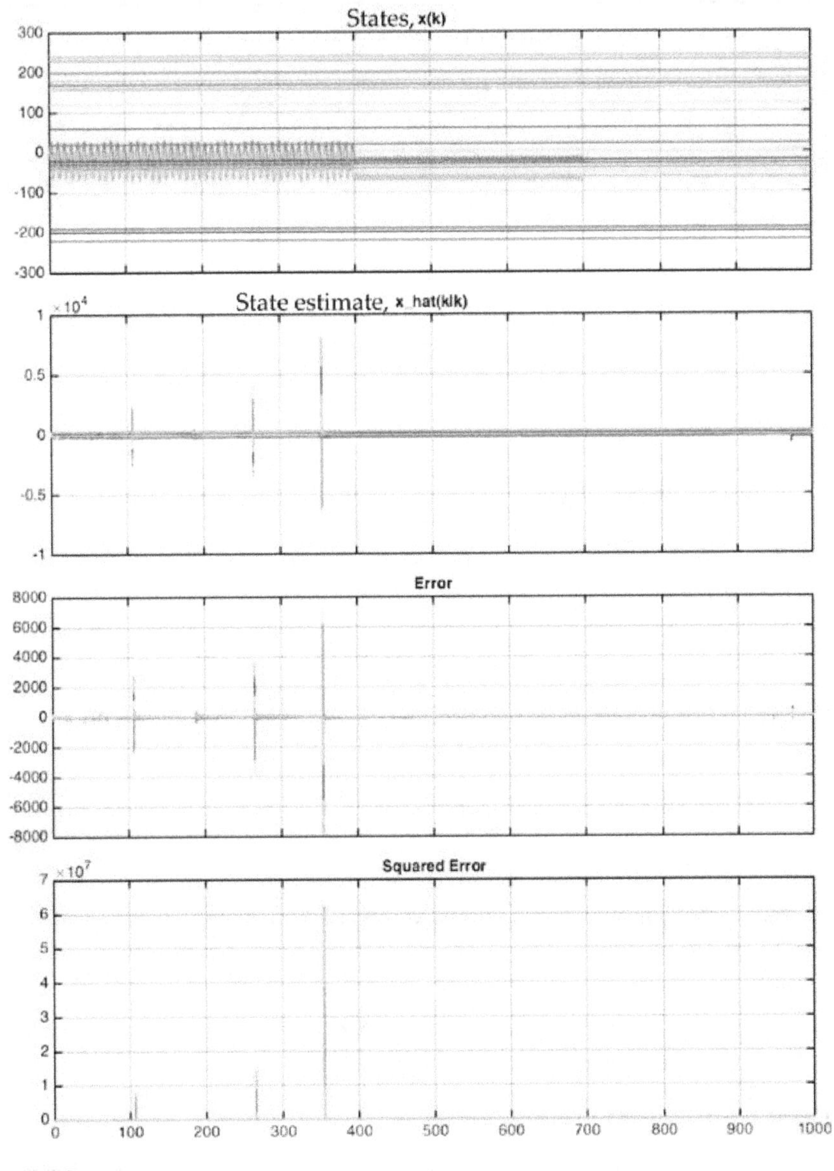

Figure 7. The state update conditions of normal HF generate a lot of estimation errors.

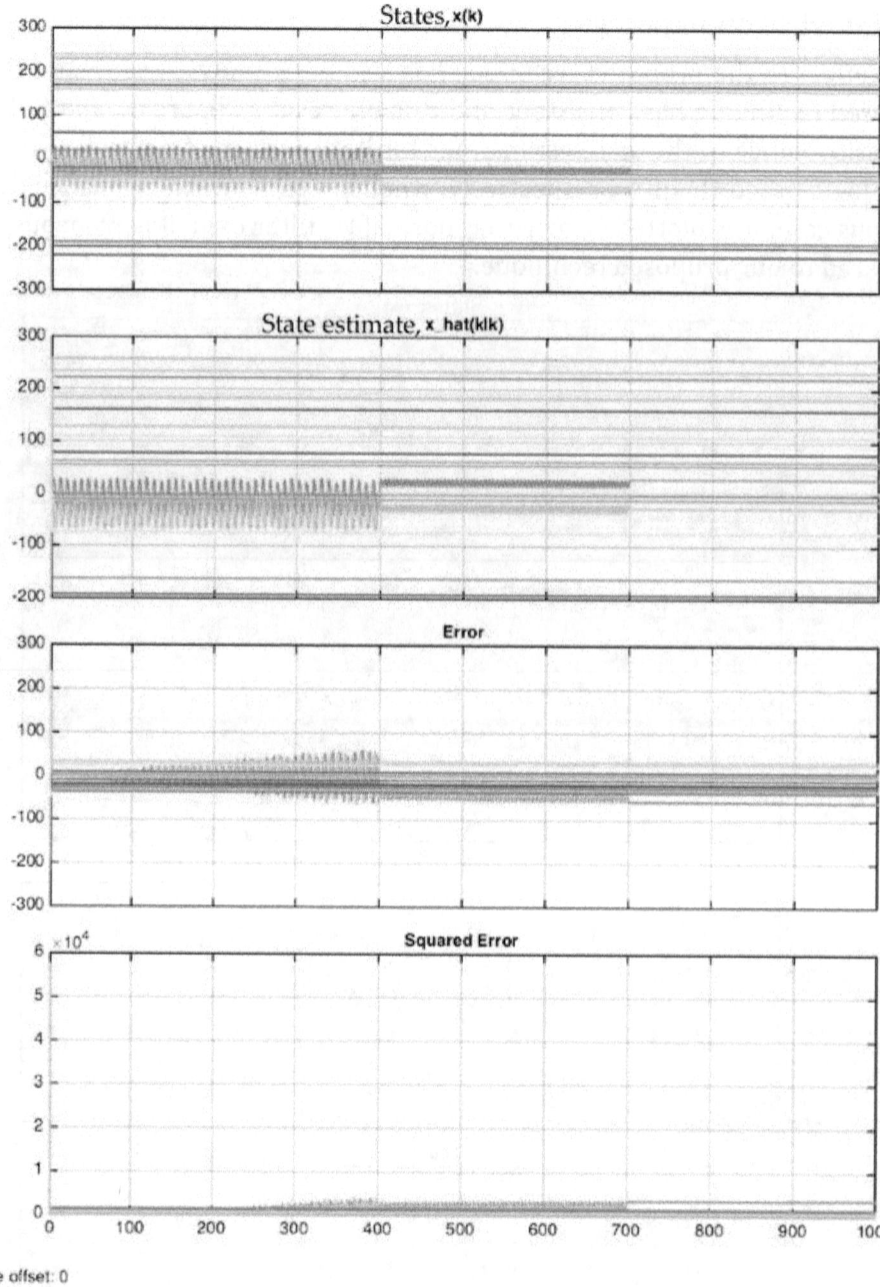

Figure 8. The state update conditions of FHF do not encounter high uncertainties or errors during mobile robot navigation.

4.2. Effect of initial state covariance

Generally, in SLAM, mobile robot does not have any prior information about its location or the environment. Therefore, the initial state covariance is designed to pose high uncertainties. Owing to these conditions, the mobile robot can probably have lost its way and is unable to navigate effectively on the environment. The analysis covers this issue by simulating the results using both normal H_∞ filter and fuzzy logic-based H_∞ filter in the next few figures.

Figure 9 demonstrates the results of estimation when mobile robot is in the above-mentioned condition. As expected, the normal H_∞ filter did not deliver good estimation results. On the other hand, the fuzzy logic-based H_∞ filter guarantees a good estimation that can still be preserved with a considerable estimation. In this analysis, the Gaussian membership is used for decision making.

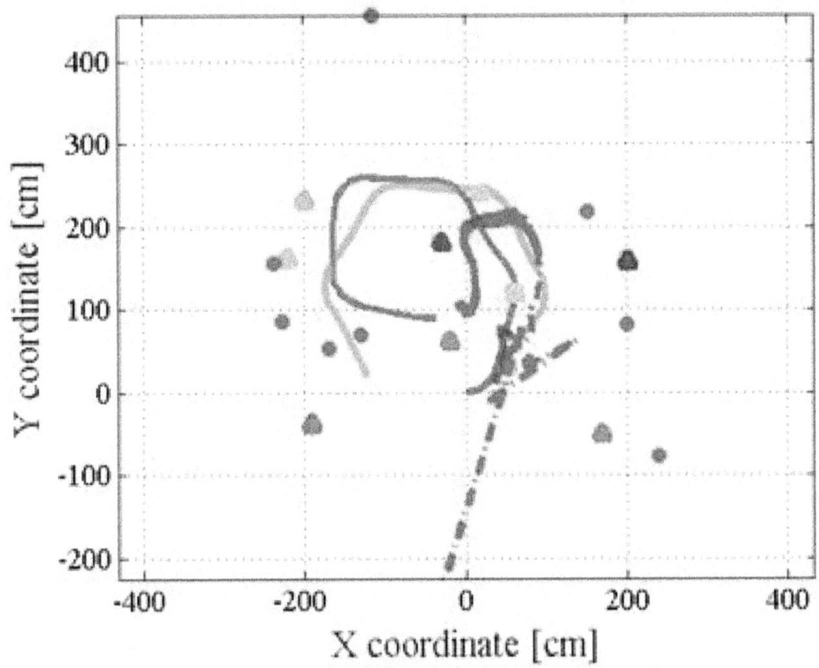

Figure 9. The normal H_∞ Filter(dotted line) shows erroneous results compared to the true mobile robot path(lighter line) and Fuzzy Logic based H_∞ Filter(darker line). Landmarks estimation of normal H_∞ Filter(round shape) is also results in erroneous estimation.

Further inspection is done by using different fuzzy membership with the similar range of fuzzy sets in the case when the mobile robot attempts to localize itself in a given environment. Other associated and related simulation parameters remain unchanged to observe any significant improvement that fuzzy logic can offer. The initial state covariance for the landmarks is now known and mobile robot does not have any information of its location. Even other types of membership functions are applied for decisions; the fuzzy logic-based H_∞ filter surpassed the normal filter performance as depicted in Figure 10.

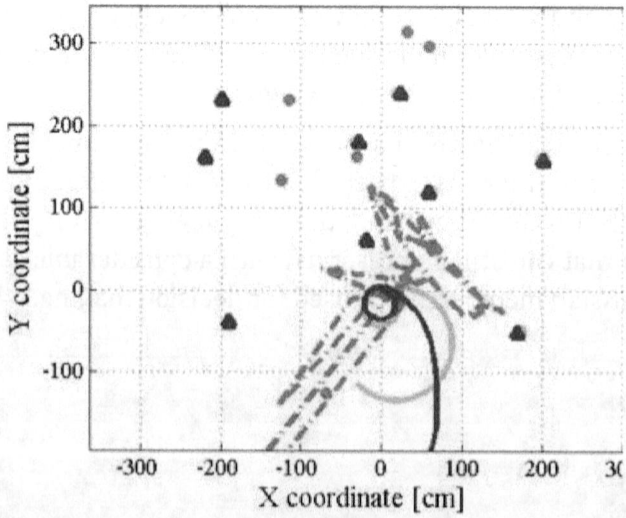

Figure 10. The normal H_∞ Filter (dotted line) cannot localize itself as well as the available landmarks even though the landmarks information is given compared to the Fuzzy Logic based H_∞ Filter (darker line) with reference to the truth locations (lighter line). Triangle shows the Fuzzy Logic based H_∞ Filter and round shape showing the normal H_∞ Filter for landmarks estimation.

Assessment on different mobile robot movements during its observations with the same simulation parameters is also conducted to evaluate the consistency of estimation. The results are not disappointing and show a reliable estimation when comparing it to the normal H_∞ filter as illustrated in Figure 11.

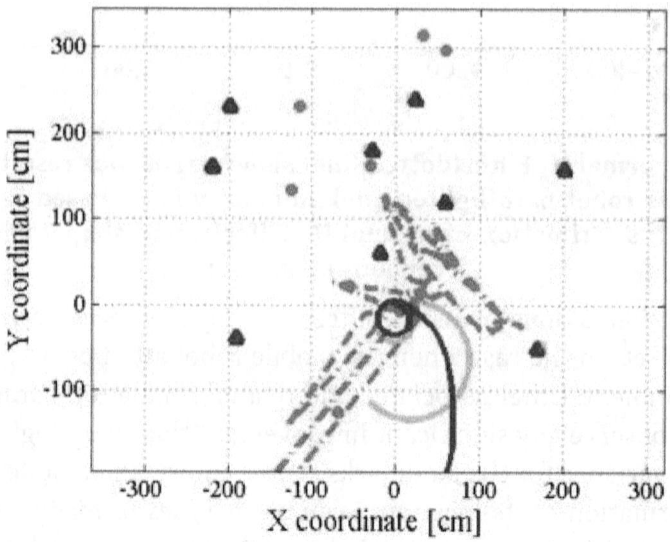

Figure 11. Normal H_∞ Filter (dotted line) produces erroneous results compared to the proposed technique (FHF in darker line) for different mobile robot movements. Triangle shows the Fuzzy Logic based H_∞ Filter and round shape showing the normal H_∞ Filter for landmarks estimation.

4.3. Effect of non-Gaussian measurement noise

H_∞ filter is known to be more robust to the extended Kalman filter (EKF) especially whenever non-Gaussian noise is available. The measurement noise is increased to observe whether the H_∞ filter with fuzzy logic still able to preserve a good estimation. Figure 12 presents the results that still define that the H_∞ filter with fuzzy logic (triangular memberships) maintains better performance than the normal H_∞ filter. Similar performance is also observed with different fuzzy memberships even though the results are not included in this chapter.

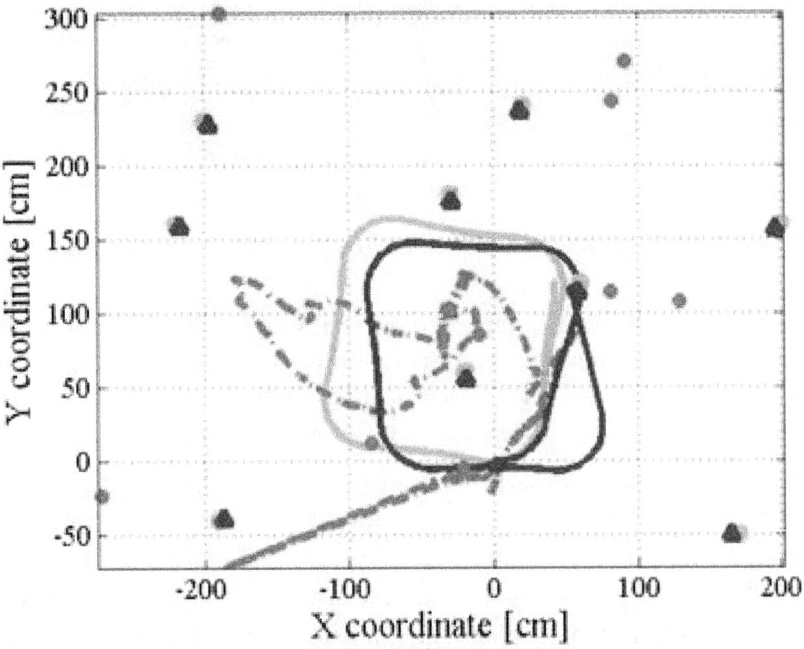

Figure 12. Normal H_∞ Filter (dotted line) performance is still low compared to the Fuzzy Logic based H_∞ Filter (darker line) with reference to the truth positions(lighter line) for bigger measurement noise. Triangle shows the Fuzzy Logic based H_∞ Filter and round shape showing the normal H_∞ Filter for landmarks estimation.

The performance analysis between fuzzy logic-based H_∞ filter and extended Kalman filter is also considered. A condition where mobile robot loses its measurement data at random time is referred in this case. Interesting results are obtained in Figure 13 showing that the H_∞ filter with fuzzy logic inference is still better compared to the normal EKF estimation in non-Gaussian noise. The landmarks estimation for H_∞ filter with fuzzy logic outperforms the EKF. Therefore, based on the figure, H_∞ filter with fuzzy logic technique offers better solutions when non-Gaussian noise as well as when measurement data is lost unexpectedly during mobile robot observations.

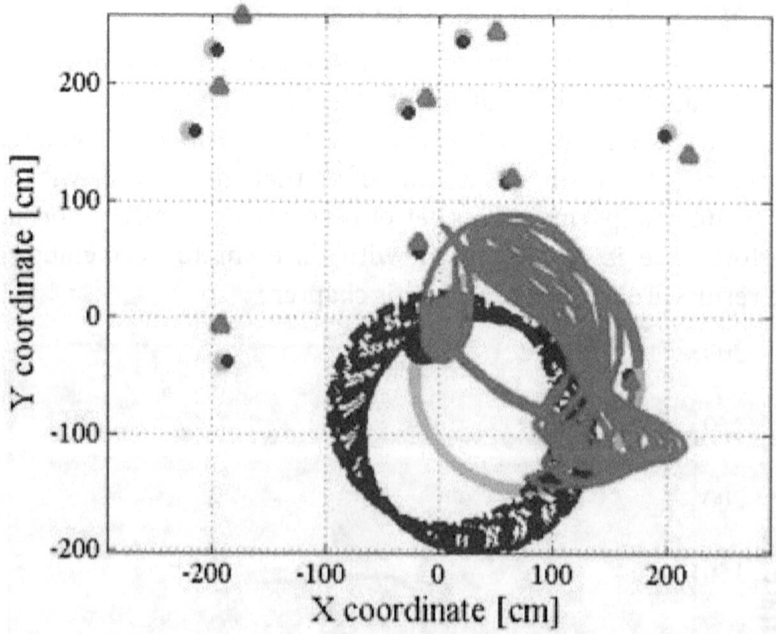

Figure 13. An analysis when measurement data is randomly unavailable. Fuzzy Logic based H_∞ Filter(darker line) performance is surpassing the EKF(triangle) especially on the landmarks estimations with reference to the truth positions(lighter line).

There are few remarks to be considered in designing the fuzzy logic control for mobile robot navigation. The measurement innovation characteristics must be first examined prior to the estimation to ensure the results achieve the desired conditions. Besides measurement innovation, the noise characteristics can also influence the estimation performance. Therefore, the designer should carefully understand and model the noise according to the environment to be observed. H_∞ filter is also sensitive to some parameters as stated by Bolzern and Maroni [7] and those parameters must be studied before conducting further analysis on the proposed technique.

5. CONCISE SUMMARY

This research has conducted an analysis and examination of the $H\infty$ filter for mobile robot navigation utilizing fuzzy logic control. The inquiry primarily concentrated on the advancement of fuzzy logic control to assess relative angle and distance measurements as inputs to minimize navigation errors. Moreover, fuzzy logic control has been identified as a viable method to circumvent the finite escape time issue in $H\infty$ filters. A series of tests have been performed on the proposed technique, examining the impacts of varying γ values, diverse noise factors, and intermittent data loss. Initial findings indicate that the fuzzy logic-based $H\infty$ filter can effectively manage the issue with a minimal number of rules and fuzzy sets. Consequently, the technique may serve as an alternative solution for navigation.

AT A GLNACE

This chapter has provided an examination of H∞ filter-based mobile robot navigation utilizing fuzzy logic to accommodate non-Gaussian noise situations. The technique utilizes information derived from H∞ filter measurement innovation to mitigate noise and uncertainty in mobile robot observations. The simulation results indicated that the proposed technique enhanced the mobile robot's estimation and the observation of any landmarks. The analysis encompasses several factors, including γ values, noise parameters, intermittent data loss, and finite escape time concerns, to assess their impact on estimation. Various fuzzy logic design configurations were examined to attain improved estimation outcomes. This study illustrates that fuzzy logic provides more trustworthy estimation results than traditional methods.

REFERENCES

1. Thrun, S. Burgard, W. Fox, D., 2000. A Real Time Algorithm for Mobile Robot Mapping with Applications to Multi-Robot and 3d Mapping, IEEE International Conference on Robotics and Automation, Vol.1, San Francisco, CA, USA, USA, pp. 321-328
2. Thrun, S. Burgard, W. Fox, D. 2005, Probabilistic Robotics, MIT Press, pp. 39-79.
3. Valavanis KP, Doitsidis L, Long M, Murphy RR. A case study of fuzzy-logic-based robot navigation. IEEE Robotics & Automation Magazine. 2006;13(3):93-107
4. Jaradat MAK, Abdel-Hafez MF. Non-linear autoregressive delay-dependent INS/GPS navigation system using neural networks. IEEE Sensors Journal. 2017;17(4):1105-1115
5. Átila V. F. M. de Oliveira; Marcelo A. C. Fernandes, 2014. Dynamic Planning Navigation Technique Optimized with Genetic Algorithm, 2014 Joint Conference on Robotics: SBR-LARS Robotics Symposium and Robot Control, Sao Carlos, Brazil, pp. 91-96
6. Ahmad H, Namerikawa T. Extended Kalman filter based mobile robot localization with intermittent measurement. System Science & Control Engineering: An Open Access Journal. 2013;1:113-126
7. Bolzern P, Maroni M. New condition for the convergence of H_∞ filters and predictors. IEEE Transactions of Automatic Control. 1999; 44:1564-1568
8. Huang S, Dissayanake G. Convergence and consistency analysis for extended Kalman filter based SLAM. IEEE Transaction on Robotics. 2007;23(5):1036-1049

9. Ahmad H, Othman N. The impact of cross-correlation on mobile robot localization. International Journal of Control, Automation and Systems. 2015;13(5):1251-1261

10. Lewis, F. L., Xie, L., & Popa, D. 2008.Optimal and Robust Estimation With an Introduction to Stochastic Control Theory. USA: CRC Press, pp. 353-385

11. Gualda, D, Urena, J. Garcia J.C, Garcia E, Ruiz D, Lindo A, 2014, Fusion of Data from Ultrasonic LPS and Isolated Beacons for Improving MR Navigation, 2014 IEEE International Conference on Instrumentation and Measurement Technology Conference, Montevideo, Uruguay, pp.1552-1555

12. Abdelnour, G. Chand, S. Chiu, S. and Kido T., 1993, On-line detection and correction of Kalman filter divergence by fuzzy logic, in Proc. Amer. Control Conf., 1993, San Francisco, CA, USA, USA, pp. 1835-1839

13. Asadian A., Moshiri, B., Sedigh A.K, 2005, A Novel Data Fusion Approach in an Integrated GPS/INS System Using Adaptive Fuzzy Particle Filters, In Proc. of the 5th IEEE Conf. on Technology and Automation (ICTA), Greece, pp. 125-130

14. Choomang R, Afzulpurkar N. Hybrid Kalman filter/fuzzy logic-based position control of autonomous mobile robot. International Journal of Advanced Robotic System. 2005;2(3):197-208

15. Ip YL, Rad AB, Wong YK, Liu Y, Ren X. A localization algorithm for autonomous mobile robots via a fuzzy tuned extended Kalman filter. Advnced Robotics. 2010;24179-206

16. Kobayashi K, Cheok KC, Watanabe K, Munekata F. Accurate differential global positioning system via fuzzy logic Kalman filter sensor fusion technique. IEEE Transactions on Industrial Electronics. 1998;45(3):510-518

17. Raimondi, F.M, Melusso, M, 2006. Fuzzy EKF Control for Wheeled nonholonomic Vehicle, 32th Anuual Conference on IEEE Indusrial Electronics, Paris, France, pp. 43-48

18. Wang JH, Song CL, Chen JB. Sigma point H infinity filter for initial alignment in marine strapdown inertial navigation system. In: 2nd International Conference on Signal Processing Systems. 2010; pp. 580-584

11 Chapter

INDOOR MOBILE LOCALIZATION UTILIZING NEURAL NETWORKS AND FUZZY LOGIC CONTROL

1. INTRODUCTION: AN OVERVIEW

The contemporary world is reliant on accurate navigation systems. Satellite navigation systems, including the Global Positioning System (GPS) and the Global Navigation Satellite System (GLONASS), are extensively utilized throughout diverse fields such as transportation navigation, engineering, surveying, and cellular communication. Consumer-grade GPS receivers are included into nearly all contemporary smartphones. In open environments, such receivers enable positioning precision within the range of 1–5 meters. Simultaneously, these receivers fail to satisfy the current need for indoor navigation. Under these conditions, GPS fails to function or delivers position data with a significant error margin of approximately 100–150 meters. Indoor navigation systems can be utilized across various domains. Incorporating navigation within expansive shopping malls, warehouses, or various systems, "smart home" technology allows for the centralized and automated management of diverse household functions, such as heating, lighting, and air conditioning. The necessity for indoor navigation is particularly evident in navigation activities inside transportation networks, such as airports and railway stations [2]:

(A) For visitors: positioning; search check-in desks/offices/storage rooms and a cafe, parking, taxi, etc.; installation of the route, taking into account number of storeys of buildings; search colleagues inside the airport/train station, social activity; service based on knowledge of the position of the visitor (location-based service).

(B) For airports, railway stations, and tenants: an additional service to visitors; analyst extensive movements of visitors and staff; advertising opportunities geo contextual advertising (location-based advertising) airlines/shops/cafes, etc., as well as products/services; promotions based on the location of the visitor.

Opportunities now exist to enhance the techniques and practical applications of indoor mobile navigation. Let us examine contemporary methods for addressing this issue and the rationale for selecting learning algorithms and neural networks based on certain characteristics of shared interests. Indoor mobile navigation systems are increasingly prevalent across various sectors, including transportation, public institutions, and logistics. Navigation utilizing fingerprint radio access points for Wi-Fi is gaining popularity. The artificial neural network (ANN) model is frequently regarded as a technique for storing and processing radio fingerprints. This chapter analyzes the primary challenges associated with developing intricate indoor mobile navigation algorithms utilizing neural networks, a component of fuzzy logic. We substantiate the selection of artificial neural network algorithms for indoor navigation on mobile devices. The data for neural network training in MATLAB, sourced from the file "train.txt," is utilized for functional testing of software navigation. To assess the efficacy of ANN training, the original file must be partitioned into two subsets (training and testing). Wireless local area networks 802.11k have gained significant popularity as precise navigation systems for various reasons. They function within unlicensed frequency ranges and require minimal effort and expense for implementation. The proliferation of several mobile devices equipped with Wi-Fi technology offers users the opportunity for greater selection and cost savings. WLAN networks possess several advantages compared to conventional wired networks.

- Much easier and cheaper to deploy a local wireless network in the new location than a traditional network.
- The subscriber does not have to be next to his desk or local network socket-outlet. He can move freely inside the area coverage.

Nonetheless, the quantity of users operating within unlicensed frequency bands is rising daily. In this context, a significant issue is the method by which the wireless device selects an access point for network connectivity. The device now selects the access point based on signal strength. This method facilitates the identification of the closest access point. The currently utilized 802.11k standard is designed for the deployment of load balancing in radio Wi-Fi networks. However, a strong signal does not necessarily indicate high network bandwidth. For instance, assume that the majority of notebooks in specific conference halls connect to the WLAN via the access point located above the entrance door. In this scenario, the number of connected users could reach tens or even hundreds, whilst other access points would remain underutilized. Consequently, the network bandwidth allocated to each user diminishes substantially, resulting in an overall decline in network productivity. Therefore, the selection of the access point in a WLAN network under conditions of high user density is quite relevant. The selection must consider both the signal strength received by the mobile device and the bandwidth width of the dedicated channel, which is contingent upon the number of subscribers connected to the access point. In this chapter, we analyzed a method that considers both signal levels and additional parameters. This chapter employs fuzzy logic to address the task. Membership functions and language norms that have been constructed are analyzed. The model's structure and simulation outcomes are presented.

2. INDOOR MOBILE POSITIONING USING NEURAL NETWORKS

2.1. Characteristics of modern approach to navigation mobile

First, we note that the modern mobile devices contain a variety of different sensors and receivers. The main of them include:

1. GPS.
2. Wi-Fi, a radio signal operates at frequencies of 2.4 and 5 GHz.
3. Bluetooth operates on the same frequencies.
4. The accelerometer and gyroscope—inertial sensors measuring linear and angular acceleration.
5. Magnetometer—a sensor that measures the intensity of magnetic fields.
6. Other sensors—humidity, light, proximity, barometer, etc.

Sensors 1–5 are well represented in many of today's mobile devices that are running operating systems iOS and Android. These systems provide programmatic access to the sensors through its own application program interfaces (API), and any application can receive data from the sensors (with certain insignificant limitations).

Currently, there are several ways to use the mobile navigation device.

1. GPS (Global Positioning System) and GLONASS (Global Navigation Satellite System) [1]: these are navigation using satellites. Well suited for positioning in the open spaces. Because of the need to be in the field of view of at least three satellites, these are poorly suited for closed premises, as they greatly impair the satellite signal. The initial search for satellites might take few minutes.
2. AGPS (Assisted GPS). Navigation using radio signals from cell towers. Typically it used in conjunction with GPS. It accelerates the initial determination of the coordinates by the fact that the use of data on the mobile device serving cell towers.
3. Navigate using radio beacons operating on technology iBeacons (Bluetooth Low Energy) [3]. This is a relatively young technology, the impetus for the development of which will serve standard Bluetooth 4.0 (+) with low power consumption. Tracker is a chip with a radio module, which is a predetermined frequency radio, sends packets with information about themselves. The receiver, knowing the map of the location beacons and signal strength to the nearest of them, calculates their relative location.

4. Navigation on the basis of fingerprints of radio access points Wi-Fi [4]. Such navigation systems are more exploratory in nature. Due to the relative novelty of this approach, ways to implement proven accurate navigation have not yet developed.
5. Inertial navigation system [5]. Navigation is based on data from the inertial sensor device, which is constructed on the basis of the model orientation in space.

Comparative characteristics of the main approaches of navigation using mobile devices [2] are given in Table 1.

Table 1. Comparative characteristics of approaches.

System	Dignity	Disadvantages
GPS	Average precision (5 m) Ease of use Good compatibility	Inability To work Indoor
GSM	Ease of use Good compatibility	Ease of use Good compatibility
Wi-Fi	Average Accuracy (5 m) Work indoors	The need for network deployment Limited compatibility High power consumption
iBeacon	High accuracy (1–2 m) Ease of use Good compatibility Low power consumption	The need to deploy BLE-network

In addition to these approaches, following approaches are also included. In Ref. [6], the integration of Wi-Fi and an inertial navigation system is considered. In Refs. [7, 8], mobile navigation services and the use of technology OpenCellID to determine the location of mobile devices are studied.

The article [9] considered a relatively new approach to the positioning of mobile devices in the premises on the basis of two-dimensional barcodes. In Ref. [10], the model of context-aware computing (context-dependent browser) based on network proximity is considered. This mobile phone is considered as a proximity sensor and geo replaced positional information network proximity. An algorithm for calculating the trajectories of mobile networks on the basis of information about network proximity is also considered.

2.2. Rationale for navigation of mobile devices using neural networks

Employment of navigation system via radio fingerprint from the access points of Wi-Fi consists of two parts:

1. Preparation of a database of known fingerprints radio with known coordinates associated with them.
2. Getting new coordinates from the database on the new print of radio signals.

As a mechanism for storing and processing the radio fingerprint has considered a model of artificial neural networks (ANNs).

This approach is interesting for the following reasons:

- The problem of determining the location based on a previously defined radio fingerprint can be viewed as a problem of classification of multidimensional data.
- Main computational load of this method accounts for the learning process, which is performed only once and can be executed on an independent computer system.
- Data on the surrounding Wi-Fi points can be supplemented by the data from other sources, such as Bluetooth LE beacons.

The possibilities of modern ANN are the subject of active research. The approach to the use of ANN to solve the above problem is considered in several publications, for example, Refs. [10–12].

2.3. Characteristics of software navigation

Here examined a complex program [13, 14] to navigate through the Wi-Fi signals using a mobile phone. It consists of two components:

1. Mobile application that
   - collects data on the surrounding Wi-Fi access points;
   - uses trained neural network to determine the current location of the mobile device based on the new data on Wi-Fi signals.
2. Desktop application that
   - prepares data radio prints for further training of the neural network;
   - training and testing neural network.

As an implementation of a neural network, we used multilayer ANN provided free library FANN (Fast Artificial Neural Network Library). To train the ANN learning algorithm, Resilient Propagation (*RProp*) has been used [15].

Unlike standard algorithm *Backprop*, *RProp* uses only partial signs for adjusting the weighting coefficients. The algorithm uses the so-called "training periods" when the correction occurs after the presentation of the balance of the network of examples from the training set.

To determine the amount of correction using the following rule [15]:

$$\Delta_{ij}^{(t)} = \begin{Bmatrix} \eta^{+}\Delta_{ij}^{(t)}, \dfrac{\partial E^{(t)}}{\partial \omega_{ij}}\dfrac{\partial E^{(t-1)}}{\partial \omega_{ij}} > 0 \\ \eta^{-}\Delta_{ij}^{(t)}, \dfrac{\partial E^{(t)}}{\partial \omega_{ij}}\dfrac{\partial E^{(t-1)}}{\partial \omega_{ij}} < 0 \end{Bmatrix}, \quad (1)$$

$$0 < \eta^{-} < 1 < \eta^{+} \quad (2)$$

If the partial derivative of the corresponding weight $\partial\omega_{ij}$ had changed its sign at the current step, it means that the latest update was great, and the algorithm passed a local minimum and therefore the value of change must be reduced by η and return the previous weight value: in other words, you must make 'roll back'.

$$\partial\omega_{ij}(t) = \partial\omega_{ij}(t) - \Delta_{ij}^{(t-1)}. \quad (3)$$

If the sign of the partial derivative is not changed, it is necessary to increase the amount of compensation η^{+} to achieve a more rapid convergence. Fixing factors η^{-} and η^{+} can be dispensed with global settings of the neural network, which can also be seen as an advantage of the algorithm to the standard algorithm *Backprop*.

Recommended values are $\eta^{-} = 0.5$, $\eta^{+} = 1.2$, but there are no restrictions on the use of other values for these parameters.

To prevent too large or small weight values, the correction value limit from above the maximum Δ_{max} and below the minimum Δ_{min} value of the correction value, which default, respectively, shall equal 50 and 1.0E–6.

The initial values for all Δ_{ij} are set to 0.1. Again, this should be seen only as a recommendation, and in the practical implementation it can specify a different value for the initialization.

The current implementation of the navigation software package is a simple one-dimensional classifier, which, by new radio-prints, is able to determine the room in which the mobile device is located.

2.4. Working with software navigation

Working with a program complex navigation is implemented using desktop and mobile software, and includes the following steps:

1. Collection of data on radio-prints in the studied areas. The output of this stage is a set of files, each of which contains a set of vectors (matrix) measuring Wi-Fi signals points (for example, Table 2).

In the first row, there are names of all the access points that were visible in the data collection process. For clarity, we have been selected to work with network names, rather than their mac address. Each line represents the measured signal strength to a point. A value 0 means that the point was not available at this time.

2. Combine all the source files into one.

   The result is a common image with the measurements in all the studied areas which is shown in Table 3.

   The first column is stored name space in which the measurement was performed. Null values (0) are replaced by -100 (very weak signal).

   In addition, a file is created "names.txt," which lists the names of all the networks in the same order in which they are placed in the file of training (step 4).

3. Represented by map matching names with their premises of formal numerical representation:

   Cabinet [0] Room [0.5] Kitchen [1]

4. On the basis of the contents of the files of the previous steps 2 and 3, create a file with the data for network training:

   60 15 1

   -86.0 -100.0 -54.0 -69.0 -100.0 -100.0 -88.0 -100.0 -100.0 -100.0 -100.0 -100.0 -100.0 -100.0 -100.0

Table 2. Set of vectors (matrix) measuring signals Wi-Fi points.

Grimas	DSL-2640U	NETGEAR	STERH76	InterZet-at-home	InetBezP	ASUS 26
−86	0	−54	−69	0	0	−88
−89	−83	−56	−66	0	0	−91
...	...	...	...	...	...	...

Table 3. Common image with the measurements in all the studied areas.

	Grimas	DSL-2640U	NETGEAR	STERH76	InterZet-at-home	InetBezP	ASUS 26	...
Study	−86	−100	−54	−69	−100	−100	−88	...
Study	−89	−83	−56	−66	−100	−100	−91	...
...	...	...	...	...	...	...	...	...

0.0

-89.0 -83.0 -56.0 -66.0 -100.0 -100.0 -91.0 -100.0 -100.0 -100.0 -100.0 -100.0 -100.0 -100.0 -100.0

0.0

-90.0 -71.0 -51.0 -67.0 -100.0 -100.0 -88.0 -100.0 -100.0 -100.0 -100.0 -100.0 -100.0 -100.0 -100.0

0.0

The first line shall include: the number of hidden layer neurons, the number of neurons in the input layer and output layer.

The following are a couple of lines:

A. The input vector.

B. The value that the National Assembly should be trained for this vector.

The result of this step is to file "network," which is a trained neural network in the library "FANN." The network name is chosen at random.

The following steps describe the workflow of software application that is running on a mobile device.

5. A mobile application launches library "FANN" and passes the file to the trained neural network.
6. Mobile application starts scanning the surrounding Wi-Fi hotspots. After the completion of the next iteration of the scanning application, an array of the currently available access points is received to which the device may try to connect. Each object of the array, among other things, contains in its structure the following two fields:

   A. Title.

   B. The strength of the signal to a predetermined point.

   With using the array of file names "names.txt" obtained in the second stage and the current array of visible access points built radio fingerprint, which, in essence, is a simple array of numbers. Create a radio footprint that includes the following steps:

   A. Create an array of numbers (float []) size which determine the number of elements in the array of names.

   B. Go through the array of names in the cycle:

      a. if the array access point is a point with the current name, the add-in radio signals strength to mark this point of the current iteration of the index;

      b. otherwise, add the current iteration index of the default value for the missing point that was made in step 2 (-100).

   C. At the output of iteration scanning radio fingerprint is obtained as a vector of values (-100, -50, -75, -100, ...).

7. The resulting vector is passed to the neural network, which returns the result vector (float []). Since the ANN has been trained on the basis of dimensional data, we are only interested in the first element of the resulting vector.
8. The resulting value is compared with the data from the map space. In the version of the software complex in the map, closest value to the current value obtained from the network is searched. After finding, it displays the name of the current premises that corresponds to the measured radio-imprint.

2.5. Training a neural network in MATLAB

The data for the neural network training in MATLAB are taken from the file "train.txt" and used for functional testing of software navigation. To verify the performance of ANN training, the original file should be divided into two sets (training and testing).

By analogy with [16], we make a comparative analysis of the main learning algorithms of the neural network used for the navigation of mobile applications for indoors, with position accuracy and complexity ratios depending on the number of neurons in the hidden layer. The corresponding results are shown in Table 4.

Table 4. Characteristics of the complexity and accuracy of learning algorithms NA.

Number of neurons	Trainlm		Trainscg		Trainbr	
	MSE	Epochs(s)	MSE	Epochs(s)	MSE	Epochs(s)
5	0.11	12(0.1)	0.08	15(0.1)	0.03	592(25)
10	0.06	16(0.1)	0.09	20(0.2)	0.02	700(20)
15	0.07	9(0.1)	0.05	13(0.2)	0.013	800(23)
20	0.05	11(0.2)	0.13	14(0.2)	0.015	269(48)
30	0.13	6(0.2)	0.19	23(0.3)	0.018	889(120)
50	0.23	13(0.2)	0.21	30(0.5)	0.02	483(187)

Table 4 uses the following symbols: *trainlm*—learning algorithm Levenberg-Markarth, *trainscg*—related gradient method scaled association, and *trainbr*—Bayesian regularization method. MSE—mean square error, epochs (s)—the number of training cycles, and in brackets the number of seconds.

Based on the analysis of Table 1, we can conclude that the optimal number of hidden layer neurons for a given set of source data is in the range 15–25. It showed the highest accuracy Bayesian regularization algorithm, but it is considerably more time-consuming compared with other algorithms. The optimal ratio of accuracy and convergence provides the Levenberg-Marquardt algorithm.

Note that low accuracy is common to all algorithms, which is associated with the original data. Probably, it reduces the dimension of the input vector (data used by

Wi-Fi 15 points, most of which are available), and increases the total number of measurements. At the same time in the original data set of vector measurements corresponded to only three different rooms, so get enough accuracy.

3. USING FUZZY LOGIC CONTROL FOR SELECTION OF THE ACCESS POINT

3.1. Introduction

Wireless local area networks of 802.11k standards become very popular due to some reasons. They operate in the unlicensed frequency bounds and they need not require large time and cost for deployment. The appearance of vast number of mobile devices supporting Wi-Fi technology gives possibilities for free choice and cost saving for the various kinds of users. WLAN networks have a number of advantages over traditional wired networks.

- Much easier and cheaper to deploy a local wireless network in the new location than with a traditional network.
- The subscriber does not need to be next to his desk or local network socket. He can move freely inside the area coverage.

However, number of users working in the unlicensed frequency bounds increases day by day. In this context, a very important problem is the way by which the wireless devices are used to select an access point to connect to the network. Now device selects the access point according with signal power [17]. This way allows determining the nearest access point. In other words, currently used 802.11k standard is aimed at the implementation of load balancing of radio Wi-Fi networks. Nevertheless, the high level of the signal does not always mean high network bandwidth. Suppose, for example, that most notebooks operating in the certain conference hall connects to WLAN using access point which is above the entrance door. In such a case, the amount of subscribers connected to it would be to tens if not hundreds, while other access points would not be fully loaded. As a result the network bandwidth per subscriber reduces to a significant degree that leads to the network reduction of productivity of the network in whole. Hence, the task of selection of the access point in WLAN network in the case of high user density is quite topical. Such selection must take into account not only the level of the signal received by the mobile device, but also a width in the dedicated channel bandwidth that depends on the number of connected subscribers to the access point.

Further, examined the method based on analysis not only signal level but also other parameters. To solve this task, fuzzy logic is used (See Ref. [18, 19]). Constructing membership functions and linguistic rules are examined. Structure of the developed model and simulation results are presented.

3.2. Choice of access points for connecting a mobile device

Signal level and the bandwidth of a system are linked by the well-known Shannon formula, which allows determining the capacity of the data transmission system:

$$C = \Delta F \times \log_2(1 + S/N). \tag{4}$$

The total bandwidth allocated in the range, is divided equally among all active subscribers [17]. If the length of the transmitted packet is *L* bits, one can determine the time required for transmission of a packet:

$$T(M, S/N) = \frac{L \times M}{F \times \log_2(1 + S/N)}, \tag{5}$$

where $\Delta F = F/M$ is the network bandwidth, and S/N is signal/noise ratio at the receiver input and *M* is the number of subscribers already connected. Thus, the functional dependence between bandwidth and the number of connected subscribers is linear. Then the task of selecting the best access point to connect can be formulated as follows:

Find the value of function (5), the T_{max} of which does not exceed a predetermined time with the following restrictions: $S/N > P_0$ and $\Delta F \geq \Delta F_{min}$, where P_0 is some ratio threshold signal/noise at which the operation of mobile subscriber receiver is possible and ΔF_{min} is a minimum possible bandwidth width. In other words, the problem is reduced to finding such a pair of values (ΔF, S/N) for which the transmission time has the minimum possible value.

This problem can be solved by various methods:

- Analytical methods that use differential and variational calculus.
- Numerical methods which use prior information in order to find improved solutions using iterative algorithms.
- Mathematical methods (linear and nonlinear programming).

In the case of choice of any method, the finding of a point of extremum of the considered function is usually required. The type of function (5) is shown in Figure 1.

Figure 1 shows that the function (5) has minimum values for various combinations of parameters and the number of connected subscribers M. Function (5) has no maximum that determines the above-mentioned formulation of the problem of choosing the access point. Table 5 shows the values of the function (5) calculated for various combinations of parameters and *M*.

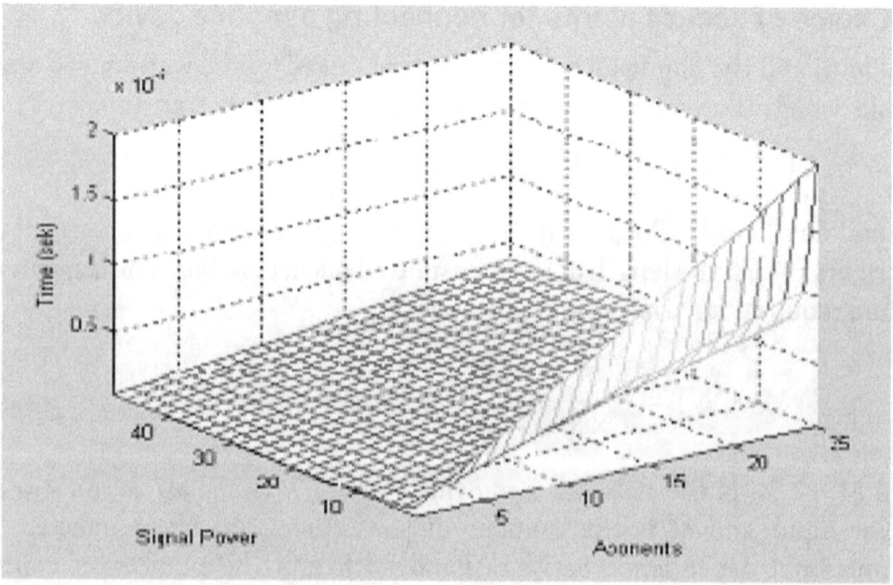

Figure 1. Dependence of the packet transmission time from the signal level and the number of subscribers.

Table 5. Values of the time of packet transmission for different ratios of parameters.

Signal power, db	The bandwidth width, hz	The transmission time, s
30	4E + 6	2.514E - 4
30	8E + 5	1.257E - 3
30	4.444E + 5	2.263E - 3

Analysis of Table 5 confirms that the high signal level does not always provide an acceptable transmission time. It's known for solve the problem of the selection the access point to connect mobile subscribers on the basis of function (5) using the above methods it is necessary and sufficient that the following conditions are met (as it's shown for instance in Ref. [20, 21]):

1. Function (5) should be continuous and differentiable at the point of extremum.
2. The Hessian matrix of function (5) must be negative definite (for a point of minimum).

Clearly, the condition 1 for the function (5) is performed. To simplify the analysis of the condition (2), it is convenient to consider function 1, taking into account that low values of transmission time are achieved with large bandwidth. Then, for the function (4), condition 2 changes to opposite. Condition 2 for the function (5) is not satisfied due to the fact that $\partial^2 C/\partial(\Delta F)^2=0$, which implies that:

$$\begin{vmatrix} \frac{\partial^2 C}{(\Delta F)^2} & \frac{\partial^2 C}{\partial(\Delta F)\partial\left(\frac{S}{N}\right)} \\ \frac{\partial^2 C}{\partial\left(\frac{S}{F}\right)\partial(\Delta F)} & \frac{\partial^2 C}{\partial(\frac{S}{N})^2} \end{vmatrix} = \begin{vmatrix} 0 & \frac{1}{\log_2 e\left(1+\frac{S}{N}\right)} \\ \frac{1}{\log_2 e\left(1+\frac{S}{N}\right)} & \left(\frac{1}{\log_2 e\left(1+\frac{S}{N}\right)}\right)^2 \end{vmatrix} \le 0. \quad (6)$$

It follows that the function in question is not concave. However, the requirement of convexity or concavity of the function is a serious restriction that is far from always performed in practical problems. This is why the concept of such functions is generalized by the introduction of pseudo-convex unimodal functions [21]. The f: $X \rightarrow R$ function is called the pseudo-unimodal in the interim $[a,]\subset X$, if $\exists$ an arbitrary interval $I^* \subset [a,b]$ such that the f function:

- Strictly increasing in the interval $[a, b]$;
- Equals to some constant, $\le \min\{f(c),f(d)\}$ on the some interval I^*;
- Strictly decreasing in the interval $[d, b]$.

The points c and d determined by the following way:

$$c = \inf_{x\subset I^*} x; \quad d = \sup_{x\subset I^*} x. \quad (7)$$

In this case, the interval I^* is the solution of the problem $\max\{f(x): x\subset[a,b]\}$. In the particular case if $c = d$, then $I^*=\{x^0\}$ and $x = c = d$, the f function is called unimodal. The example of unimodal function is shown in Figure 2.

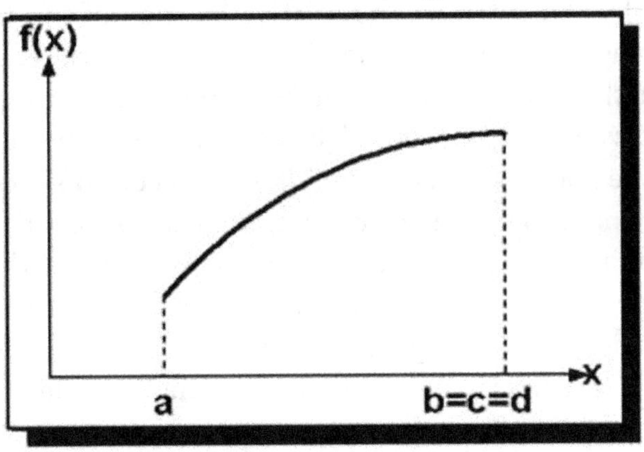

Figure 2. The example of unimodal function.

Then, if we set

$$a = [\Delta F_{min}, P_0]; \quad b = [\Delta F_{min}, P_{max}]. \quad (8)$$

Here P_{max} is some maximum possible value of S/N ratio, then unimodality of the function (4) and, consequently, function (5) derived from their definitions. Consequently, the optimization problem has a solution in the following formulation: find the minimum of function (4) in the interval (8). This solution can be found by any of the methods listed above. However, their use is associated with a large number of calculations (solution of the corresponding equations), or with a large amount of stored data, requiring constant modification (various search methods).

In Ref. [18], to solve above-mentioned problem is proposed to use the apparatus of fuzzy logic, free from above disadvantages. Fuzzy logic operators are very similar to conventional Boolean operators and allow simplified algorithms for solving this problem. Complicated mathematical modeling can be replaced by evaluation of membership functions and rules of fuzzy logic [22]. Various approaches to solving this type of problems are considered in Refs. [23, 24]. In Ref. [25] there is an example of controlling the operation of the charger using intelligent controller which applies various algorithms of fuzzy inference. One of the most powerful tools for solving such problems is the MATLAB system, which provides users a various species of software, including the visual ones. With the help of visual programming, the necessary model can be built, and then can run the simulation in program mode.

3.3. Model of algorithm of determination of the best access point

In accordance with the 802.11 standard, the mobile station scans the individual channels for the detection of the best signal from the access point. APs periodically send a Beacon signal in a broadcast mode. Mobile network station accepts these beacon signals and takes note of the relevant signal level. Thus, the received signal level actually characterizes the relative location of the subscriber and the access point. During this process, a mobile subscriber searches such access point. To determine whether channel is free, the well-known algorithm, which is named clear channel assessment (CCA), is used. Its essence lies in the measurement of the signal power at the antenna and determining the received signal strength (RSSI). If the received signal strength is below a certain threshold, then the channel is declared free, and the MAC layer receives the CTS status. If the power is above the threshold, the data transfer is delayed in accordance with the protocol rules. The standard provides another opportunity to determine the idle channel, which can be used either separately or together with the RSSI measurement—method of checking of the carrier. The most appropriate method to use depends on the level of interference in the work area. Various 802.11 standards use one of the five possible CCA modes:

- solution that channel is free and is based on the detection of channel power which exceeds a certain threshold value;

- the decision that the channel is free based on the detection of a carrier signal corresponding to the 802.11 standard;
- carrier signal detection and discovering of power (combination of modes 1 and 2);
- carrier signal detection with message that the ether is free, if neither signal is detected during 3.5 ms timeslot;
- detection of the power corresponding to the increased transmission rate in a physical layer and carrier detection at mode 3, but with reference to the ERP.

802.11 MAC layer is responsible for the way in which the subscriber is connected to the access point. When a subscriber enters into the coverage area of one or more access points, it selects the access point based on signal power values and observed number of errors, selects one of them and connects thereto. Once the subscriber receives confirmation that it is accepted by the access point, it tunes to a radio channel in which it operates. From time to time, it checks all the channels to see if there is another point which provides access services of higher quality. If such an access point is found, then the subscriber connects to it and readjusts to its frequency.

In accordance with this principle, the diagrams of signal levels and levels of channel diagrams of signal levels and levels of channel congestion can be constructed for each access point. These diagrams look like shown in Figure 3.

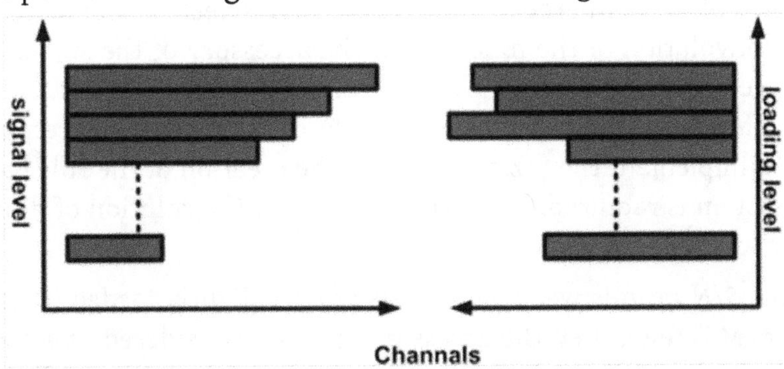

Figure 3. Diagrams of signal levels and levels of channel congestion.

Analysis of these diagrams and subject area makes it possible to apply fuzzy logic to make a decision on selecting the best access point to connect. As shown above, a signal power can be characterized through signal/noise ratio, and a congestion level of AP can be described with the help of number of connected subscribers. Then it becomes obvious that proposed model must have two input linguistic variables (S/N ratio and the degree of loading of the access point). As the set of terms of the first linguistic variable "signal/noise" the following set is used:

$$S/N = \{\text{'weak'}, \text{'medium'}, \text{'high'}\}. \tag{9}$$

As the set of terms of the second linguistic variable "degree of loading" the following set is used:

$$\text{NUMUSERS} = \{\text{'few'}, \text{'medium'}, \text{'many'}\}. \tag{10}$$

As the set of terms of the output linguistic variable the following set is used:

$$\text{RESULT} = \{\text{'is a candidate for connection'}, \text{'is not a candidate for connection'}\} \tag{11}$$

Since the algorithm Mamdami is used for fuzzy inference, the following methods of execution had been chosen for the stages of composition:

1. To perform a logical conjunction in conditions of the fuzzy rules (And method), the minimum value (min) method was chosen;
2. To perform a logical disjunction in conditions of the fuzzy rules (Or method), the maximum value (max) method was chosen;
3. To perform a logical conclusion in the every of fuzzy rules (Implication), the minimum value (min) method was chosen;
4. For aggregation of membership function values of each input variable in conclusions of fuzzy rules (Aggregation), the maximum (max) method was chosen;
5. To perform defuzzification of output variables (Defuzzification), the method of center of gravity for a discrete set of values of the membership functions (centroid) was chosen.

In the process of fuzzy inference, it is necessary to select such an access point for which the convolution of the functions of the accessory of the signal/noise ratio and the degree of congestion of the access point give the best result. To implement the fuzzy inference, the creation of the rule base of fuzzy inference system is required. To implement the fuzzy inference, the creation of the rule base of fuzzy inference system is required. Such base of rules for the solution of discussed task looks like:

- If the S/N ratio is weak and access point is lightly loaded (the number of users M is few), then the access point is not considered as a candidate for connection.
- If the S/N ratio is weak and access point is moderately loaded (the number of users M is medium), then the access point is not considered as a candidate for connection.
- If the S/N ratio is weak and access point is heavily loaded (the number of users M is many), then the access point is not considered as a candidate for connection.

- If the S/N ratio is medium and access point is lightly loaded (the number of users M is few), then the access point is considered as a candidate for connection.
- If the S/N ratio is medium and access point is moderately loaded (the number of users M is medium), then the access point is not considered as a candidate for connection.
- If the S/N ratio is medium and access point is heavily loaded (the number of users M is many), then the access point is not considered as a candidate for connection.
- If the S/N ratio is high and access point is lightly loaded (the number of users M is few), then the access point is considered as a candidate for connection.
- If the S/N ratio is high and access point is moderately loaded (the number of users M is medium), then the access point is considered as a candidate for connection.
- If the S/N ratio is high and access point is heavily loaded (the number of users M is many), then the access point is not considered as a candidate for connection.

The following notations are in use here to simplify the formalization of fuzzy productions:

1. if (S/N is A1) and (Numusers is B1) then (Result is E1) (1);
2. if (S/N is A1) and (Numusers is B2) then (Result is E1) (1);
3. if (S/N is A1) and (Numusers is B3) then (Result is E1) (1);
4. if (S/N is A2) and (Numusers is B1) then (Result is E2) (1);
5. if (S/N is A2) and (Numusers is B2) then (Result is E1) (1);
6. if (S/N is A2) and (Numusers is B3) then (Result is E1) (1);
7. if (S/N is A3) and (Numusers is B1) then (Result is E2) (1);
8. if (S/N is A3) and (Numusers is B2) then (Result is E2) (1);
9. if (S/N is A3) and (Numusers is B3) then (Result is E1) (1).

A1 means the S/N ratio is weak, A2 means the S/N ratio is medium, A3 means the S/N ratio is high, B1 means the number of users is few, B2 means the number of users is medium, B3 means the number of users is many, E1 means that the access point is not a candidate for connection, E2 means that the access point is a candidate for connection. Thus, the connection to the access point occurs only when the access point is not overloaded and has an acceptable signal level. The numbers in round brackets mean the weight of rule (0 or 1). The next step in the construction of the model is to determine the membership functions of input and output variables. Kind of membership functions for the signal/noise ratio, the number of subscribers, and the output variable are shown in Figure 4.

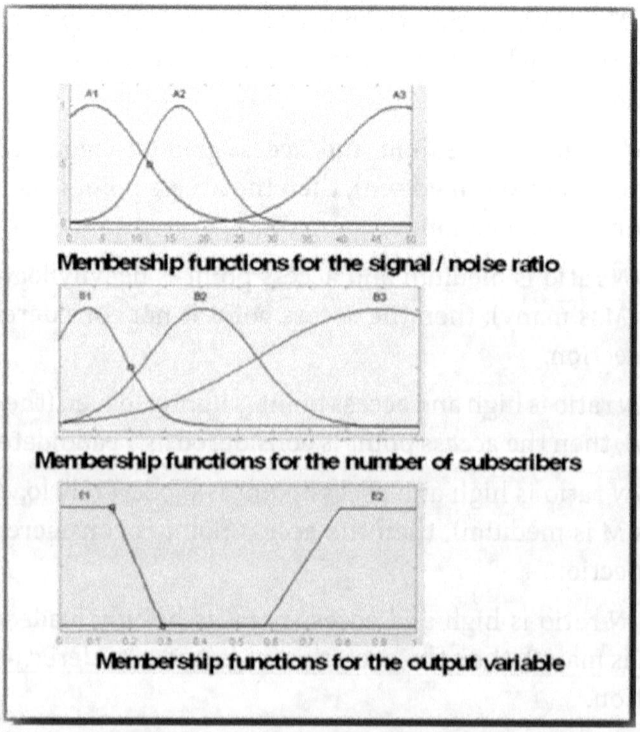

Figure 4. Membership functions for input and output variables.

For analyzing the adequacy of the developed fuzzy model, the surface fuzzy inference can be quite useful. It allows assessing the impact of changes in the values of the input fuzzy variables on the values of the output fuzzy variables. This surface is shown in Figure 5. A comparison of this figure with Figure 1 allows concluding that the nature of the surface of the fuzzy inference coincides in general with the frame transmission time dependent on the signal level and the number of subscribers. This confirms the adequacy of the proposed model.

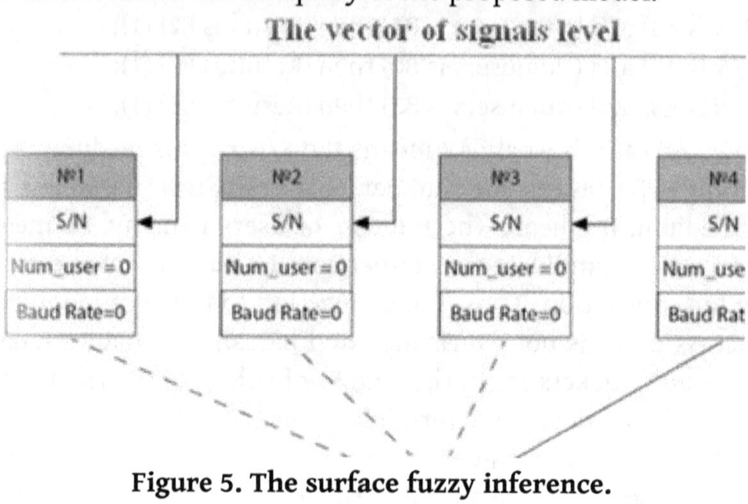

Figure 5. The surface fuzzy inference.

3.4. Simulation of the algorithm for determining the best access point

For checking the correctness and efficiency of the proposed algorithm, its styling is first carried out. MATLAB provides the ability to download a model developed with the help of visual aids into the MATLAB—program and perform simulation in program mode. The data structure of a software model is shown in Figure 6.

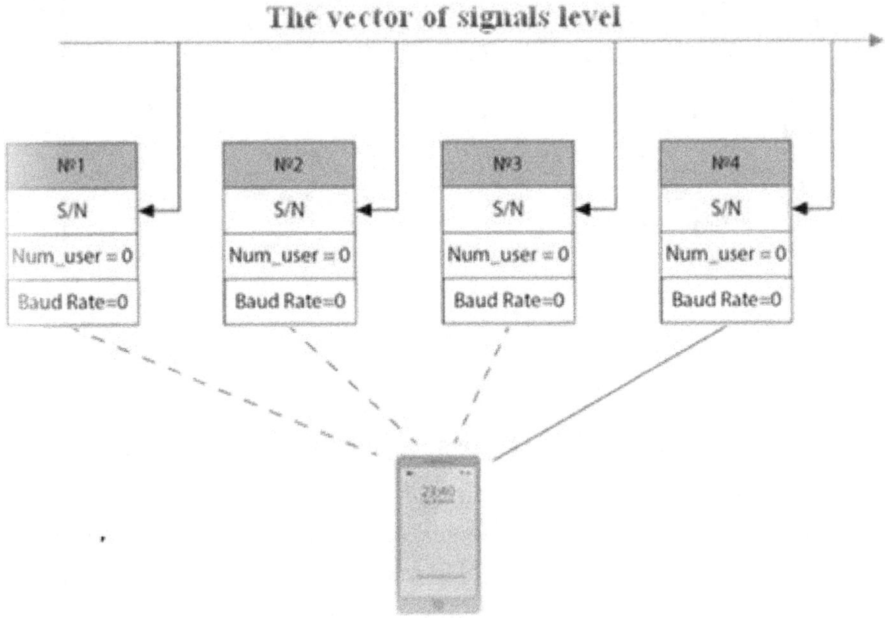

Figure 6. The data structure of model.

Each access point is represented by the data structure containing the following data fields:

- Index of the access point (its number in a vector of access points).
- The current value of the signal level at that point.
- The current number of connected subscribers at this point.
- The current time of the frame transmission for this point.

At initialization of this model, these structures combine into a vector. Thus, the sequential number of structure in the vector imitates the Beacon signal. All other fields of structures are set to zero. Also, the vector of signal levels is forming. This vector simulates a signal power at which every access point radiates, and thereby, the location of subscriber is relative to access points in the moment of its connection. The signal levels are forming according with normal distribution and exponential distribution. This adequately reflects the real situation when connecting subscribers grouped in a small area, for example, when entering a room or in some corner of the room, as it is shown in Figure 7.

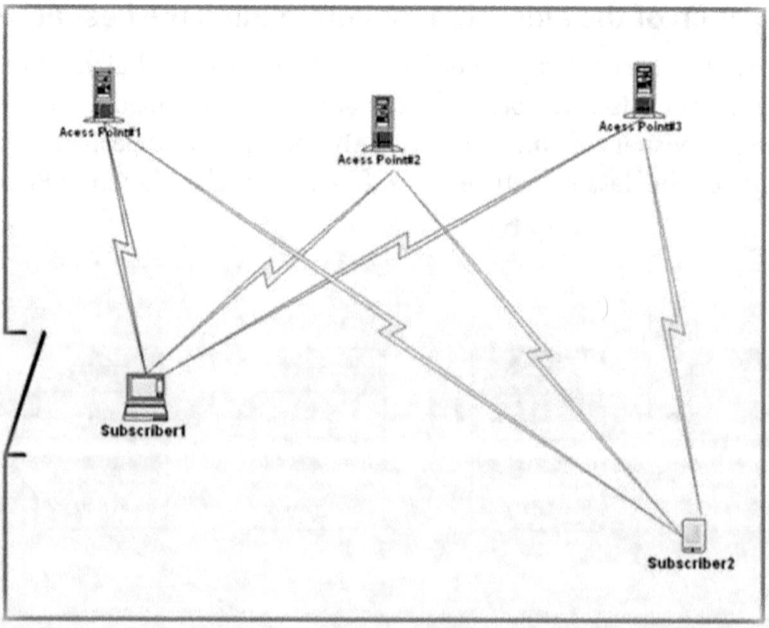

Figure 7. The example of the subscriber's location indoors.

Simulation process starts with the loading of model into the program. Further, the current values of signal level out of the vector of signal levels are moving into the fields of current values of the signal level in every access point. After this, the values of membership functions are calculated for every access point. Further, the values of membership functions are calculated for every access point and their maximum value is looked for. The access point for which this value is maximal, the value of the field current number of connected subscribers to this point is increased by 1. This simulates the RSSI level correction for the next simulation step. Furthermore, the current time of the frame transmission for this point is calculated according with function (5). This process is repeated for all the elements of the vector of signal levels. Thus, after the completion of simulation process, total loading and the frame transmission time in every point would be determined.

The modeling process is based on a method that was considered in Ref. [26]. In this case the sufficient characteristics are current values of congestion of access point and frame transmission time on every simulation step; the final characteristics are the distributions of load of access points, transmission time, mean value, and dispersion of the transmission time. For the confirmation of efficiency of the proposed algorithm, its result was compared with the result of simulation of classic, which is based on the measuring of maximal signal power. The results of the comparison of the analyzed algorithms are shown in Figure 8. In this picture, "Sample Model" means the model is based on the measuring of maximal signal power. This figure shows that the algorithm that uses fuzzy logic (the right-hand side of Figure 8), taking into account the current load on each access point, ensures a more even loading of the entire LAN as a whole. The frame transmission time for the second algorithm is approximately same for all access points (the dispersion is equal to 0.0799). For the first algorithm, the transmission time fluctuates strongly (the dispersion is equal to 0.3129). Hence, it follows that accounting the degree of congestion of access points increases the efficiency of the wireless LAN.

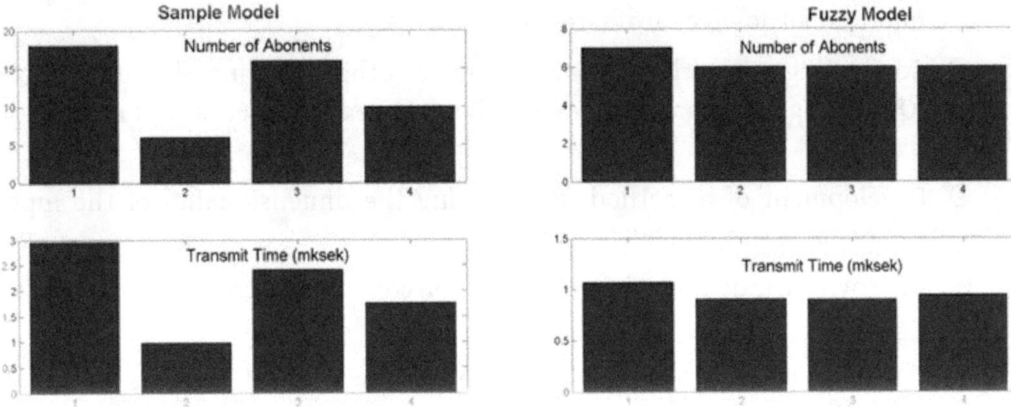

Figure 8. The results of the comparison of the analyzed algorithms.

Developed in the framework of this chapter, MATLAB—the program cannot be used directly in some hardware for the following reasons:

- Developed program is a model.
- MATLAB—program runs in interpreted mode and cannot be loaded directly into the controller.

However, there are various possibilities to convert MatLab-program into equivalent program on one of the high-level programming languages. In particular, there is the possibility of constructing an equivalent program on high level language such as C# or C++. Such program can be installed on the controller or other device that will automatically determine the best access point to connect to the local network.

4. CONCISE SUMMARY

- We have evaluated the sophisticated solutions for indoor navigation and conducted a brief assessment of the program complex for navigation utilizing radio fingerprints. The navigation software package employs the ANN model. Furthermore, the justification for selecting the ANN learning algorithm and determining the optimal number of neurons in the hidden layer for the specified number of Wi-Fi networks was executed using MATLAB. It is advisable to utilize the MAC address of an access point rather than its name, as the latter is not unique. The resultant collision may disrupt the entire system. This strategy offers the benefit that the software package may operate on the current infrastructure of Wi-Fi networks installed in diverse locations, including residential buildings and shopping centers. The drawbacks of this strategy stem from the necessity of manually mapping the radio fingerprint prior to its utilization. We recommend that additional research be conducted in the following areas.

- The implementation multivariate classifier for converting radio prints multidimensional coordinates (*x*, *y*, height).
- Conducting research on the influence of other learning algorithms ANN (Quickprop, Batch, and Incremental) the speed and accuracy of navigation training.
- Development of a method for reducing the dimensionality of the input vector radio fingerprint.
- Improving the quality of navigation when combined with inertial navigation system.
- The addition of processing signals from Bluetooth LE beacons to signals from Wi-Fi access points.

The task of selection of the access point in WLAN network in the case of high user density is quite topical. Such selection must take into account not only the level of the signal received by the mobile device, but also a width in the dedicated channel bandwidth that depends on the number of connected subscribers to the access point.

- Uneven loading of access points in a wireless LAN tends to reduce the overall effectiveness of its work. This is manifested in the fact that the frame transmission time for the overloaded access point may increase significantly.
- Connecting the mobile subscriber to the access point in accordance with the maximum received signal level leads to that some access points may be overloaded.
- Accounting of degree of current workload of access points when a subscriber connects to them provides better balancing of total workload in the network.
- To find the best access point for connecting a subscriber with account of its workload is necessary to solve the optimization task.
- The most effective method for solving this task is to use the apparatus of fuzzy logic.
- The developed model can be used in controllers or in embedded devices by converting into an executable program on high-level programming language.

AT A GLANCE

Indoor mobile navigation systems are increasingly common across various sectors, including transportation, public institutions, and logistics. Interior navigation utilizing access points organized by radio fingerprints is gaining popularity. The artificial neural network (ANN) model is frequently employed as a method for the storage and processing of radio fingerprints. The selection of the access point in a WLAN network under conditions of high user density is a pertinent issue. The selection must consider both the signal strength received by the mobile device and the width of the allocated channel bandwidth. This chapter discusses the primary challenges associated with developing a program complex for mobile interior navigation utilizing neural networks, as well as the strategy for access point selection based on an analysis of not only signal strength but also additional parameters. Fuzzy logic is employed to address this task.

REFERENCES

1. Yuri S. Satellite Navigation Systems. Moscow: ECO Trendz; 2000. p. 267. http://www.indoorsnavi.com/
2. Hatami A, Pahlavan K. A comparative performance evaluation of RSS-based positioning algorithms used in WLAN networks. In: Proceedings of the IEEE Wireless Communications and Networking Conference (WCNC '05); Vol. 4. March; New Orleans, LA, USA; 2005. pp. 2331-2337
3. Kukolj D, Vuckovic M, Pletl S. Indoor location fingerprinting based on data reduction. International Conference on Broadband and Wireless Computing, Communication and Applications (BWCCA); 26-28 October; 2011. pp. 327-332
4. Butakov NA. Applicability inertial navigation in mobile devices. International Journal of Open Information Technologies. 2014;2(5):24-32. ISSN: 2307-8162.
5. Evennou F, Marx F. Advanced integration of wi-fi and inertial navigation systems for indoor mobile positioning. EURASIP Journal on Applied Signal Processing. 2006;2006:1-11. Article ID 86706.
6. Wang S, Min J, Yi BK. Location based services for mobiles: Technologies and standards. LG Electronics Mobile Research, USA. 2008. 122 pp.
7. Dworkina NB, Namiot DE, Dvorkin BA. Mobile navigation services and the use of technology to determine the location OpenCellID. Geomatics. 2010:2;80-87. http://sovzond.ru/upload/iblock/fbf/2010_02_014.pdf.
8. Abdrakhmanova AM, Namiot DE. Using two-dimensional bar code to create a system of positioning and navigation inside. Applied Informatics. 2013; 1(43):31-39

9. Namiot DE, Shneps-Shneppe MA. Analysis of the trajectories in mobile networks on the basis of information about network proximity. Automation and Computer Science. 2013;3:S48-S60

10. Mok E, Cheung Bernard KS. An improved neural network training algorithm for wi-fi fingerprinting positioning. ISPRS International Journal of Geo-Information. 2013;2:854-868

11. Mehmood H, Tripathi NK, Tipdecho T. Indoor positioning system using artificial neural network. Journal of Computer Science. 2010;6(10):1219-1225

12. Fang SH, Lin TN. Indoor location system based on discriminant-adaptive neural network in IEEE 802.11 environments. IEEE Transactions on Neural Networks. 2008;19(11):1973-1978

13. Novikov PA, Khomonenko AD, Yakovlev EL. Justification of the choice of neural networks learning algorithms for indoor mobile positioning. Proceeding CEE-SECR '15 Proceedings of the 11th Central & Eastern European Software Engineering Conference in Russia. Moscow, Russian Federation; 22-24 October 2015; ACM New York, NY, USA©; 2015. Article No. 9. DOI: 10.1145/2855667.2855677

14. Novikov PA, Khomonenko AD, Yakovlev EL. Software for mobile indoor navigation using neural networks. Informatsionno-Upravliaiushchie Sistemy, [Information and Control Systems]. 2016;1:32-39. (In Russian). DOI: 10.15217/issn1684-8853.2016.1.32.

15. Akobir S. The learning algorithm RProp – mathematical apparatus. http://basegroup.ru/community/articles/rprop.

16. Khomonenko AD, Yakovlev EL. Neural network approximation of characteristics of multi-channel non-Markovian queuing systems. SPIIRAS Proceedings. 2015;4(41):81-93. (In Russian). DOI: http://dx.doi.org/10.15622/sp.41.4.

17. IEEE. Wireless LAN medium access control (MAC) and physical layer (PHY) specifications. The Institute of Electrical and Electronics Engineers, IEEE 802.11. 2006.

18. Krasnovidow AV. Model of algorithm to determine a best access point to connect a mobile device to the LAN. Intellectual Technologies on Transport. 2016;2(6):36-42 (In Russian).

19. Larsson D, Merty R. Adaptivniy Podhod k optimizacii proizvoditelnosty besprovodnyh setey/ Dylan Larsson, Ravi Merty//[Adaptive approach to optimize the performance of wireless networks] Technology@Intel Magazine, - 2004, No 8, pp. 24–27.

20. Himmelblaw D. Prikladnoe nelineinoe programmirovanie. [Applied Nonlinear Programming]. Moscow: Mir; 1976. p. 448

21. Elter K-H, Reinhardt R, Schauble M, Donath G. Vvedenie v Nelineinoe Programmirovanie. [Introduction to the Nonlinear Programming]. Moscow: Nauka; 1985. p. 264

22. Leonenko A. Nechetkoie modelirovanie v srede matlab i fuzzyTECH. [Fuzzy Modelling in MATLAB Environment and FuzzyTECH]. Saint-Petersburg: BHV – Petersburg; 2005. p. 736

23. Terano T, Asai K, Sugeno M, editors, Prikladnye nechetkie sistemy. [Applied Fuzzy Systems]. Moscow: Mir; 1993. p. 386

24. Ihmig M. Dynamic channel allocation for self-managing WLAN access points in chaotic wireless networks [Thesis]. Diplomarbeit, Master's Thesis, Technische Universität München, Mar 2006.

25. Leith DJ, Clifford P. A self-managed distributed channel selection algorithm for WLANs. 4th International Symposium on Modelling and Optimization in Mobile, Ad Hoc and Wireless Networks. 2006; pp. 1-9. DOI: 10.1109/WIOPT.2006.1666484.

26. Krasnovidow AV. An approach to the construction of algorithms for the statistical analysis of error flows in digital communications channels. Intellectual Technologies on Transport. 2015;2(2):20-25. (In Russian).

12
Chapter

A FUZZY LOGIC METHOD FOR SEPARATION ASSURANCE AND COLLISION AVOIDANCE IN UNMANNED AERIAL SYSTEMS

1. INTRODUCTION: AN OVERVIEW

Recently, significant advancements have occurred in the capabilities of mobile robots across several aerospace applications. These advancements encompass autonomous intelligence, surveillance and reconnaissance (ISR) initiatives, aerial firefighting, and airborne delivery services. Nonetheless, despite the prospective advantages, these innovations are presently underutilized because to numerous unsolved safety concerns regarding their integration into the National Airspace System (NAS). Consequently, there is a necessity to create algorithms enabling a heterogeneous team of small unmanned aerial systems (sUAS) to autonomously communicate and execute time-sensitive tasks in intricate situations. Given the expanding applications and functionalities of sUAS, it is essential to ensure their safe integration into the National Airspace System (NAS). The majority of efforts in this domain focus on separation assurance, as it generally takes precedence in NAS dispute resolution situations [4, 5]. Nonetheless, the majority of techniques require the exchange of state information among vehicles to effectively determine resolution actions [6]. Numerous intelligent systems have been created for collision avoidance, yielding promising outcomes [7–12], although only a limited number have demonstrated behavioral verification through formal methods [13, 14]. A fuzzy logic method was employed to enable real-time control of several sUAS. This method was employed to reduce the probability of separation losses and, consequently, collisions between the sUAS. A realistic simulation environment was established to generate situations for testing the sUAS's collision avoidance capabilities. This simulation environment was constructed modularly, allowing for the implementation of alternative algorithms to coordinate sUAS operations. This facilitates the testing and evaluation of many vehicle platforms, sensor models, software packages, and traffic management approaches. The primary objective of this research is to formulate a comprehensive operational concept for a UAS traffic management (UTM) system. This system must tackle the issues of collision avoidance and separation assurance. All of these platforms will include fuzzy logic controllers (FLCs) to facilitate real-time decision-making and dynamic control. Moreover, the assurance in accurate decision-making and avoidance control outcomes must be exceptionally strong. Consequently, formal approaches were utilized for behavioral verification. The subsequent sections of the paper are organized as follows. Section 2 delineates the background information regarding various approaches and technologies employed in this study. Section 3 delineates the suggested solution, encompassing the separation assurance and collision avoidance methodologies. This encompasses comprehensive development protocols for the decision-making and fuzzy avoidance controllers. Section 4 delineates the technique for the implementation and assessment of the decision-making and fuzzy avoidance controllers through formal methods. Section 5 delineates the test cases, their implementations, and provides an overview of the simulation environment and its limitations. Section 6 delineates the outcomes of the formal techniques assessments and simulation executions, while Section 7 articulates the findings and prospects for future endeavors.

2. BACKGROUND

2.1. Hybrid fuzzy systems

Most fuzzy inference systems (FISs) encompass several operations that correlate inputs to outputs through many if-then rules, which are ultimately resolved to a singleton. The output space is generally nonlinear and challenging to characterize as a function of the input variables. By imposing specific features and association procedures on the FIS, an explicit statement can be more readily identified. Hybrid systems are characterized by segments of continuous behavior interspersed with discrete transitions [15]. This is comparable to a subset of FISs that has membership functions limited to a finite domain. Such FISs can be characterized as hybrid systems if an explicit expression is derived. This expression associates an input set with an output set through a collection of mathematical functions. This is advantageous because of the numerous low-level tools created for the analysis of hybrid systems. Included in this category are formal techniques tools [16], which are detailed in the subsequent section.

2.2. Formal methods

In systems such as UTM collision avoidance algorithms, the level of confidence that they will always behave as intended needs to be extremely high. Typical methods for evaluating these algorithms usually involve simulation, but simulation and other numerical methods can miss critical cases that result in undesired behavior. To increase the confidence that the avoidance algorithms presented work as intended, formal methods were employed. Formal methods are defined by NASA as "mathematically rigorous techniques and tools for the specification, design, and verification of software and hardware systems." [17] There are numerous types of tools that fall under this definition, but in this work, satisfiability modulo theories (SMT) solvers and model checkers were used.

SMT solvers are tools that extend the Boolean Satisfiability (SAT) problem to first order logic (FOL) sentences and incorporate other theories for evaluating the truth assignments of variables (real values, bitvectors, etc.). If a behavior can be described in FOL, it can be encoded and evaluated by an SMT solver to find truth assignments that violate this behavior. If a behavior is found, it is returned as satisfying the behavioral specification. If there are no possible assignments to the variables that render the specification true, it is then said to be unsatisfiable. Therefore, "safety" properties, properties which should always hold true, can be evaluated by negating a specification that encapsulates the respective behavior. If a satisfying case is found for the negated specification, this means that there are conditions that violate the original specification. If no satisfying cases are found, then the original specification will hold for all possible conditions.

Model checkers are tools that exhaustively check the states of a system to search for combinations of variable assignments that violate behavioral specifications. In finite state systems, they use deductive proofs, and in infinite state systems, they can use inductive methods. These tools can also use SAT or SMT solvers in conjunction with their own search methods for finding counterexample cases. Encoding safety properties in model checkers is slightly different, however, as model checkers typically use some type of temporal operator in conjunction with logical sentences. However, there are methods for relating quantified FOL sentences for safety properties to temporal representations for use in model checkers [18].

In this work, an infinite-state model checker named JKind [19, 20] was used. JKind is a Java implementation of the Kind model checker which uses *k*-induction. To evaluate the truth assignments for variables within each state, JKind employs SMT solvers. The SMT solver used for this work is Z3 [21], a state-of-the-art SMT solver developed by Microsoft.

3. PROPOSED SOLUTION

To ensure that two or more sUAS do not collide with one another, an intruder avoidance system was developed. This avoidance problem is broken down into two sub-systems: strategic (separation assurance) and tactical (collision avoidance).

The strategic separation assurance platform uses a centralized approach to coordinate trajectory modifications for sUAS to ensure that vehicles do not get too close to one another. This separation assurance technique is employed when two or more vehicles come within 0.4 nmi laterally of one another (separation alert threshold), 100 ft. vertically, and are predicted to have a loss of separation (LOS), defined by when vehicles come within 0.1 nmi laterally and 100 ft. vertically of one another. This LOS threshold was determined based on the characteristics of the vehicle platforms and feasible sensing abilities of sUAS. Based on the system constraints, the avoidance platform would have roughly 2 sec to resolve maximum closure rate encounters. If the separation assurance system fails to prevent an LOS, the vehicles will employ their onboard sense and avoid systems to prevent a mid-air collision. A mid-air collision occurs when two vehicles come within 60 m of one another. This collision threshold is intentionally conservative to introduce a notion of spatial uncertainty. Since the sensor models provide perfect state information, as described in Section 5, all vehicle locations are precisely known. For this study, the collision avoidance platform uses a de-centralized approach, that is, all vehicles attempt to avoid intruding vehicles independently. Thus, no communication between aircraft is available (i.e., uncoordinated maneuvers). In Table 1, the various distance thresholds used to describe the separation boundaries are shown.

Table 1. Separation threshold values.

Threshold label	Lateral distance	Vertical distance
Separation alert	0.4 nmi	100 ft.
LOS	0.1 nmi	100 ft.
Collision	60 m	50 ft.

Prior to presenting the details of each avoidance sub-system, Section 3.1 provides an overview of the avoidance system architecture. This overarching logic is used to determine when a vehicle should perform an avoidance maneuver. When deemed necessary, the system will activate the appropriate avoidance platform. In each avoidance platform, a set of heuristics are used to determine the appropriate action to resolve a conflict. These details are presented in Sections 3.2 and 3.3. Once the appropriate action has been decided, an FLC is used to control the vehicle's turn rate in the desired direction. The details of each FLC are shown in Section 3.4.

Finally, it is important to note that these two sub-systems use different approaches when trying to resolve conflicts between aircraft. This is primarily due to the overall purpose each sub-system serves, and the information that is available to each. If vehicles are reporting their state information to a ground-based station, a centralized separation assurance platform could be used to coordinate a trajectory modification to one or more of the vehicles. Thus, coordinated maneuvers are possible. However, when two vehicles are within seconds away from a collision, minimizing the time between sensing the vehicle and performing an action is critical. Therefore, when the collision avoidance system is activated, the vehicles must independently choose the appropriate action using onboard processors. In these collision avoidance scenarios, there is no communication between aircraft. Thus, each sub-system requires a different set of rules to determine the appropriate action.

3.1. Overarching control logic

The overarching control logic determines whether to perform a separation assurance maneuver, activate the collision avoidance system, or allow vehicles to continue along their desired trajectories. This logic is shown in flow chart form in Figure 1. First, the system will find the distance separating all aircraft pairs. With this information, a calculation is made to see how much time can pass prior to two vehicles coming within 0.4 nmi. To calculate this value, the current separation, minus the 0.4 nmi threshold, is divided by the maximum closure rate of the aircraft pair. Therefore, if both vehicles moved directly toward one another at their maximum allowable speeds, this is the time it would take them to reach the 0.4 nmi separation threshold. This future time is known as the "time threshold", as shown in Figure 1. Using this time threshold, if two vehicles cannot possibly be within 0.4 nmi of one another, the system will not unnecessarily check if the two aircraft are in conflict. Rather, it remains idle between checks to improve the performance of the system.

Once enough time has passed and an aircraft pair reaches their assigned time threshold, the system will again check their separation. If the two aircraft are still more than 0.4 nmi apart, a new time threshold is calculated and set. However, if the aircraft pair has reached the 0.4 nmi threshold, it will next check to see if an LOS has occurred. If the vehicles have violated the 0.1 nmi LOS threshold, the collision avoidance system is enabled. Otherwise, the separation assurance system may be needed to ensure that two vehicles do not have an LOS. This decision is based on two criteria: if the predicted closest point of approach (pCPA) creates an LOS and if the time to LOS (tLOS) is within 2 min.

If both criteria are met, a final check is used to see if the aircraft pair has already been assigned a separation assurance maneuver. If neither vehicle has been assigned a maneuver to avoid an impending LOS, a resolution advisory is sent from the centralized system to one of the sUAS. However, if the sUAS was already assigned a maneuver and is currently in the middle of its resolution, a check is used to see if turning back toward its preferred heading will cause another predicted

LOS. If resuming its originally intended mission will not cause an LOS, it will do so, otherwise, the sUAS will continue on its current bearing.

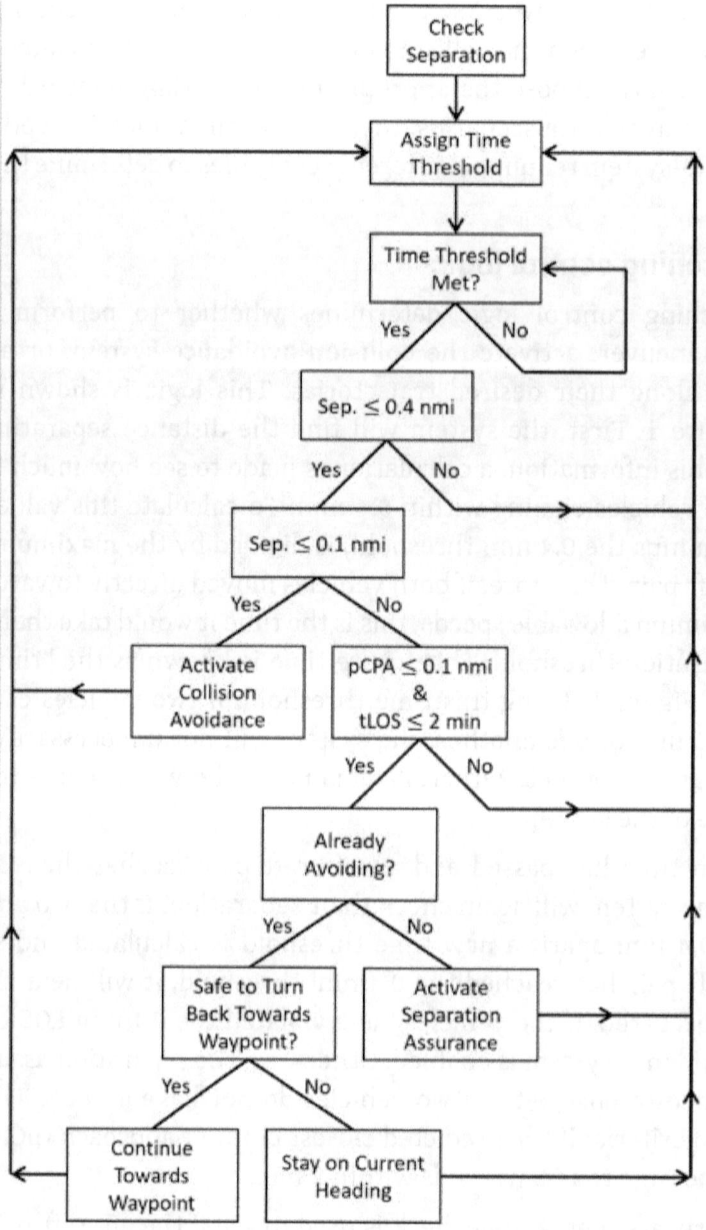

Figure 1. Control logic flow chart.

If neither criterion (pCPA and tLOS) is met, then no separation assurance command is given and the aircraft will continue toward its respective target using its navigation controller. After determining the appropriate separation assurance action, or deciding that no action is needed, the algorithm calculates another time threshold for each aircraft pair.

3.2. Separation assurance logic

The centralized separation assurance system observes only the current position, heading, and velocity of each vehicle. With this limited information, vehicle intent is unknown. Therefore, the system must be robust to dynamic scenarios and resolve conflicts without the knowledge of other vehicles' goals.

The separation assurance platform will be enabled if three criteria are met: separation less than 0.4 nmi, the pCPA is less than 0.1 nmi, and the tLOS is within 2 min. If two aircraft meet all three of these criteria, the separation assurance platform will activate and assign one of the vehicles a new trajectory in an effort to avoid the predicted LOS.

To determine what action the separation assurance platform should use to avoid a potential LOS, a series of conflict classification techniques are used. For this study, three pieces of information are used to classify all conflict scenarios: relative heading, relative angle, and crossing time. These parameters can be described using the more common parameters: location, speed, and heading. In Figure 2, a sample conflict scenario is shown. Here, the triangular objects each represent an sUAS, the arrows represent the velocities of each aircraft (both magnitude and direction), and the "x" represents the heading intersection point location. The heading intersection point is not to be confused with the pCPA. It is simply the point where the projected headings intersect with one another. For this example, the vehicle with the small circle represents aircraft 1, and the other represents aircraft 2.

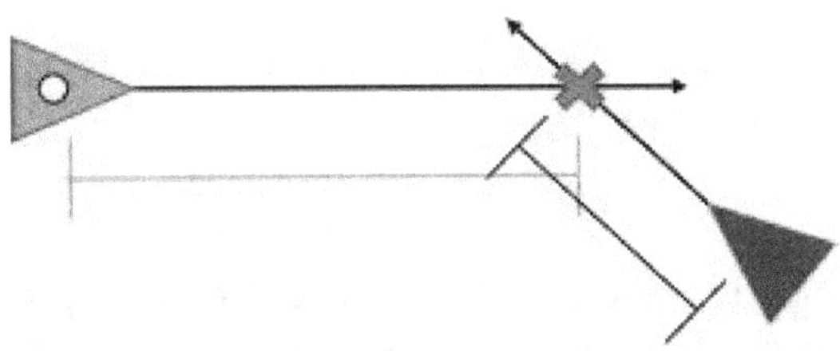

Figure 2. Conflict scenario classification information.

The relationship used to describe the heading of vehicle 2 relative to vehicle 1's perspective is shown in Eq. (1).

$$R_{H_1} = H_2 - H_1 \tag{1}$$

where H_1 is the heading of vehicle 1, H_2 is the heading of vehicle 2, and R_{H1} is the relative heading from vehicle 1's perspective. In this study, $0° \le R_{hi} < 360°$ for all vehicles, where i represents the index for each vehicle. Therefore, if $H_1 > H_2$, a 360° phase shift must be added to R_{H1} to remain within the constrained range. Computing the relative heading for the example shown in Figure 2, R_{H1} would be 140° (i.e., moving to the left with respect to vehicle 1), whereas, from vehicle 2's perspective, R_{H2} would be roughly 220° after applying the 360° phase shift (i.e., moving to the right with respect to vehicle 2).

Similarly, the relative angle between two vehicles describes the relative position of one vehicle with respect to the other. This relationship has been shown in Eq. (2).

$$R_{A_1} = \tan^{-1}\left(\frac{y_2 - y_1}{x_2 - x_1}\right) - H_1 \tag{2}$$

where x_2 and y_2 are the two-dimensional coordinates of vehicle 2 in the global frame, x_1 and y_1 are the two-dimensional coordinates of vehicle 1 in the global frame, H_1 is the heading of vehicle 1, and R_{A1} is the relative angle of vehicle 2 from vehicle 1's perspective. Like the relative heading, the relative angle is constrained. Here, we constrain the relative angle by the following relationship: $-180° \le R_{A1} \le 180°$. Therefore, if the relative angle is less than -180°, a 360° phase shift is added to meet this constraint, or if the angle is greater than 180°, a 360° phase shift is subtracted. Again, using Figure 2 as an example, R_{A1} would be roughly -15° after subtracting a 360° phase shift (i.e., vehicle 2 is to the right of vehicle 1 from vehicle 1's perspective), and R_{A2} would be roughly 25° (i.e., vehicle 1 is on the left of vehicle 2 from vehicle 2's perspective).

Lastly, the crossing time, t_1, can be defined by the relationship shown in Eq. (3). This relationship defines how long it would take for vehicle 1 to reach the heading intersection point, as shown in Figure 2, if they remained on their current trajectory.

$$t_1 = sign\left(\vec{V_1} \cdot \vec{C_1}\right)\frac{c_1}{v_1} \tag{3}$$

where *sign* is a function used to determine the sign (positive or negative) of an expression, $\vec{V_1}$ is the velocity of vehicle 1, $\vec{C_1}$ is the vector used to describe the location of the heading intersection point relative to the vehicle's current position, v_1 is the speed of vehicle 1, and c_1 is the magnitude of the vector $\vec{C_1}$.

With these relationships, all possible encounter scenarios can be described. To aid in understanding what each of these parameters represent, Table 2 describes, in linguistic terms, what each range of values represents in the physical conflict scenarios. Here, the term crossing point is synonymous to the heading intersection point. Each of these descriptions has been listed to describe the motion, location, or crossing time of vehicle 2 from vehicle 1's perspective.

Table 2. Linguistic descriptions of encounter scenarios.

Parameter	Range	Meaning
Relative heading	$0°<R_{H1}<180°$	Moving from right to left
	$180°<R_{H1}<360°$	Moving from left to right
	$R_{H1}=0°$	Same direction
	$R_{H1}=180°$	Head-on
Relative angle	$R_{A1}>0°$	On left
	$R_{A1}<0°$	On right
	$R_{A1}=0°$	Straight ahead
Time to crossing point	$t_1 > 0$	Crossing point in front
	$t_1 < 0$	Crossing point behind
	$t_1 > t_2$	Farther from crossing point
	$t_1 < t_2$	Closer to crossing point

Although the primary goal of this system is to ensure safe separation of vehicles, it is also important to try and limit the number of unnecessary flight adjustments. This is particularly important when operating sUAS due to their typical limitations in power and endurance. Because vehicle intent is unknown in this study, a predicted LOS does not guarantee an LOS is imminent. Therefore, there is a tradeoff between using strict and relaxed criteria when determining if a trajectory modification is necessary. The criteria should be relaxed to ensure sUAS do not repetitively perform unnecessary adjustments but strict enough to ensure safe operation.

Aside from optimizing this time to predicted LOS threshold, a second way to limit the number of vehicles that divert from their desired flight paths is to assign vehicles priority. This priority assignment ranks all vehicles in conflict from highest priority to lowest priority. Therefore, the vehicle with the highest priority will continue on its preferred trajectory without modification. However, all vehicles with a lower priority must avoid all other vehicles with a higher priority.

To determine which aircraft has a lower priority, a series of evaluations are made. First, the system will be checked to see if the two aircraft are moving in a similar direction. If two vehicles have a heading within 5° of one another, that is, $355°R_{H1}<360°$ or $0°\leq R_{H1}\leq 5°$, the trailing aircraft will have lower priority. This encounter scenario can be seen in Figure 3. If several aircraft have similar headings, the aircraft furthest behind will be assigned the lowest priority so must avoid all other aircraft. However, the vehicle in the front of the group will have the highest priority and will disregard all other aircraft.

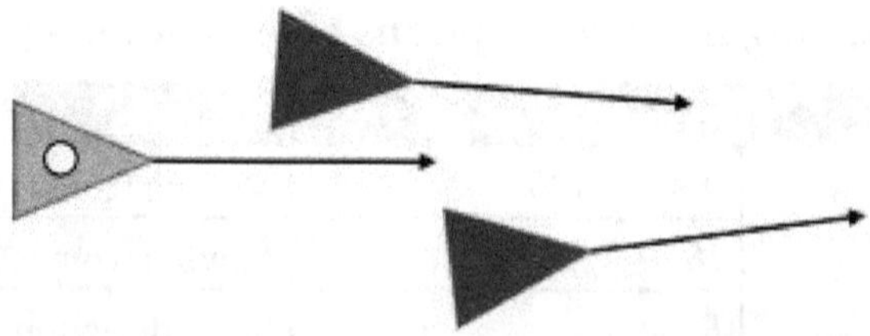

Figure 3. Encounter example with relative heading within 5°.

If the vehicles in conflict do not have similar headings (i.e., more than a 5° difference), the vehicle closest to its next waypoint is given priority. Since the separation assurance system logic does not use the location of a vehicle's next waypoint (i.e., intent is unknown), this priority assignment was simply a means to an end. In practice, the priority of each vehicle in these scenarios would be randomly assigned.

To predict whether an LOS will occur, the separation assurance algorithm uses the current location and velocity of each aircraft to calculate a projected flight path for each. Using these projected trajectories, the pCPA between the aircraft is found. If the pCPA will result in an LOS within the next 2 min, a resolution is calculated and employed to prevent the predicted LOS.

Recalling the sample encounter scenario shown in Figure 2, the definitions described by Eqs. (1) through (3), and the constraints shown in Table 2, all encounter scenarios can be described. Figure 4 depicts all the possible conflict scenarios when vehicle 2 is located to the right of vehicle 1. In each diagram, the aircraft with the small circle represents vehicle 1 and the other is vehicle 2. Thus, all parameters used to describe a particular conflict scenario are from the perspective of vehicle 1. Within the scope of the separation assurance system, vehicle 2 is classified as the vehicle that has been assigned the higher priority. Thus, vehicle 1 (lower priority) must perform a maneuver to prevent an LOS.

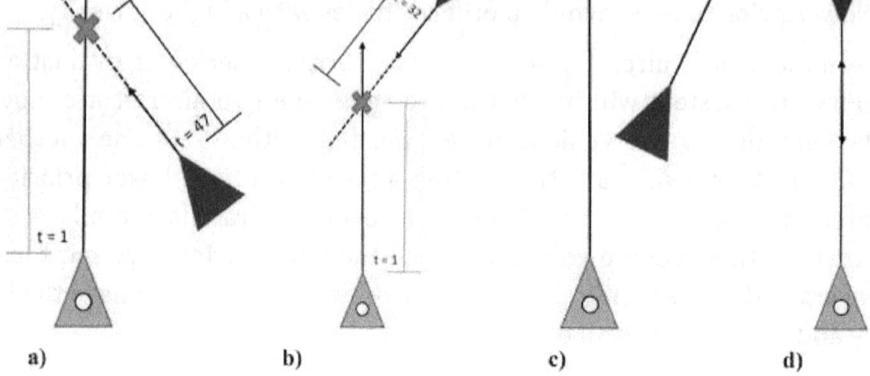

Figure 4. Separation assurance conflict scenario classifications: (a) from behind, (b) coming toward, (c) diverging, and (d) head-on.

In Figure 4(a), vehicle 2 is moving to the left, vehicle 1 is approaching from behind, and vehicle 1 is more than 45 sec closer to the heading intersection point. In this case, vehicle 1 would decide to go in front of vehicle 2. However, if vehicle 1 is not at least 45 sec closer, it will go behind.

In Figure 4(b), vehicle 2 is still going to the left, but in this case, it is coming toward vehicle 1. In these scenarios, vehicle 1 must be more than 30 sec closer to the intersection point to go in front. In Figure 4(c), vehicle 2 is to the right of vehicle 1 but is also going to the right. In these instances, the logic will determine vehicle 1 should turn left to avoid a potential LOS. This also holds for when vehicle 2 is located directly in front of vehicle 1. If, however, vehicle 2 is located on the left of vehicle 1 and going left, it will be instructed to turn right. (NOTE: The 30 and 45 sec buffers were selected after testing a handful of design iterations. Optimizing these buffer thresholds is left to future work.)

In Figure 4(d), vehicle 2's heading is parallel and coming toward vehicle 1. If vehicle 2 is directly in front of, or to the left of, vehicle 1, the logic will instruct vehicle 1 to turn right. However, if vehicle 2 is located to the right of vehicle 1, it will be instructed to turn left.

To prevent an aircraft from prematurely exiting an avoidance maneuver, the system checks if reverting to the navigation controller generates another predicted LOS. Since the avoidance controller has only local sensor knowledge (i.e., $-90° \leq R_{A1} \leq 90°$), switching back to the navigation controller can result in a turning action that generates another predicted LOS. If turning back to its desired target would create another predicted LOS, the avoiding aircraft will continue with its trajectory until authorized to resume its desired mission.

A summary of the separation assurance logic can be found in Table 3. For all cases, the crossing time is strictly positive. That is, the absolute value of the true crossing time found using Eq. (3) is used for all separation assurance logic.

Table 3. Summary of separation assurance logic.

Condition	Relative angle or time to cross	Connector	Relative heading	Action
IF	Any angle within sensor range ($-90° \leq R_{A1} \leq 90°$)	AND	Similar direction ($R_{H1} < 5°$ OR $R_{H1} > 355°$)	Go behind
ElseIF	On right OR straight ahead ($R_{A1} \leq 0°$)	AND	Going right ($180° < R_{H1} < 360°$)	Go behind
ElseIF	On left ($R_{A1} > 0°$)	AND	Going left ($0° < R_{H1} < 180°$)	Go behind
ElseIF	On right ($R_{A1} < 0°$)	AND	Head-on ($R_{H1} = 180°$)	Turn left
ElseIF	On left OR straight ahead ($R_{A1} \geq 0°$)	AND	Head-on ($R_{H1} = 180°$)	Turn right

Condition	Relative angle or time to cross	Connector	Relative heading	Action
ElseIF	I'm more than 45+ seconds closer ($t_1 < (t_2 - 45)$)	AND	Approaching from behind (R_{H1}<90° OR R_{H1}>270°)	Go in front
ElseIF	I'm NOT 45+ seconds closer ($t_1 \geq (t_2 - 45)$)	AND	Approaching from behind (R_{H1}<90° OR R_{H1}>270°)	Go behind
ElseIF	I'm more than 30+ seconds closer ($t_1 < (t_2 - 30)$)	AND	Coming towards (90°≤R_{H1}≤270°)	Go in front
ElseIF	I'm NOT 30+ seconds closer ($t_1 \geq (t_2 - 30)$)	AND	Coming towards (90°≤R_{H1}≤270°)	Go behind

3.3. Collision avoidance logic

When two vehicles have an LOS, are converging, and are within one another's sensor ranges, the collision avoidance system will be activated. In this study, each sUAS will attempt to avoid all intruders within its sensor range; therefore, no vehicle priority will be assigned. Like the approach used for the separation assurance platform to classify conflict scenarios, the collision avoidance system will use the same inputs to decide the appropriate action (i.e., relative angle, relative heading, and time to crossing point). Here, it is important to note that no two vehicles can communicate with one another. Therefore, the same logic is used onboard each system independently. This means that from each vehicle's perspective, we need to ensure that both vehicles will choose complementary actions, that is, the action will not force the vehicles to turn toward one another.

In Figure 5, all possible encounter scenarios are shown. Here, the black triangle and arrow represent the "ownship" (vehicle 1) location and heading respectively, the filled circle represents the "intruder" (vehicle 2) location, the dashed line connecting the two vehicles represents the relative position, and the other two dashed lines represent intruder headings that are either parallel or perpendicular to the ownship's heading. In this figure, the intruder can have any heading between 0° and 360°.

Using these dashed lines to divide the possible intruder heading into cases, the geometry of each encounter scenario can be broken down into twelve cases, provided that two of the three dashed lines do not coincide with one another. In Figure 5(a), the twelve cases are depicted: two cases where the intruder has a parallel relative heading (vertical line), two cases where the headings are perpendicular (horizontal line), two cases where the intruder heading is directly toward or away from the ownship position (line connecting the two vehicles), and all headings that lie in between these angles each count as one case (i.e., six angle ranges in between the lines). If two of the three dashed lines coincide, this possible geometric space reduces to eight possible cases, as shown in Figure 5(b) and c).

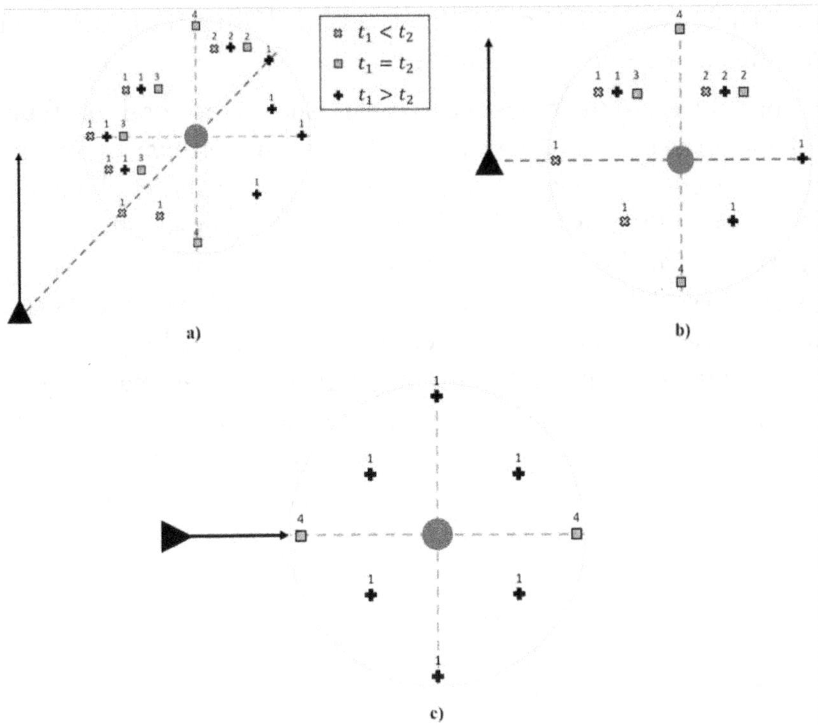

Figure 5. Classification of all possible encounter scenarios: (a) intruder not straight ahead or on side, (b) intruder on side, (c) intruder straight ahead.

Now that the geometric configurations have been defined, let us now introduce the third characteristic, time to heading intersection point. Unlike the separation assurance platform, the time to the heading intersection point, or "crossing time," can be either positive or negative. Therefore, if the crossing point lies behind the vehicle, the crossing time becomes negative. Using the crossing time, there are up to three possible new situations for each of the twelve cases (or eight cases): the times are equal, the ownship crossing time is greater than that of the intruder, or the intruder crossing time is greater than that of the ownship. All possible crossing time scenarios based on the relative heading and angle have been shown. All instances where the ownship can have a crossing time less than the intruder crossing time have been marked by an "x." The black cross represents scenarios where the intruder crossing time can be less than the ownship crossing time. Finally, the square represents the situations where the two vehicles can have equal crossing times.

Using these three designations, all pairwise encounters where an intruder is not directly in front of or beside the ownship can be described by 20 possible cases, as shown in Figure 5(a). In Figure 5(b), cases where the intruder is directly beside the ownship are shown, resulting in 12 possible cases. Lastly, if the intruder is directly in front of the ownship, 8 additional cases can be attained, as shown in Figure 5(c). Thus, a total of 40 cases can be attained using these three parameters to describe the pairwise encounter space.

Knowing that 40 possible cases have been shown, a total of four conflict classifications can be used to solve all possible encounter scenarios. In Figure 5, the number above each icon represents which of the four conflict classifications the scenario belongs. In Figures 6 and 7, the four conflict scenarios have been shown. In each figure, the ownship (vehicle 1) location is marked by a black triangle and its heading is designated by the black arrow. The small circle represents the location of the intruder (vehicle 2). This intruder can have any heading, but the different headings have been separated into different sections, as designated by the shaded regions. In each figure, the shaded regions represent the intruder headings that are excluded by that particular encounter scenario, whereas, the unshaded regions are the possible intruder headings allowed by that encounter scenario. In addition, like Figure 5, there is a dashed line connecting the two vehicles to represent the relative position. Lastly, if a dashed black line is on the boundary of the included and excluded regions, this represents an inclusive boundary (i.e., that heading is included in the possible headings allowed), whereas, the solid black line between the regions represents an exclusive boundary.

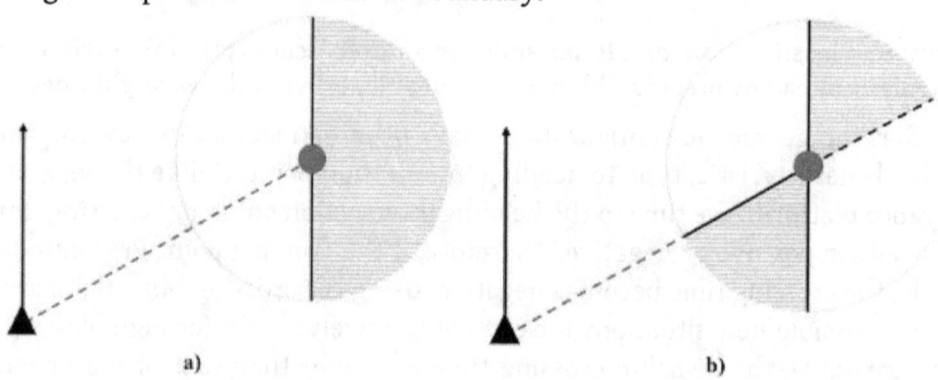

Figure 6. Conflict classification #1 where $(t_1 > 0) \vee (t_2 > 0)$ and: (a) $t_1 < t_2$, (b) $t_1 > t_2$.

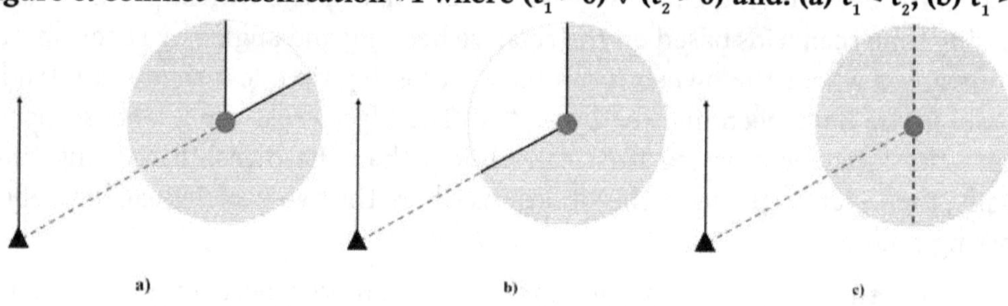

Figure 7. Conflict classifications #2, #3, and #4: (a) $(t_1 < 0) \wedge (t_2 < 0)$, (b) $(t_1 = t_2) \wedge (t_1 \neq \infty) \wedge (t_1 > 0)$, (c) $(t_1 = t_2) \wedge (t_1 = \infty)$.

Figure 6 shows the first conflict classification scenario. Here, at least one of the vehicles must have a positive crossing time (t_i), and they are not equal to one another. As shown in Figure 6(a), the ownship crossing time must be strictly less than that of the intruder. Thus, the intruder crossing time can be negative, or equal to zero (i.e., pointed directly at the ownship). If the intruder heading passes into the excluded region, either both values are negative, or the ownship value must be greater than the intruder value. For all the cases that satisfy this relationship, the ownship would determine to go in front of the intruder. Including the instances where the intruder is to the left of the ownship.

In Figure 6(b), the ownship crossing time must be strictly greater than that of the intruder crossing time. As seen in the figure, the boundary along the relative position line is excluded when the intruder is pointing toward the ownship (i.e., the ownship crossing time is zero, but the intruder crossing time is positive), violating the relationship. However, when the intruder is pointed directly away from the ownship, the boundary is inclusive. In all possible instances, the ownship will determine to go behind the intruder. This includes the cases when the intruder is located to the left of the ownship or is directly in front of the ownship (i.e., t_2 = 0). For all cases represented by conflict classification #1, if one vehicle decides to go in front, the other will decide to go behind, given that they are both within one another's sensor field of view.

In the second conflict classification, as shown in Figure 7(a), both the intruder and the ownship have crossing times less than zero. These negative crossing times, and the fact that the vehicles cannot have the same heading, result in the intruder never sensing the ownship. In these scenarios, if the intruder is on the left, the ownship will turn right. But, if the intruder is on the right, the ownship will turn left.

In Figure 7(b), the scenarios where both the ownship and the intruder have the same time to the crossing point are shown. In this classification case, the intruder cannot have the same heading as the ownship, thus it must be crossing the ownship's path. In addition, this case is restricted to instances where both vehicles have strictly positive crossing times. For these instances, regardless of whether the vehicle is on the left, or on the right, both vehicles will decide to turn right. Since the crossing times are equivalent, and there is no coordination of intent with the other vehicle, both vehicles must choose the same action.

There is one remaining aircraft orientation in the encounter space. This occurs when an intruder is parallel to the ownship, either traveling in the same or opposite direction. This scenario has been shown in Figure 7(c). As seen from this figure, only intruder headings that lie on the dashed line are included. If the intruder is to the right of the ownship, each vehicle will turn to the left (given that the intruder can see the ownship as well). However, if the intruder is directly in front of, or to the left of the ownship, both vehicles will be instructed to turn right.

Although the above classifications describe all possible encounters between moving aircraft, the quad-rotor vehicles can stop and hover. Therefore, a final classification must be described. If an intruder is stationary, the intruder crossing time is set to negative infinity. When this is the case, if the intruder is to the left or directly in front of the ownship, the controller will instruct the vehicle to turn right. However, if the intruder is to the right of the ownship, the logic will instruct the vehicle to turn left.

Table 4 describes the different encounter scenarios using both linguistic and mathematical descriptions, as well as the decided actions. Each of the above conflict classification numbers numerically matches the respective cases in this table. However, the encounter scenario where the intruder vehicle is stationary is referred to as case 0. Like Table 3, the linguistic descriptions in Table 4 represent how the ownship perceives the intruder. Furthermore, all values with the subscript 1 represent the ownship, whereas, all values with the subscript 2 represent the intruder. (NOTE: the * designation indicates that the crossing time is negative infinity for that vehicle.)

Table 4. Summary of collision avoidance logic.

Case	Position	Direction	Time to cross	Crossing point location	Action
0	On left or straight ($0° \le R_{A1} \le 90°$)	N/A	More ($t_1 > t_2{}^*$)	N/A	Turn right
	On right ($-90° \le R_{A1} < 0°$)	N/A	More ($t_1 > t_2{}^*$)	N/A	Turn left
1	Any ($-90° \le R_{A1} \le 90°$)	Not parallel ($R_{H1} \ne \{180, 0\}$)	Less ($t_1 < t_2$)	Not behind both ($t_1 \ge 0$ OR $t_2 \ge 0$)	Go in front
	Any ($-90° \le R_{A1} \le 90°$)	Not parallel ($R_{H1} \ne \{180, 0\}$)	More ($t_1 > t_2$)	Not behind both ($t_1 \ge 0$ OR $t_2 \ge 0$)	Go behind
2	On left ($R_{A1} > 0°$)	Going left away ($R_{H1} < 90°$)	Any	Behind both ($t_1 < 0$ AND $t_2 < 0$)	Turn right
	On right ($R_{A1} < 0°$)	Going right away ($R_{H1} > 270°$)	Any	Behind both ($t_1 < 0$ AND $t_2 < 0$)	Turn left
3	On left ($R_{A1} > 0°$)	Going right ($180° < R_{H1} < 360°$)	Equal ($t_1 = t_2$)	In front of both ($t_1 > 0$ AND $t_2 > 0$)	Turn right
	On right ($R_{A1} < 0°$)	Going left ($0° < R_{H1} < 180°$)	Equal ($t_1 = t_2$)	In front of both ($t_1 > 0$ AND $t_2 > 0$)	Turn right

Case	Position	Direction	Time to cross	Crossing point location	Action
4	On left or straight (R_{A1}≥0°)	Head-on or same (R_{H1}={180,0})	Equal (t_1 = t_2)	N/A	Turn right
	On right (R_{A1}<0°)	Head-on or same (R_{H1}={180, 0})	Equal (t_1 = t_2)	N/A	Turn left

3.4. Fuzzy inference systems

Using the methodology described in Sections 3.2 and 3.3, four possible actions can be used to avoid an intruding sUAS: go behind, go in front, turn right, and turn left. When the command action is either go in front or go behind, the corresponding FIS is activated to execute the maneuver. Therefore, two FISs were developed for this study: go in front and go behind. If a turn left or turn right command is selected, the turn rate of the vehicle will always be a constant value, either positive or negative, depending on which way it should turn. This constant value is one half of the maximum turn rate for collision avoidance maneuvers and one eighth of the maximum turn rate for separation assurance maneuvers.

Both the go in front and go behind fuzzy systems are of Mamdani-type and were constructed in such a way that the input-output relationship can be described using a simple mathematical representation. By using a hybrid representation, as described in Section 2.1, the fuzzy system can easily be expressed mathematically. In this study, the fuzzy systems have a common architecture: triangular membership functions, normalized inputs and outputs, membership function partitioning, product "and" method, minimum implication method, sum aggregation, and mean of maximum defuzzification. If more than one membership function exists for a particular input or output, membership functions are partitioned such that the endpoints of one membership function coincides with the center points of the neighboring membership functions.

Each FIS was developed using a three-input one-output structure. Each FIS uses the distance separating the two aircraft, their relative heading, and their closure rate as inputs to determine the appropriate turn rate output. Since a heterogeneous system is used in this study, the FIS must provide a sufficient turn rate output to avoid a collision for all vehicle type combinations. By considering the separation and closure rate, the conflict can be solved without expelling more energy than required.

In order to use the FISs for both the separation assurance and the collision avoidance platforms, all inputs and outputs must be normalized. Regardless of the avoidance platform being used, the relative heading and closure rate inputs are always normalized by the same values. The relative heading which falls between 0° and 360° is divided by 360°. Thus, a normalized relative heading of 0.5 would represent a head-on encounter. The closure rate is normalized by dividing the true closure rate by the maximum possible closure rate between two vehicles (i.e., 61.762 m/sec in this study). This maximum closure rate would be a result of two fixed wing vehicles approaching one another in a head-on scenario.

The third input, distance, is normalized by 0.4 nmi for separation assurance cases and 0.1 nmi for collision avoidance cases. Lastly, the turn rate output is always between -1 and 1. This output is then scaled by the respective vehicle platform's maximum turn rate. In the case of the collision avoidance system, the output is also multiplied by 1.58. This is to compensate for the fuzzy output not providing a sufficient turn rate command to avoid a collision, especially in head-on scenarios.

Figure 8 shows the structure of each avoidance FIS. Here, the number of membership functions and the corresponding classification can be seen for each input and output. Both the relative heading and the turn rate are partitioned such that the endpoints of the center membership functions coincide with the center points of the adjacent membership functions. As a result of this, and using a product "and" connector, for all possible inputs, at most two of the three rules will be active at one time. This drastically reduces the possible solution space, making it much easier to represent the system mathematically.

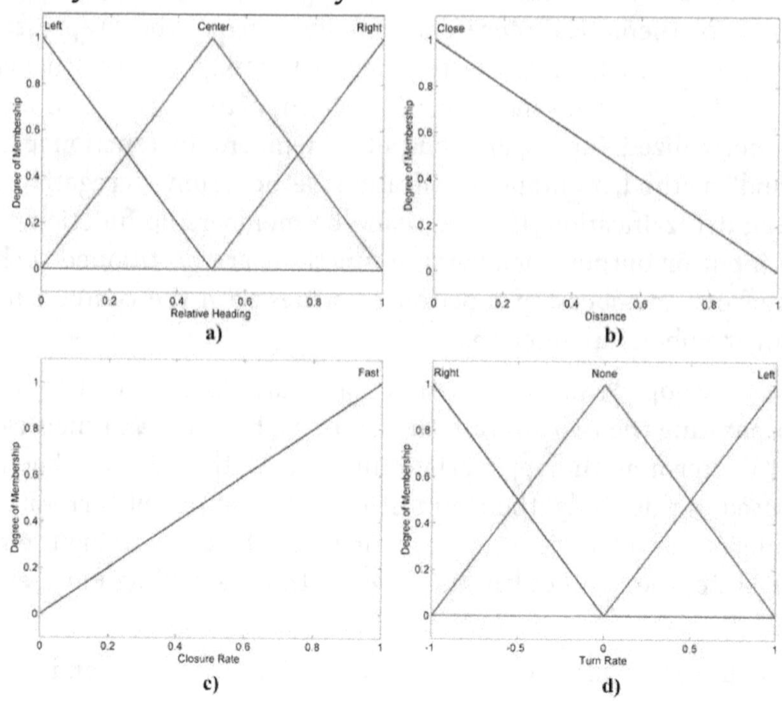

Figure 8. Avoidance FIS structure: (a) Input 1, (b) Input 2, (c) Input 3, (d) Output.

The input and output membership function sets for both the go behind and go in front FISs are identical. However, the rule bases associating the inputs and outputs are opposite, therefore, when one FIS outputs "turn right" (i.e., negative turn rate), the other FIS would output "turn left" (i.e., positive turn rate), and vice versa. In Table 5, the respective rule bases can be seen. Because the other two inputs only have one membership function, they have been excluded from the table.

Table 5. FIS logic.

Rule #	Input 1: relative heading	Go in front output	Go behind output
1	Left	Left	Right
2	Center	Center	Center
3	Right	Right	Left

To map the fuzzy system to a set of nonlinear expressions, the following process was used. First, the system was discretized into three possible modes based on the domain of the ith input, D_i. These modes are described in Table 6. Here, let a square bracket represent an inclusive bound and a round bracket represent an exclusive bound. Based on the structure of the FIS, if input 1 is exactly 0.5 (Mode 2 in Table 6), the output will yield a turn rate of 0. It remains to find the explicit input-to-output mappings for modes 1 and 3.

Table 6. Input domains.

Mode #	D_1	D_2	D_3
1	[0,0.5)	[0,1]	[0,1]
2	[0.5]	[0,1]	[0,1]
3	(0.5,1]	[0,1]	[0,1]

When the FIS is in mode 1, only the "Left" and "Center" membership functions are active. Thus, the output will be independent of rule 3. Therefore, given that a product "and" method and a mean of maximum aggregation technique is used, the following relationship describes how the output is calculated.

$$\dot{\psi} = \left[\left(\prod_{i=1}^{3} h_{i_1}\right) + \left(1 - \prod_{i=1}^{3} h_{i_2}\right)\right]/2 \tag{4}$$

where, Ψ' , is the turn rate, h_{i1} is the degree of membership for the ith input membership function when using rule 1 (i.e., left, close, fast membership functions), and hi2$hi2$ is the degree of membership for each input when using rule 2. When substituting the equation of each membership function into Eq. (4), the expression shown in Eq. (5) is found, which can be reduced to the polynomial shown in Eq. (6).

$$\dot{\Psi} = \{[(-2\mu_1+1)(-\mu_2+1)(\mu_3)] + [1-(2\mu_1)(-\mu_2+1)(\mu_3)]\}/2 \tag{5}$$

$$\dot{\Psi} = (4\mu_1\mu_2\mu_3 - 4\mu_1\mu_3 - \mu_2\mu_3 + \mu_3 + 1)/2 \tag{6}$$

where μ_1, μ_2, and μ_3 are inputs 1, 2, and 3, respectively. The polynomial shown in Eq. (6) can be used to map any combination of inputs that belong to D_1 to an output for the go in front FIS.

Using the same methodology for mode 3, Eqs. (7) through (9) can be found.

$$\dot{\Psi} = \left[\left(-\prod_{i=1}^{3} h_{i_3}\right) + \left(\prod_{i=1}^{3} h_{i_2} - 1\right)\right]/2 \tag{7}$$

$$\dot{\Psi} = \{[-(2\mu_1+1)(-\mu_2+1)(\mu_3)] + [(-2\mu_1+2)(-\mu_2+1)(\mu_3) - 1]\}/2 \tag{8}$$

$$\dot{\Psi} = (4\mu_1\mu_2\mu_3 - 4\mu_1\mu_3 - 3\mu_2\mu_3 + 3\mu_3 - 1)/2 \tag{9}$$

The polynomial shown in Eq. (9) will map any combination of inputs belonging to D_3 to an output for the go in front FIS.

This same approach was used to map any combination of inputs for the go behind FIS to an output using a polynomial function. Since the rule bases are exactly opposite to one another, the output is the negation of Eqs. (6) and (9). Eqs. (10) and (11) describe the input-output relationships for modes 1 and 3, respectively, for the go behind FIS.

$$\dot{\Psi} = (-4\mu_1\mu_2\mu_3 + 4\mu_1\mu_3 + \mu_2\mu_3 - \mu_3 - 1)/2 \tag{10}$$

$$\dot{\Psi} = (-4\mu_1\mu_2\mu_3 + 4\mu_1\mu_3 + 3\mu_2\mu_3 - 3\mu_3 + 1)/2 \tag{11}$$

It is important to note that if three or more vehicles are in conflict, each pair of vehicles will be evaluated separately. Thus, for a single ownship, several turn rate outputs will be obtained. Once all respective turn rates are calculated for each intruder, the values are averaged into a single value. This mean turn rate serves as the final vehicle turn rate command.

4. AVOIDANCE ALGORITHM VERIFICATION

To verify that the avoidance algorithm performed as intended, two levels of avoidance control were evaluated. The first was the high-level, decision-making logic that determined which action a vehicle would take during potential collision scenarios as shown in Table 4. The second that was evaluated was the low-level logic in the avoidance FLCs. These FLCs determine the actual vehicle turn rate output after being selected by the decision-making heuristics. For each of these levels, specifications about their behavior were developed and then translated into FOL sentences that could be evaluated by an SMT solver. For the first cases that deal with the avoidance decision-making logic, the specifications and system model were implemented in JKind, with Z3 being used as the SMT solver. The evaluation of the avoidance FISs was performed directly in Z3. The difference was purely a practical one, as the language JKind uses, Lustre [22], is more conducive to easily create a more detailed environment model.

4.1. Collision avoidance decision-making logic

As shown in Table 4, there are several conditions for which there are different output actions for avoidance. The specifications were first translated into FOL sentences. The sentences are shown in the following equations. Note that *RAL* is a predicate that acts on the relative angle limit to represent the vehicle's sensors detecting an intruder. Since they can only sense intruders between ±90°, the relative angle values were limited in the specifications such that being in that range implied that the output action would be the correct one. If the intruder is outside of this range, the specifications do not say what the output action should be. Note that the variables *out*1 and *out*2 are the output actions for the two vehicles. Since both vehicles pick actions based on the states and have no knowledge of the other vehicle's selected action, the specifications need to go both ways. Also, the actions are represented by integer values 1–4 such that {1,2,3,4} = {go behind, go in front, turn left, turn right}, respectively.

Although several avoidance actions could be taken by each vehicle, only cases 1, 3, and 4 required formal verification. In these cases, both vehicles can sense one another and will perform some type of avoidance maneuver. Thus, their actions must be such that they do not move closer to one another. In cases 0 and 2, however, there was no need to write requirements because only one vehicle performs an avoidance action. In case 0, vehicle 2 is stationary, so it does not try to avoid vehicle 1. In case 2, vehicle 2 cannot sense vehicle 1, thus, there is no need to check that the two vehicles' actions produce a converging result.

Eqs. (12) through (15) are the specifications for case 1 in Table 4 and are intended to ensure that the vehicles do not both choose to go in front, or go behind.

$$S_{1_1} = \forall R_{A_1} \forall R_{A_2} \forall R_{H_1} \forall R_{H_2} (RAL \rightarrow (out1 = 1 \rightarrow out2 = 2)) \tag{12}$$

$$S_{1_2} = \forall R_{A_1} \forall R_{A_2} \forall R_{H_1} \forall R_{H_2} (RAL \rightarrow (out1 = 2 \rightarrow out2 = 1)) \tag{13}$$

$$S_{1_3} = \forall R_{A_1} \forall R_{A_2} \forall R_{H_1} \forall R_{H_2} (RAL \rightarrow (out2 = 1 \rightarrow out1 = 2)) \tag{14}$$

$$S_{1_4} = \forall R_{A_1} \forall R_{A_2} \forall R_{H_1} \forall R_{H_2} (RAL \rightarrow (out2 = 2 \rightarrow out1 = 1)) \tag{15}$$

Eqs. (16) and (17) are the specifications for case 3 and ensure that the vehicles turn the same way when resolving these particular conflicts. Recall that in this case, both vehicles are at the same time from the crossing point. This specification ensures that they will then be forced into new positions such that the crossing times are not equal and the FISs are then selected.

$$S_{3_1} = \forall R_{A_1} \forall R_{A_2} \forall R_{H_1} \forall R_{H_2} (RAL \rightarrow ((t_1 = t_2 \wedge t_1 > 0 \wedge t_2 > 0 \wedge ((R_{H_1} \geq 0 \wedge R_{H_1} < 180) \vee (R_{H_1} > 180 \wedge R_{H_1} < 360))) \rightarrow (out1 = 4 \wedge out2 = 4))) \tag{16}$$

$$S_{3_2} = \forall R_{A_1} \forall R_{A_2} \forall R_{H_1} \forall R_{H_2} (RAL \rightarrow ((t_1 = t_2 \wedge t_1 > 0 \wedge t_2 > 0 \wedge ((R_{H_2} \geq 0 \wedge R_{H_2} < 180) \vee (R_{H_2} > 180 \wedge R_{H_2} < 360))) \rightarrow (out2 = 4 \wedge out1 = 4))) \tag{17}$$

Finally, Eqs. (18) and (19) are for case 4. These two specifications are for cases where the vehicles are head-on, or traveling next to each other in the same direction, respectively. The first specification ensures that while in a head-on encounter, both vehicles turn the same direction (i.e., both left, or both right, forcing them to diverge). The second specification ensures that while traveling in the same direction, the vehicles turn in opposite directions.

$$S_{4_1} = \forall R_{A_1} \forall R_{A_2} \forall R_{H_1} \forall R_{H_2} (RAL \rightarrow ((t_1 = t_2 \wedge (R_{H_1} = 180 \wedge R_{H_2} = 180)) \rightarrow ((out1 = 3 \wedge out2 = 3) \vee (out1 = 4 \wedge out2 = 4)))) \tag{18}$$

$$S_{4_2} = \forall R_{A_1} \forall R_{A_2} \forall R_{H_1} \forall R_{H_2} (RAL \rightarrow ((t_1 = t_2 \wedge (R_{H_1} = 0 \wedge R_{H_2} = 0)) \rightarrow ((out1 = 3 \wedge out2 = 4) \vee (out1 = 4 \wedge out2 = 3)))) \tag{19}$$

where t_1 is the time until the crossing point, R_{H1} is the relative heading from vehicle 1 to vehicle 2, and R_{A1} is the relative angle from vehicle 1 to vehicle 2. Similarly, t_2, R_{H2} , and R_{A2} are for vehicle 2. In addition to these specifications, another specification was created to ensure that the vehicles always chose one of the valid actions. This specification is not shown but is similar to Eqs. (12) through (19) such that the output actions are one of the possible outcomes (1–4). These specifications were then translated to a temporal representation for evaluation in

JKind using previously developed methods [18].

4.2. Avoidance fuzzy inference systems

The method for verifying the avoidance FISs is similar, but the specifications were left in FOL and then implemented directly into Z3. The main reason for this was that less of the system model was needed to check these outputs. The behavior that the specifications needed to encapsulate was that the turn rate output for each FIS needs to always be in the correct direction. The correct direction for each of these cases is shown in Table 7.

Table 7. Avoidance FIS outputs.

Mode	Action	Intruder position	Intruder direction	Turn direction
1	Go behind	On right	Going left	Right (negative)
3	Go behind	On left	Going right	Left (positive)
1	Go in front	On right	Going left	Left (positive)
3	Go in front	On left	Going right	Right (negative)

These can then be encoded in FOL sentences using the polynomial representation of the FISs shown in Section 3.4. These sentences are then negated to show that there are no possible variable values that make the negated sentences true. The negated FOL sentences are shown in Eqs. (20) through (23). Note that μ_1, μ_2, μ_3, 'Ψ' , and modes 1 and 3 are the same as detailed in Section 3.4.

$$S_{behind_1} = \exists\mu_1\exists\mu_2\exists\mu_3(\dot{\Psi}(\mu_1,\mu_2,\mu_3)\geq 0) \tag{20}$$

$$S_{behind_3} = \exists\mu_1\exists\mu_2\exists\mu_3(\dot{\Psi}(\mu_1,\mu_2,\mu_3)\leq 0) \tag{21}$$

$$S_{front_1} = \exists\mu_1\exists\mu_2\exists\mu_3(\dot{\Psi}(\mu_1,\mu_2,\mu_3)\leq 0) \tag{22}$$

$$S_{front_3} = \exists\mu_1\exists\mu_2\exists\mu_3(\dot{\Psi}(\mu_1,\mu_2,\mu_3)\geq 0) \tag{23}$$

These negated sentences were then implemented in Z3. If Z3 finds that these sentences were all unsatisfiable, there are no possible real-valued assignments to μ_1, μ_2, or μ_3 that allow the FISs to turn the vehicles in an undesired direction.

5. SIMULATION ENVIRONMENT DESCRIPTIONS

5.1. Testing environment

Prior to integrating the controllers presented in Section 3 into the full simulation environment, each avoidance platform was tested using pairwise encounter scenarios. This component testing was used to ensure that each was operating as desired. In this section, the methods used to test each controller are described.

5.1.1. Separation assurance

A testing environment was created to evaluate a considerable amount of pairwise encounters between aircraft. This testing environment was used to identify potential controller failures. To accomplish this, various initial relative headings and relative angles were tested. In all cases, the initial location and heading of one aircraft was held constant. Then, by placing the intruding vehicle at a relative angle of -90° and just outside the vehicle's sensing radius, as shown in Figure 9, we evaluated the interactions for 720 initial intruder heading values. This was then repeated four more times by changing the initial relative angle between -90° (to the right) and 0° (straight ahead). A visualization of these scenarios is shown in Figure 9. Note that the radius of the sensing semi-circle for the separation assurance tests was set at 0.4 nmi.

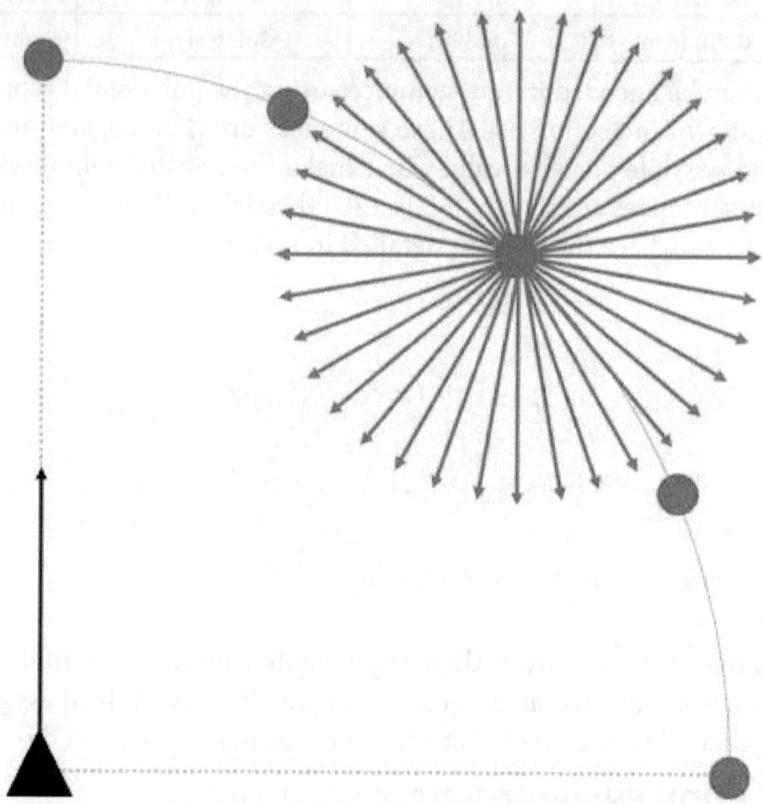

Figure 9. Separation and avoidance testing scenarios between ownship (triangle) and intruder UAS for multiple different initial positions (circles).

Although intruders can approach a vehicle from the left, that is, a relative angle between 0° and 90°, symmetry allows us to limit the initial relative angles to lie between -90 and 0°. To verify, a sample scenario was tested for the full field of view. In all cases, the trajectories of the vehicles were symmetric to one another (i.e., when reflected across the vehicle's initial heading).

5.1.2. Collision avoidance

The collision avoidance platform was tested in the same manner as the separation assurance. The difference being that the sensing radius for the interaction tests as shown in Figure 9 is 0.1 nmi. Thus, the initial position of each intruder scenario was just outside of this sensing radius.

5.2. Full simulation environment

5.2.1. Airspace description

The simulation environment created for this study models a portion of the US airspace over central Ohio. A depiction of the selected airspace can be seen in Figure 10. This airspace covers approximately 2500 square miles where sUAS can operate at a maximum altitude of 400 ft.

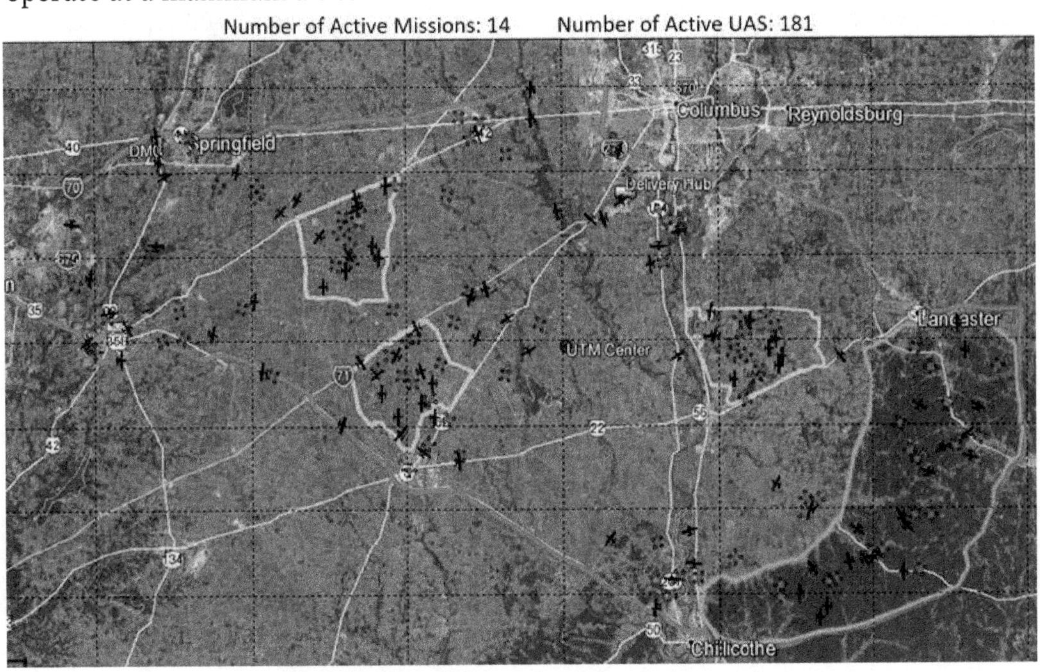

Figure 10. Simulation platform example.

5.2.2. Mission types and objectives

Many sUAS will be active throughout this airspace, each with individual missions, such as, precision agriculture, forest monitoring, roadway surveillance, disaster management, and package delivery. During simulation runs, the UAS will travel to various waypoints to fulfill their assigned missions. After visiting all waypoints for a given mission, the aircraft will return to their respective starting locations. For more on the mission types, please refer to Refs. [11, 12].

5.2.3. Scalability and operations

The full simulation environment can accommodate any number of sUAS and missions. However, these numbers were constant throughout testing. A maximum of 184 sUAS can be airborne at any given time. This is a result of having centralized control of the aircraft. In a realistic environment, some of the computationally heavy components can be handled in parallel onboard each individual sUAS.

5.2.4. Aircraft models

Two vehicle platforms were used in all simulations: fixed wing and quad-rotor. Kinematic models were developed for each within the constraints of the environment. These constraints include maximum turn rate, maximum climb/decent rate, minimum/maximum speed, and maximum altitude.

For this study, each aircraft is assumed to climb at 4 ft./sec upon takeoff. The vehicles will continue to climb at this rate until an altitude of 400 ft. is reached. When vehicles are in conflict and are required to make trajectory modifications, all adjustments are constrained to lateral deviations in flight path (i.e., speed and altitude modifications are not used). Therefore, UAS are limited to level, two-dimensional flight during conflict resolution scenarios. Using this assumption, the maximum turn rate for each fixed wing vehicle is described in Eq. (24).

$$\dot{\psi}_{max} = \frac{g\sqrt{n^2 - 1}}{V} \tag{24}$$

All fixed wing vehicles travel at 60 knots and have a maximum load factor of 3.5. Therefore, they are constrained to a maximum turn rate of 61.06 deg/sec. The multi-rotor systems, however, can travel at a maximum airspeed of 38 knots. Due to the nature of the quad-rotor sUAS, Eq. (24) does not accurately model the maximum turn rate for this vehicle type and it is assumed that the aircraft can yaw at a maximum rate of 45 deg/sec. Also, each aircraft is assumed to have the capability to detect an intruding aircraft at a distance of 0.1 nmi if within a 180° field of view in front of the aircraft (i.e., $-90° \leq R_{A1} \leq 90°$).

As previously described in Table 1, the collision threshold is defined by aircraft coming within 60 m laterally and 50 ft. vertically of one other. To track the vehicles, ground-based sensors for detecting aircraft have been dispersed in the airspace. These sensors, along with telemetry data, provide continuous and reliable vehicle state information (i.e., position and velocity). This capability allows the implementation of global separation assurance practices. More details about the simulation environment and its properties can be found in Refs. [11, 12]. (NOTE:

The collision buffer value was used to accommodate for position uncertainties while tracking the sUAS. In this study, the avoidance system has perfect knowledge of all vehicle state information acquired from ground based and onboard sensors.)

6. RESULTS

6.1. Avoidance systems testing

For each pairwise encounter shown in Figure 9, the closest point of approach (CPA) was found. The CPA is defined as the minimum recorded distance between the two vehicles throughout the entire encounter. For both the collision avoidance and separation assurance platforms, a total of four different vehicle platform trials were conducted: fixed vs. fixed, quad vs. quad, quad vs. fixed, and fixed vs. quad. The first vehicle type designation represents the ownship vehicle's type in Figure 9 (i.e., has the same starting position and heading for all tested cases), whereas, the second vehicle platform designation represents the intruder vehicle's type (i.e., initial conditions change for each tested scenario).

To measure the effectiveness of the avoidance logic, the minimum CPA was recorded for all initial relative angles tested. In addition, the total number of collisions for each case were tallied, and for the separation assurance case, the total number of LOSs. As a reminder, a collision is deemed by two vehicles coming within 60 m of one another, and an LOS is when two vehicles come with 0.1 nmi of one another.

The results for the separation assurance testing have been shown in Table 8. Here, the "angle case" refers to the various initial relative angles, from -90 to 0°, respectively. Thus, case 1 represents an intruder directly to the right, and case 5 represents an intruder directly in front of the ownship.

Table 8. Separation assurance testing results.

Angle case	Fixed vs. Fixed		Quad vs. Quad		Fixed vs. Quad		Quad vs. Fixed	
	Min CPA (m)	LOS	Min CPA (m)	LOS	Min CPA (m)	LOS	Min CPA (m)	LOS
1	185.3	0	185.1	0	574.3	0	185.8	0
2	185.6	0	185.5	0	351.4	0	158.6	12
3	185.3	0	185.4	0	185.6	0	133.2	24
4	186.1	0	131.2	9	175.1	24	137.3	31
5	98.4	116	137.0	8	128.1	161	132.0	28

Although all vehicle platform combinations had more than one LOS, no LOS resulted in two vehicles colliding (i.e., a CPA less than 60 m). To try and understand where the separation assurance platform was breaking down to allow an LOS to occur, the CPA results were plotted for each initial intruder heading. When evaluating the results of each vehicle case, it was found that although several LOSs occurred, the failures tended to lie in groups near the same intruder angle depending on the initial position of the intruder. For example, in the fixed vs. quad case, it can be seen that no LOSs occurred in the first three trials. However, once the intruder was positioned such that it was nearly in front of or directly in front of the ownship, the separation assurance platform began to fail. This was caused by the fixed wing vehicle traveling at higher speeds than the quad-rotor vehicle. Therefore, when the intruder heading was away from the ownship, the ownship tended to approach the intruder from behind and begin to pass the vehicle. When approaching the vehicle from behind, it proved difficult for the avoidance platform to solve the conflict prior to an LOS in all vehicle configurations, as shown by angle case 5.

Table 9 shows the results of the collision avoidance testing. Each case had a minimum CPA greater than 60 m, implying no collisions were found throughout testing. Although not all possible encounter scenarios have been tested, this shows that the avoidance system logic is quite robust. As seen from the results, the homogeneous quad-rotor and the fixed vs. quad cases showed promising results. They consistently had a higher minimum CPA than the other two cases. The closure rates in the fixed vs. fixed cases were higher than cases involving quad-rotors. This resulted in consistently lower minimum CPA values. A noteworthy result is in the fixed vs. fixed scenario for angle case 5. The intruder being head-on and directly in front of the ownship resulted in a CPA of 60.8 m. Although this is close to the collision boundary, this shows that even in the highest closure rate scenario, the collision avoidance system was able to resolve the conflict.

Table 9. Collision avoidance testing results.

Angle Case	Fixed vs. Fixed Min CPA (m)	Quad vs. Quad Min CPA (m)	Fixed vs. Quad Min CPA (m)	Quad vs. Fixed Min CPA (m)
1	132.1	138.9	154.7	125.3
2	111.9	125.2	133.3	121.0
3	101.9	113.5	122.9	111.1
4	82.9	104.4	103.5	97.8
5	60.8	96.7	98.9	99.0

6.2. Formal verification

6.2.1. Avoidance logic

After evaluating all the specifications outlined in Eqs. (12) through (19), JKind returned that they always held. This means that for all possible real-valued assignments to the variables (within the sensor domain limitations), the vehicles will always select the desired output action.

Although the final version of the avoidance logic adhered to all of the specifications, during development there were several cases where JKind found values that violated one or more of the specifications. These counterexamples are invaluable as they identify exact cases that result in undesired behavior. This gives way to corrections based on the counterexample conditions.

As an example of this, one of the conditions that violated a specification found during development is shown in Table 10.

Table 10. Avoidance logic counter-example values.

Specification	R_{H1}	R_{A1}	$out1$	R_{H2}	R_{A2}	$out2$
S_{32}	180	$-5 \cdot 10^{-15}$	3	180	$-5 \cdot 10^{-15}$	4

These conditions mean that one vehicle is heading in the exact opposite direction of the other and there is a slight position offset between them as shown in Figure 11. One vehicle selects the turn left action (3) while the other selects the turn right action (4). This implies they are not turning away from each other. The reason for this was that the range of angles that would force vehicle 1 into the correct action was not inclusive on one of its boundaries. This then meant that the conditions forced vehicle 1 into a different action and generated this counterexample.

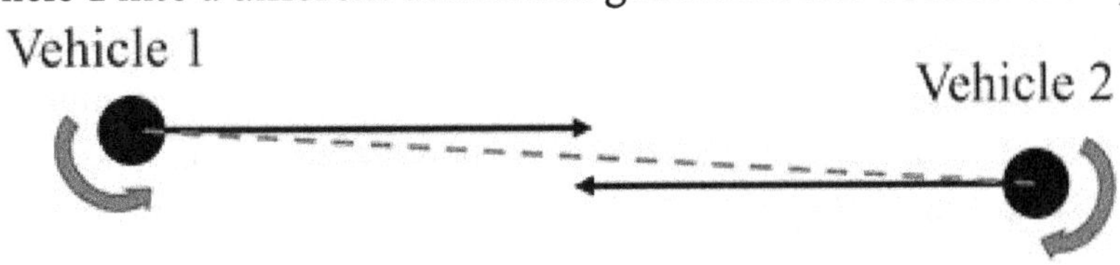

Figure 11. Counterexample showing head-on vehicles turning into each other.

6.2.2. Avoidance FISs

Similarly, after evaluating the specifications in Eqs. (20) through (23), Z3 showed that they were all unsatisfiable. This shows that the FLCs will always make the ownship turn away from an intruder.

6.3. Full simulation results

To test the algorithms in a dynamic environment, simulations were run both with the separation assurance mitigations and then without. The number of LOSs and number of collisions were recorded in order to directly compare the mitigated and unmitigated cases.

6.3.1. Unmitigated study

For this unmitigated study, the separation assurance system was disabled. The results of this study are shown in Table 11.

Table 11. Results for simulation without separation assurance.

Separation less than 0.4 nmi	LOSs	Collision avoidance maneuvers	Collisions	Collision avoidance success rate	Number of flight hours	LOSs per flight hour	Collisions per flight hour
26,576	8263	8252	2	99.98%	25,116	0.33	$7.96 \cdot 10^{-15}$

Over the span of 25,116 flight hours, there were 26,576 recorded violations of the 0.4 nmi separation threshold. These resulted in 8263 LOSs. The collision avoidance algorithm was employed for all except for 11 LOS occurrences. In those cases, the vehicles were outside of one another's field of view, thus the collision avoidance system was not used. The collision avoidance system was 99.98% successful at resolving conflicts.

The only collisions that occurred throughout simulation can be attributed to the restriction on the detection sensor field of view and having no memory of state time histories. Therefore, if two vehicles were nearly parallel and directly beside one another, they would turn to resolve the conflict (i.e., turn away from one another). However, this turning again puts each intruder outside of the other vehicle's field of view. The lack of state memory combined with no sensor input caused a switch back to their navigation controllers. The navigation controller caused them to go back toward one another. Since the navigation controller had a higher turn rate output than the avoidance output, this cycle would continue (each vehicle turning away then toward) until they converged and were within 60 m of one another.

6.3.2. Mitigated study

For this study, the separation assurance features were enabled to help mitigate the risk of having an LOS. Although fewer LOSs were expected with the mitigations enabled, some LOSs were expected due to sub-optimal performance in head-on and trailing situations. The results of this mitigated study are shown in Tables 12 and 13.

Table 12. Separation assurance results.

Separation less than 0.4 nmi	Separation assurance maneuvers	LOSs	Separation assurance success rate	Number of flight hours	LOSs per flight hour
33,550	14,750	3579	75.74%	9277	0.39

Table 12 presents the results of the separation assurance platform. In this mitigated study, only 9277 flight hours were recorded. Thus, sUAS were able to complete their respective missions in a shorter period of time. Throughout the simulation aircraft came within 0.4 nmi on 33,550 occasions. However, of those instances, the separation assurance system predicted an LOS to occur within 2 min only 14,750 times. Of these resolution advisories, only 75.74% were successful, resulting in 3579 LOSs and 0.39 LOSs per flight hour. Although this number is slightly larger than the number of LOSs per flight hour in the unmitigated study, this additional layer of avoidance kept vehicles from entering any scenarios that resulted in collision. Thus, the overall safety of the UTM system has been improved.

If an LOS occurred and the vehicles were within one another's sensor ranges, the sense and avoid software would activate. The results of the collision avoidance software can be seen in Table 13. Of the 3568 encounters, the collision avoidance software was 100.00% successful at resolving conflicts.

Table 13. Collision avoidance results.

Collision avoidance maneuvers	Collisions	Collision avoidance success rate	Number of flight hours	Collisions per flight hour
3568	0	100.00%	9277	0

7. CONCISE SUMMARY

This study employed numerous fuzzy logic controllers and decision-making systems collaboratively to avert potential loss of separation in a congested three-dimensional airspace. This simulation environment facilitated comprehensive encounter scenarios among diverse vehicles to evaluate the two dispute resolution technologies. A sense-and-avoid system was created to avert probable collisions by utilizing only current state information, without inter-vehicle communication. A separation assurance platform was subsequently built to further reduce the risk of a potential collision. This platform utilizes worldwide aircraft state data to forecast if two aircraft will experience a loss of separation during a specified look-ahead period. If a loss of separation (LOS) was anticipated, the system would provide requisite resolution advisories to the appropriate aircraft to avert a LOS. Upon the development of the controllers, numerical simulations and formal approaches were employed to validate their projected performance. By employing a formal techniques approach, we could demonstrate that the controller output consistently aligned with the intended direction. Furthermore, we confirmed that in all pairwise encounter situations involving sUAS, the maneuvers of each vehicle were executed in a manner that precluded any turning towards one another during collision avoidance. Following the verification of the control logic behavior and fuzzy logic controller outputs through a formal methods approach, numerical simulations were performed. In all simulations, the avoidance system possessed complete awareness of all vehicle state variables (i.e., velocity, orientation, and position). The fuzzy system effectively resolved all potential conflicts in the studied collision avoidance scenarios for both homogeneous and heterogeneous cases. Nonetheless, the separation assurance platform had difficulties in addressing specific sorts of encounter scenarios. Consequently, it occasionally fails to avert a line of sight obstruction between vehicles. Nevertheless, when a loss of separation transpired, the collision avoidance system once more averted any mid-air crashes. A number of comprehensive simulation environment missions were conducted to assess the efficacy of the avoidance algorithms. These missions encompassed scenarios in which the separation assurance mitigations were either activated or deactivated. The experiment's results aligned with expectations. No collisions between planes transpired in the mitigation research. Upon the removal of mitigations, vehicles faced situations in which the collision avoidance system failed to avert a collision. The collisions were not caused by the collision avoidance logic or the fuzzy logic controllers, but rather by the restricted performance of the car sensors and insufficient memory. In future endeavors, we intend to enhance separation assurance strategies to avert a loss of separation in all encounter conditions. Furthermore, employing a higher fidelity model of the environment within the formal methods tools would provide more comprehensive specifications (e.g., vehicles maintaining separation) and then discover instances that contravene these specifications. Furthermore, given that the avoidance system possessed flawless vehicle state information, we aim to incorporate a degree of uncertainty into the sensor models. Ultimately, we intend to integrate the proposed avoidance software into hardware testing environments.

AT A GLANCE

In the forthcoming years, activities in low-altitude airspace will significantly escalate as the functionalities and uses of small unmanned aerial systems (sUAS) proliferate. Consequently, identifying solutions for managing sUAS in densely populated airspace will enhance sUAS operations. This study employed a fuzzy logic-based methodology to decrease the risk of aircraft collisions through separation assurance and collision avoidance strategies. The technology was assessed for its efficacy in reducing the risk of mid-air collisions between aircraft. This system employs solely current state information and may address possible conflicts without understanding the intruder's intent. The avoidance logic was validated by formal methods and demonstrated to choose the appropriate action in every case. Furthermore, the fuzzy logic controllers consistently directed the vehicles accurately. Numerical testing indicated that the avoidance system successfully prevented mid-air collisions between two sUAS in all examined scenarios. Simulations were conducted in a three-dimensional environment featuring a heterogeneous fleet of sUAS executing various realistic missions. Simulations demonstrated that the technology achieved 99.98% efficacy in averting mid-air collisions when separation assurance was disabled (unmitigated scenario) and 100% efficacy when enabled (mitigated scenario).

REFERENCES

1. Cook K. The silent force multiplier: The history and role of UAVs in warfare. Proceedings of the IEEE Aerospace Conference; 3–10 March 2007; Big Sky, MT, USA. IEEE. 2007:1-7. DOI: 10.1109/AERO.2007.352737
2. Ollero A, Merino L. Unmanned aerial vehicles as tools for forest-fire fighting. Forest Ecology and Management. 2006;234(1):S263
3. Bamburry D. Drones: Designed for product delivery. Design Management Review. 2015;26(1):40-48. DOI: 10.1111/drev.10313
4. Prevot T, Homola J, Mercer J. Human-in-the-loop evaluation of ground-based automated separation assurance for NEXTGEN. In: The 26th Congress of ICAS and 8th AIAA ATIO. 2008 Sep. 8885
5. Erzberger H, Heere K. Algorithm and operational concept for resolving short-range conflicts. Proceedings of the Institution of Mechanical Engineers, Part G: Journal of Aerospace Engineering. 2010 Feb 1;224(2):225-243
6. Lauderdale TA, Erzberger H. Automated separation assurance with weather and uncertainty. Air Traffic Management and Systems. 2014:35-47. Springer, Japan
7. Temizer S, Kochenderfer M, Kaelbling L, Lozano-Pérez T, Kuchar J. Collision avoidance for unmanned aircraft using Markov decision processes. In: AIAA Guidance, Navigation, and Control Conference. 2010 Aug 2. 8040
8. Durand N, Alliot JM, Noailles J. Collision avoidance using neural networks learned by genetic algorithms. In: IEA-AEI 1996, 9th International Conference on Industrial and Engineering Applications of Artificial Intelligence and Expert systems. 1996 Jun 1
9. Hromatka M. A fuzzy logic approach to collision avoidance in smart UAVs. Honors Thesis. Collegeville, MN: College of Saint Benedict and Saint John's University; 2013
10. Kuchar JE, Drumm AC. The traffic alert and collision avoidance system. Lincoln Laboratory Journal. 2007 Nov 2;16(2):277
11. Cook B, Cohen K, Kivelevitch EH. A fuzzy logic approach for low altitude uas traffic management (UTM). In: AIAA Infotech@ Aerospace. 2016:1905. DOI: 10.2514/6.2016-1905
12. Cook BM. Multi-Agent Control Using Fuzzy Logic. Electronic Thesis or Dissertation. University of Cincinnati. 2015. OhioLINK Electronic Theses and Dissertations Center. pp. 90-166

13. Peled D. Software Reliability Methods. New York, NY: Springer Science & Business Media; 2013. 332

14. Baier C, Katoen J, Larsen K. Principles of Model Checking. Cambridge, MA: MIT press; 2008. 984

15. Clark M, Rattan K. Piecewise affine hybrid automata representation of a multistage fuzzy pid controller. In: Proceedings of the 2014 AAAI Spring Symposium Series: Formal Verification and Modeling in Human-Machine Systems; 24–26 March 2014

16. Ross T. Fuzzy Logic with Engineering Applications. 3rd ed. Hoboken, NJ: John Wiley & Sons; 2009. 606

17. Butler RW. "What is Formal Methods?" NASA LaRC Formal Methods Program [Internet]. 2001. Available from: https://shemesh.larc.nasa.gov/fm/fm-what.html. [Accessed: 03-02-2017]

18. Wagner L, Fifarek A, DaCosta D, Gross K. SpeAR: Specification and Analysis of Requirements. InS5 Symposium. 2014

19. Hagen GE. Verifying safety properties of Lustre programs: An SMT-based approach. Ann Arbor, MI: ProQuest; 2008

20. Ghassabani E, Gacek A, Whalen MW. Efficient generation of inductive validity cores for safety properties. In: Proceedings of the 2016 24th ACM SIGSOFT International Symposium on Foundations of Software Engineering. 2016 Nov 1:314-325. ACM

21. De Moura L, Bjørner N. Z3: An efficient SMT solver. In: International Conference on Tools and Algorithms for the Construction and Analysis of Systems. 2008 Mar 29. 337-340. Berlin Heidelberg: Springer

22. Halbwachs N, Caspi P, Raymond P, Pilaud D. The synchronous data flow programming language LUSTRE. Proceedings of the IEEE. 1991 Sep;79(9):1305-1320

13 Chapter

ENHANCEMENT OF PRECISION IN MINIATURIZED INERTIAL NAVIGATORS THROUGH FUZZY LOGIC DENOISING OF SENSOR SIGNALS

1. INTRODUCTION: AN OVERVIEW

The enduring transformation in sensor fabrication technologies over the past two decades, particularly focused on miniaturization, has significantly impacted the field of navigation and positioning systems, specifically regarding strap-down inertial navigation systems (SDINS). Consequently, an increasing number of applications have been developed and are presently being developed based on this notion, utilizing smaller inertial measurement devices and compact data processing systems. The contemporary technologies, including MEMS (micro-electro-mechanical systems), NEMS (nano-electro-mechanical systems), MOEMS (micro-opto-electro-mechanical systems), and NOEMS (nano-opto-electro-mechanical systems), are presently employed effectively in inertial sensors, which are integral to numerous SDINS applications. Due to downsizing, the application range of SDINS systems has significantly expanded beyond traditional navigation systems used in aerospace vehicles, encompassing land vehicle applications, robotics, medicine, and assisted living. The categorization of inertial sensor fabrication technologies based on the accuracy classes of future applications can be structured as illustrated in Figure 1 [1]. The accuracy levels of NEMS technologies, potentially including NOEMS, remain at the commercial application stage; specifically, IOG—integrated optics gyro, IFOG—interferometric fiber optic gyro, HRG—hemispherical resonant gyro, and PCF-FOG—photonic crystal fiber-fiber optic gyro.

<table>
<tr><th>Very Precise Navigation</th><th>Long-Range Guidance</th><th>1 nm/hr Navigator</th><th>Tactical Weapons</th><th>Comercial Consumer</th></tr>
<tr><td colspan="5">Accelerometer Technology</td></tr>
<tr><td rowspan="3">Cold Atom</td><td rowspan="2">Cold Atom</td><td rowspan="6">MEMS</td><td rowspan="6">MEMS</td><td rowspan="6">MEMS / NEMS</td></tr>
<tr></tr>
<tr><td rowspan="2">MEMS</td></tr>
<tr><td rowspan="3">Mechanical</td></tr>
<tr><td rowspan="2">Mechanical</td></tr>
<tr></tr>
<tr><td colspan="5">Gyro Technology</td></tr>
<tr><td rowspan="2">Cold Atom</td><td>Cold Atom</td><td rowspan="2">IOG</td><td rowspan="4">MEMS</td><td rowspan="4">MEMS / NEMS</td></tr>
<tr><td>IFOG</td></tr>
<tr><td rowspan="2">IFOG</td><td>HRG</td><td rowspan="2">PCF FOG</td></tr>
<tr><td>IOG</td></tr>
</table>

Figure 1. Inertial sensor technologies on the application accuracy classes.

Included in the dead-reckoning navigation systems category, the SDINS was succeeded on the market due to its availability to provide the vehicle speed and position with abundant dynamic information, due to the possibility to calculate at an extremely high rate comparatively with the GPS system, and due to its excellent short-term performance. Having characteristics complementary to satellite positioning systems, in time, it became an important component of the modern navigation systems near the GPS [5–6]. If the two systems are used in an integrated configuration, for a short time, derived from the quality of the used inertial sensors, the inertial navigator can overcome the nonavailability of GPS signals and can maintain in this way the navigation solution at a high quality; a well-designed INS/GPS navigator provides improved performance in terms of accuracy, availability, and reliability in front of a simple GPS system.

From another point of view, the inertial navigators are subjected to various error sources, their main deficiency residing in the great accuracy degradation over time mainly due to the quality of the used inertial sensors [7–9]. New miniaturization technologies plays an important role in the reduction of the cost and the volume of the inertial sensors, which, apparently, offers an important advantage for the miniaturized inertial sensors, but, in fact, the fabrication processes of such sensors make these very sensitive to the changes of the environmental conditions: temperature, pressure, electric and magnetic fields, and vibrations. Therefore, the sensors output can vary quickly, widely, and sometimes randomly and is very hard to be modeled. In many cases, this sensitivity leads to the decreased sensor performance, adding more error types, and possibly, with higher values than those of classical sensors [3, 9].

In this context, the need to develop low-cost, small-size, and high-precision INS/GPS integrated navigators, which are able to be used also in GPS challenging environments, generated few research directions in the field; the main two being [4–6, 10–14]: (1) to develop standalone accurate SDINS structures with miniaturized IMUs; (2) to develop new INS/GPS data-fusion methods based on artificial intelligence algorithms, having objectives to overcome the limitations in terms of model dependency, prior knowledge dependency, and linearization dependency.

The increase of the SDINS accuracy in standalone configuration targets, in fact, the improvement of the quality for the signals provided by the IMUs to the navigation processor through the obtaining of high quality miniaturized sensors or through the development of various IMU hardware architectures accompanied by right numerical algorithms able to produce sensor errors estimation and compensation [9, 13, 14]. Two categories of errors parasitize the inertial sensors: deterministic errors and stochastic errors. For the deterministic errors, the specialists conceived various calibration procedures, which can easily estimate and eliminate it. Unfortunately, the stochastic error estimation necessitates a more complex process and they cannot be fully removed from the sensors data. For both categories of inertial sensors (accelerometers and gyros), the most important stochastic errors are caused by noise and by instability of bias and of scale factor [7].

The main difficulty with the noise is related to the impossibility to apply a direct filtering procedure by using classical methods, because, in terms of frequency spectrum, it is superimposed on the band 0–100 Hz of the navigation useful signal [7]. As a consequence, many researches have been initiated to find alternative solutions to limit its impact on the solution of navigation. On the other way, it must not forget that at the level of the noise there are present two components, long term noise (low frequency noise) and short term noise (high frequency noise), each one with its own spectral characteristics and with its own compensation mechanism: optimal low pass filtering or using algorithms that fusing the data from multiple sensors [13, 15–19].

Our identified way to improve the quality of the inertial sensor signals is based on the use of redundant detection clusters with the sensors disposed in linear configurations along each detection axis in IMU (Figure 2) [9, 13]. For each cluster, the measured data are subsequently fused to provide to the navigation processor a better measured signal of acceleration or angular speed for the respective axis in IMU. Simultaneously, this structure provides the advantage to have a redundant inertial navigator in terms of the detection unit components, i.e., when one or more sensors in a detection cluster break down they are removed from calculation process but the system still remains operational. The inertial sensors in the same cluster have the sensitivity axes parallel and oriented in the same sense with the detection axis.

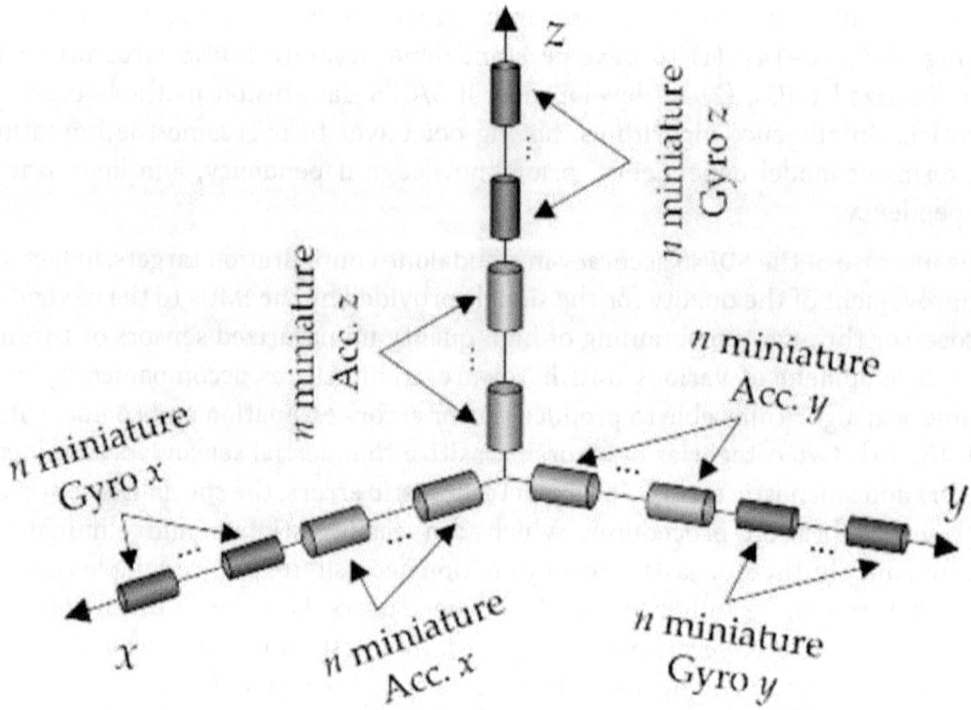

Figure 2. Inertial sensors disposed in linear configurations along each detection axis in IMU.

The proposed fuzzy logic approach was primarily intended to mitigate sensor noise, while it also yielded some advantages in reducing bias effects. This work pertains to a research project conducted in Romania, funded by the Executive Unit for the Financing of Higher Education and Scientific Research, which sought to enhance the precision of strap-down inertial navigators through the utilization of redundant sensor networks for their IMUs and innovative adaptive numerical algorithms for data fusion. This chapter presents the information in the following manner: Section 1 comprises a brief introduction; Section 2 delineates the fundamental components and architecture of the algorithm, encompassing the fuzzy logic mechanism; Section 3 presents the numerical simulation outcomes derived from the developed algorithm for detecting clusters of varying sizes and sensor types; Section 4 details the results obtained from experimental data collected using detection clusters, each equipped with six MEMS sensors (aided by an integrated GPS/INS navigator as a reference positioning system) and processed for a bidimensional INS navigator; and Section 5 underscores the advantages conferred by the algorithm in the form of conclusions.

2. DATA FUSION ALGORITHM BASIC ELEMENTS AND STRUCTURE

To describe the generalized form of the proposed algorithm, it is denoted with n the number of collinear sensors included in a detection cluster, as in Figure 2. On the other way, if x_i ($i = 1 \div n$) are the measurements of the x quantity provided by the n sensors, characterized by the standard deviations σ_i ($i = 1 \div n$), then the weighted mean obtained by applying the data fusion algorithm conducts to the x_e estimate of the x quantity under the next form [14]:

$$x_e = w_1 \cdot x_1 + w_2 \cdot x_2 + \cdots + w_n \cdot x_n = \sum_{i=1}^{n} w_i \cdot x_{i'} \left(\sum_{i=1}^{n} w_i = 1 \right); \tag{1}$$

w_i ($i = 1 \div n$) are the weights of the x_i ($i = 1 \div n$) measurements. For each sensor, the standard deviation is evaluated starting from the last m samples acquired from it. The algorithm has an adaptive character, the data frame used to estimate the standard deviations of the sensors signals being permanently updated with the new sensor measurements, and, in this way, the weights associated with each sensor are reestimated at each measurement step. The estimation of the new weight set is realized based on a fuzzy logic mechanism. The necessary data frame used in the estimation of sensors standard deviations is built using a FIFO (first in first out) buffer with n channels and able to memorize m successive samples on each channel. In this way, at each measurement time, the last column in the data frame goes out from the buffer and a new one comes in; two consecutive data frames are superposed with (m - 1) samples [13]. The standard deviations of the n-independent channels result with the next equation [13, 14]:

$$\sigma_i(k) = \sqrt{\frac{1}{m} \sum_{j=1}^{m} [x_i(j) - \overline{x}_i(k)]^2}, \; (i = \overline{1, n}); \tag{2}$$

$x_i(j)$ is the jth measurement provided by the ith sensor in the kth data frame; $\sigma_i(k)$ is the standard deviation for the kth data frame, which characterize the ith sensor; and $\overline{x}_i x'_i(k)$ is the arithmetic mean of the ith sensor data for the kth data frame [14]. The value of $\sigma_i(k)$ is used to fuse the data at the next sensor reading.

According to the proposed algorithm, a fuzzy logic controller is used in each channel of a detection cluster, and, based on the standard deviation, $\sigma_i(k)$ of the m consecutive samples acquired from the sensor in the kth data frame provides a weight $w_i^{\text{fuzzy}}(k)(i = \overline{1, n}1, n)$ for the respective sensor. Based on the fact that in a weighted sum, the sum of all parts weight should be equal to 1, the sensors weights are recalculated and result with the equations [13, 14]):

$$w_1(k) = \frac{w_i^{\text{fuzzy}}(k)}{\sum_{i=1}^{n} w_i^{\text{fuzzy}}(k)}, \cdots, w_n(k) = \frac{w_n^{\text{fuzzy}}(k)}{\sum_{i=1}^{n} w_n^{\text{fuzzy}}(k)}. \tag{3}$$

As a consequence, by calculating the weighted sum from the data fusion equation, at the time $t_{k+1,}$ the x_e estimate of x is [13, 14]:

$$x_e(k+1) = w_1(k) \cdot x_1(k+1) + w_2(k) \cdot x_2(k+1) + \cdots + w_n(k) \cdot x_n(k+1), \tag{4}$$

i.e.,

$$x_e(k+1) = \sum_{i=1}^{n} w_i(k) \cdot x_i(k+1), \left(\sum_{i=1}^{n} w_i(k) = 1\right). \tag{5}$$

Based on the previous considerations, the architecture of the data fusion algorithm results as in Figure 3 [14]. The chosen fuzzy logic controllers are based on rules set that created a proportional dependence between the quality of the sensor signals and their associated weights, e.g., a "P" fuzzy logic controller. Therefore, the implemented fuzzy logic controllers established an inverse proportionality between the calculated standard deviations and sensor-associated weights.

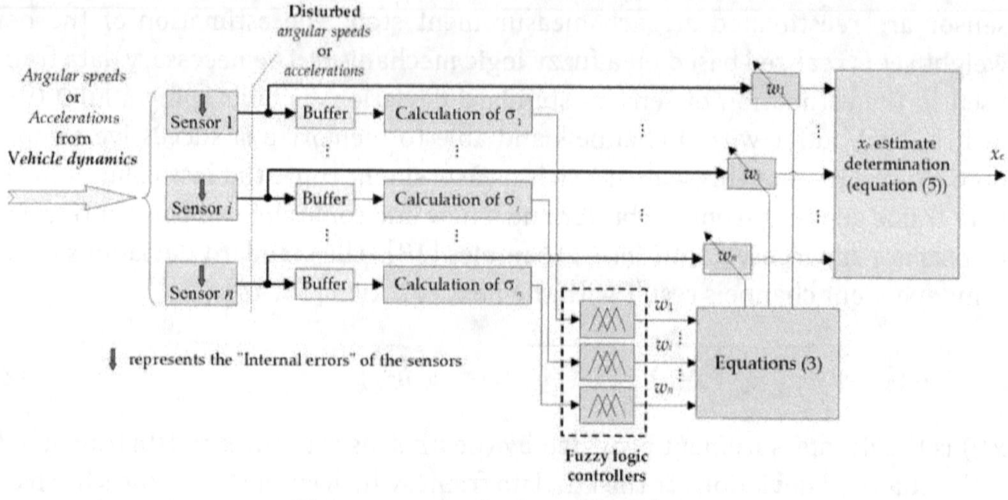

Figure 3. Data fusion algorithm.

The aim is to obtain a simpler fuzzy controller in order to reduce at minimum the computing time at each sample in order to implement the controller on a cheaper microcontroller. The way to reduce the computing time is to reduce on one hand the number of the membership functions and on the other hand to simplify these functions. The structure of fuzzy controller realized in Matlab/Simulink is presented in Figure 4 [20, 21].

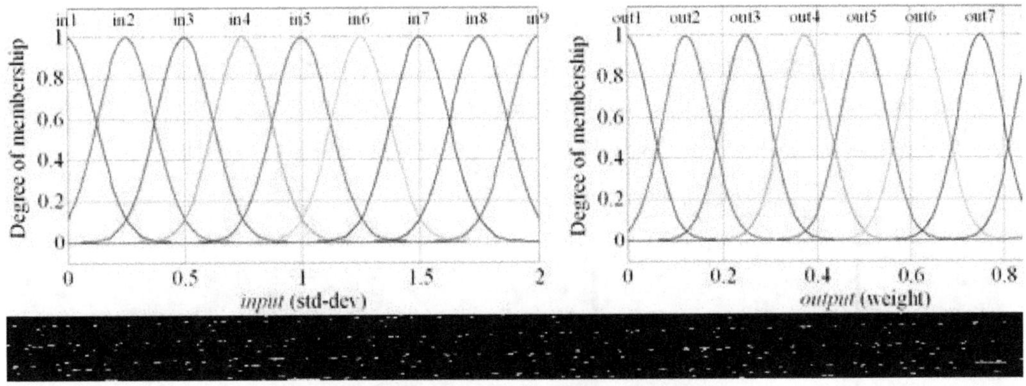

Figure 4. Fuzzy controller structure.

The universe of discourse for the controller input has considered a [0, 2] interval, where nine membership functions (*in1* to *in9*) with Gaussian shapes were uniformly distributed. In the same time, nine membership functions (*out1* to *out9*) with similar shapes were considered in the [0, 1] interval, taken as universe of discourse for the output. Figure 5 presents the input/output membership functions.

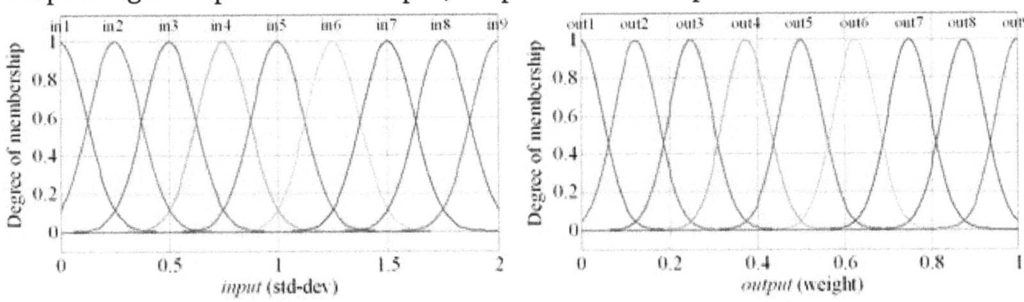

Figure 5. Membership functions (mfs) for input and output.

In the fuzzification process, the next nine rules were defined as: (1) If (*input* is *in1*) then (*output* is *out9*); (2) If (*input* is *in2*) then (*output* is *out8*); (3) If (*input* is *in3*) then (*output* is *out7*); (4) If (*input* is *in4*) then (*output* is *out6*); (5) If (*input* is *in5*) then (*output* is *out5*); (6) If (*input* is *in6*) then (*output* is *out4*); (7) If (*input* is *in7*) then (*output* is *out3*); (8) If (*input* is *in8*) then (*output* is *out2*); and (9) If (*input* is *in9*) then (*output* is *out1*). Widely accepted for capturing expert knowledge, a Mamdani controller type was used due to its simple structure of "min-max" operations [21]; and for defuzzification, the centroid method was applied.

3. NUMERICAL SIMULATION RESULTS

To simulate the algorithm, some models in Matlab/Simulink for detection clusters with various sizes and sensors types were realized. An example of such model is shown in Figure 6, being realized for a cluster with four sensors by the same type, with errors software modeled by the blocks "Model Acc" placed at the input of the simulation model. These models, realized by the authors, are based on the sensor data sheets and on the IEEE equivalent models for the inertial sensors [22, 23]. Accelerometers were modeled as in Figure 7, the obtained model having inputs, such as acceleration a_i, applied along the sensitive axis, and the cross-axis acceleration a_c, and as output the perturbed acceleration a [22, 23].

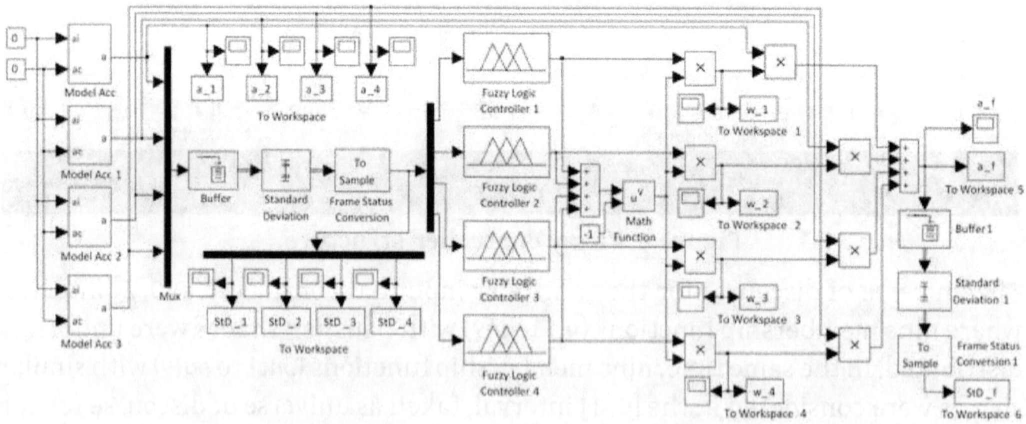

Figure 6. Simulation model with four sensors in cluster.

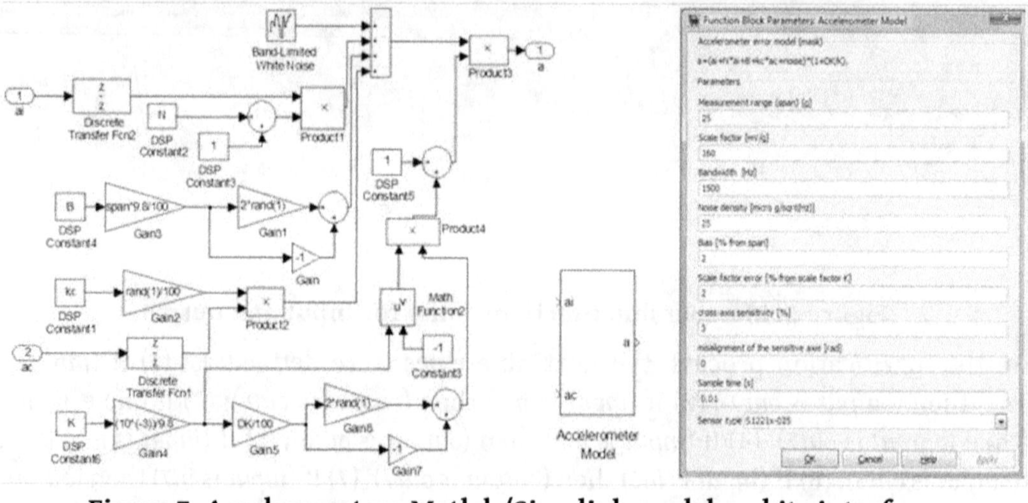

Figure 7. Accelerometers Matlab/Simulink model and its interface.

Studying the data sheets for various acceleration sensors, it was observed that a part of the included parameters is provided by using an interval inside which they can vary arbitrary. For example, for the bias it is provided a maximum absolute value (*B*) which is considered as percent from span, for the cross-axis sensitivity it is provided a maximal value (k_c) considered as a percent from the cross-axis acceleration (a_c), for the scale factor calibration error the producers give a maximum absolute value (ΔK) under the form of a percent from the scale factor (*K*), and for the noise is provided the maximum value of its density. Therefore, according to Figure 7, in the model, the "rand(1)" MATLAB function is used three times to generate a random value in the (-*B*, *B*) interval for the bias, a random value in the (0, k_c) interval for the cross-axis sensitivity, and a random value in the ($-\Delta K$, ΔK) interval for the scale factor calibration error. The noises are realized using the SIMULINK block "Band-Limited White Noise" and MATLAB function "RandSeed" through the generation of a random value of the density in the ($80\%\cdot\nu_d$, ν_d) interval [24].

In a simulation example at null accelerations as inputs, graphical characteristics resulted are shown in Figures 8–10. Figure 8 shows the sensor measurements from "Acc-1" to "Acc-4" and also the results of the accelerometer data fusion "Acc-f". In the simulation, there was considered just the sensor noise overlapped on an ideal null input for four sensors with a span of 18 g, a bandwidth of 2500 Hz each, and with various noise densities taken aleatory between 280 and 380 $\mu g/Hz^{1/2}$. The noise pattern of the data fusion resulted a signal that proves an important reduction of the noise level, a similar observation resulting from Figure 9 also depicting the evolution in time of the standard deviations for the sensors signals and for the estimate x_e. On the other way, Figure 9 proves that the adaptive character of the algorithm is maintained, the standard deviations values being updated at each simulation time step. The best sensor in the cluster was sensor #3, the standard deviation of the fused signal being two times smaller (approx. $3\cdot10^{-3}$ m/s^2) than the raw values of the standard deviations of the sensors signals (approx. $6\cdot10^{-3}$ m/s^2). Figure 10, depicting the values of the sensor weights estimated at each calculation step, also proves that the best sensor in the cluster was the sensor #3, and it receives the biggest values from the four weights during the entire numerical simulation.

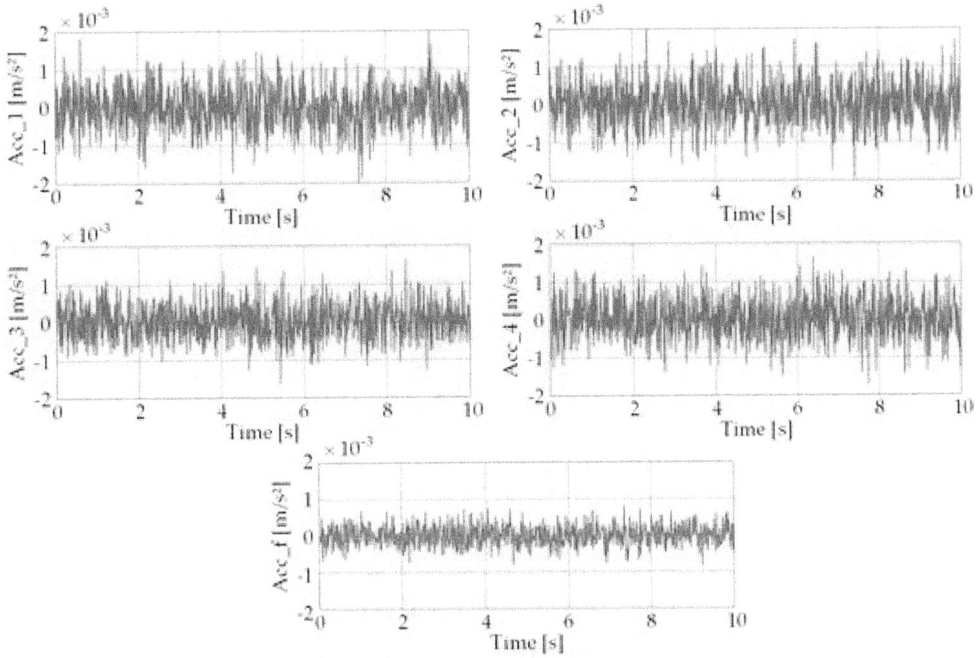

Figure 8. The sensor measurements and the data fusion results.

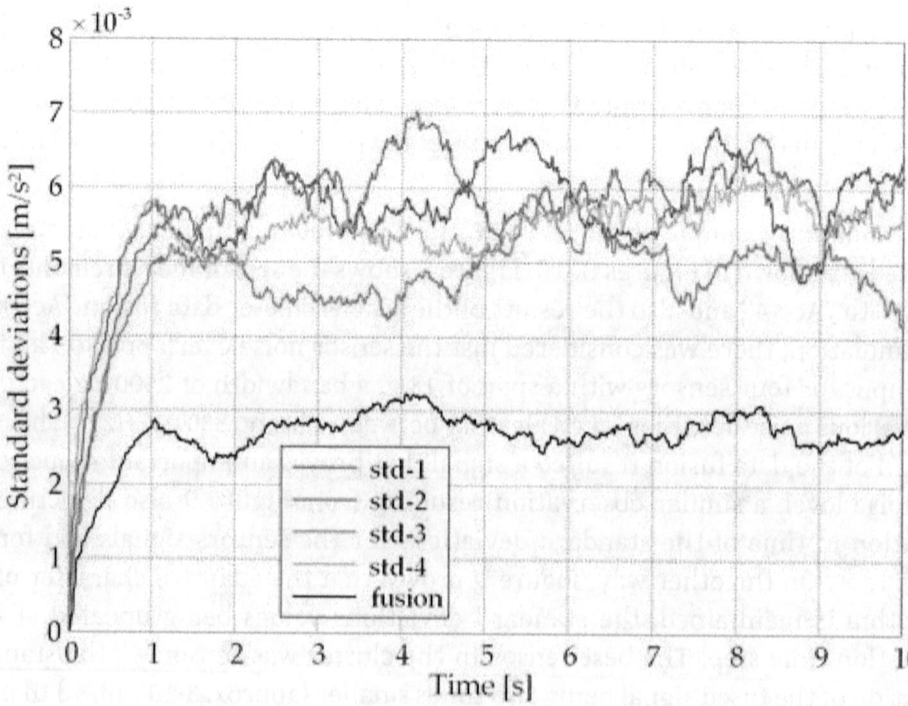

Figure 9. Standard deviations for the sensor signals and for the data fusion signals.

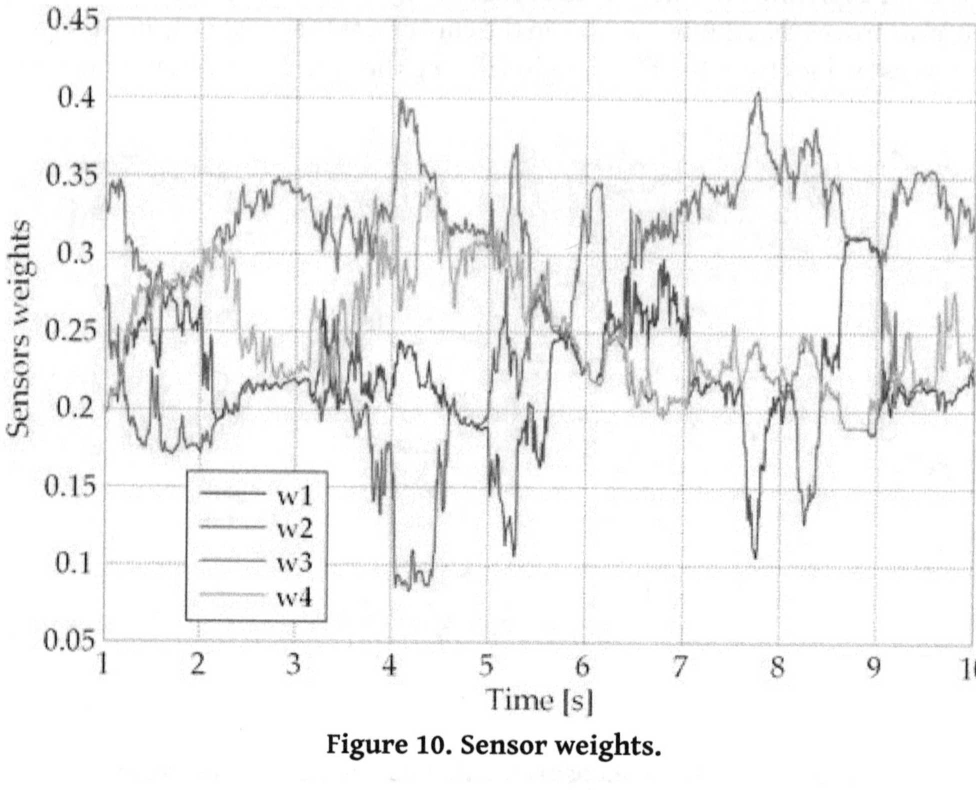

Figure 10. Sensor weights.

The sample time for the numerical simulation was 0.01 s, which means a data processing rate of 100 samples/s. Moreover, the data frame size for one detection channel was established to be m = 100, i.e., the buffers allow to be obtained data frames of 100 consecutive samples for each channel. In this way, any two consecutive data frames for one channel are overlapped with 99 samples. From Figure 9, the initialization phase of the algorithm with a duration of 1 s, time necessary for the first buffer to be full with 100 samples can be easily observed. All simulation results proved a good functioning of the algorithm which provided an important reduction of the sensor noise level, with a higher potential in the improvement of the solution of navigation accuracy when it is used in an inertial navigator.

4. RESULTS

The proposed algorithm was also tested using some experimental data acquired with an IMU equipped with detection clusters, which included six MEMS sensors each (assisted by an integrated GPS/INS navigator as reference positioning system). For the used IMU structure, with six sensors in each detection cluster, the Matlab/Simulink implementation of the data fusion algorithm resulted as shown in Figure 11 [14]. Grouping the algorithm model resulted in "Fuzzy-logic data fusion" block from the right side of the figure. The data fusion block inputs are the measured data obtained from the sensors in the cluster ("Si" correspond to the measurements x_i, i = 1 ÷ 6), whereas its outputs are the fusion signal "S_f" (the estimate x_e), the sensor weights "w1–w6" (w_i, i = 1 ÷ 6), the standard deviation of the fusion signal "Std_f", and the standard deviations of the sensor data "Std1–Std6".

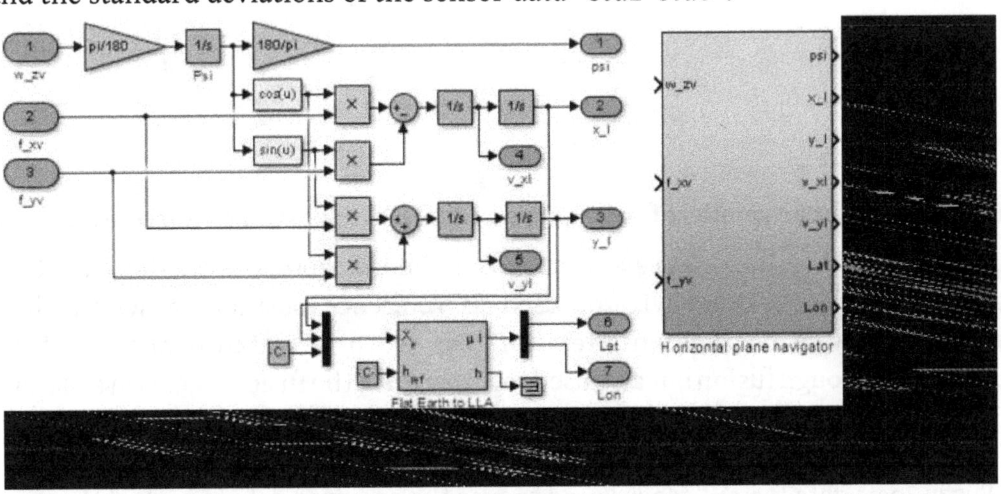

Figure 11. Matlab/Simulink implementation of the data fusion algorithm for experimental data.

After the fusion, the obtained data were processed in a bidimensional INS navigator in horizontal plane, software implemented as in Figure 12 [14]. Therefore, in this testing mechanism, three detection clusters were implied: two accelerometers clusters along the longitudinal and lateral axes of the vehicle and a gyro cluster along the vertical axis of the vehicle. The model included the "Flat Earth to LLA" Matlab/Simulink block in order to obtain the vehicle coordinates also in terms of latitude and longitude. The block "Horizontal plane navigator" in the right side of the figure has been obtained by grouping the navigator software model in the left side of the figure. The inputs of the navigator are the inertial measurements provided by the accelerometers and by the gyro: longitudinal (*f_xv*) and lateral (*f_yv*) accelerations in vehicle reference frame, and the angular speed (*w_zv*) along the vertical axis of the vehicle reference frame. As outputs, it provides information related to the angular orientation of the vehicle in horizontal plane (yaw angle—*psi*), to the vehicle speed and position components in local horizontal frame, *v_xl* (North speed), *v_yl* (East speed), *x_l* (position in North direction), and *y_l* (position in East direction), and to the latitude and longitude coordinates of the vehicle (*Lat* and *Lon*).

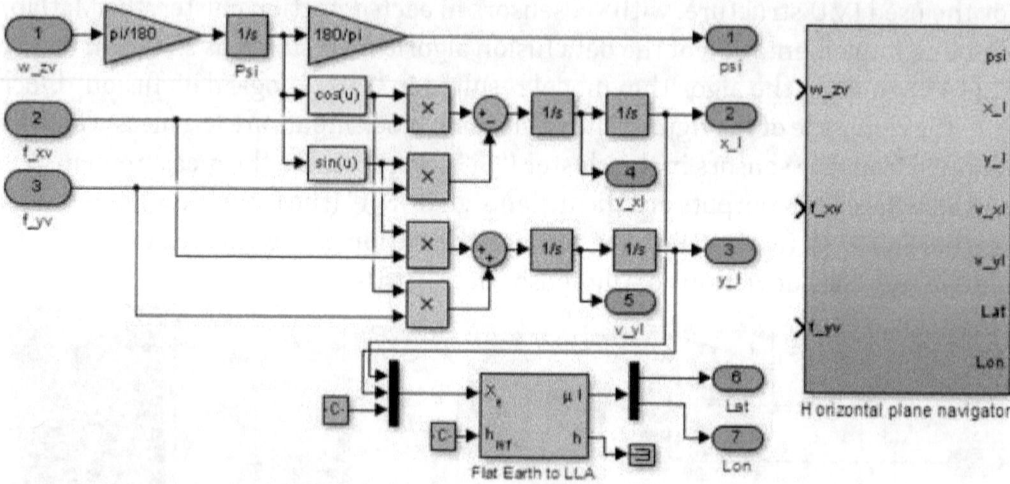

Figure 12. Software model of the bidimensional INS navigator in horizontal plane.

During tests, the IMU data were acquired simultaneously with the data from an integrated GPS/INS navigator used as reference positioning system; both equipments were boarded on a test car. The sensor acquired data were offline processed through fusion but also used together with the fusion signals in a Matlab/Simulink software model implementing seven bidimensional INS navigators ("Horizontal plane navigator" blocks) as in Figure 13 [14]. It also includes three "Fuzzy-logic data fusion" blocks used to fuse the acquired data in each of the three detection clusters.

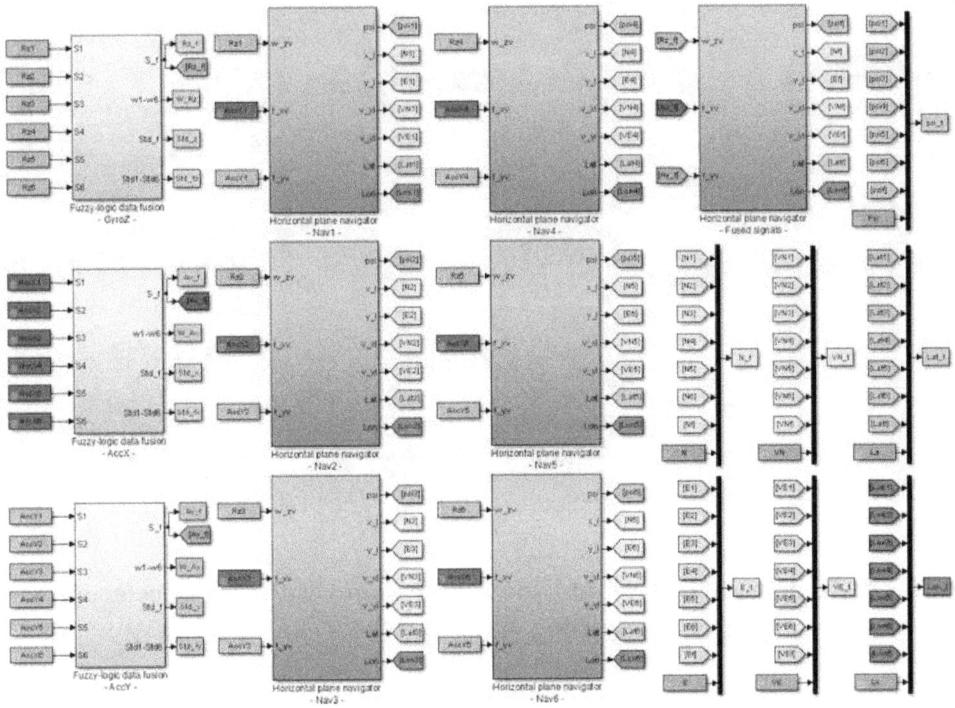

Figure 13. The evaluation Matlab/Simulink model.

The sensors in IMU were organized in six groups, with three sensors each (two accelerometers and one gyro), one sensor from each of the three detection clusters. To avoid the complications in the data processing, the detection groups included sensors with the same number in the detection clusters. Six of the seven navigators in Figure 13 ("Nav1" to "Nav6") processed data from the six detection groups, whereas the seventh navigator ("Horizontal plane navigator – Fused signals") processed data resulted from the fusion procedure applied on each of the three detection clusters. In this way, the evaluation model allows the obtaining of seven solutions of navigation, six solutions based on inertial sensor raw data (nonredundant INSs) and one solution based on inertial sensor fused data. All these solutions are further compared with the navigation solution provided by the integrated GPS/INS navigators which are used as reference positioning system.

For an evaluation situation, the experimental data acquired in each of the three detection clusters and the data fusion results for each cluster are shown in Figure 14 for the accelerometers in the *x*-axis cluster, in Figure 15 for the accelerometers in the *y*-axis cluster, and in Figure 16 for the gyros in *z*-axis cluster. The data fusion signals are the last ones in each figure.

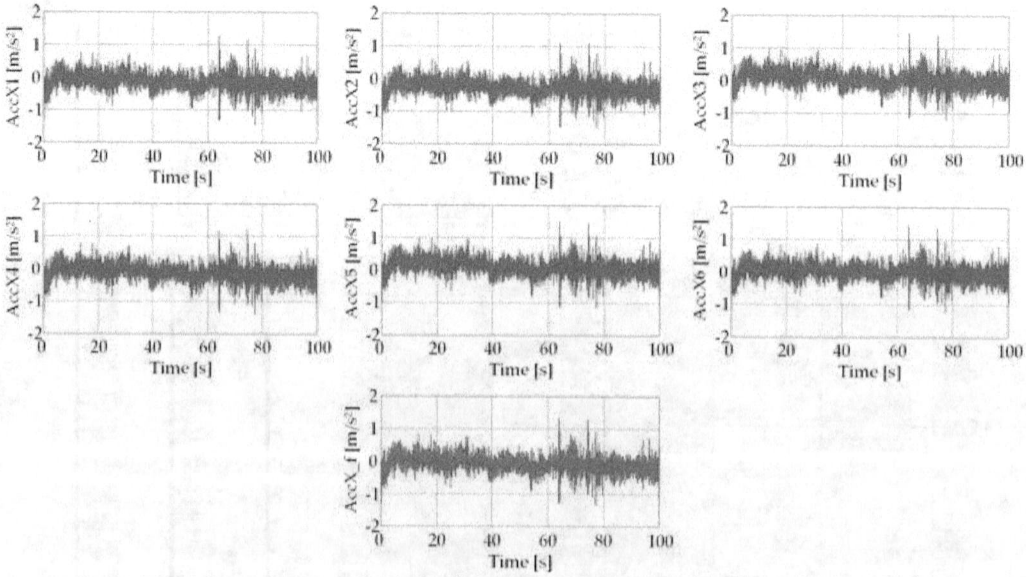

Figure 14. Experimental data and the data fusion results for accelerometers in the x-axis cluster.

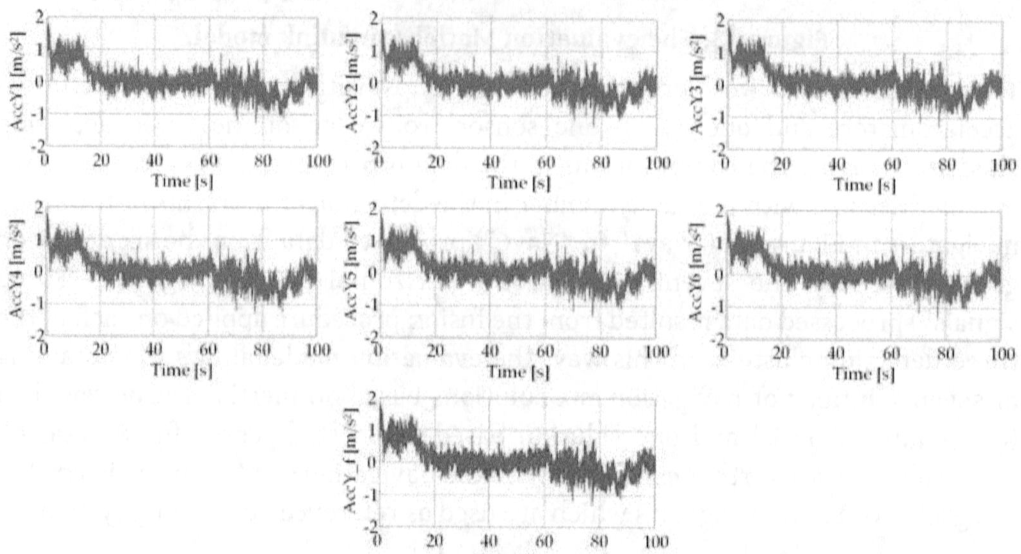

Figure 15. Experimental data and the data fusion results for gyros in the z axis cluster.

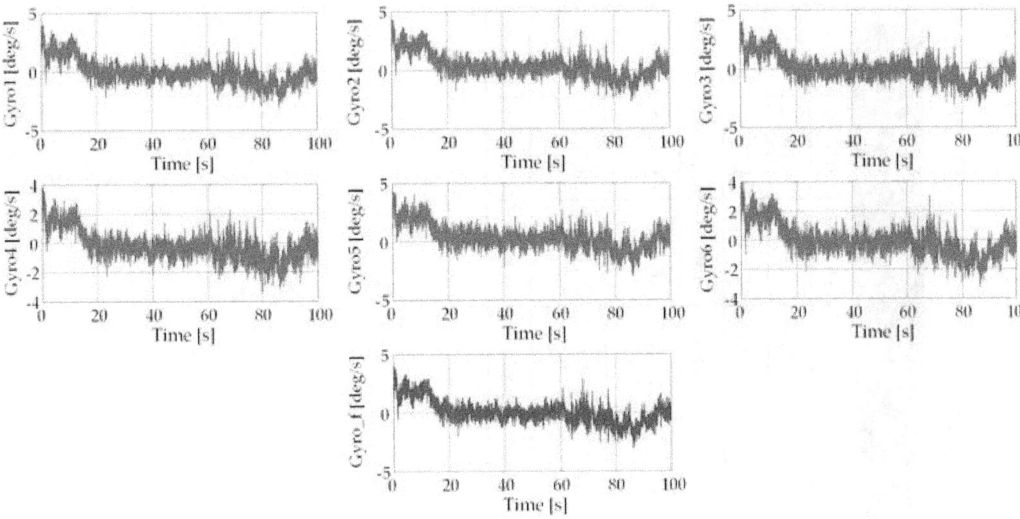

Figure 16. Experimental data and the data fusion results for gyros in the z axis cluster.

The processing of the acquired data with the model in Figure 13 conducted at the navigation solutions components shown in the next figures. Together with the navigation solution, components have shown the associated errors, evaluated vis-à-vis of the reference navigation solution provided by the GPS/INS integrated navigator. Table 1 centralizes the absolute maximal values of the positioning and speed errors for all six nonredundant INSs ("INS1" to "INS6" columns) and for the redundant INS ("Fusion" column), and also the mean values obtained for all these errors ("Mean value" column). In the same time, by calculating the mean of the absolute maximal values of the errors for the six nonredundant INSs in each line of Table 1 and dividing it with the corresponding value in the "Fusion" column, the ratios in the "Mean/Fus" column have been obtained. Similarly, the "Max/Fus" and "Min/Fus" columns show the ratios between the maximal value of the tabled errors for the six nonredundant INSs in each line of Table 1 and the corresponding value in the "Fusion" column, respectively, between the minimum value of the tabled errors for the six nonredundant INSs in each line of Table 1 and the corresponding value in the "Fusion" column.

Table 1. Error analysis for redundant INS and for nonredundant INSs.

Abs. max error	INS1	INS2	INS3	INS4	INS5	INS6	Fusion	Mean value	Mean/ Fus	Max/ Fus	Min/ Fus
North [m]	411.3583	**990.5235**	599.5493	308.5221	692.1193	266.7808	**32.6736**	544.8089	16.6742	30.3156	8.1650
East [m]	73.0415	**318.1701**	262.5502	107.2303	310.4679	132.3827	**39.2776**	200.6405	5.1082	8.1005	1.8596
V-North [m/s]	9.0599	**19.4083**	11.1598	7.1313	14.0987	7.0160	**4.0333**	11.3123	2.8046	4.8119	1.7395
V-East [m/s]	10.9701	9.8672	3.4987	**13.6819**	**1.5752**	5.5067	**7.1445**	7.5166	1.0520	1.9150	0.2204
Latitude [degrees]	$37.019 \cdot 10^{-4}$	$\mathbf{89.141 \cdot 10^{-4}}$	$53.955 \cdot 10^{-4}$	$27.765 \cdot 10^{-4}$	$62.286 \cdot 10^{-4}$	$24.008 \cdot 10^{-4}$	$\mathbf{2.940 \cdot 10^{-4}}$	$49.029 \cdot 10^{-4}$	16.6742	30.3156	8.1659
Longitude [degrees]	$9.156 \cdot 10^{-4}$	$\mathbf{39.886 \cdot 10^{-4}}$	$32.914 \cdot 10^{-4}$	$13.442 \cdot 10^{-4}$	$38.921 \cdot 10^{-4}$	$16.595 \cdot 10^{-4}$	$\mathbf{4.923 \cdot 10^{-4}}$	$25.152 \cdot 10^{-4}$	5.1082	8.1005	1.8598
Yaw [degrees]	14.4621	25.2716	18.7894	**31.1004**	24.9227	19.4309	**5.7402**	22.3295	3.8899	5.4179	2.5194

At the level of the vehicle attitude, the characteristics in Figure 17 were obtained for the yaw angle. Also, from the point of view of the horizontal positioning, Figure 18 presents the covered distances in the North direction, and Figure 19 presents the covered distances in East direction. The North speed evolutions during time are depicted in Figure 20, whereas the speed evolutions during time in the East direction are shown in Figure 21. For each navigator, the incorporated "Flat Earth to LLA" block allows the calculus of the vehicle Latitude and Longitude as shown in Figures 22 and 23. Combining the positioning data in a horizontal plane, it resulted the vehicle trajectories (Latitude versus Longitude) and the deviations from the reference trajectory (Latitude errors versus Longitude errors), established by all seven navigators, as in Figure 24.

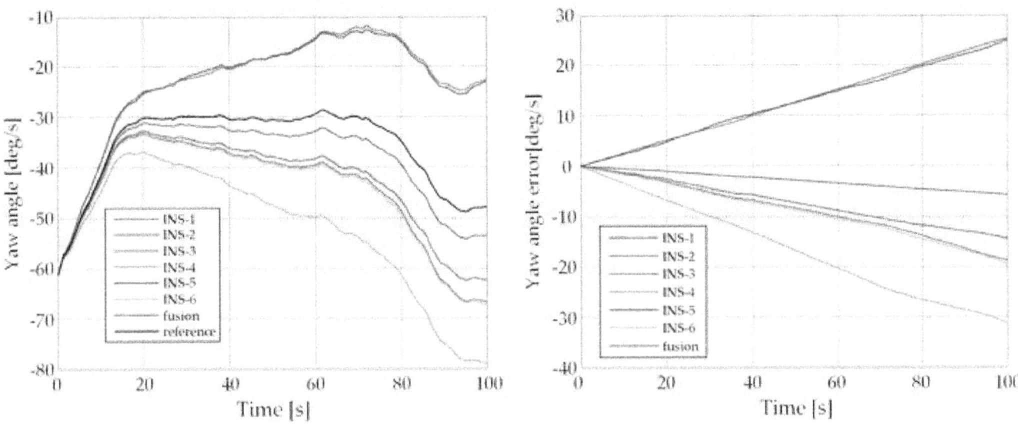

Figure 17. Yaw angle values and errors.

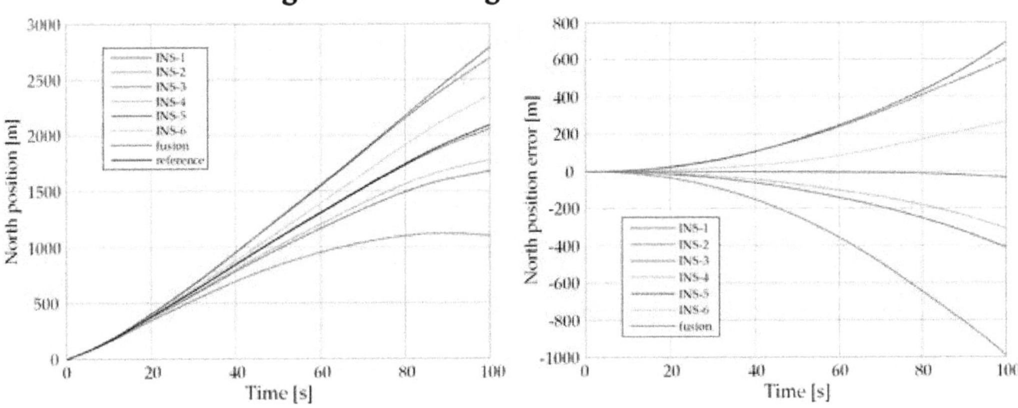

Figure 18. North positions and errors.

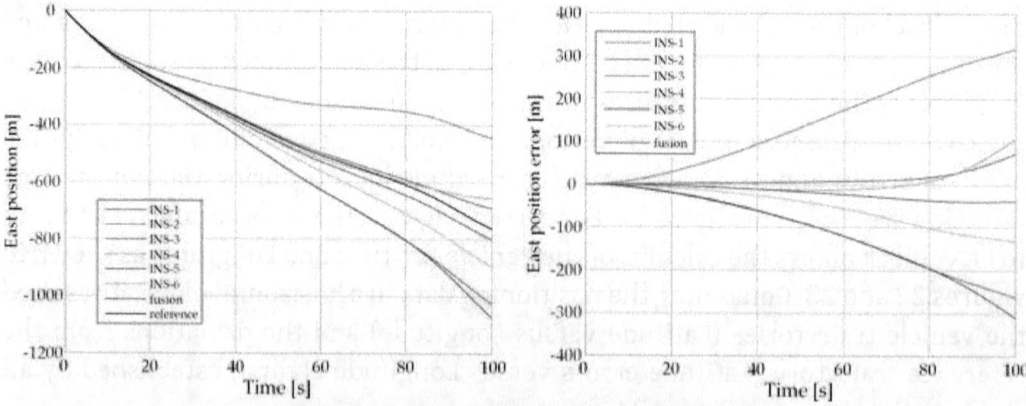

Figure 19. East positions and errors.

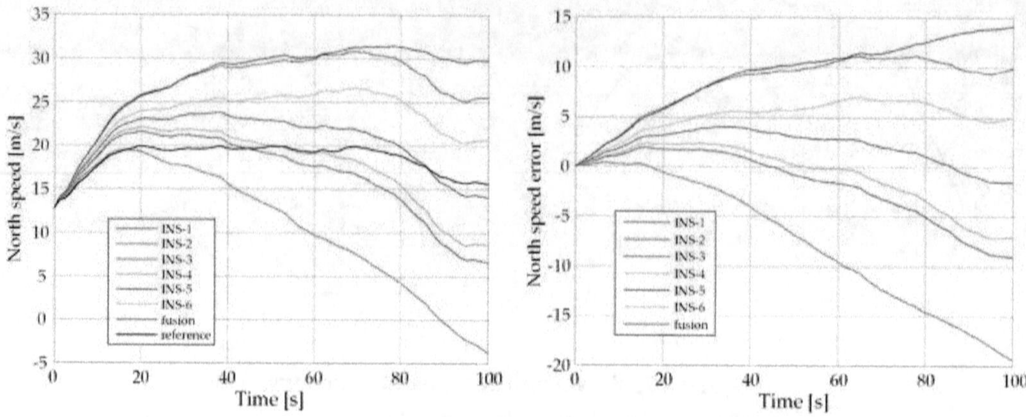

Figure 20. North speed evolutions during time.

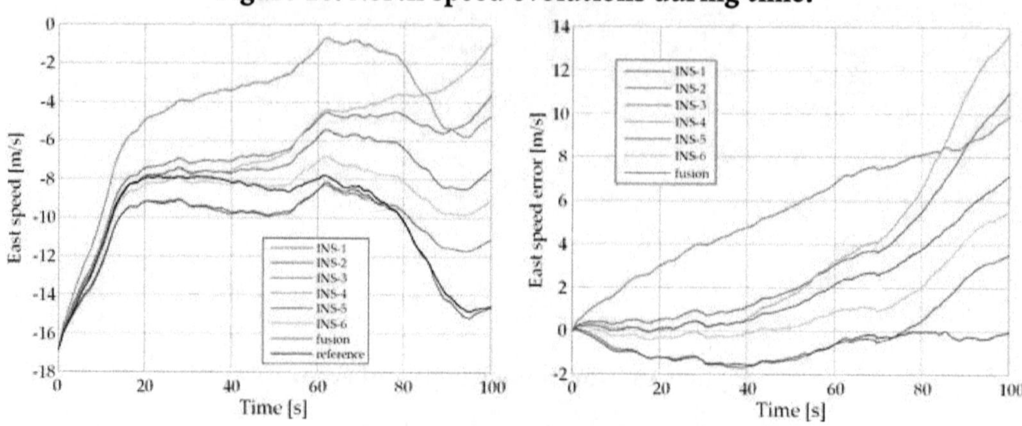

Figure 21. East speed evolutions during time.

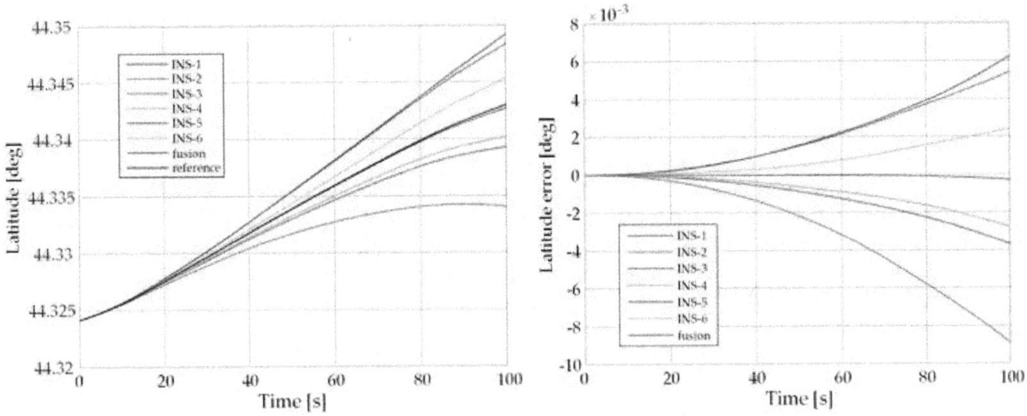

Figure 22. Latitude and errors.

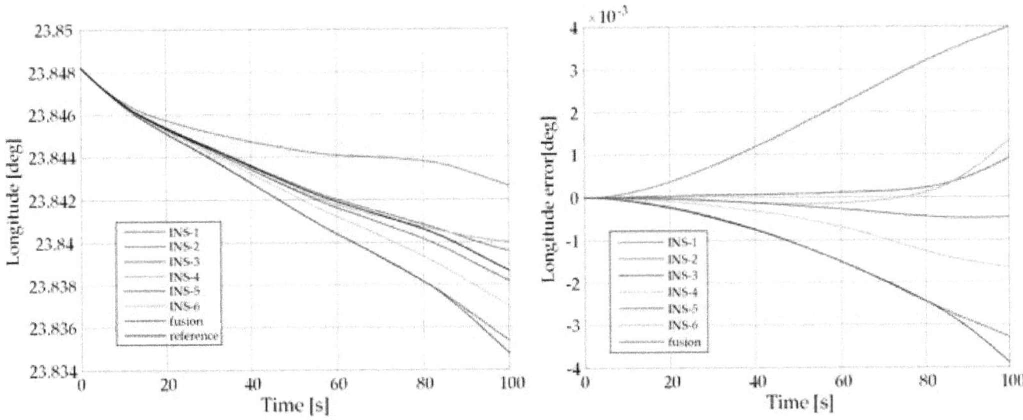

Figure 23. Longitude and errors.

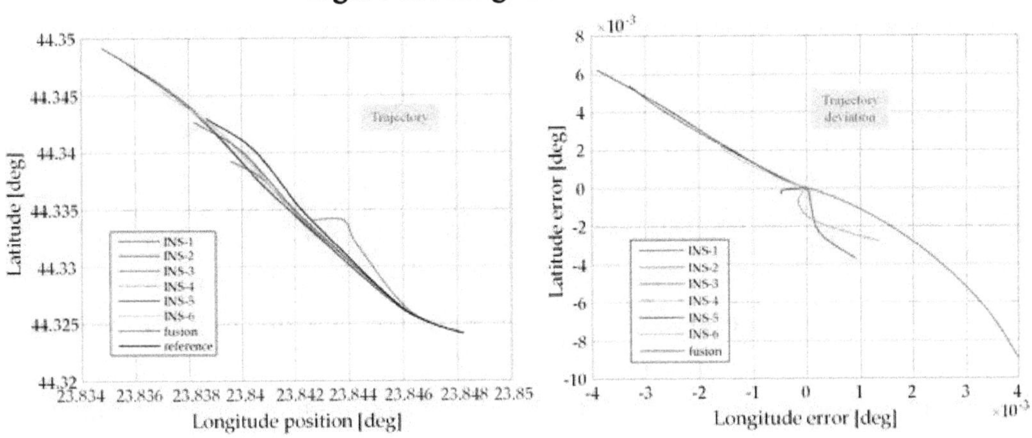

Figure 24. Vehicle trajectories and deviations from the reference trajectory in horizontal plane.

Figure 25 depicts the time evolution of the vehicle trajectories estimated with all seven navigators, whereas Figure 26 presents their deviations from the reference trajectory.

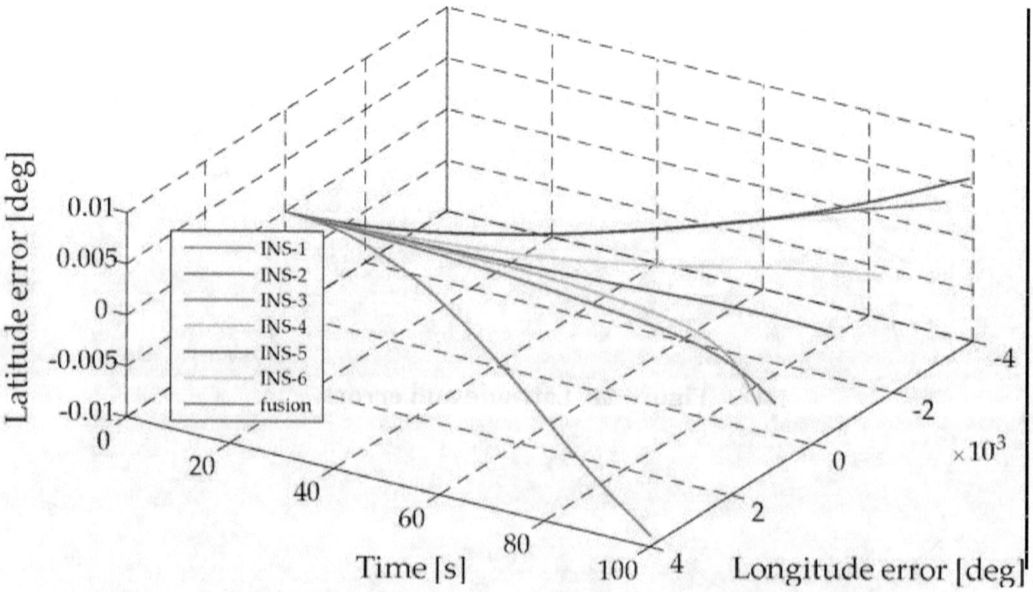

Figure 25. The time evolution of the vehicle trajectories for all seven INSs.

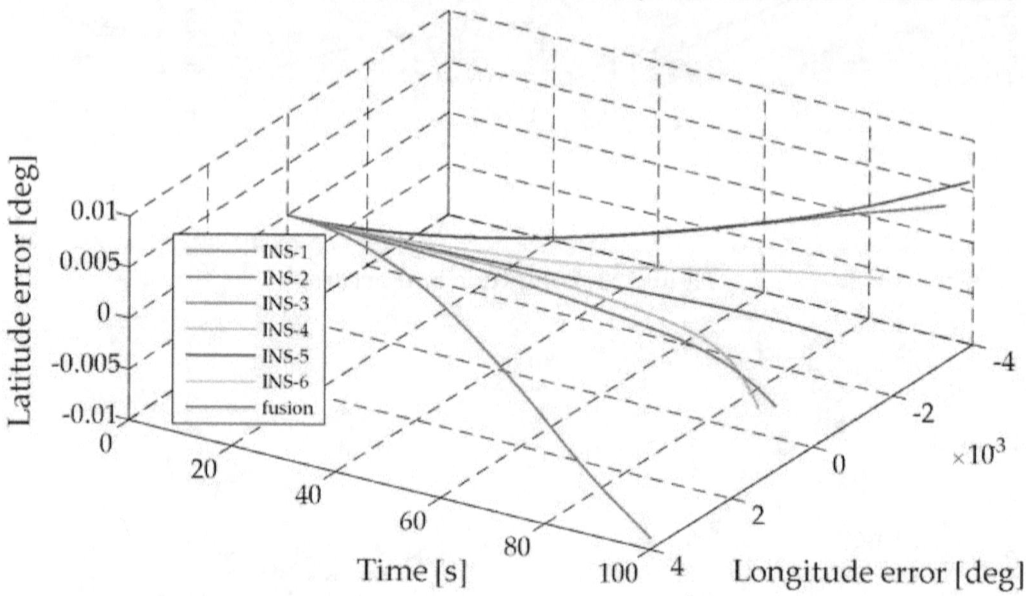

Figure 26. Time evolution of the deviations of the vehicle from the reference trajectory.

Both graphical results and numerical values presented in Table 1 prove an important positioning precision improvement by using the proposed data fusion mechanism. The errors of the navigation solution are substantially reduced comparatively with the nonredundant navigation solutions. The numerical data in the "Mean/Fus" column of Table 1 highlighted a decrease of the mean of the absolute maximal values of the errors for the six nonredundant INSs of approximately 16.6 times for the positioning in the North direction, 5.1 times for the positioning in the East direction, 2.8 times for the speed component in North direction, 1.05 times for the speed component in the East direction, and 3.8 times in the angular positioning in the horizontal plane (yaw angle). It is very important to be mentioned that the inertial sensor outputs in the experimental model of the inertial measurement unit were not previously corrected with the biases values or with other errors values. The values in the "Fusion" column of Table 1, related to the positioning and speed errors for the redundant navigator after 100 s have shown us that the absolute maximal values of these errors were 32.67 m for the linear positioning in North direction, 39.27 m for the linear positioning in East direction, 2.03 m/s for the speed component in North direction, 7.14 m/s for the speed component in East direction, $2.940 \cdot 10^{-4}$ degrees for the Latitude positioning, $4.923 \cdot 10^{-4}$ degrees for the Longitude positioning, and 5.74 degrees for the angular positioning in the horizontal plane (Yaw angle).

The nonredundant inertial navigation system's best configuration was switched between the detection groups associated with the first sensor in each detection cluster (East channel and Yaw angle channel) and with the sixth sensor in each detection cluster (North channel). The worst configurations for the nonredundant INSs are for the second sensor in each detection cluster, the absolute maximal values of errors in North position channel being 990.52 m, 318.17 m in East position channel, 19.40 m/s in North speed channel, $89.141 \cdot 10^{-4}$ degrees in the Latitude position channel, and $39.886 \cdot 10^{-4}$ degrees in the Longitude channel. From the point of view of East speed channel and yaw angle channel, the worst nonredundant INS is INS4, the East speed error being 13.68 m/s and the yaw angle error being 31.10 degrees.

5. CONCISE SUMMARY

The chapter revealed an innovative method for diminishing noise in inertial sensor data utilizing a fuzzy logic technique. The devised system requires a redundant inertial measurement unit with inertial sensors arranged in linear detection clusters, all measuring the same quantity. The suggested technique was initially subjected to numerical simulation using models developed in Matlab/Simulink for the detection of clusters of varying sizes and including diverse sensor types. All simulation results demonstrated the effective operation of the method, which significantly reduced sensor noise levels and had enhanced potential for improving navigation accuracy when employed in an inertial navigator. Secondly, the algorithm was incorporated into a redundant Inertial Navigation System (INS) framework, with an Inertial Measurement Unit (IMU) comprising three detection clusters, each containing six miniaturized sensors. A combined GPS/INS navigator served as the reference positioning system to assess the inaccuracies of the redundant INS in each navigation solution channel. The data obtained from the redundant IMU were utilized in a software model in both non-redundant configuration to compute navigation solutions for six detection groups, which comprised sensors with identical quantities in the detection clusters, and in the INS redundant configuration when our data fusion methodology was initially applied. The graphical findings and tabulated numerical values demonstrated a significant enhancement in positioning precision with the use of the proposed data fusion mechanism. The mistakes in the navigation solution were significantly diminished in comparison to the nonredundant navigation solutions. The numerical data indicated a reduction in the mean of the absolute maximal error values for the six nonredundant INSs, approximately 16.6 times for North positioning, 5.1 times for East positioning, 2.8 times for the North speed component, 1.05 times for the East speed component, and 3.8 times for angular positioning in the horizontal plane (yaw angle). It is crucial to note that the inertial sensor outputs in the experimental model of the inertial measurement unit were not previously adjusted for bias or other error values.

AT A GLANCE

The chapter introduces a novel technique to enhance the accuracy of inertial navigators by processing the inputs from tiny sensors in a fuzzy manner within their inertial measurement units (IMU). The application of the established technique necessitated the design of the hardware component of the inertial measurement units to have redundant clusters of inertial sensors arranged in linear configurations along the measurement axes. The algorithm operates at the level of each detection cluster, which is intended to quantify acceleration or angular velocity along an IMU axis by integrating the data acquired from the sensors inside the particular cluster. Utilizing the standard deviations of the sensor signals calculated for a data frame of a predetermined size, the fuzzy logic mechanism generates a set of weights corresponding to each sensor within the cluster, which are subsequently employed to amalgamate the data obtained from the sensors at the present moment. The technique has an adaptive nature, with the data frame utilized for estimating the standard deviations of the sensor signals being continuously updated with fresh sensor measurements. Consequently, the weights assigned to each sensor are re-evaluated at each measurement interval.

REFERENCES

1. Barbour, N.M. Inertial navigation sensors. In: NATO RTO-EN-SET-116; 2010.
2. Barbour, N.M., Schmidt, G. Inertial sensor technology trends. IEEE Sensors Journal. 2001;1(4):332-339.

3. Barbour, N.M. et al. Inertial MEMS Systems and Applications. In: NATO RTO-EN-SET-116-2011, Low-Cost Navigation Sensors and Integration Technology; Bagneux, France, 28-29 March, 2011.
4. Schmidt, G.T. INS/GPS technology trends. In: NATO RTO-EN-SET-116-2011, Low-Cost Navigation Sensors and Integration Technology; Bagneux, France, 28-29 March, 2011.
5. Farrell, J. Aided Navigation. GPS with High Rate Sensors. McGraw-Hill, New York; 2008.
6. Mohinder,S., Lawrence, R., Angus, P. Global Positioning Systems, Inertial Navigation, and Integration. John Wiley & Sons, Inc., Hoboken, New Jersey, 2001.
7. Titterton, D.H., Weston, J. Strapdown Inertial Navigation Technology. 2nd ed. IET, Stevenage Herts, United Kingdom; 2004.
8. Grigorie, T.L., Sandu, D.G. Navigation Systems Synergic Architectures With Strap-Down Inertial Components. SITECH, Craiova, Romania; 2013.
9. Grigorie, T.L., Edu, I.R. Inertial Navigation Applications With Miniaturized Sensors. SITECH, Craiova, Romania; 2013.
10. He, X., Hu, X., Wu, M. Trends in GNSS/INS Integrated Navigation Technology. Coordinates. 2007; III(3):26-28.
11. Al-Faiz, M.Z., Ismaeel, S.A. Design of Kalman filter for augmenting GPS to INS systems. In: International Conference on Advanced Remote Sensing for Earth Observation. 2005; 14-20.
12. Lorinda, S., Aboelmagd, N. Bridging GPS outages using neural network estimates of ins position and velocity errors. Measurement of Science & Technology. 2006; 17:2783-2798.
13. Grigorie, T.L., Edu, I.R., Corcau, J.I. Fuzzy logic denoising of the miniaturized inertial sensors in redundant configurations. In: 33rd International Conference on Information Technology Interfaces (ITI 2011); June 27-30; Cavtat, Croatia. 2011. p. 521-526.
14. Grigorie, T.L., Botez, R.M. Positioning monitoring improvement in a horizontal plane ins by using fuzzy logic data fusion for denoising of inertial sensors in redundant clusters. International Journal of Fuzzy Systems And Advanced Applications. 2015; 2:33-40.
15. Hasan, A.M., Samsudin, K., Ramli, A.R., Azmir, R.S. Wavelet-based pre-filtering for low cost inertial sensors. Journal of Applied Science. 2010;10:2217-2230.

16. Mao, B., Wu, J.W., Wu, J.T., Zhou, X.M. MEMS gyro denoising based on second generation wavelet transform. In: First International Conference on Pervasive Computing; China. 2010. pp. 255-258.

17. Sabat, S.L., Giribabu, N., Nayak, J., Krishnaprasad, K. Characterization of fiber optics gyro and noise compensation using discrete wavelet transform. In: Second International Conference on Emerging Trends in Engineering & Technology; Indian. 2009. pp. 909-913.

18. Skaloud, B., Bruton, A.M., Schwartz, K. Detection and filtering of short-term (1/F) noise in inertial sensors. Journal of the Institute of Navigation. 1999; 46(2):97-107.

19. Grigorie, T.L., Negrea, P., Edu, I.R., Adochiei, F.C. Assistive positioning system based inertial techniques and wavelet denoising. In: 8th International Conference on PErvasive Technologies Related to Assistive Environments (PETRA); July 1-3, 2015; Corfu, Greece. 2015

20. Kovacic, Z., Bogdan, S. Fuzzy Controller Design – Theory and Applications. Taylor and Francis Group, USA; 2006.

21. Castellano, G., Fanelli, A.M., Mencar, C. Design of transparent mamdani fuzzy inference systems. In: Design and Application of Hybrid Intelligent Systems. IOS, Amsterdam; 2003. pp. 468-476.

22. Grigorie, T.L., Lungu, M., Edu, I.R., Obreja, R. Concepts for error modeling of miniature accelerometers used in inertial navigation systems. Annals of the University of Craiova, Electrical Engineering Series. 2010; 34:212-219.

23. Grigorie, T.L., Botez, R.M. Modelling and simulation based Matlab/Simulink of a strap-down inertial navigation system' errors due to the inertial sensors. In: Kelly Bennett, editor. Matlab Applications for the Practical Engineer. InTech, Rijeka, Croatia; 2014. p. 305-338. DOI: 10.5772/57583

24. Grigorie, T.L., Botez, R.M. Modeling and numerical simulation of an algorithm for the inertial sensors errors reduction and for the increase of the strap-down navigator redundancy degree in a low cost architecture. Transactions of the Canadian Society of the Mechanical Engineering (CSME). 2010; 34(1):1-16.

14 Chapter

MODELING AGENT-BASED IMAGE ANALYSIS WITH FUZZY BELIEFS, DESIRES, AND INTENTIONS

1. INTRODUCTION: AN OVERVIEW

The examination of remote sensing data is intrinsically linked to image processing and image analysis methodologies. Unlike other imaging techniques, remote sensing is defined as a way for obtaining information about the Earth's surface by detecting and analyzing its reflected or emitted electromagnetic radiation without direct touch. In addition to the radiation of the visible spectrum, infrared (optical data) and microwave radiation (RADAR) are utilized to generate remote sensing images. Remote sensing devices may be deployed on spacecraft (often satellites) or aerial vehicles (such as airplanes and drones). To obtain geo-information from remote sensing data, the generated images must be evaluated, namely through preprocessing and classification. In this context, picture classification refers to the allocation of pixels to significant item categories on the earth's surface, with the identified and categorized objects ultimately stored in a geographic information system (GIS) as polygons, lines, or points (vector model). The ongoing enhancement of spatial and radiometric resolution in remote sensing images has rendered image analysis increasingly complicated. Until the late 1990s, most remote sensing data was evaluated using categorization methods that considered the radiation captured in each individual pixel. Simultaneously, advanced techniques in pattern analysis, artificial intelligence, and computer vision are utilized. The emergence of very high resolution (VHR) satellite imagery has rendered traditional image classification algorithms ineffective, as the objects of interest in VHR data are depicted by multiple spectrally heterogeneous pixels. Furthermore, attributes like as shape, texture, and spatial context are significant in recognizing and defining items of interest within this data [1–3]. Consequently, almost concurrently with the emergence of VHR satellite imagery, object-based image analysis (OBIA) has developed into a recognized paradigm for the interpretation of remote sensing data. Unlike pixel-based analysis methods, OBIA utilizes picture objects as the fundamental components for analysis. Image objects are first created using a random image segmentation, succeeded by a preliminary categorization of these segments. The feature space for classification may possess great dimensionality, encompassing attributes such as color, shape, texture, or spatial context relevant to the target item classes. A variety of classifiers can be utilized, including simple thresholding, Support Vector Machines (SVM), Bayesian Network Classifiers (BNC), and Artificial Neural Networks (ANN). Fuzzy set assignments are also feasible. The definition of fuzzy sets and the corresponding fuzzy classification rules must align with the ontology of the intended object classes [4–6]. The standard workflow of OBIA commences with initial segmentation and classification, subsequently followed by an iterative process aimed at enhancing segmentation and classification through knowledge-based methods. The latter illustrates the task-ontology that delineates the requisite expert knowledge in image processing, which may be preserved as an OBIA rule set and subsequently reapplied. Nevertheless, as the analysis of remote sensing data requires greater precision and reliability, the methods and rule sets become increasingly complex. The former ultimately diminishes the transferability of the rule sets. To attain satisfactory outcomes for various picture datasets, increased manual intervention, including the modification of individual rules or the human adjustment of object boundaries and class allocations, is required [7, 8]. Therefore, to leverage the benefits of OBIA for multiple images or entire image archives, intelligent and adaptable technologies are required that can autonomously adjust to image variability.

2. AGENT-BASED AND MULTI-AGENT SYSTEMS

Agent-based and multiagent systems (MAS) recently show a variety of applications: they range from simulation of complex systems such as social systems [9] and ecosystems [10, 11] to the automation and optimization of complex production systems such as industrial processes [12–14]. In software development, meanwhile the agent-oriented paradigm has evolved as a new paradigm that extends the classic object-oriented approach. Simply spoken, the general differences are: (1) objects behave rather passive than agents, that is, objects only change once they receive an appropriate signal while agents behave proactive and collaborative and (2) agents can be mobile while objects are static. Thus, agents have (individual) goals they intend to achieve; they have sensors and effectors that enable them to become aware about their current status and to interact with their environment. Agents can decide autonomously about their potential next action. The environment agents are interacting with can be of arbitrary complexity ranging from other (human) agents to factory plants, sports fields to traffic situations, etc. When embedded in (collaborative) MAS individual agents often have different roles but common goals.

All these abilities allow software agents and MAS to react flexible but robust on unforeseeable changes of their environment.

Since each agent needs to have certain situation awareness, each agent must be capable to appraise its current situation. That is, evaluating the grade of goal achievement and the acting opportunities supporting the agent's goal achievement. This sort of situation awareness is commonly known as the belief-desire-intention (BDI) model [15–19]. Simply spoken, the BDI model allows an agent to analyze its current situation and to choose from a predefined list of plans the most promising one in order to achieve its goals. It is obvious that for the design of software agents and MAS ontologies are required, which are capable to formally describe an agent's or MAS' environment and which allow individual agents to infer their current situation. Further, ontologies are necessary to describe an agent's goals and to infer the most promising action for goal achievement [20]. Casali et al. [21] extend the classic BDI model to a graded BDI (g-BDI) model, which allows each agent to express its preferences among its acting opportunities, while Shen et al. [22] introduce an agent fuzzy decision making (AFDM) approach, which extends the classic BDI model "by making decisions based on quantified fuzzy judgment." Zarandi and Ahmadpour [23] present a fuzzy agent-based expert system for the steel making process that uses a fuzzy described knowledge base.

2.1. Agent-based image analysis

Although meanwhile matured, the agent-based paradigm and potential applications based on it are not very widespread in the image analysis community, yet. The most applications can be found in the field of image coregistration and image fusion [24–26]. In the field of object detection and delineation reported applications are still rare [27–31]—even more in the field of remote sensing image analysis [32–35].

2.2. Software agents and multiagent systems in OBIA

As already mentioned, in order to fully exploit the advantages of OBIA there is a strong need for more robustness and transferability of methods. The limiting factors are the rule sets' complexity and the unpredictable variations of the image objects' appearance in remote sensing data. At this background recently the integration of the agent-based paradigm with OBIA are investigated in order to improve its degree of automation, its robustness, and its transferability. MAS seem to have the potential to overcome OBIA's obstacles [33, 36–38]. Especially their ability to react flexibly on environmental perturbations, which is given in the remote sensing domain by varying illuminations, seasons, locations, sensors, and atmospheric conditions, are promising aspects to be investigated. Consequently, Hofmann et al. [38] developed a conceptual framework for agent-based image analysis (ABIA), which suggests two principle ways of integration: (1) adapting already existing OBIA rule sets(e.g., thresholds of single rules) by means of a MAS built by respective rule set adaptation agents (RSAAs) and (2) evolving OBIA image objects to image object agents (IOAs). In the first approach different RSAAs adapt a rule set's rules in order to improve its classification results. As constraint, adaptations must not violate the underlying ontology of the original rule set. The latter is controlled by one or more control agents (CAs), which also give feedback whether a to-be-defined minimum classification quality has been achieved after rule set adaptation (Figure 1).

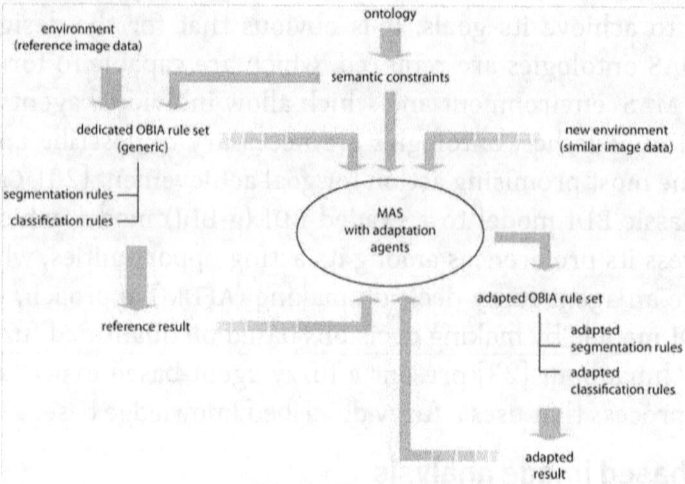

Figure 1. Principle workflow for OBIA rule set adaptation by means of a MAS built by RSAAs (Source: [38]).

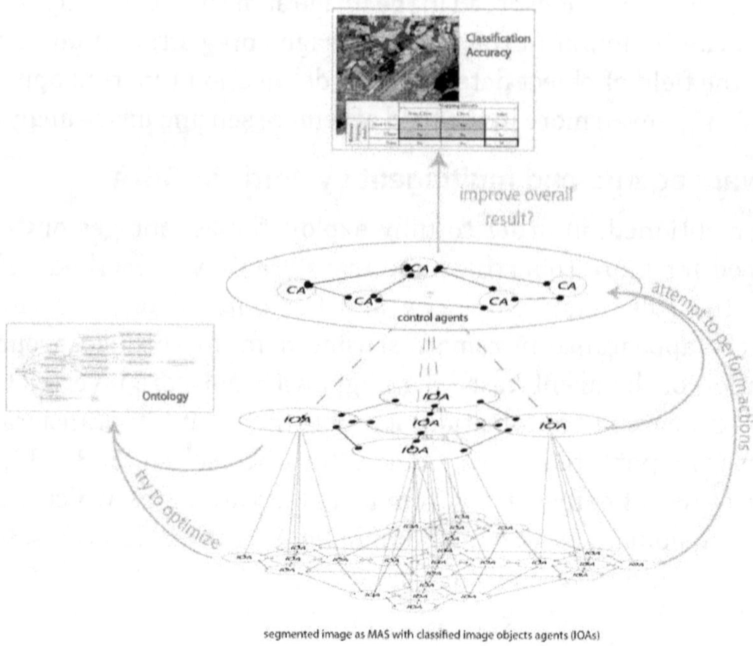

Figure 2. Scheme of a MAS built up by hierarchically organized IOAs and CAs (Source: [38]).

In the second approach after initial segmentation and (fuzzy) classification IOAs build a hierarchical network of IOAs. Each IOA intends to become a best possible member of its assigned class (goal, desire). To achieve this goal, every IOA can change its shape by a number of predefined methods (effectors, intention). Further, every IOA is aware about its topological situation and can communicate within other IOAs in the hierarchical net. The underlying ontology for this approach is given by the (fuzzy) class descriptions (Figure 2).

The ABIA framework has been implemented in a typical environment for agent-based modeling (REPAST [39]) as well as in a typical OBIA environment (eCognition) [40], realizing the IOA approach. In both implementations the real world objects to be delineated and identified were described as fuzzy sets based on an appropriate ontology. As test scene a very high resolution digital orthophoto (0.08m) has been used together with an appropriate digital surface model (DSM) together with the calculated slope and curvature (slope of slope) per pixel (Figure 3).

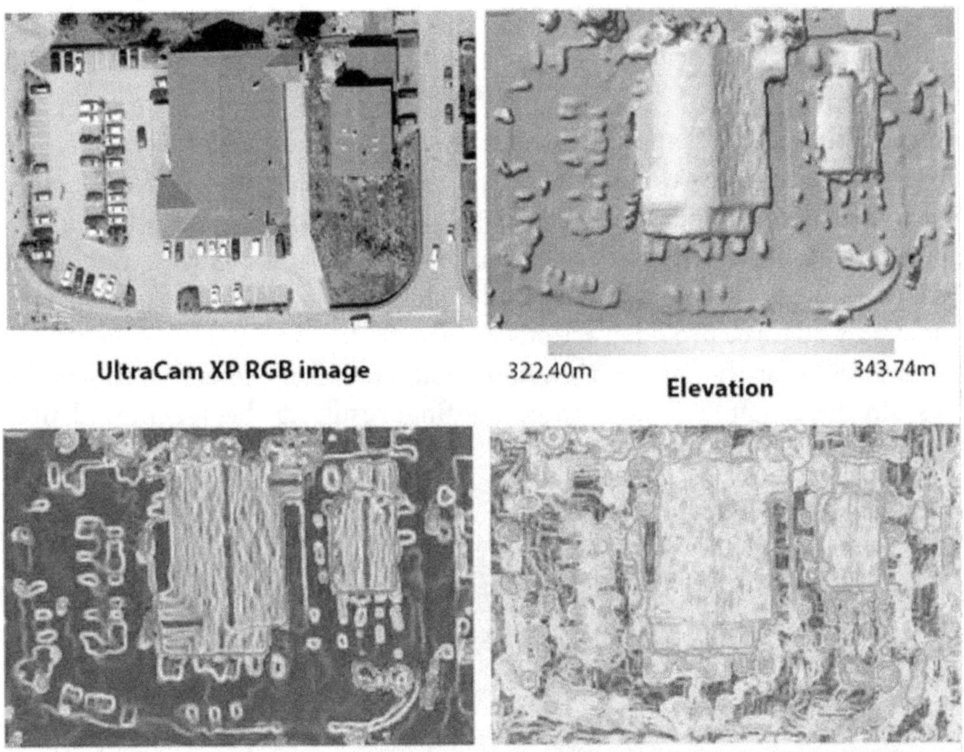

Figure 3. Used image and DSM data for first implementation of the ABIA framework (Source: [40]).

The rule set was intentionally designed to delineate buildings in that particular scene following the ontology as outlined in Figure 4.

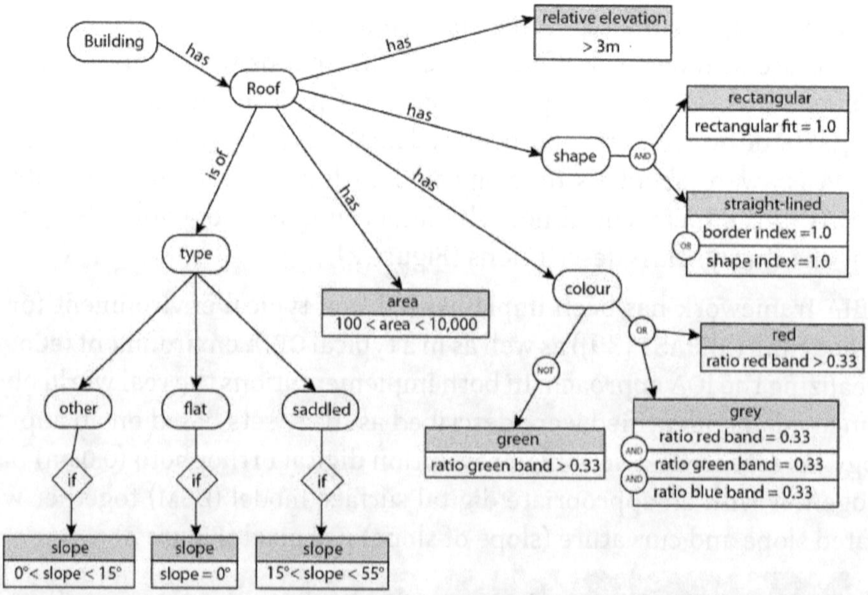

Figure 4. Ontology describing buildings and their appearance in the given data (Source: [38]).

However, if the rule set is applied without any further adaptations it creates a rather over-segmented image—which would be a typical OBIA use case after reapplying a given rule set on similar images. The BDI model to solve this problem has been implemented as a hybrid model. That is, the class definitions were implemented as fuzzy sets, whereas the next action's decision rules were designed crisp. For the latter simply all three provided actions were virtually executed for each IOA. Every IOA then opted for that action that improved its class membership to "building" at best. In the example demonstrated the final result has been achieved after 17 iteration steps already (Figure 5).

Figure 5. Segmentation and classification result before (left) and after (right) applying the agent-based optimization approach according to the ABIA framework. Numbers indicate the membership degree to "building" (Source: [38]).

3. A FUZZY BELIEVE-DESIRE-INTENTION MODEL FOR AGENT-BASED IMAGE ANALYSIS

Based on the already achieved results with the relative simple crisp BDI model it has been investigated here whether the IOAs' intentions could also be expressed in fuzzy manner and whether this is of advantage compared to the above described BDI model. For this purpose the existing rule set, which has been developed in a software environment dedicated for OBIA (here: the commercial software eCognition), has been extended by the necessary components. In a standard fuzzy control-system control is given implicitly through the membership functions, e.g., "the colder room temperature the more open heater's valve." In this particular case the used software only allows fuzzy sets (alias classes) to be described in fuzzy manner. Consequently, the agents' acting opportunities had to be expressed as fuzzy sets whereas the membership degree to an "action-class" can be interpreted as the "intention-degree" or the willingness of an agent to perform that particular action. Another difference of the fuzzy BDI model developed here is that it only evaluates the current situation. That is, there is no virtual test for each potential action if and how it would improve an agent's situation (here: its class membership).

3.1. Components of the fuzzy BDI model

In order to fuzzify each IOA's believe its current status quo after segmentation and classification is analyzed. That is, each IOA had to be enabled not only to know its current class membership degree (degree of goal/desire achievement), but also for each classification rule the degree of its contribution to that particular classification result. The latter would allow each IOA to select an action that improves one of those conditions.

3.1.1. Fuzzy beliefs

For this purpose, the conditions that build a "building" as described in the ontology were separated into the four categories: color conditions, DSM conditions, and shape conditions. The class "roof"—which represents buildings—consequently was now described through these aggregated conditions whereas the property "Classification Value of ..." expresses the membership degree μ for this particular class or the degree of fulfillment (DOF) of that particular condition. Similarly, the condition classes were further deconstructed to classes describing the DOF's property conditions (feature-2-class) or operator conditions (operator-2-class). Table 1 shows the cascaded classification scheme. With this decomposition every IOA now can determine the grade each of the fuzzy classification rules or rule groups (color conditions, shape conditions, or context conditions) it contributes to its final class assignment result (= grade of goal achievement). In the example displayed in Figure 6 the **red** outlined IOA fulfills the criteria for "roof" only by

0.432, whereas the color conditions are fully fulfilled, the shape conditions are fulfilled by 0.555 and the DSM conditions are fulfilled by 0.432. The shape conditions are not fulfilled, because the IOA's rectangular fit value of 0.76 leads to a grade of fulfillment for that condition of 0.555 only.

Table 1. Cascaded fuzzy classification scheme for the class "roof" as described in Figure 4.

Class	Properties	Operator	Membership function type	Function values	
				Lower bound	Upper bound
Roof	Classification value of 'Color conditions'	AND	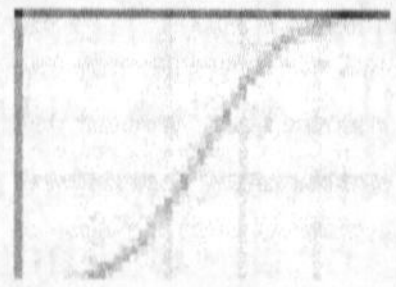	0.00	1.00
	Classification value of 'DSM conditions'		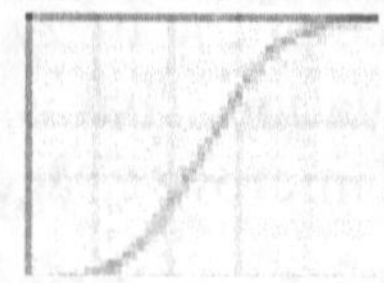	0.00	1.00
	Classification value of 'Shape conditions'		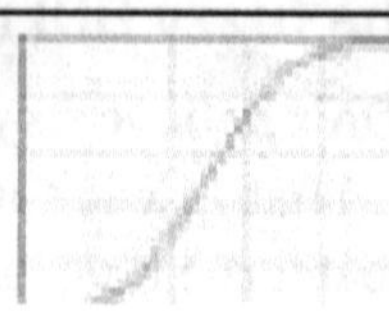	0.00	1.00
Color conditions	Classification value of 'feature-2-class "Ratio green"'	AND	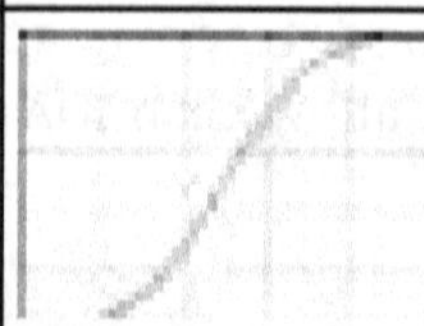	0.00	1.00
	Classification value of 'operator-2-class "Red-OR-Grey"'		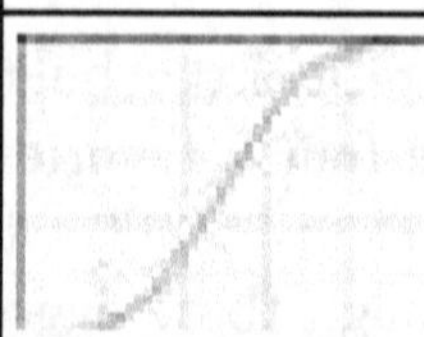	0.00	1.00
DSM conditions	Classification value of 'feature-2-class "Mean difference to neighbors DSM"'	AND	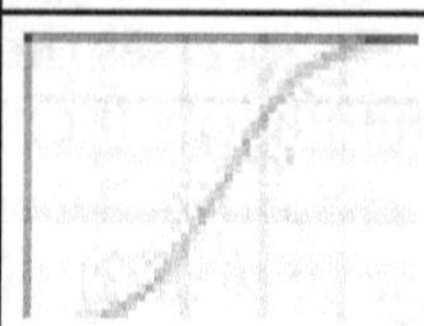	0.00	1.00
	Classification value of 'feature-2-class "Mean Slope-of-Slope"'		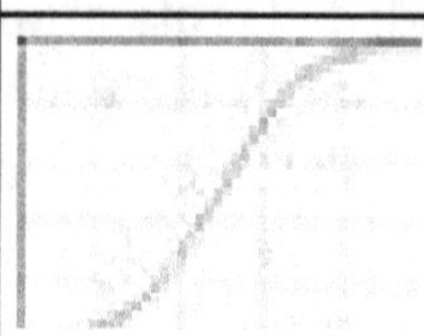	0.00	1.00

Class	Properties	Operator	Membership function type	Function values	
Shape conditions	Classification value of 'feature-2-class "Lower Bound of Area"'	AND	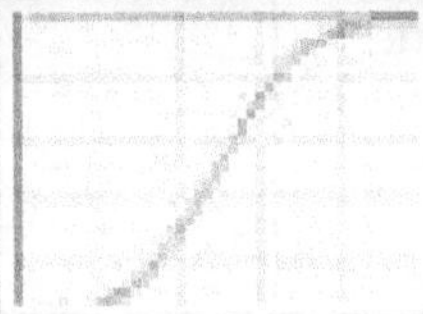	0.00	1.00
	Classification value of 'feature-2-class "Upper Bound of Area"'		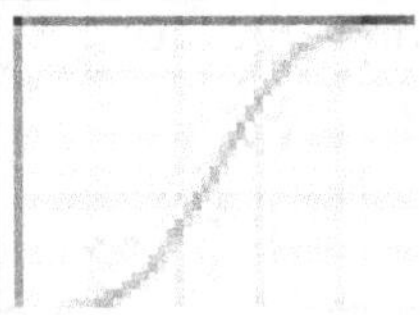	0.00	1.00
	Classification value of 'feature- 2-class "Rectangular fit"'			0.00	1.00
	Classification value of 'operator-2-class "Border Index-OR-Shape Index"'		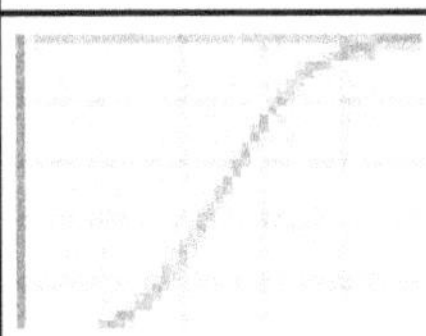	0.00	1.00
Feature-2-class "Ratio green"	Ratio green	–	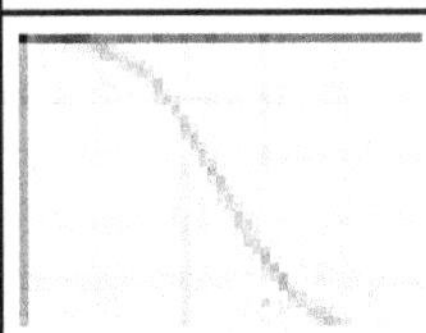	0.32	0.34
Feature-2-class "Ratio red"	Ratio red	–	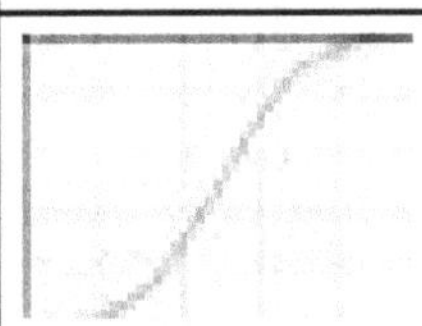	0.32	0.34
Feature-2-class "grey"	Ratio green	AND	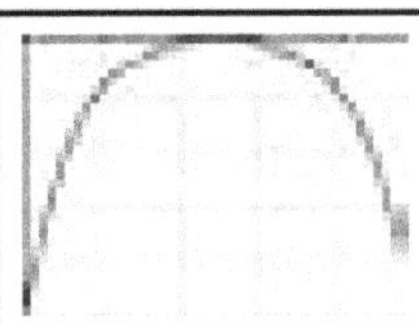	0.31	0.35
	Ratio red		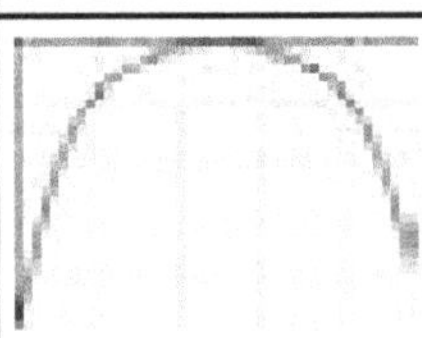	0.31	0.35

Class	Properties	Operator	Membership function type	Function values	
	Ratio blue			0.31	0.35
Operator-2-class "Red-OR-Grey"	Classification value of 'feature-2-class "grey"'	OR	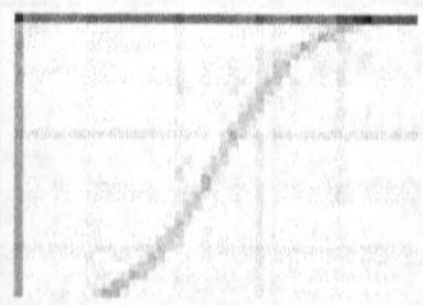	0.00	1.00
	Classification value of 'feature-2-class "Ratio red"'		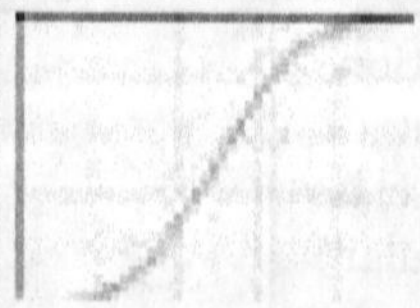	0.00	1.00
Feature-2-class "Mean difference to neighbours DSM"	Mean difference to neighbors DSM	–	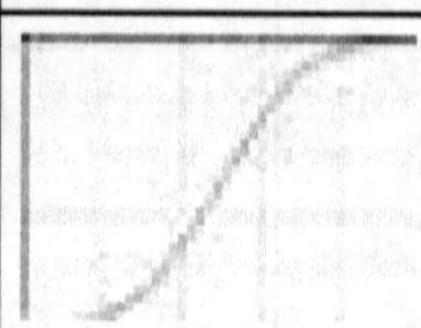	0.00	3.50
Feature-2-class "Mean Slope-of-Slope"	Mean Slope-of-Slope	–	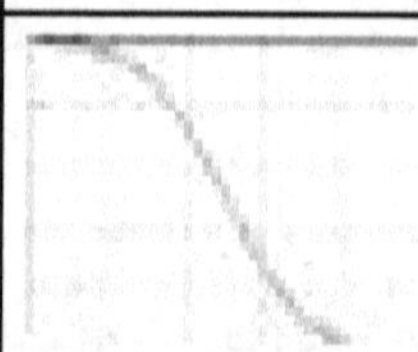	88.00	90.00
Feature-2-class "Lower Bound of Area"	Area	–	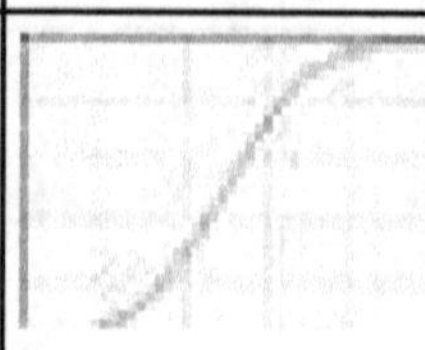	10.00	100.00
Feature-2-class "Upper Bound of Area"	Area	–	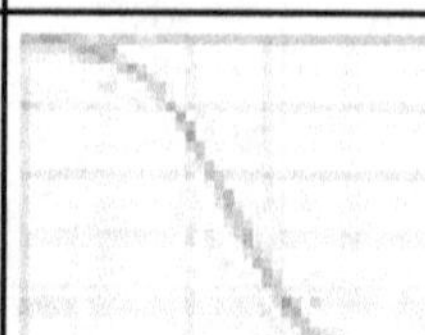	7391.60	7391.70
Feature-2-class "Border Index"	Border Index	–	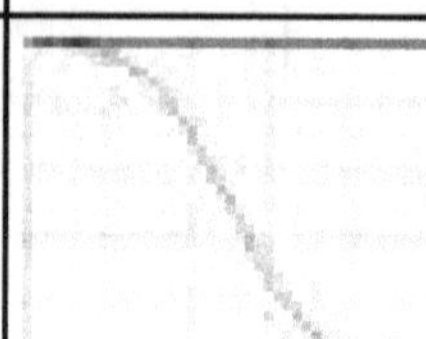	1.20	3.00

Class	Properties	Operator	Membership function type	Function values	
Feature-2-class "Rectangular fit"	Rectangular fit	–		0.50	1.00
Feature-2-class "Shape Index"	Shape Index	–		1.30	3.00
Operator-2-class "Border Index-ORShape Index"	Classification value of 'feature-2-class "Border Index"'	OR	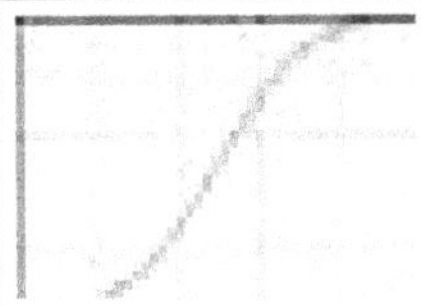	0.00	1.00
	Classification value of 'feature-2-class "Shape Index"'		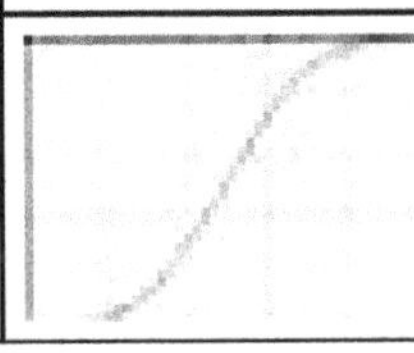	0.00	1.00

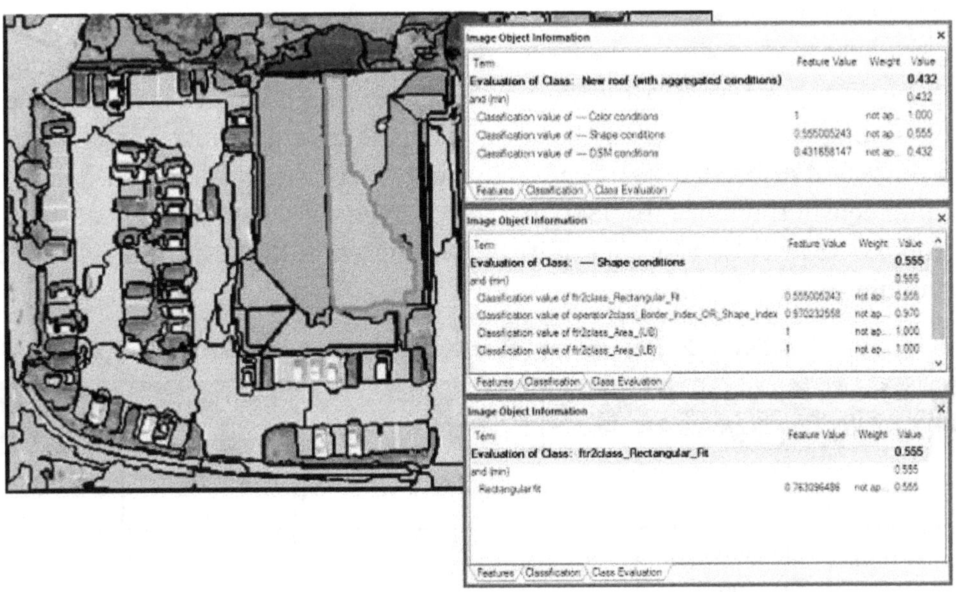

Figure 6. Evaluation of classification conditions of an IOA.

Similarly, the IOA's DSM conditions show a DOF of 0.432 because the mean difference to neighbors in the DSM is at 1.631 translating to a DOF for that feature of 0.432.

3.1.2. Fuzzy intentions

Based on these fuzzy beliefs an intended next action can be determined in fuzzy manner based on appropriate fuzzy decision rules—again expressed as fuzzy sets. In the example present the following possible actions were implemented: "grow," "shrink," "smooth," "merge," and "do nothing." While the latter action is obvious and applies only if an IOA has achieved its goal already, the former actions point to procedures eCognition offers and can be easily exchanged or adapted if necessary. In this particular example the actions translate to:

- "grow": grow the IOA of concern by one pixel into neighbor IOAs, which are unclassified.
- "shrink": shrink the IOA of concern by one pixel.
- "merge": merge the IOA of concern with its unclassified neighbors.
- "smooth": perform a grow-and-shrink sequence by one pixel each starting with shrink.

The appropriate intentions are defined as the fuzzy sets: "want grow," "want shrink," "want smooth," "want merge," or "do nothing." The degree or the intensity an agent wants to execute one of these actions can depend on the prior determined classification conditions or on the spatial situation in which IOA is embedded in, or a combination of both. Each of the action intensities is expressed gradually, that is, through an action-membership (Table 2). In the example given, an IOA wants to shrink if the situation is similar to that of "want grow;" it wants the more shrink the closer its size is at the upper bound of the area-rule for "building" ("Upper bound of Area"). An IOA wants to do nothing the more it fulfills already the "building" criteria. It wants the more grow, the lower its contrast and elevation difference is at its border. Its intention to merge increases, the lower its area is and simultaneously the more its DSM and color criteria are similar with those of its neighbors. Its intention to smooth its border increases the less the shape criteria for building are fulfilled.

Table 2. Definition of intentions as fuzzy sets.

Intension	Properties	Membership function	Operator	Function values	
				Lower bound	Upper bound
Do nothing	Classification value of "Color conditions"		AND	0.00	1.00

	Classification value of "DSM conditions"	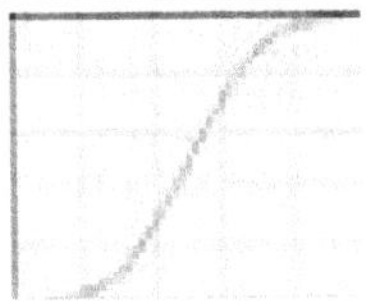		0.00	1.00
	Classification value of "Shape conditions"	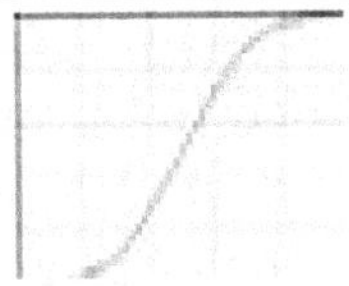		0.00	1.00
Want grow	Border contrast Blue	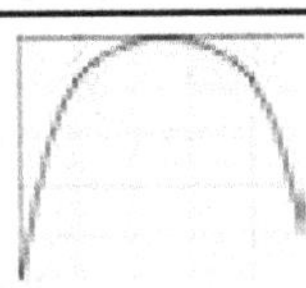	AND	Minimum/10	25.00
	Border contrast Green	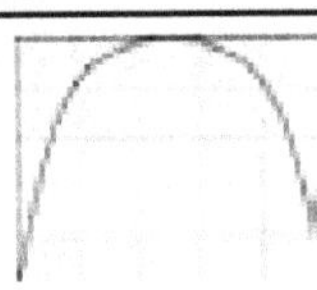		Minimum/10	25.00
	Border contrast Red	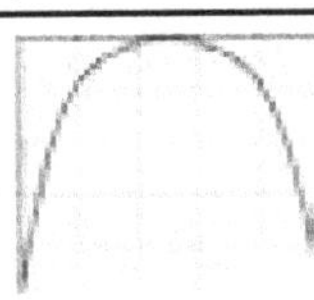		Minimum/10	25.00
	Border contrast DSM	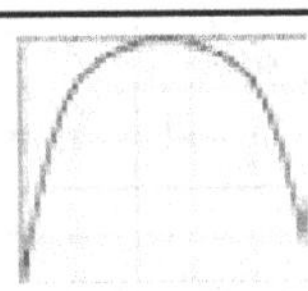		-200	2.00
Want merge	Area	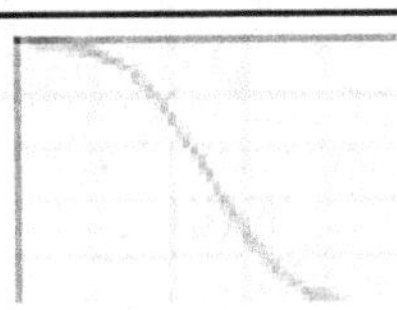	AND	10.00	2500.00
	Classification value of "Similarity to neighbors (DSM)"	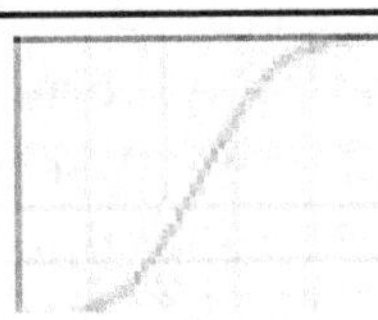		0.00	1.00
	Classification value of "Similarity to neighbors (Color)"	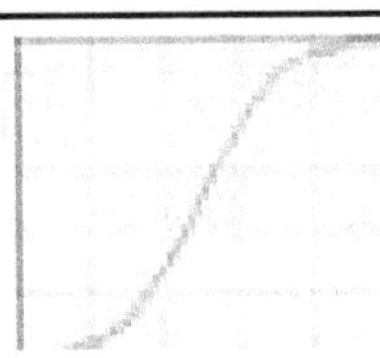		0.00	1.00
Want shrink	Want grow		AND		
	Feature-2-class "Upper bound of Area"				

Want smooth	NOT(feature-2-class “Rectangular Fit”)	AND		
	NOT(operator-2-class “Border Index OR Shape Index”			

As displayed in Figure 7, the **red** outlined IOA prefers to grow. However, similar to the ambiguity of class assignments of objects, each IOA can have ambiguous intentions in terms of a favorite, a second favorite, etc. action. Further, as with fuzzy class assignments [41], a minimum intention value should be defined (sensibly not less than 0.5) below which an intended action must be seen as not clearly enough wanted and thus not further considered.

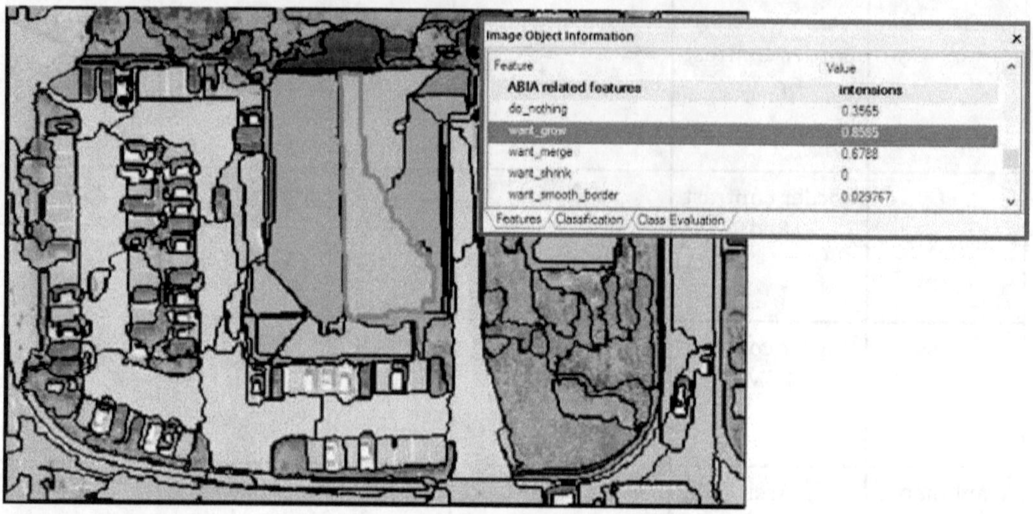

Figure 7. Grades of intentions for actions an IOA wants to perform for its improvement.

In the example given in Figure 7 although the IOA’s most wanted action is to grow, this action is not the only one it intends to perform. Since the intention value for “want_grow” is not clearly around 1.0 the IOA seems to be not fully convinced about this action to achieve its goal. Merging (intention = 0.68) seems to be an option although it is second-best. In other words, the IOA could also be satisfied if it merges. The only thing that is clear, is, that the IOA does not want to shrink (intention = 0.0). And its willingness to smooth its border (intention = 0.0298) is even lower than that for doing nothing (intention = 0.356).

4. CONCISE SUMMARY

The volume of remote sensing data stored in archives is perpetually escalating. In light of the growing demand for dependable, accurate, and prompt geoinformation, the exploration of these archives and the analysis of the stored picture data necessitate methods for trustworthy and automated image analysis. Although OBIA is a recognized and very precise technique for assessing particularly VHR remote sensing data, its durability, transferability, and level of automation remain limited. Significant challenges in automating OBIA and related methodologies include its sensitivity to perturbations, specifically the variability of pictures and objects. ABIA, as an extension of the OBIA paradigm, possesses the capacity to surmount these challenges due to its enhanced flexibility and robustness in responding to unforeseen perturbations. Nonetheless, research in this domain remains in its nascent stages.

5. OUTLOOK

The established fuzzy BDI model serves as the foundation for a negotiation framework applicable to ABIA agents: leveraging their distinct action priorities, IOAs can negotiate their collective subsequent action(s) and thereby enhance the overall classification outcome of multiple yet varied images of the same category, as is customary in remote sensing. The fuzzy BDI model, although built for IOAs, is also fundamentally relevant to rule set adaptation agents (RSAAs).

AT A GLANCE

Contemporary image analysis techniques in remote sensing exhibit inadequate robustness and transferability. Techniques like object-based image analysis (OBIA) yield satisfactory outcomes on individual photos. Nonetheless, the foundational rule sets for OBIA are typically too intricate to be directly implemented on diverse picture data without modifications or human intervention. Consequently, new research endeavors explore the feasibility of merging the agent-based paradigm with Object-Based Image Analysis (OBIA). Agent-based systems exhibit significant adaptability and, thus, robustness, even amongst fluctuating environmental conditions. In the realm of image analysis, this implies that even minor variations in the image data (e.g., due to seasonal effects, diverse locations, atmospheric conditions, or slightly different sensors) enable agent-based methods to autonomously adjust existing analytical rules or segmentation outcomes in response to evolving imaging circumstances. The foundation for the behavior of individual software agents is the belief-desire-intention (BDI) paradigm. The BDI delineates each agent's objectives, its presumed current circumstances, and several action rules that may assist the agent in attaining its goals. The chapter presents a belief-desire-intention (BDI) model founded on fuzzy rules inside the framework of agent-based image analysis, thereby enhancing the traditional object-based image analysis (OBIA) paradigm through the agent-based approach.

REFERENCES

1. U. C. Benz, P. Hofmann, G. Willhauck, I. Lingenfelder, and M. Heynen, "Multi-resolution, object-oriented fuzzy analysis of remote sensing data for GIS-ready information," ISPRS Journal of Photogrammetry and Remote Sensing, vol. 58, pp. 239-258, 2004.
2. T. Blaschke, M. Kelly, P. Hofmann, et al, "Geographic object-based image analysis - towards a new paradigm," ISPRS Journal of Photogrammetry and Remote Sensing, vol. 87, pp. 180-191, 2014.
3. T. Blaschke and J. Strobl, "What's wrong with pixels? Some recent developments interfacing remote sensing and GIS," GeoBIT/GIS, vol. 6, pp. 12-17, 2001.
4. P. Hofmann, "Detecting informal settlements from IKONOS image data using methods of object oriented image analysis-an example from Cape Town (South Africa)." In: Jürgens, C. (Ed.): Proceedings of the 2nd International Symposium on Remote Sensing of Urban Areas, Regensburg, Germany, June 22-23 2001, pp. 41-42, 2001.
5. M. Belgiu, B. Hofer, and P. Hofmann, "Coupling formalized knowledge bases with object-based image analysis," Remote Sensing Letters, vol. 5, pp. 530-538, 2014.
6. M. Belgiu and J. Thomas, "Ontology based interpretation of very high resolution imageries–grounding ontologies on visual interpretation keys," AGILE 2013—Leuven, pp. 14-17, 2013.
7. P. Hofmann, "Übertragbarkeit von Methoden und Verfahren in der objektorientierten Bildanalyse-das Beispiel informelle Siedlungen," ed: Dissertation, University of Salzburg, 2005.
8. P. Hofmann, T. Blaschke, and J. Strobl, "Quantifying the robustness of fuzzy rule sets in object-based image analysis," International Journal of Remote Sensing, vol. 32, pp. 7359-7381, 2011.
9. A. Koch and D. Carson, "Spatial, temporal and social scaling in sparsely populated areas–geospatial mapping and simulation techniques to investigate social diversity," in Proceedings of the Geoinformatics Forum Salzburg, Offenbach, 2012, pp. 44-53.
10. H. R. Gimblett, Integrating geographic information systems and agent-based modeling techniques for simulating social and ecological processes: Oxford University Press, Oxford, 2002.
11. P. Göhner, Agentensysteme in der Automatisierungstechnik: Springer-Verlag, Heidelberg, 2013.

12. W. Shen, D. H. Norrie, and J.-P. Barthès, Multi-agent systems for concurrent intelligent design and manufacturing: Taylor & Francis, London, New York, 2005.

13. M. Wooldridge, An introduction to multiagent systems: 2nd Ed., John Wiley & Sons, Chichester, U.K., 2009.

14. M. Wooldridge, An introduction to multiagent systems: John Wiley & Sons, 2009.

15. M. Wooldridge and N. R. Jennings, "Intelligent agents: theory and practice," The knowledge Engineering Review, vol. 10, pp. 115-152, 1995.

16. M. Wooldridge, "Agent-based software engineering," IEEE Proceedings-software, vol. 144, pp. 26-37, 1997.

17. M. Georgeff, B. Pell, M. Pollack, M. Tambe, and M. Wooldridge, "The belief-desire-intention model of agency," in International Workshop on Agent Theories, Architectures, and Languages, 1998, pp. 1-10.

18. A. S. Rao and M. P. Georgeff, "BDI agents: From theory to practice". In: Proceedings of the First International Conference on Multiagent Systems, June 12-14, 1995, San Francisco, California, USA. The MIT Press 1995 pp. 312-319.

19. M. Bratman, Intention, plans, and practical reason, Center for the Study of Language and Information. 1987.

20. K. Sycara and M. Paolucci, Ontologies in agent architectures, In: S. Staab, R. Studer (Eds.): Handbook on Ontologies, Springer, Berlin, Heidelberg, 2004, pp. 343-363.

21. A. Casali, L. Godo, and C. Sierra, "A graded BDI agent model to represent and reason about preferences," Artificial Intelligence, vol. 175, pp. 1468-1478, 2011.

22. S. Shen, G. M. O'Hare, and R. Collier, "Decision-making of BDI agents, a fuzzy approach," in Computer and Information Technology, 2004. CIT'04. The Fourth International Conference on, September 14-16, 2004, pp. 1022-1027.

23. M. H. F. Zarandi and P. Ahmadpour, "Fuzzy agent-based expert system for steel making process," Expert Systems with Applications, vol. 36, pp. 9539-9547, 2009.

24. J. Beyerer, M. Heizmann, J. Sander, and I. Gheta, "Bayesian methods for image fusion," In: T. Stathaki (Ed.): Image Fusion: Algorithms and Applications, Elsevier, Amsterdam, pp. 157-192, 2008.

25. C. Elmas and Y. Sönmez, "A data fusion framework with novel hybrid algorithm for multi-agent Decision Support System for Forest Fire," Expert Systems with Applications, vol. 38, pp. 9225-9236, 2011.

26. F. Castanedo, J. García, M. A. Patricio, and J. M. Molina, "A multi-agent architecture based on the BDI model for data fusion in visual sensor networks," Journal of Intelligent & Robotic Systems, vol. 62, pp. 299-328, 2011.

27. G. P. Bovenkamp, J. Dijkstra, J. G. Bosch, and J. H. C. Reiber, "Multi-agent segmentation of IVUS images," Pattern Recognition, vol. 37, pp. 647-663, 2004.

28. N. Richard, M. Dojat, and C. Garbay, "Automated segmentation of human brain MR images using a multi-agent approach," Artificial Intelligence in Medicine, vol. 30, pp. 153-176, 2004.

29. V. Rodin, A. Benzinou, A. Guillaud, P. Ballet, F. Harrouet, J. Tisseau, et al., "An immune oriented multi-agent system for biological image processing," Pattern Recognition, vol. 37, pp. 631-645, 2004.

30. K. E. Melkemi, M. Batouche, and S. Foufou, "A multiagent system approach for image segmentation using genetic algorithms and extremal optimization heuristics," Pattern Recognition Letters, vol. 27, pp. 1230-1238, 2006.

31. J. Nolan, A. Sood, and R. Simon, "Agent-based, collaborative image processing in a distributed environment". In Proceedings of the 5th International Conference on Autonomous Agents (ACM Agents 2001), Montreal, QC, Canada, May 28 - June 01,2001, pp. 228-235.

32. Samadzadegan, F. T. Mahmoudi, and T. Schenk, "An agent-based method for automatic building recognition from lidar data," Canadian Journal of Remote Sensing, vol. 36, pp. 211-223, 2010.

33. K. Borna, A. Moore, and P. Sirguey, "A vector agent approach to extract the boundaries of real-world phenomena from satellite images," in Proceedings of Research at Locate'14 (R@Loc-2014), Canberra, Australia, 2014.

34. K. Borna, A. Moore, and P. Sirguey, "Towards a vector agent modelling approach for remote sensing image classification," Journal of Spatial Science, vol. 59, pp. 283-296, 2014.

35. Tabib Mahmoudi, F. Samadzadegan, and P. Reinartz, "Object oriented image analysis based on multi-agent recognition system," Computers & Geosciences, vol. 54, pp. 219-230, 2013.

36. J. Liu and Y. Y. Tang, "Adaptive image segmentation with distributed behavior-based agents," IEEE Transactions on Pattern Analysis and Machine Intelligence, vol. 21, pp. 544-551, 1999.

37. R. D. Labati, V. Piuri, and F. Scotti, "Agent-based image iris segmentation and multiple views boundary refining." In: Proceedings IEEE 3rd International Conference on Biometrics: Theory, Applications, and Systems 28-30 Sept., Washington DC, 2009, pp. 1-7.

38. P. Hofmann, P. Lettmayer, T. Blaschke, M. Belgiu, S. Wegenkittl, R. Graf, et al., "Towards a framework for agent-based image analysis of remote-sensing data," International Journal of Image and Data Fusion, vol. 6, pp. 115-137, 2015.

39. C. M. Macal and M. J. North, "Tutorial on agent-based modelling and simulation," Journal of Simulation, vol. 4, pp. 151-162, 2010.

40. P. Hofmann, V. Andrejchenko, P. Lettmayer, M. Schmitzberger, M. Gruber, I. Ozan, et al., "Agent based image analysis (ABIA) - preliminary research results from an implemented framework." In: GEOBIA 2016 : Solutions and Synergies., 14 September 2016 - 16 September 2016, University of Twente Faculty of Geo-Information and Earth Observation (ITC), 2016.

41. P. Hofmann, "Defuzzification strategies for fuzzy classifications of remote sensing data," Remote Sensing, vol. 8(6), doi:10.3390/rs8060467, 2016.

15 Chapter

DEFINITION OF FOCAL LENGTH FOR ZOOM LENS USING FUZZY LOGIC FUNCTIONS IN ANFIS

1. INTRODUCTION: AN OVERVIEW

A fuzzy inference system (FIS) is a methodology for transforming specified inputs into outputs utilizing the principles of fuzzy set theory [1]. FIS generates an output utilizing an inference engine founded on IF-THEN rules. Two established types of FIS exist. Mamdani FIS employs defuzzification of a fuzzy output, whereas Sugeno FIS utilizes a weighted average to derive the crisp result [4]. The adaptive neuro-fuzzy inference system (ANFIS) is structured as a network that incorporates the neural learning capabilities of Sugeno-type fuzzy systems. This network consists of a collection of nodes arranged in layers to execute certain functions [5, 6]. The adaptive neuro-fuzzy inference system (ANFIS), introduced by J.S. Roger in his 1992 Ph.D. thesis, is one of the most utilized neuro-fuzzy models today [7–9]. ANFIS adjusts its settings based on training data through the application of a hybrid learning algorithm. The approach employs gradient descent for optimizing the non-linear antecedent parameters and least squares for adjusting the linear consequent parameters [1]. The success of ANFIS primarily derives from its implementation within the MATLAB Fuzzy Logic Toolbox, featuring an exceptional graphical interface developed by J.S. Roger in collaboration with N. Gulley, which also includes various fuzzy logic pattern classification algorithms for defining and dimensioning the input membership functions (mf) [9, 10]. The literature presents numerous instances of ANFIS, demonstrating its significant performance across various scientific applications. The beneficial attributes of an ANFIS encompass straightforward implementation, rapid and precise learning, robust generalization capabilities, superior explanatory mechanisms via fuzzy rules, and the seamless integration of both verbal and numerical knowledge for problem-solving. ANFIS models have recently gained significant popularity for calibrating non-linear relationships due to their advantages over traditional modeling techniques.

These advantages include the capacity to manage substantial volumes of noisy data from dynamic and non-linear systems, particularly when the underlying physical relationships are not completely understood, as well as for addressing linear systems, including interpolation modeling such as time series. Research indicates that numerous ANFIS and FIS methodologies have been presented for assessing uncertainty in pattern recognition, image matching, and three-dimensional (3D) positioning studies within computer vision applications. The study on the color recognition ANFIS model for a robust vision system is detailed in Ref. [12]. An alternative application of the ANFIS method is proposed for the effective calibration of a stereo camera system utilized for non-intrusive distance measurement in [13]. Another instance of the ANFIS approach in remote sensing research is the utilization of functions for analysis and prediction from satellite imagery of Prionace glauca for pattern identification, including Backpropagation networks, Radial Basis Function (RBF) networks, functional separability networks, and neuro-diffuse networks (ANFIS) [14]. Reference [15] presents a novel hybrid approach for executing eye-to-hand coordination and manipulation, resulting in the development of a functional robot named COERSU. The technique is an enhanced amalgamation of two neuro-fuzzy methodologies devised by the authors: direct fuzzy servoing and fuzzy correction. The adaptive neuro-fuzzy inference system (ANFIS) calibrates the fuzzy approaches. Human action recognition is a significant study domain within computer vision, with numerous practical applications. Reference [16] proposes the ANFIS controller approach for path tracking, a virtual field strategy for obstacle avoidance and path planning, and employs various sensors (an ultrasonic array, a thermal sensor, and a video streaming system) to gather environmental information. Reference [17] describes the advancement of a virtual robot teleoperation platform utilizing hand gesture detection derived from visual data through ANFIS and support vector machines (SVM). Reference [18] introduces a multi-view action identification framework that utilizes human silhouette cues from various cameras, employing a fuzzy rule-based approach to analyze scene dynamics and interpret human movements. The interaction tool has been positioned in 3D space, and the 3D model has been acquired by a structured light system calibrated just with the vanishing points derived from a basic planar surface. An immersive interaction strategy was employed to modify the 3D model via a fuzzy approach, which offers benefits such as minimal memory consumption, real-time functionality, and reduced positioning mistakes in contrast to traditional techniques referenced in [19]. A quantitative analysis of the error in reconstructing a 3D scene, captured using a synthetic aperture integral imaging system, has been conducted for two-dimensional (2D) images obtained under unknown camera parameters utilizing an adaptive neuro-fuzzy inference system (ANFIS) in [20]. This chapter presents an ANFIS methodology for identifying effective camera focal distance parameters, along with supplementary parameters that facilitate the measurement of image coordinates essential for three-dimensional modeling derived from two-dimensional images, akin to studies [19, 20]. Consequently, photogrammetric investigations utilizing a self-calibrating conventional model will be more applicable across various cameras and temporal contexts. This chapter primarily aims to process the chosen radial distances using ANFIS routines to derive a focal distance value for each image. The efficacy of ANFIS functions pertaining to optical distortion law can be substantiated through their impact on data sets. This further demonstrates that ANFIS functions corroborate a physical reality and articulate it with considerable reliability. The research indicates that fuzzy logic can be employed for the calibration process of digital cameras. Consequently, it is determined that further use and examination of the fuzzy logic approach in photogrammetric applications will be advantageous [21]. The second section of this chapter delineates information regarding fuzzy systems. The third section of this chapter pertains to the experimental application. The test field was captured from five distinct perspectives using a Nikon Coolpix E8700 digital camera, featuring a maximum focal distance of 71.2 mm and a minimum of 8.9 mm. The input and output data are established based on the resultant 10 photos; a spatial resection methodology is developed and analyzed utilizing ANFIS capabilities in MATLAB 7.0 software. The Trimf, Trapmf, Gbellmf, Gaussmf, Gauss2mf, Pimf, Dsigmf, and Psigmf functions in the Fuzzy Interface System (FIS), utilized for establishing relationships between fuzzy sets under the anfisedit menu in MATLAB 7.0, are analyzed through a spatial resection approach to identify the most suitable functions for focal distance detection. The fourth section delineates the findings and their interpretations.

2. FUZZY LOGIC AND ITS FUNDAMENTAL PRINCIPLES

Fuzzy logic is based on the logic of clustering and determination of membership degrees depending of this clustering. Membership degrees generate rule-based work systematic which constitutes the rules of fuzzy systems. Fuzzy cluster sections which are placed in the inlets and outlets of the fuzzy rules express an approximation for each. In this respect, all expressions like “approximately 3,” “nearly 9,” “over 5 and approximate” always express a fuzzy number. Each of these approximations corresponds to a fuzzy cluster. It cannot be possible to solve mathematical operations with these fuzzy numbers. The operations are made with fuzzy numbers by defining some restrictions. In order for a fuzzy number to exist, fuzzy cluster of this needs to have an interval of normal, convex, limited support and at section of membership degree, closed and finite. For the fuzzy numbers to be normal, membership degree of at least one of the real numbers in the fuzzy expression must be 1.

Generally, two fuzzy numbers as a triangle and trapezoid are in question. Mathematical expression of a triangle fuzzy number showed with a fuzzy cluster is given as Eq. (1):

$$mf(x) = mf(x; a, b, c) = \begin{cases} (x-a)/(b-a) & a \le x \le b \\ (c-x)/(c-b) & b \le x \le c \\ 0 & x > c \quad \text{or} \quad x < a \end{cases} \tag{1}$$

In the expression of mf(x;a,b,c), a and c show the lower and upper limit values relatively and b shows the single number with full membership. Similarly, trapezoid fuzzy numbers are expressed with four full whole numbers as a, b, c, d. Here a and d show lower and upper limit values of trapezoid fuzzy numbers; b and c show the limits of the cluster of the trapezoid numbers whose membership degree is full between these two numbers. Mathematical indication of trapezoid fuzzy number is like Eq. (2):

$$mf(x) = mf(x; a, b, c, d) = \begin{cases} (x-a)/(b-a) & a \le x \le b \\ 1 & b \le x < c \\ (d-x)/(d-c) & c < x \le d \\ 0 & x > d \quad \text{or} \quad x < a \end{cases} \tag{2}$$

If we pay attention, when b = c, trapezoid fuzzy number is transformed into triangle fuzzy number. Graphical indication of these fuzzy numbers is as shown in Figure 1 .

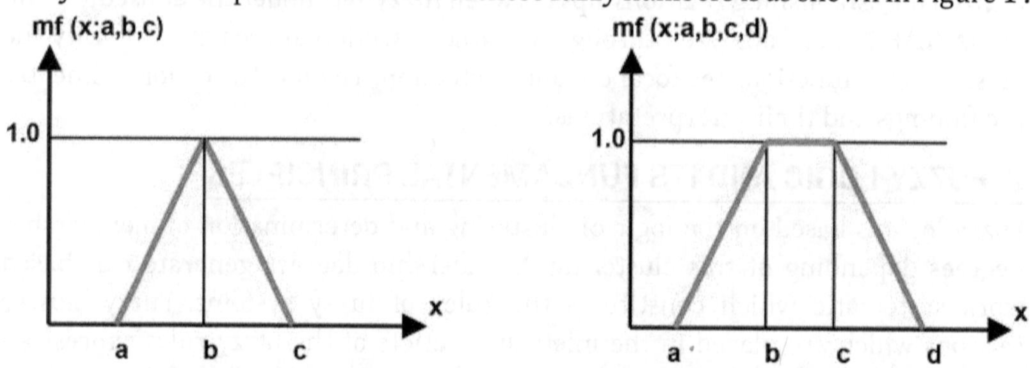

Figure 1. Fuzzy numbers: (a) triangle and (b) trapezoid.

Generally, there are two reasons for the researchers to use fuzzy systems:

Since the real-world incidents are very complicated, these incidents are not possible to be taken under control by being defined with specific equations. As a natural result of this, the researcher always prefers applying to the methods that have approximate solubility even if not certain. As Einstein said, if it can be said that real-life incidents can be indicated with mathematical equations, either the accuracy of the equation cannot be mentioned or if the result that mathematical equations exactly depicts the reality is taken, then real-life incidents cannot be mentioned. So, the solutions are approximate to a certain extent in all studies conducted. Otherwise, many non-linear equations must be solved simultaneously which is known to lead to chaotic unspecified problems according to current information.

All theories and equations in engineering express the real world approximately. Even though many real systems are not linear, every effort is given in order to accept the linearity in examination of these with classic methods. In order for the verbal data presented by people to be converted into numerical data and calculated by being perceived by computers and algorithms, fuzzy systems are needed.

Almost all of mathematical, stochastic, or conceptual systems so far are made up of three units given in Figure 2 .

Figure 2. Conventional system.

These are input unit, a transition function which converts this input to output and is called system behavior and output unit. In all of the units here, numerical data are processed. The difference of the fuzzy systems from the conventional systems is that the system behavior section is divided into two and there are four connected units between each other as shown in Figure 3 .

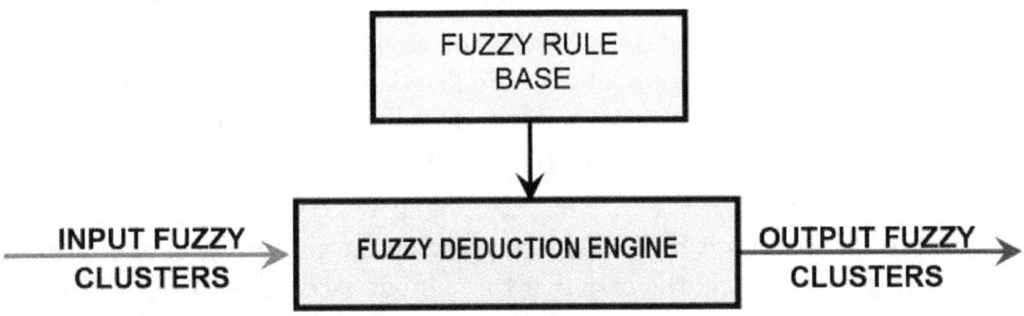

Figure 3. General fuzzy system.

Each of the units here has tasks, which are different but can be related to each other:

- **General information base unit:** It includes input variants to which the incident to be examined is subjected and all information about these. Because the it includes both numerical and/or textual data, the database is called as 'general database'.
- **Fuzzy rule base unit:** It includes all the rules which can be written in the type of logical **IF-THEN**, connecting the inputs in data base to output

variants. In writing of these rules, the whole interval connections (fuzzy cluster) which can only be between input and output data are considered. Thus, each rule connects one piece of input space to the output space logically. All of these contexts form the base rule.

- **Fuzzy deduction engine unit:** It is a mechanism including the operations, which enable a system to behave as having one output by bringing the relations formed between input and output fuzzy clusters in a fuzzy rule base unit together. This engine serves for determination of what kind of output the whole system will give between the inputs by bringing deductions of each rule together.
- **§Output unit:** It indicates the output values obtained as a result of the interaction of information and fuzzy rule bases with deduction engine.

Figure 3 represents a general fuzzy system. The point to be paid attention to here is information and outputs in the input; in other words, databases are fuzzy values. So each unit in Figure 3 is made up of fuzzy clusters entirely. The most significant deficiency of basic fuzzy system is that numerical database cannot enter into such a fuzzy system and the outputs are not numerical, so they cannot be directly used in engineering designs.

In order to certainly eliminate the deficiencies of the general fuzzy system is extented to a new system which is proposed in Refs. [3, 22] This is called Takagi-Sugeno-Kank (TSK) fuzzy system. For each inputs and the outputs obtained as a result of operation of fuzzy rule and deduction engine are in the way of a function of inputs which are from input database. So like the input variants in rule base, it was thought that these variants are reflected to rule result section after THEN word as a linear function of these variants [21]. According to this, the rule is presented in the following part.

IF the speed of your car is high, **THEN** stepping on the gas force can be expressed as **y = ax.** For example, in the case that three input variants exist (x_1, x_2, and x_3), in one of fuzzy system, y as output variant generally; **IF** x_1 is little and x_3 is wide, **THEN** it can be expressed as $\mathbf{y = a_0 + a_1x_1 + a_2x_2 + a_3x_3}$. Result sections of all rules are made up of a polynomial linear equation. Since in the fuzzy system which has such a structure output variants are not used in the deduction of fuzzy clustering, instead of fuzzy deduction unit in Figure 3 , weighted deduction calculation unit comes as mainly the membership degrees calculated from the input section of each rule. This system is shown in Figure 4 .

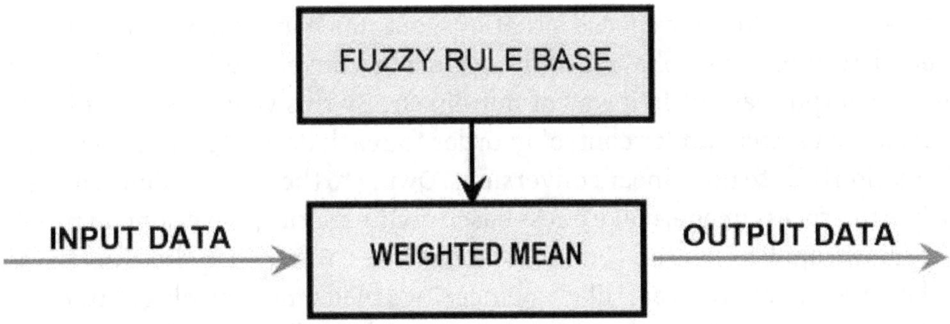

Figure 4. TSK fuzzy system.

Actually in such a fuzzy system, output space is represented as a function of the inputs for being a rule valid in each sub-space. Even in the case that the output surface is not linear, with TSK approach they are understood as modeled in the way of plain pieces of input variant kind on sub-spaces.

However, TSK fuzzy systems have deficiencies as there is a mathematical relation after **IF** part; output sections of the rules cannot model the verbal data given by human, and output sections of all rules, which are possible to be written between input and output variants cannot be written because they are fuzzy. In order to eliminate these deficiencies, fuzzy system, which is respectively blur, and clarification units are used in input and output units given in Figure 5 [21].

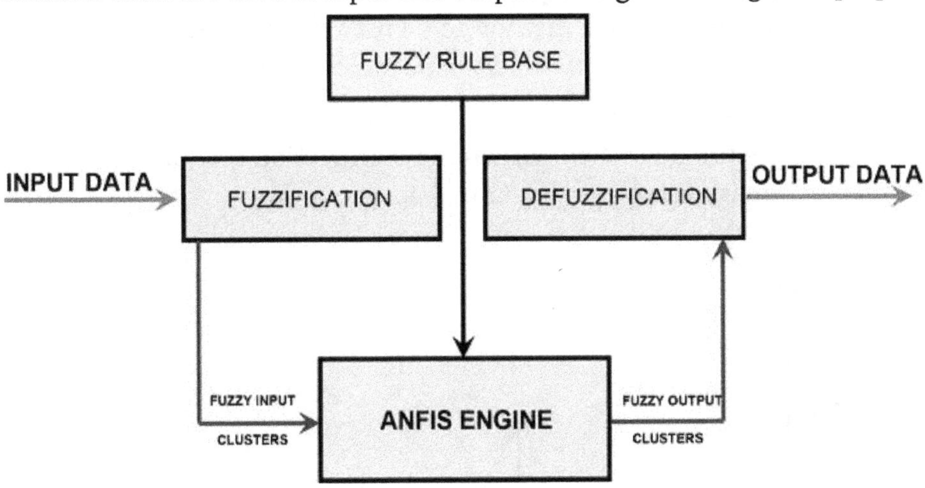

Figure 5. Fuzzy system with blurring-clarifying unit.

Here, fuzzy rule base and deduction engine in a general fuzzy system remain as they are. In the case that the inputs are numerical, there are blurring agent unit, which is for blurring them with a process, and clarifying agent unit, which is for digitizing the fuzzy outputs. Blurring (fuzzification) and clarifying (defuzzification) mean, respectively, blurring the input numbers and digitizing the fuzzy numbers.

Among the main properties of fuzzy systems, multiple inputs, rule base, and conversion as single output by processing with deduction engine come as the most important issues. In some special situations, outputs may be more than one. Fuzzy system determines the behavior of the system by converting variants forming the inputs to output variant in a way of non-linear. By this way, it is possible to take the examined system under control in order to reach desired results by subjecting information base to non-linear conversions. Owing to the fuzzy systems, it becomes possible to process image, make guess based on time series, solve control problems, and perform applications on communication issues. Other than this, fuzzy systems can be used in many areas like engineering, medicine, sociology, psychology, business management, expert systems, artificial intelligence, signal processing, transportation, and signalization.

Exact numerical values are needed in practical applications particularly for sizing in engineering designs. In such cases, fuzzy information must be clarified in order for the required answers to be given by benefiting from information obtained or given as fuzzy. All processes made for conversion of fuzzy information to exact results are called clarification.

There are many clarification methods [23]. A few of them are presented below. In these processes, **z** indicates fuzzy deduction cluster, indicates components, and z * indicates clarified value.

- **The highest membership principle:** Another name of this process is height method. For this method to be used, fuzzy deduction clusters with peaks are needed. This method can be indicated with Figure 6 .
- **Centroid method:** Another name of this method is center of gravity method. It could be the most common one of clarification methods. Mathematical expression is as shown in Eq. (3):

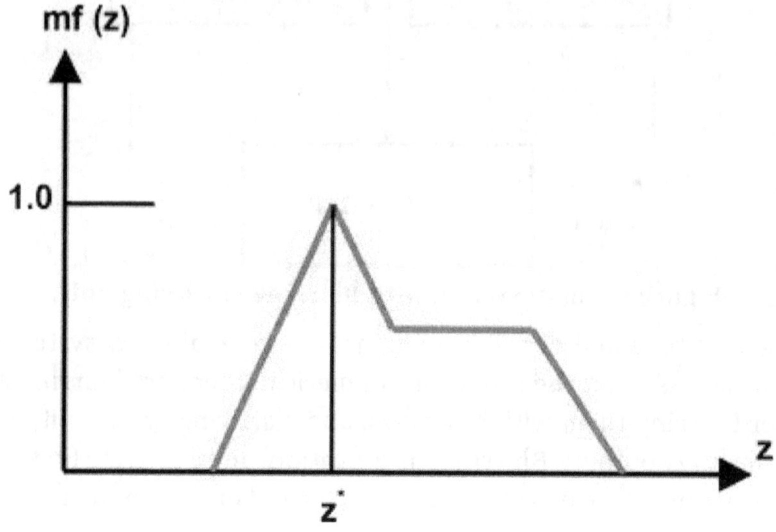

Figure 6. Clarification of the highest membership degree.

$$z^{*} = \frac{\int mf_{C}(z).zdz}{\int mf_{C}(z)dz} \tag{3}$$

Clarification is performed according to the center of gravity of fuzzy expression.

- **Weighted mean method:** In order for this method to be used, symmetric membership function (mf) is needed. Mathematical explanation is as shown in Eq. (4):

$$z^{*} = \frac{\sum mf_{C}(z).z}{\sum mf_{C}(z)} \tag{4}$$

Here Σ sign indicates algebraic sum. Graphical indication of this method is shown in Figure 7 . Here, each membership function of fuzzy cluster constituting the output is multiplied with the highest membership degree value it has and their weighted mean is calculated. For instance, weighted mean (clarified value) of two fuzzy clusters in Figure 7 is calculated as shown in Eq. (5).

$$z^{*} = \frac{a(0.5) + b(0.9)}{0.5 + 0.9} \tag{5}$$

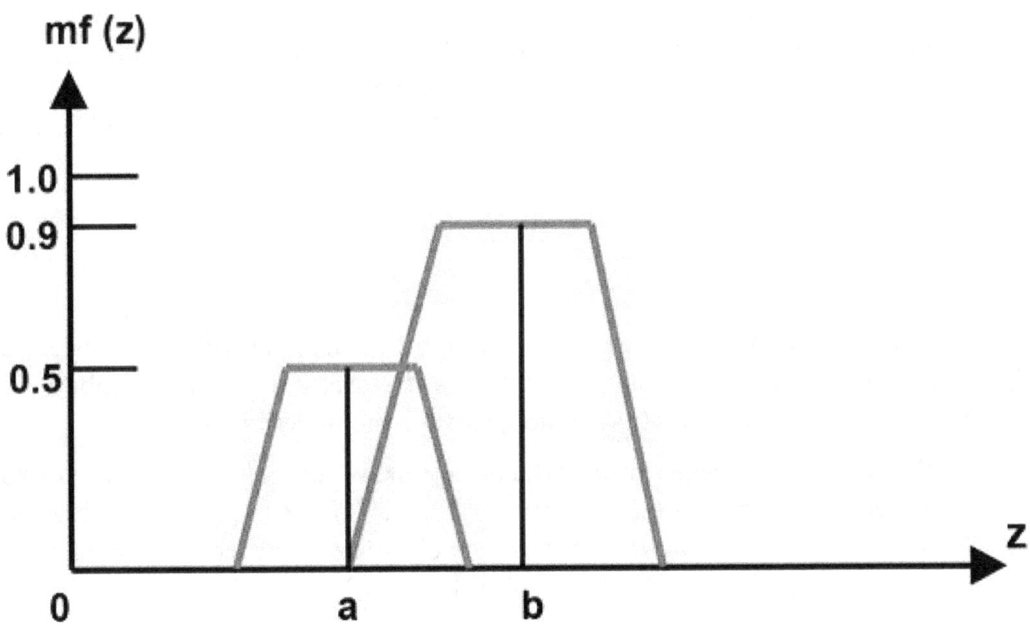

Figure 7. Clarification with weighted mean method.

Since this clarifying process is valid only for symmetric membership functions, a and b values are the mean of the figures they represent.

- **Averagely the highest membership degree method:** This method is also known as the mean of the highest ones. In this respect, it is very close to the first clarification principle. However, the position of the highest membership may not be singular. This means that the one having the highest membership degree in the membership function, mf (z) = 1, may exist in a plain section instead of a point.

Clarified value according to this method in graphical indication in Figure 8 is as shown in Eq. (6):

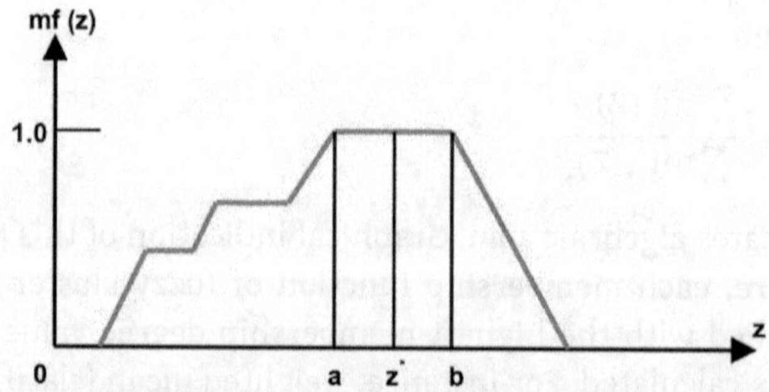

Figure 8. Averagely the highest membership clarification.

$$z^* = \frac{a+b}{2} \tag{6}$$

- **Central management of the sums:** It is the fastest one among the clarification methods used. In this method, instead of combination of two fuzzy clusters, their algebraic sum is used. A deficiency of this is overlapping sections entering into the sum twice. Clarified value can be calculated as Eq. (7).

$$z^* = \frac{\int_z z\sum_{k=1}^{n} mf_c(z)\,dz}{\int_z \sum_{k=1}^{n} mf_c(z)\,dz} \tag{7}$$

In a way, this calculation form resembles weighted mean clarification. However, in the method of center of sums, weights are the areas of related membership functions. In the average weights method, this is a membership degree.

- **Center of the largest area:** If the outlet fuzzy cluster includes at least two convex fuzzy sub-clusters, center of gravity of the one with the biggest area of convex fuzzy clusters is used in clarification process. Mathematical indication of this method is as shown in Eq. (8):

$$z^* = \frac{\int mf_{Cm}(z)z\,dz}{\int mf_{Cm}(z)\,dz} \tag{8}$$

Here, $ü_{ebç}(z)$ indicates the sub-region where the convex fuzzy cluster with the biggest area dominates. This circumstance is used when all fuzzy deduction clusters are not convex; however, in the case that all deductions are convex, the result is the same with the one obtained with z* centroid method.

- **The highest first and last membership degree method:** In this method, the lowest (or the highest) fuzzy cluster value with the highest membership degree in the fuzzy cluster that comes up as a combination of all outputs is selected. The equations below are valid for z* value to be given by calculations. At first, the biggest height y_{eb} is determined in the combination of fuzzy cluster deduction, B.

$$y_{eb}(B) = \max\left|m\, f_B(z)\right| \tag{9}$$

Then, first highest value, z* is found. Another way of this is to find the last highest fuzzy cluster value instead of the first one.

ANFIS name, which means fuzzy deduction system based on open adaptive networks or adaptive neural fuzzy detection system, is made of the first letters of adaptive network-based fuzzy interface system or adaptive neuro-fuzzy interface system. This network is the combination of the nodes, which are placed as layers for performing specific functions [5, 21]. ANFIS consists of six layers. This system is presented in Figure 9 .

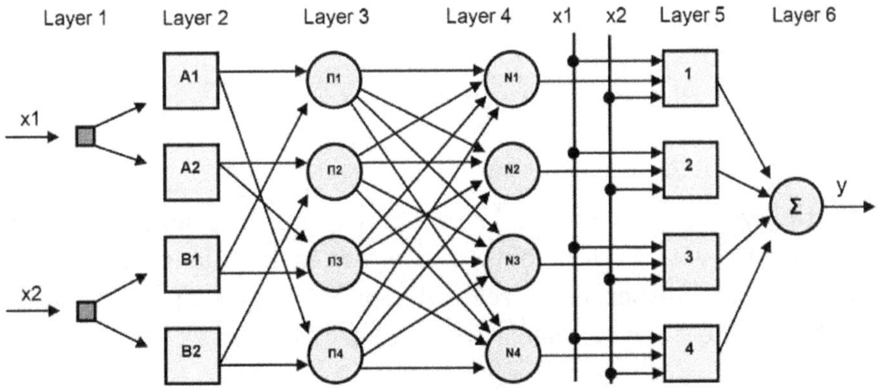

Figure 9. ANFIS network system.

Node functions of each layers on ANFIS structure and the operations of layers are briefly explained [23, 24]:

- **First layer:** This is called the input layer. Input signals, which are taken from each of the nodes on this layer, are transferred into other layers.
- **Second layer:** This is called the fuzzification layer. Jang's ANFIS model uses the current Bell activation function as the membership function in order to divide input values into fuzzy sets. Here, output of each node is formed of input values, and membership degrees related to membership function and membership degrees, which are obtained from the second layer are shown as $\mu_{Aj}(x)$ and $\mu_{Bj}(y)$.

- **Third layer:** This is the rule layer. Each node in this layer expresses the rules and number of Sugeno fuzzy logic deduction system. Output of each rule node μ_i is the multiplication of membership degrees coming from second layer, and obtaining μ_i values are such as (j=1,2) and (i=1,....,n);

$$y_i^3 = \prod i = \mu_{A_j}(x)\, x\, \mu_{Bj}(y) = \mu_i \tag{10}$$

Here, y_i^3 represents the output values of the third layer and n represents the number of nodes in this layer.

- **Fourth layer:** This is the normalization layer. Each node in this layer assumes all the nodes coming from rule layer as the input value, and it calculates the normalized ignition level of each rule.

Calculation of normalized ignition level i μ is as shown in Eq. (11):

$$y_{i^4} = N_i = \frac{\mu_i}{\sum_{i=1}^{n} \mu_i} = \overline{\mu}, \; (i = 1, n) \tag{11}$$

- **Fifth layer:** This is the debugging layer. In this layer, weighted result values of a rule, which is given in each node, are calculated. Output value of i node in fifth layer is as shown in Eq. (12):

$$y_{i^5} = \overline{\mu}_i \left[p_i x_1 + q_i x_2 + r_i\right], \; (i = 1, n) \tag{12}$$

Here, (p_i, q_i, r_i) variables are result parameter set for i rule.

- **Sixth layer:** This is the sum layer. There is only one node in this layer, and it is tagged with Σ. Here, output values of each node in the fifth layer are added to each other, and a real value of ANFIS system is obtained. System output value y can be calculated with Eq. (13):

$$y = \sum_{i=1}^{n} \overline{\ } \left[p_i x_1 + q_i x_2 + r_i\right] \tag{13}$$

3. APPLICATION

MATLAB is a software development instrument designed for technical calculations and the solutions of mathematical problems. MATLAB, which is the abbreviation of the words "MATrix LABoratory," works using matrixes as understood from its name or, in other words, using arrays. MATLAB, which is estimated to be used by over 500,000 academician, researcher, and students, is also described as the most advanced technical and scientific problem solving and application development instrument of the computer world with many interfaces it includes. MATLAB, particularly used in the analysis of the systems in engineering area, can perform the operations of data analysis and examination, visibility and image processing, generating algorithm prototype, modeling and simulation, programming and application development [25].

Application study consists of two basic steps. These are:

- Geometric camera calibration with conventional method
- Focal length calculation with ANFIS functions

In the application phase, geometric camera calibration study has been performed with bundle block adjustment method by using conventional method as the first step. In this study made by using zoom lens, camera calibration study has been preferred in both minimum and maximum focal lengths for Nikon Coolpix-E8700 camera whose focal length has been selected. The image of this camera is shown in Figure 10 .

Figure 10. Nikon coolpix-E8700.

Images of test fields on which there are 145 points have been taken with Nikon Coolpix-E8700 digital camera from five angles in the situations of maximum (71.2 mm) and minimum (8.9 mm) focal length of the objective in the Photogrammetry Laboratory of Department of Geodesy and Photogrammetric Engineering of Gebze Technical University. This process is shown in Figure 11 .

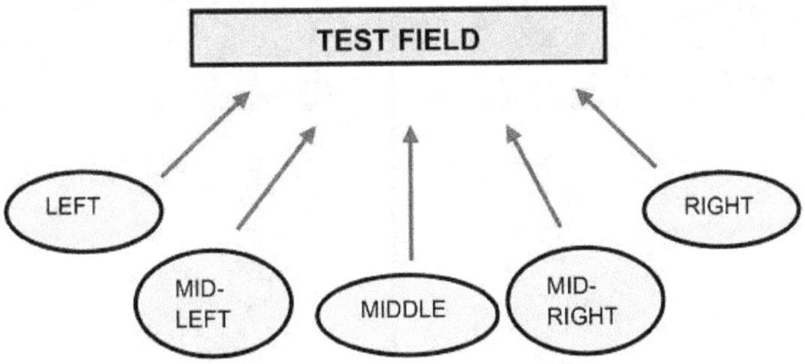

Figure 11. Test field shooting plan.

Coordinates of 145 points on test field were determined locally, and these values have been regarded as field coordinate (ANNEX-1). After that, 10 images obtained have been subjected to the evaluation in Topcon Pl-3000 software, and calibration parameters (inner orientation factors and distortion parameters) have been determined in situations of maximum focal length (71.2 mm) and minimum focal length (8.9 mm) of digital camera. These steps have been explained in Figures 11 and 12 for both focal distances.

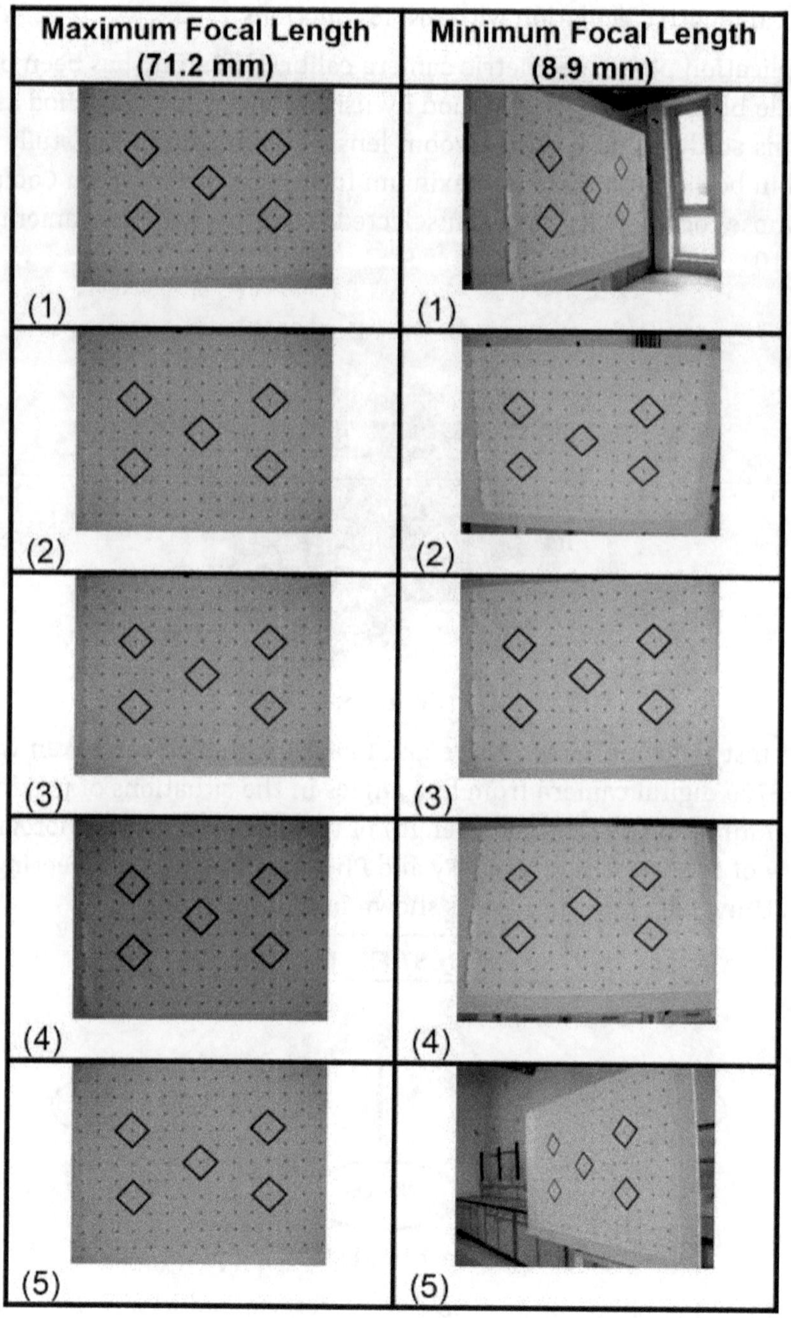

Figure 12. Camera calibration test field.

ANNEX-1

1 1.0000000e-002 1.0000000e-002 1.0000000e-002	73 4.3500000e-001 6.0500000e-001 1.0000000e-002
2 9.5000000e-002 1.0000000e-002 1.0000000e-002	74 4.3500000e-001 7.7500000e-001 1.0000000e-002
3 1.8000000e-001 1.0000000e-002 1.0000000e-002	75 4.3500000e-001 8.6000000e-001 1.0000000e-002
4 2.6500000e-001 1.0000000e-002 1.0000000e-002	76 4.3500000e-001 9.4500000e-001 1.0000000e-002
5 3.5000000e-001 1.0000000e-002 1.0000000e-002	77 4.3500000e-001 1.0300000e+000 1.0000000e-002
6 4.3500000e-001 1.0000000e-002 1.0000000e-002	78 4.3500000e-001 1.1150000e+000 1.0000000e-002
7 5.2000000e-001 1.0000000e-002 1.0000000e-002	79 4.3500000e-001 1.2000000e+000 1.0000000e-002
8 6.0500000e-001 1.0000000e-002 1.0000000e-002	80 5.2000000e-001 1.0000000e-002 1.0000000e-002
9 6.9000000e-001 1.0000000e-002 1.0000000e-002	81 5.2000000e-001 9.5000000e-002 1.0000000e-002
10 7.7500000e-001 1.0000000e-002 1.0000000e-002	82 5.2000000e-001 1.8000000e-001 1.0000000e-002
11 8.6000000e-001 1.0000000e-002 1.0000000e-002	83 5.2000000e-001 3.5000000e-001 1.0000000e-002
12 9.4500000e-001 1.0000000e-002 1.0000000e-002	84 5.2000000e-001 4.3500000e-001 1.0000000e-002
13 1.0300000e+000 1.0000000e-002 1.0000000e-002	85 5.2000000e-001 5.2000000e-001 1.0000000e-002
14 1.1150000e+000 1.0000000e-002 1.0000000e-002	86 5.2000000e-001 6.9000000e-001 1.0000000e-002
15 1.2000000e+000 1.0000000e-002 1.0000000e-002	87 5.2000000e-001 7.7500000e-001 1.0000000e-002
16 9.5000000e-002 1.0000000e-002 1.0000000e-002	88 5.2000000e-001 8.6000000e-001 1.0000000e-002
17 9.5000000e-002 9.5000000e-002 1.0000000e-002	89 5.2000000e-001 1.0300000e+000 1.0000000e-002
18 9.5000000e-002 1.8000000e-001 1.0000000e-002	90 5.2000000e-001 1.1150000e+000 1.0000000e-002
19 9.5000000e-002 2.6500000e-001 1.0000000e-002	91 5.2000000e-001 1.2000000e+000 1.0000000e-002
20 9.5000000e-002 3.5000000e-001 1.0000000e-002	92 6.0500000e-001 1.0000000e-002 1.0000000e-002
21 9.5000000e-002 4.3500000e-001 1.0000000e-002	93 6.0500000e-001 9.5000000e-002 1.0000000e-002
22 9.5000000e-002 5.2000000e-001 1.0000000e-002	94 6.0500000e-001 2.6500000e-001 1.0000000e-002
23 9.5000000e-002 6.0500000e-001 1.0000000e-002	95 6.0500000e-001 4.3500000e-001 1.0000000e-002
24 9.5000000e-002 6.9000000e-001 1.0000000e-002	96 6.0500000e-001 5.2000000e-001 1.0000000e-002
25 9.5000000e-002 7.7500000e-001 1.0000000e-002	97 6.0500000e-001 6.0500000e-001 1.0000000e-002
26 9.5000000e-002 8.6000000e-001 1.0000000e-002	98 6.0500000e-001 6.9000000e-001 1.0000000e-002
27 9.5000000e-002 9.4500000e-001 1.0000000e-002	99 6.0500000e-001 7.7500000e-001 1.0000000e-002
28 9.5000000e-002 1.0300000e+000 1.0000000e-002	100 6.0500000e-001 9.4500000e-001 1.0000000e-002
29 9.5000000e-002 1.1150000e+000 1.0000000e-002	101 6.0500000e-001 1.1150000e+000 1.0000000e-002
30 9.5000000e-002 1.2000000e+000 1.0000000e-002	102 6.0500000e-001 1.2000000e+000 1.0000000e-002
31 1.8000000e-001 1.0000000e-002 1.0000000e-002	103 6.9000000e-001 1.0000000e-002 1.0000000e-002
32 1.8000000e-001 9.5000000e-002 1.0000000e-002	104 6.9000000e-001 9.5000000e-002 1.0000000e-002
33 1.8000000e-001 1.8000000e-001 1.0000000e-002	105 6.9000000e-001 1.8000000e-001 1.0000000e-002
34 1.8000000e-001 3.5000000e-001 1.0000000e-002	106 6.9000000e-001 3.5000000e-001 1.0000000e-002
35 1.8000000e-001 4.3500000e-001 1.0000000e-002	107 6.9000000e-001 4.3500000e-001 1.0000000e-002
36 1.8000000e-001 5.2000000e-001 1.0000000e-002	108 6.9000000e-001 5.2000000e-001 1.0000000e-002

37 1.8000000e-001 6.0500000e-001 1.0000000e-002	109 6.9000000e-001 6.0500000e-001 1.0000000e-002
38 1.8000000e-001 6.9000000e-001 1.0000000e-002	110 6.9000000e-001 6.9000000e-001 1.0000000e-002
39 1.8000000e-001 7.7500000e-001 1.0000000e-002	111 6.9000000e-001 7.7500000e-001 1.0000000e-002
40 1.8000000e-001 8.6000000e-001 1.0000000e-002	112 6.9000000e-001 8.6000000e-001 1.0000000e-002
41 1.8000000e-001 1.0300000e+000 1.0000000e-002	113 6.9000000e-001 1.0300000e+000 1.0000000e-002
42 1.8000000e-001 1.1150000e+000 1.0000000e-002	114 6.9000000e-001 1.1150000e+000 1.0000000e-002
43 1.8000000e-001 1.2000000e+000 1.0000000e-002	115 6.9000000e-001 1.2000000e+000 1.0000000e-002
44 2.6500000e-001 1.0000000e-002 1.0000000e-002	116 7.7500000e-001 1.0000000e-002 1.0000000e-002
45 2.6500000e-001 9.5000000e-002 1.0000000e-002	117 7.7500000e-001 9.5000000e-002 1.0000000e-002
46 2.6500000e-001 2.6500000e-001 1.0000000e-002	118 7.7500000e-001 1.8000000e-001 1.0000000e-002
47 2.6500000e-001 4.3500000e-001 1.0000000e-002	119 7.7500000e-001 2.6500000e-001 1.0000000e-002
48 2.6500000e-001 5.2000000e-001 1.0000000e-002	120 7.7500000e-001 3.5000000e-001 1.0000000e-002
49 2.6500000e-001 6.0500000e-001 1.0000000e-002	121 7.7500000e-001 4.3500000e-001 1.0000000e-002
50 2.6500000e-001 6.9000000e-001 1.0000000e-002	122 7.7500000e-001 5.2000000e-001 1.0000000e-002
51 2.6500000e-001 7.7500000e-001 1.0000000e-002	123 7.7500000e-001 6.0500000e-001 1.0000000e-002
52 2.6500000e-001 9.4500000e-001 1.0000000e-002	124 7.7500000e-001 6.9000000e-001 1.0000000e-002
53 2.6500000e-001 1.1150000e+000 1.0000000e-002	125 7.7500000e-001 7.7500000e-001 1.0000000e-002
54 2.6500000e-001 1.2000000e+000 1.0000000e-002	126 7.7500000e-001 8.6000000e-001 1.0000000e-002
55 3.5000000e-001 1.0000000e-002 1.0000000e-002	127 7.7500000e-001 9.4500000e-001 1.0000000e-002
56 3.5000000e-001 9.5000000e-002 1.0000000e-002	128 7.7500000e-001 1.0300000e+000 1.0000000e-002
57 3.5000000e-001 1.8000000e-001 1.0000000e-002	129 7.7500000e-001 1.1150000e+000 1.0000000e-002
58 3.5000000e-001 3.5000000e-001 1.0000000e-002	130 7.7500000e-001 1.2000000e+000 1.0000000e-002
59 3.5000000e-001 4.3500000e-001 1.0000000e-002	131 8.6000000e-001 1.0000000e-002 1.0000000e-002
60 3.5000000e-001 5.2000000e-001 1.0000000e-002	132 8.6000000e-001 9.5000000e-002 1.0000000e-002
61 3.5000000e-001 6.9000000e-001 1.0000000e-002	133 8.6000000e-001 1.8000000e-001 1.0000000e-002
62 3.5000000e-001 7.7500000e-001 1.0000000e-002	134 8.6000000e-001 2.6500000e-001 1.0000000e-002
63 3.5000000e-001 8.6000000e-001 1.0000000e-002	135 8.6000000e-001 3.5000000e-001 1.0000000e-002
64 3.5000000e-001 1.0300000e+000 1.0000000e-002	136 8.6000000e-001 4.3500000e-001 1.0000000e-002
65 3.5000000e-001 1.1150000e+000 1.0000000e-002	137 8.6000000e-001 5.2000000e-001 1.0000000e-002
66 3.5000000e-001 1.2000000e+000 1.0000000e-002	138 8.6000000e-001 6.0500000e-001 1.0000000e-002
67 4.3500000e-001 1.0000000e-002 1.0000000e-002	139 8.6000000e-001 6.9000000e-001 1.0000000e-002
68 4.3500000e-001 9.5000000e-002 1.0000000e-002	140 8.6000000e-001 7.7500000e-001 1.0000000e-002
69 4.3500000e-001 1.8000000e-001 1.0000000e-002	141 8.6000000e-001 8.6000000e-001 1.0000000e-002
70 4.3500000e-001 2.6500000e-001 1.0000000e-002	142 8.6000000e-001 9.4500000e-001 1.0000000e-002
71 4.3500000e-001 3.5000000e-001 1.0000000e-002	143 8.6000000e-001 1.0300000e+000 1.0000000e-002
72 4.3500000e-001 4.3500000e-001 1.0000000e-002	144 8.6000000e-001 1.1150000e+000 1.0000000e-002
	145 8.6000000e-001 1.2000000e+000 1.0000000e-002

In Figures 13 and 14 , focal length obtained with bundle block adjustment (conventional method), inner orientation parameters, and radial distortion parameters were summarized from Topcon PI-3000 software interface.

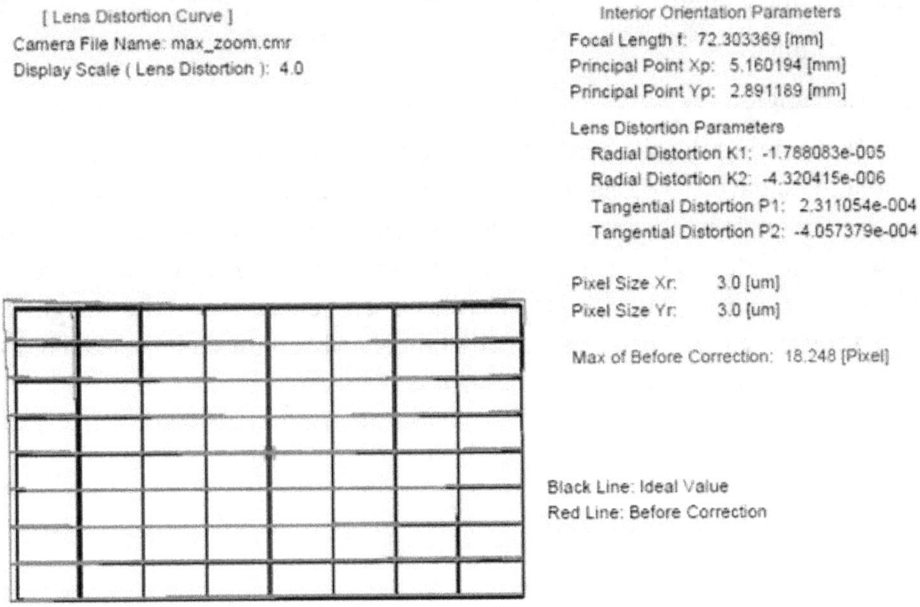

Figure 13. The one obtained by Topcon PI-3000 program in maximum focal length.

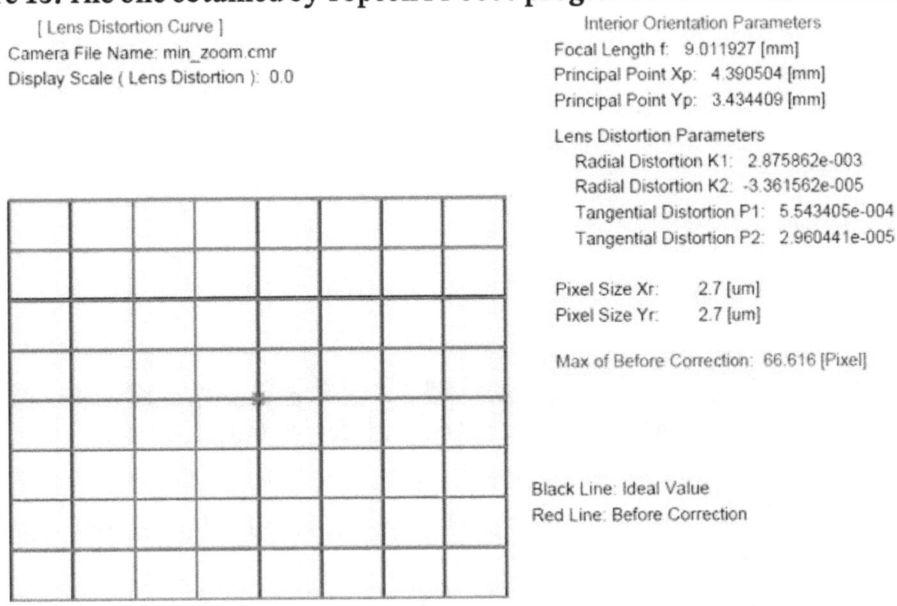

Figure 14. The one obtained by Topcon PI-3000 program in minimum focal length.

As the second step of the application, image coordinates measured over images have been indicated radially with Eq. (14) for maximum focal length and minimum focal length values of digital camera.

$$c_k = \frac{r'}{\tan \tau} \tag{14}$$

c_k: calibrated focal length

r^1: radial distance

τ: distortion angle

Twenty-three points have been selected and arranged in the format of training data (smalldata.dat) and remaining 122 points have been arranged in the format of test data (bigdata.dat) as shown in Figure 15 . ANFIS data set which are used in fuzzification and defuzzification have been composed from equation process with radial distance and angle. This process has been repeated for the images taken from 10 different angles for each zoom distance. ANFIS data set has been composed from equation process with radial distance and angles.

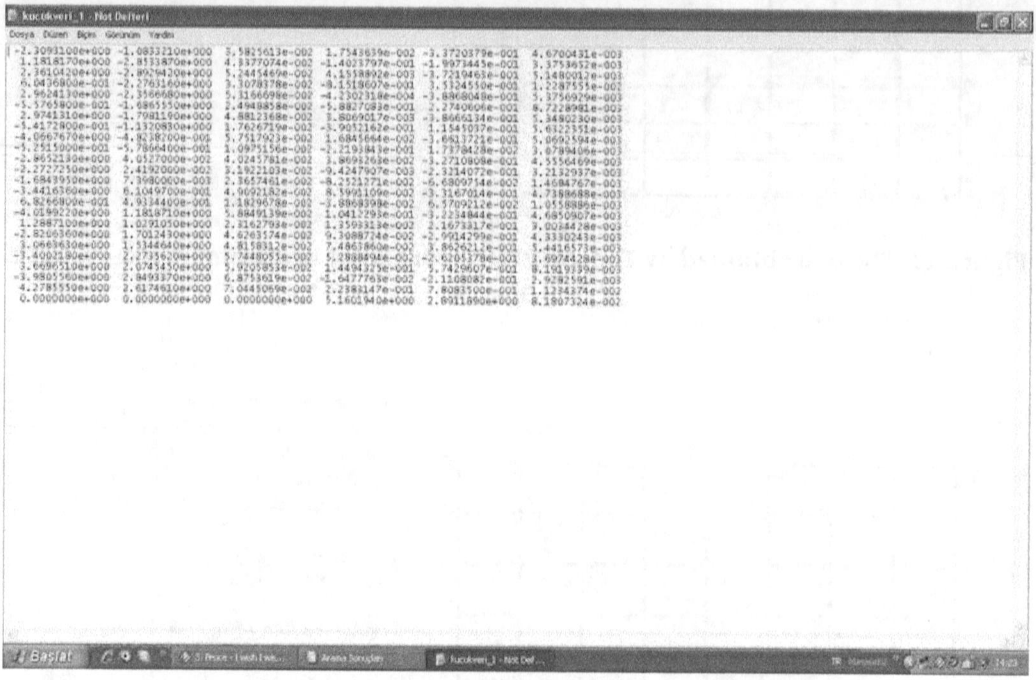

Figure 15. ANFIS data set structure for radial distances.

In Figure 15 , first and fourth columns indicate x values of image coordinates, second and fifth columns indicate y values, and third and sixth columns indicate focal lengths calculated radially.

After ANFIS data set has been generated, an ANFIS application stated below has been performed by using MATLAB software. The basic logic in this application is radial distances selected as input is processed with the help of ANFIS functions, and as a result, a focal length value for each image has been obtained. Algorithm figure determined for this method is indicated below. Here, it is seen that training algorithm determined for generation of ANFIS artificial neural network operates iteratively. Artificial neural network structure shaped with training data described as training data and consisting of 23 points gives result with the last iteration where 122 points are also included for getting the exact result. Thus, a calculation, which includes again the same quantity of measurement (145 radial values) corresponding to focal length distance, is determined with conventional method. For this reason, a logical measurement number equality has been enabled in comparison of ANFIS results with conventional method. In brief, the measurement values in the first and second steps are in the same quantity. This process has been explained in Figure 16 .

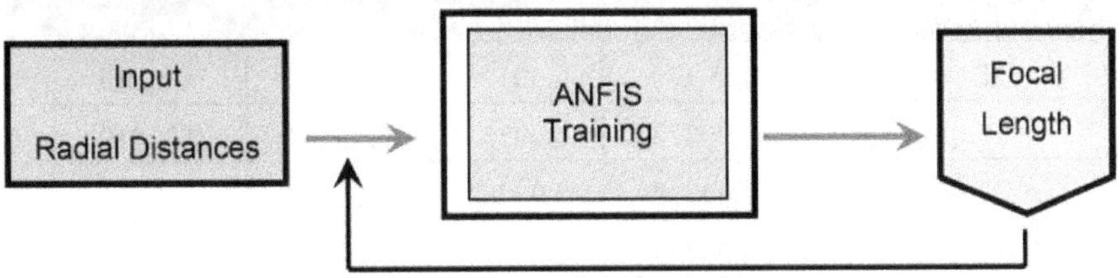

Figure 16. ANFIS algorithm.

Each data set has been subjected to process with eight different ANFIS functions (trimf, trapmf, gbellmf, gaussmf, gauss2mf, pimf, dsigmf, and psigmf) as indicated above and quadratic average error values of Artificial Neural Network and ANFIS functions have been obtained. Artificial neural network structure has been shown in Figure 17 from MATLAB ANFIS interface.

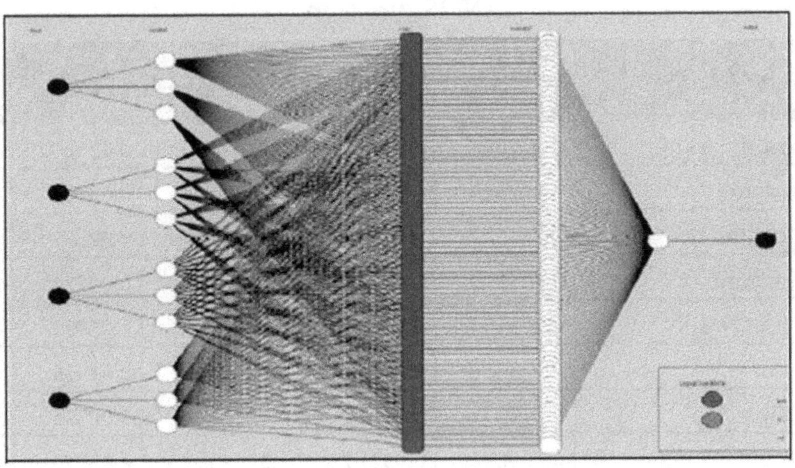

Figure 17. Artificial neural network structure generation.

Measured radial values and ANFIS artificial neural network structure have been indicated in Figure 17 . Quadratic mean errors (standard deviation) of focal length differences related to radial lengths obtained for function structures used in forming artificial neural network have been generated for 10 images in total according to maximum focal length data as shown in Table 1 and minimum focal length data as shown in Table 2 . While there are standard deviation values here, focal length value obtained (with bundle block adjustment) in the first step has been used as exact value. At the end of the first step, inner orientation parameters obtained by conventional method will be considered as true, and they will constitute reference data to be used in comparison for the results obtained with second step in other word ANFIS.

Table 1. Standard deviation values of maximum focal length obtained with ANFIS functions.

		Maximum focal length (71.9 mm)				
Num.	ANFIS function	Image 1	Image 2	Image 3	Image 4	Image 5
1	Trimf	0.00179	0.00173	0.00210	0.00235	0.00190
2	Trapmf	0.00373	0.00382	0.00381	0.00319	0.00441
3	Gbellmf	0.00243	0.00261	0.00283	0.00218	0.00219
4	Gaussmf	0.00234	0.00229	0.00237	0.00210	0.00210
5	Gauss2mf	0.00347	0.00530	0.00515	0.00321	0.00307
6	Pimf	0.00512	0.00344	0.00488	0.00356	0.00342
7	Dsigmf	0.00306	0.00288	0.00452	0.00327	0.00277
8	Psigmf	0.00306	0.00288	0.00452	0.00327	0.00308

Table 2. Standard deviation values of minimum focal length obtained with ANFIS functions.

		Minimum focal length (8.9 mm)				
Num.	ANFIS function	Image 1	Image 2	Image 3	Image 4	Image 5
1	Trimf	0.01814	0.02217	0.01626	0.02560	0.02326
2	Trapmf	0.02204	0.03220	0.02181	0.02701	0.04293
3	Gbellmf	0.01637	0.02047	0.01736	0.02276	0.03822
4	Gaussmf	0.01429	0.01982	0.01745	0.02221	0.03269
5	Gauss2mf	0.02195	0.03180	0.02524	0.02729	0.04476
6	Pimf	0.02528	0.03589	0.02648	0.02965	0.04609
7	Dsigmf	0.02323	0.03274	0.02448	0.02774	0.04494
8	Psigmf	0.02323	0.03274	0.02448	0.02774	0.04494

The result of determining focal distance with eight ANFIS functions, which are used in the application and square mean errors which are obtained with these functions, are compared to camera calibration values which are calculated with the conventional method (bundle block adjustment), and the differences of focal distances are illustrated image by image.

The results that eight ANFIS functions have given standard deviations for first image to fifth image depending on the maximum and minimum focal lengths have been detailed in the graphs in Figures 18 and 19 . When ANFIS method is examined in terms of reliability and inner reliability parameters between selected functions with this application realized, it has been seen that it gives the most suitable results in determination of focal length of trimf and gaussmf functions.

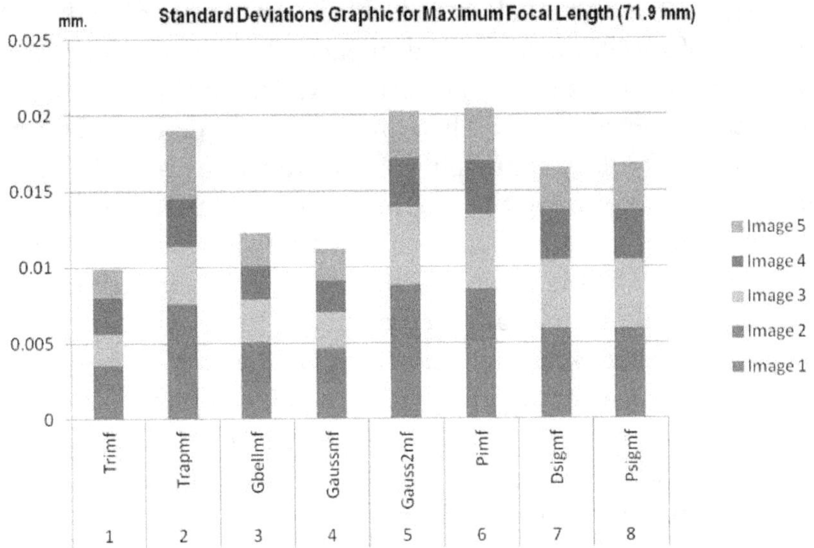

Figure 18. Standard deviations graphic for maximum focal length (71.9 mm).

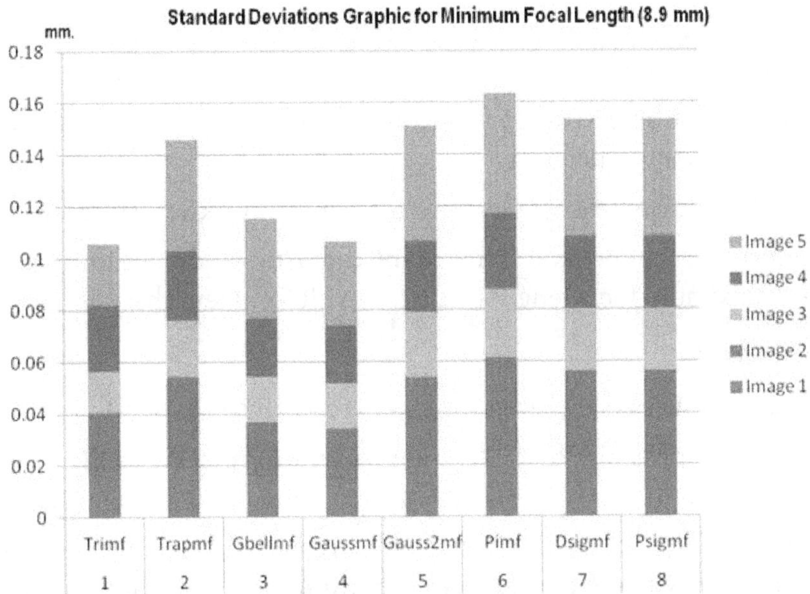

Figure 19. Standard deviations graphic for minimum focal length (8.9 mm).

4. CONCISE SUMMARY

Upon obtaining the findings at the conclusion of the application, it was seen that data with extended focal lengths yielded lower quadratic mean errors, whereas test data generated using short focal lengths resulted in elevated quadratic mean error values within the identical ANFIS functions. Analysis of the mathematical framework concerning radial distortion in focal length graphics reveals that a shorter focal length correlates with an increased radial distortion mistake. This circumstance reveals that radial distortion error directly affects the implementation of ANFIS functions and impacts the overall error magnitude in the dataset. The optical distortion law can be validated through the application of ANFIS functions on data sets. This simultaneously demonstrates a physical truth in the general table, confirming ANFIS functions and illustrating them appropriately. Upon examining the graphical representations of the functions in relation to one another, it is evident that the most suitable (exhibiting the least quadratic mean error) or the most effective ANFIS functions for modeling in an ANFIS structure pertaining to spatial resection are "trimf, gbellmf, and gaussmf." The ANFIS functions that represent balancing as the least effective (exhibiting the highest quadratic mean error) are the "pimf, dsigmf, and psigmf" functions within this framework. When evaluating the overall implications of this work on spatial resection utilizing the ANFIS system, the following advantages and disadvantages will arise for the calibration investigations.

Advantages:

- There will be no need for generating a separate unreal algorithm for spatial resection step in bundle block adjustment.
- Study will be able to be made without the need of approximate rotation and reciprocation parameters.
- It will decrease iteration number for bundle block adjustment.
- Calibration will be able to be enabled during study not only in minimum and maximum focal lengths but also in all focal lengths.

Disadvantages:

- Data sets must be prepared for ANFIS sets. This requires additional time.
- It is not possible to reach approximate rotation and reciprocation parameters.
- Since the system will directly give result, focal length necessary for calibration will be known only as approximately depending on data number.
- It is verified for more than sufficient images that focal length found with additional transaction can be used as calibrated for certain.

- All images to be determined for the project must be taken from the same focal length.

The completed investigations indicate that the fuzzy logic approach is applicable in the calibration of digital cameras. The accuracy can be enhanced by augmenting the dataset of the estimation model developed using the ANFIS method, and estimations can also leverage other artificial intelligence techniques. Consequently, it was determined that the utilization and exploration of fuzzy logic approaches would be more advantageous in photogrammetric applications.

AT A GLANCE

The digital cameras have been affected by systematic flaws that diminished the metric quality of the images. The digital cameras have been affected by systematic flaws that diminish the metric quality of images. This chapter aims to examine the application of fuzzy logic in the calibration of digital cameras. A 145-point planar test field has been established in the facilities of the Department of Geodesy and Photogrammetric Engineering at Gebze Technical University. The test field has been captured from five perspectives using the Nikon Coolpix-E8700 digital camera, with a maximum focal length of 71.2 mm and a minimum focal length of 8.9 mm. The input-output data were derived from ten calibration photos acquired for the fuzzy logic procedure. These data have also been utilized and structured for the space resection procedure. The adaptive neuro-fuzzy inference system (ANFIS) functions have been utilized for fuzzy processing in MATLAB 7.0, and the outcomes of these two separate methodologies have been contrasted. Ultimately, the most convenient (least squares average error) or the most effective ANFIS functions, namely "Trimf, trapmf, gbellmf, gaussmf, gauss2mf, pimf, dsigmf, and psigmf," are identified and evaluated for the space resection approach within the typical bundle adjustment process.

REFERENCES

1. Kajornrit J, Wong KW, Fung CC. A comparative analysis of soft computing techniques used to estimate missing precipitation records. In: Proceedings of the 19th ITS Biennial Conference; 18-21 November 2012; Bangkok, Thailand.
2. Mamdani EH, Assilian S. An experiment in linguistic synthesis with a fuzzy logic controller. International Journal of Human-Computer Studies. 1999;51(2):135-147. doi:10.1006/ijhc.1973.0303.
3. Takagi T, Sugeno M. Fuzzy identification of systems and its applications to modeling and control. IEEE Transactions on Systems, Man and Cybernetics. 1985;15:116-132. WOS:A1985AFX0900011.

4. Hamam A, Georganas ND. A comparison of mamdani and sugeno fuzzy ınference systems for evaluating the quality of experience of hapto-audio-visual applications. In: Proceeding of the IEEE International Workshop on Haptic Audio Visual Environments and their Applications 18-19 October 2008; Ottawa, Canada. IEEE.
5. Tsoukalas LH, Uhrig RE. Fuzzy and neural approaches in engineering (1st ed.). New York: John Wiley & Sons; 1996, 600 p. ISBN:0471160032.
6. Cıtak E. Fuzzy logic approach to determine the factors affecting hand dexterity [thesis]. Unıversıty of Cukurova; 2008.
7. Roger JS, Sun C. Functional equivalence between radial basis function networks and fuzzy inference systems. IEEE Transactions on Neural Networks. 1993;4:156-159. doi:10.1109/72.182710.
8. Roger JS. ANFIS: adaptive network based fuzzy inference systems. IEEE Transactions on Systems Man and Cybernetics. 1993;23(3):665-685. doi:10.1109/21.256541.
9. Perez JA, Gonzalez M, Dopico D. Adaptive neurofuzzy ANFIS modeling of laser surface treatments. Neural Computing & Applications. 2010;19:85-90. doi:10.1007/s00521-009-0259-x.
10. Roger JS, Gulley N. Matlab fuzzy logic toolbox. User#x2019;s guide. Natick: The MathWorks Inc.; 2007.
11. Jang JSR, Sun CT, Mizutani E. Neuro-fuzzy and soft computing: a computational approach to learning and machine intelligence. Englewood Cliffs, NJ: Prentice-Hall; 1997.
12. Sabzevari R, Masoumzadeh S, Ghahroudi MR. Employing anfis for object detection in robo-pong. In: Proceeding of 5th International Symposium on Mechatronics and its Applications (ISMA08); 27-29 May 2008; Amman, Jordan, pp. 707-712.
13. Orghidan R, Danciu M, Vlaicu A, Oltean G, Gordan M, Florea C. Fuzzy versus crisp stereo calibration: a comparative study. In: Proceeding of 7th International Symposium on Image and processing and analysis (ISPA 2011); 4-6 September 2011; Dubrovnik, Croatia, pp. 627-632.
14. Nuno AI, Arcay B, Cotos JM, Varela J. Optimisation of fishing predictions by means of artificial neural networks, anfis, functional networks and remote sensing images. Expert Systems with Applications. 2005;29:356-363. doi:10.1016/j.eswa.2005.04.008.
15. Jafari S, Jarvis R. Robotic hand-eye coordination: from observation to

manipulation. In: Proceedings of the Fourth International Conference on Hybrid Intelligent Systems (HIS'04); 5-8 December 2004; Kitakyushu, Japan. IEEE. doi:10.1109/ICHIS.2004.82.

16. Gallardo A, Taylor J, Paolini C, Lee HK, Lee GK. An ANFIS-based multi-sensor structure for a mobile robotic system. In: Proceeding of Computational Intelligence in Control and Automation; 11-15 April 2011; Paris, France. IEEE. doi:10.1109/CICA.2011.5945755.
17. Li C, Ma H, Yang C, Fu M. Teleoperation of a virtual iCub robot under framework of parallel system via hand gesture recognition. In: Proceeding of International Conference on Fuzzy Systems (FUZZ-IEEE); 6-11 July 2014; Beijing, China, pp. 1469-1474.
18. Orazio TD, Leo M, Mazzeo PL, Spagnolo P. Soccer player activity recognition by a multivariate features integration. In: Proceeding of Seventh IEEE International Conference on Advanced Video and Signal Based Surveillance; 29 August 2010-1 September 2010; Boston, USA, pp. 32-39.
19. Orghidan R, Gordan M, Danciu M, Vlaicu A. A prototype for the creation and interactive visualization of 3D human face models. Advanced Engineering Forum. 2013; 8-9:45-54. doi:10.4028/www.scientific.net/AEF.8-9.45.
20. Navarro H, Orghidan R, Gordan M, Saavedra G, Martinez-Corral M. Fuzzy integral imaging camera calibration for real scale 3D reconstructions. Journal of Dısplay Technology. 2014;10(7): 601-608. doi:0.1109/JDT.2014.2312236.
21. Ozbilge EU. An approach of matlab about calibration of digital cameras [thesis]. Gebze Institute of Technology; 2010.
22. Sugeno M, Kang G. Structure identification of fuzzy model. Fuzzy Sets and Systems. 1988;26(1):15-33. doi:10.1016/0165-0114(88)90113-3.
23. Ross TJ. Fuzzy logic with engineering applications. Chichester: McGraw-Hill; 1995, pp. 134-147. ISBN: 0-07-053917-0.
24. Hocaoğlu FO, Kurban M. Adaptif ağ tabanlı bulanık mantık çıkarım sistemi ile Eskisehir bolgesi icin güneşlenme süreleri tahmini. Elektrik-Elektronik-Bilgisayar Mühendisliği 11. Ulusal Kongresi ve Fuarı, 22-25 Eylul 2005; Istanbul, Turkiye.
25. Mathworks. 2017. Retrieved 25 January 2017, from https://www.mathworks.com/products/matlab.html.

16 Chapter

ELECTROMYOGRAPHY-CONTROLLED PROSTHETIC HAND UTILIZING FUZZY LOGIC CLASSIFICATION ALGORITHM

1. INTRODUCTION

Individuals lose limbs as a result of accidents and medical disorders. Robotic devices that replicate the form and function of a lost limb are produced for individuals who have experienced limb loss in various circumstances. In recent years, researchers have focused on the design and control of multifunctional prosthetic hands [1–7]. The intricacy of movement, specifically the quantity of separate motions, escalates in relation to the amount of joints. The adult skeletal system comprises 206 bones. The 90 bones of the skull and face are interconnected by non-immobilized joints, while the 33 vertebrae of the spine are linked by semi-movable joints. Movable joints are exclusively found between the bones of the arm (25) and leg (25), with the exception of the metacarpal bones. Considering this knowledge, the human hand possesses 15 separate joints, excluding the wrist joints, with three joints in each finger. Consequently, the biological hand movement necessitates the autonomous control of these joints. Consequently, the regulation of the hand is notably intricate. Consequently, of all human appendages, the hand is the most intricate regarding kinetic analysis [8]. Two primary characteristics facilitate the utilization of the functional and visual prosthesis akin to a human hand:

- Prosthetic hand mechanical design and modeling [9, 10] and
- Perform the position and speed controls of each joint efficiently and precisely [11–19].

Nonetheless, regardless of the excellence in the design and fabrication of the prosthetic hand, its functionality is contingent upon the right design of the cognitive interaction, namely the control algorithm, such as the type of movement and coordination among the fingers. If information is not conveyed to the prosthetic hand swiftly enough, the prosthesis will fail to attain the intended position. Cognitive connection is the paramount aspect for users to engage effectively. Numerous studies exist about cognitive interaction between humans and robotic technologies [20–25]. All voluntary muscular movements in humans come from bioelectrical signals transmitted from the brain via the muscle nerves. Bioelectrical electromyogram (EMG) signals relayed to the muscles convey information regarding the nature of movement, velocity, and extent of muscular contraction or relaxation. The biological hand executes fundamental functions of grasping and gripping, necessitating diverse finger movements. The wrist movements fundamentally serve as the axis and facilitate grasping and holding actions. The primary aspect enhancing the usefulness of the prosthetic hand is the movement of the fingers. As the quantity of independent movements executed by the prosthetic hand escalates, its ability to replicate the biological hand improves significantly. This study develops a bioelectrical signal control method and expands the bioelectrical signal database to enhance the finger motion capabilities of prosthetic hands controlled by bioelectrical signals. Figure 1 illustrates bioelectrical signals in relation to muscle movement activities (e.g., flexion, relaxation force), as depicted in the block flow diagram. Electromyography (EMG) can detect signals from the flexor pollicis longus, flexor carpi radialis, brachioradialis, extensor carpi radialis, extensor digiti minimi, and extensor carpi ulnaris muscles. Bioelectrical signals were captured using four surface electrodes and underwent a number of preprocessing and classification procedures to elucidate the correlations between EMG signals and hand and finger movements. The signals were subsequently applied to the prosthetic hand (space and simulator) as a reference motion signal. The devised controllers enable the regulation of the prosthetic hand finger joints' positions. Consequently, a cognitive interface and communication network are created between the user and the prosthetic hand. The study establishes a bioelectrical database documenting the actions of hand muscles and the interaction network between the human and prosthetic hand, utilizing this database and interface to make a simulator and formulate a control algorithm.

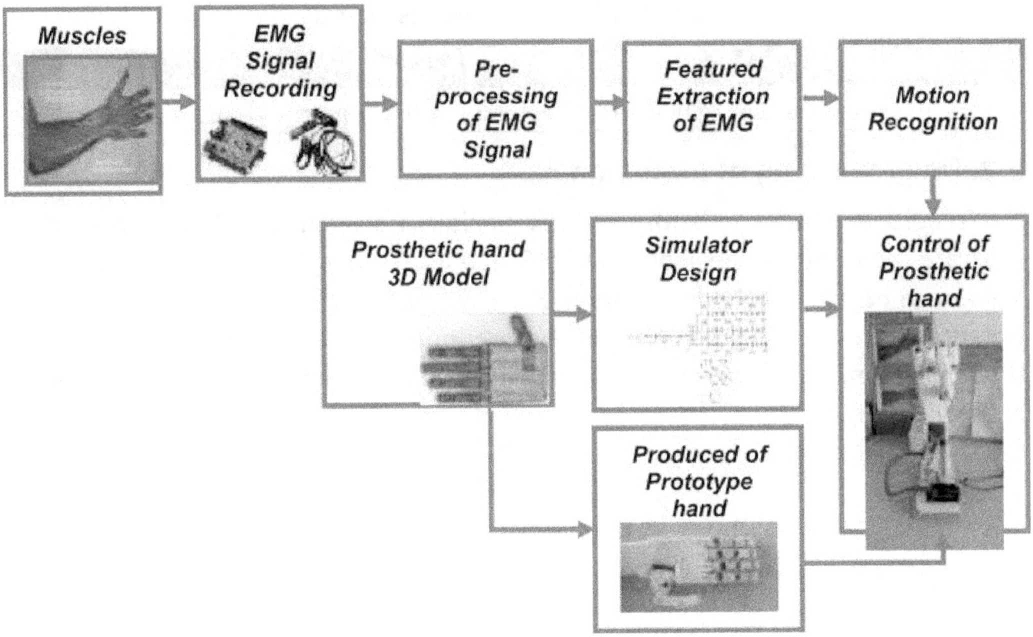

Figure 1. Control of multifunctional prosthetic hand simulator and prototype with EMG signals.

2. RECORDING, PREPROCESSING, AND FEATURED EXTRACTIONS OF EMG SIGNAL

2.1. Recording of EMG signals

EMG signals were recorded from the forearm muscles (the flexor pollicis longus, flexor carpi radialis, brachioradialis, extensor carpi radialis, extensor digiti minimi, and extensor carpi ulnaris) with the help of four surface electrodes. Electrode placements are shown in Figure 2. Electrode layout was chosen according to the protocol [26–28].

Figure 2. Placement of surface electrodes.

The signals, which support movements of the thumb, middle, ring, index, and pinkie fingers, were recorded separately for each of the respective muscles. Channels and finger relations are shown in Table 1.

Table 1. Channel finger relations.

Channel 1	Channel 2	Channel 3	Channel 4
Pinkie finger muscle	Ring finger muscle	Middle finger muscle	İndex finger muscle

2.2. Preprocessing of EMG signals

The recorded EMG signals also include various noise signals. It is necessary to separate the noise signals from the EMG signals, so that the characteristics of the signal can be accurately determined. For this reason, the raw EMG signal is first preprocessed. The block diagram of the preliminary preparation stage, including the separation, rectification, and sampling of the recorded EMG signals from noise, is shown in Figure 3.

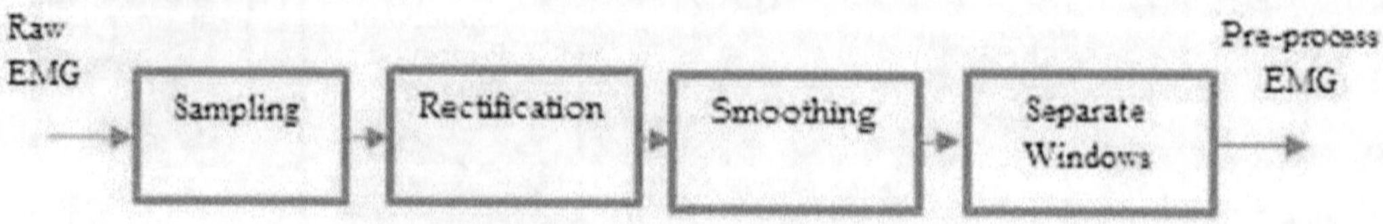

Figure 3. Preprocessing steps.

2.2.1. Numerical sampling

EMG signals are analog voltage signals. Their amplitudes change constantly over the voltage range. Analog-to-digital conversion is the process by which the amplitude of the analog signal voltage is represented by a number sequence at specific time points [29–31]. The EMG voltage signals used in this study are converted into a number sequence by sampling with a period of 0.001 s.

2.2.2. Rectification process

Rectification is the evaluation of only the positive parts of the signal. This is done either by half-wave or full-wave rectification of the signal. A full-wave rectification method was applied to preserve the energy of the signal [25, 29–34], and the expression for the method is given in Eq. (1).

$$X\_training = |x\ (t)| \tag{1}$$

2.2.3. Smoothing of signal

A bandpass filter (50–500 Hz) was designed to soften the signal by eliminating high-frequency components.

2.2.4. Separate the signal into windows

Before the attributes of the obtained EMG signals are calculated, the frame is processed by the method adjacent to the signal. Experiments in the study of Englehart [18, 19] for framing and optimal framing values (R = 256, r = 32 ms) reached with calculations were used.

2.3. Featured extractions of EMG signal

The EMG signal is a non-stationary, time-varying signal that varies in amplitude by random negative and positive values [25, 31, 32]. Bioelectrical signals have certain characteristic values, i.e., information. Features in time domain have been widely used in medical and engineering practices and researches. Time domain features are used in signal classification due to its easy and quick implementation. Furthermore, they do not need any transformation, and the features are calculated based on raw EMG time series. Moreover, much interference that is acquired through the recording because of their calculations is based on the EMG signal amplitude. However, compared to frequency domain and time-frequency domain, time domain features have been widely used because of their performances of signal classification in low noise environments and their lower computational complexity [29]. In this study, five time domain features methods widely used in the literature have been utilized to obtain the features of the EMG signal.

2.3.1. Signal energy

Mathematically, the energy of the signal m (t) is calculated as in Eq. (2), where t_j and t_i denote the lower and upper bounds of the part of the signal to be integrated, respectively. The above expression represents the area below the absolute value of the signal curve at time $T = t_i - t_j$ [30–35].

$$E = \int_{t_i}^{t_j} |m(t)|dt \tag{2}$$

2.3.2. Maximum value of signal

The maximum value of the signal represents the largest of the sampled signal values in each packet divided by windows [29].

2.3.3. Signal average value

Mathematically, the average of the signal m (t) is calculated as Eq. (3) [30, 31], where t_i and t_j denote the upper and lower bounds of the part of the signal to be integrated, respectively. The above expression represents the overall average of the signal at time interval $T = t_i - t_j$.

$$AVR = \frac{1}{t_j - t_i} \int_{t_i}^{t_j} |m(t)|dt \tag{3}$$

2.3.4. Effective value of the signal

Effective value is a commonly used signal analysis method in the time domain, such as average rectification [29–32]. The effective value of the *m*(*t*) signal is calculated as Eq. (4).

$$RMS = \left(\frac{1}{T} \int_0^t m^2(t)dt \right)^{\frac{1}{2}} \quad (4)$$

2.3.5. Variance of signal

The variance value of the signal represents the amount of deviation from the mean of the sampled signal values in each packet divided by windows [30]. $p(t)$ is the variance of the signal to represent the probability density function of t:

$$VAR = \left(\frac{1}{T} \int_0^t (x - ORT)^2 p(t)dt \right) \quad (5)$$

3. PATTERN RECOGNITION WITH FUZZY LOGIC ALGORITHM

A classifier's function should be able to map different patterns, match them appropriately, and, in this case, select different hand grip postures. The extracted features were then fed into the fuzzy logic (FL) classifier for the developed control system. FL developed by Lofty Zadeh [35–41] provides a simple way to arrive at a definite conclusion based solely on imprecise input information. A summary of the feature extraction process from the forearm muscles is shown in Table 2 according to motion.

Table 2. Summary of the feature extraction process from the forearm muscles.

	Signal	Hand closure	Hand opening	Index-thumb touch	Middle-thumb touch	Ring-thumb touch	Pinky-thumb touch
Energy	Channel 1	16,41091	9,949203	5,853087	5,405963	5,354211	**12,84222**
	Channel 2	12,48169	10,92331	7,334108	6,46115	**13,25441**	5,029002
	Channel 3	12,02946	9,254157	8,313991	**12,82708**	7,183281	4,252198
	Channel 4	14,59524	7,548085	**11,22431**	6,920272	9,376161	4,381767
Maximum value	Channel 1	2,378095	1,398911	0,822295	0,61429	0,725287	**2,255524**
	Channel 2	1,674114	1,183987	1,126519	0,961061	**1,90971**	0,609637
	Channel 3	1,606747	1,351835	1,163335	**1,60762**	1,147475	0,666139
	Channel 4	1,990469	0,844166	**1,437937**	0,906574	1,485923	0,532234
Average value	Variance	0,656436	0,397968	0,234123	0,216239	0,214168	**0,513689**
	Channel 1	0,499268	0,436932	0,293364	0,258446	**0,530176**	0,20116
	Channel 2	0,481178	0,370166	0,33256	**0,513083**	0,287331	0,170088
	Channel 3	0,58381	0,301923	**0,448973**	0,276811	0,375046	0,175271

	Signal	Hand closure	Hand opening	Index-thumb touch	Middle-thumb touch	Ring-thumb touch	Pinky-thumb touch
RMS value	Channel 4	0,474695	0,273057	0,163739	0,134428	0,148438	**0,387735**
	Channel 1	0,325763	0,25909	0,207215	0,173207	**0,370618**	0,124443
	Channel 2	0,316673	0,25826	0,223731	**0,339657**	0,213173	0,122159
	Channel 3	0,383453	0,188114	**0,295885**	0,180392	0,269928	0,10675
Variance	Channel 4	0,72476	0,223357	0,08254	0,045411	0,066981	**0,508143**
	Channel 1	0,293061	0,15076	0,133987	0,086676	**0,422607**	0,038505
	Channel 2	0,281122	0,204654	0,145503	**0,326644**	0,150682	0,047588
	Channel 3	0,410777	0,089351	**0,246002**	0,089669	0,232966	0,027352

In total, there are 20 features of EMG signal for four channels. In order to make relations easier, a featured function, which occurs at RMS, AVR, MAX, VAR, and E values, is defined for each channel. Finally, the number of inputs is reduced by four. The featured function is calculated as follows in Eq. (6).

$$F_i = E_i + AVR_i + MAX_i + VAR_i + RMS_i \tag{6}$$

For the FL classification analysis, the triangular shape of the membership function (MF) for the inputs (F_i) and output and the centroid method for defuzzification are used. The rules are created based on information from the states of contraction. FLC rules are shown in Table 3. Recorded SEMG signals have been used to initial testing. Then real time data implemented to Prosthetic hand model.

Table 3. FL rules.

Rules	F1	F2	F3	F4	Result
1	BIG	BIG	BIG	BIG	Hand closure
2	MEDIUM	MEDIUM	MEDIUM	MEDIUM	Hand opening
3	MEDIUM	MEDIUM	MEDIUM	BIG	Index-thumb touch
4	MEDIUM	MEDIUM	BIG	MEDIUM	Middle-thumb touch
5	MEDIUM	BIG	MEDIUM	MEDIUM	Ring-thumb touch
6	BIG	MEDIUM	MEDIUM	MEDIUM	Pinky-thumb touch
7	SMALL	SMALL	SMALL	SMALL	Relax-no motion

Fi Featured functions were inputs to the FL. The limits of F were set to [0, 20]. The three linguistic variables used were Small (S), Medium (M), and Big (B). The outputs of FL were Hand closure, Hand opening, Index-thumb contact, Middle-thumb contact, Ring-thumb contact, and Pinky-thumb contact. Figure 4 shows the flow diagram of FL classification process from four SEMG signals for six hand patterns [35].

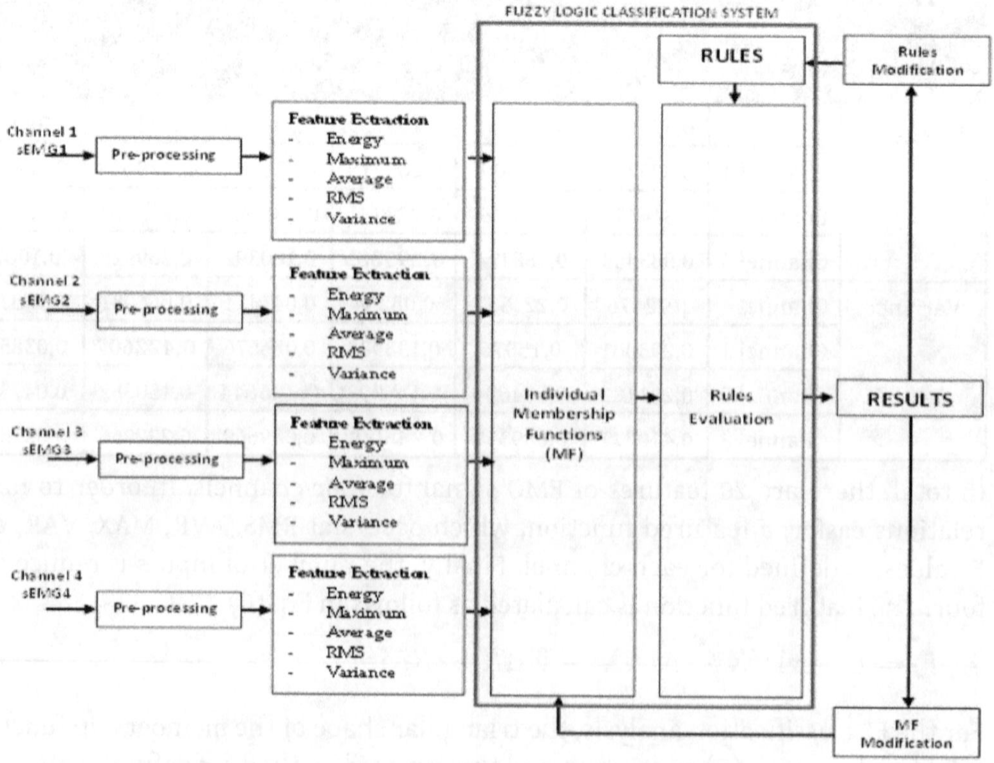

Figure 4. The flow diagram of the control system with FL classification components.

Performance of FL tested 200 hand motions. Classification performance value for the six motions is shown in Table 4.

Table 4. Classification achievement percentages.

Hand pattern	Pattern number	Tested total number of motion (A + B)	Number of true classified motion (A)	Number of wrong classified motion (B)	Average percentage of success (%)
Hand closure	MOTION 1	84	84	0	100
Hand opening	MOTION 2	84	84	0	100
Index-thumb touch	MOTION 3	84	76	8	90.476
Middle-thumb touch	MOTION 4	84	66	18	78.57
Ring-thumb touch	MOTION 5	84	72	12	85.714
Pinky-thumb touch	MOTION 6	84	76	8	90.476

In the medical decision-making process, ROC analysis method is used to determine the discrimination of the test or classification algorithm. In this study, performance of FLC algorithm for six motion class are demonstrated in Table 5 via ROC analysis.

Table 5. ROC analysis.

ROC analysis	Motions					
Classification algorithm result	Hand closure	Hand opening	Index-thumb touch	Middle-thumb touch	Ring-thumb touch	Pinky-thumb touch
Hand closure	84	0	0	0	0	0
Hand opening	0	84	0	0	0	0
Index-thumb touch	0	0	76	6	6	4
Middle-thumb touch	0	0	1	66	2	0
Ring-thumb touch	0	0	4	10	72	3
Pinky-thumb touch	0	0	2	0	1	76
No motion	0	0	1	2	1	1

Performance values calculated as Eqs. (7)–(10) for each hand motion

$$Accuracy(ACC) = \Sigma True\ positive + \Sigma True\ negative / \Sigma Total\ population \tag{7}$$

$$Positive\ predictive\ value(PPV), Precision = \Sigma True\ positive / \Sigma Test\ out\ comepositive \tag{8}$$

$$True\ positive\ rate(TPR), Sensitivity = \Sigma True\ positive / \Sigma Condition\ positive \tag{9}$$

$$False\ positive\ rate(FPR) = \Sigma False\ positive / \Sigma Condition\ negative \tag{10}$$

The four outcomes can be formulated in a 2 × 2 contingency table. All contingency matrixes for each motion are shown in Table 6.

Table 6. Contingency matrixes.

Hand closure

TP=84	FN=0	84
FP=0	TN=420	420
84	420	504

TPR= 1.00
FPR= 0.00
PPV=1.00
ACC=1.00

Hand opening

TP=84	FN=0	84
FP=0	TN=420	420
84	420	504

TPR= 1.00
FPR= 0.00
PPV=1.00
ACC=1.00

Index -thumb touch

TP=76	FN=16	92
FP=8	TN=404	412
84	420	504

TPR= 0.826
FPR= 0.0194
PPV=0.904
ACC=0.952

Middle -thumb touch

TP=66	FN=4	70
FP=18	TN=416	434
84	420	504

TPR= 0.942
FPR= 0.041
PPV=0.785
ACC=0.956

Ring -thumb touch

TP=72	FN=17	89
FP=12	TN=403	415
84	420	504

TPR= 0.808
FPR= 0.028
PPV=0.857
ACC=0.942

Pinky -thumb touch

TP=76	FN=4	80
FP=8	TN=416	424
84	420	504

TPR= 0.95
FPR= 0.018
PPV=0.904
ACC=0.976

4. 3D MODELING AND MANUFACTURING OF PROSTHETIC HAND

4.1. 3D modeling of prosthetic hand via SolidWorks

In order to develop a multifunctional prosthetic hand model, the structural characteristics of the human hand must first be determined. In other words, it is necessary to determine the number of joints, the number of links, the fingers and the length and width parameters of each finger. In order to obtain a prosthetic hand the same size as a human hand, the hand characteristics of an adult male were recorded as in Table 7 for the purposes of this study [42–44].

Table 7. Part of the hand.

	First link		Second link		Third link	
	Length (mm)	Width (mm)	Length (mm)	Width (mm)	Length (mm)	Width (mm)
Thumb	70	30	45	30	40	30
Index	55	30	40	25	30	25
Middle	55	30	50	25	40	25
Ring	55	30	40	25	30	25
Pinky	30	30	40	25	30	25
Palm	130	120				

Using the parameter values in Table 5, the prosthetic hand 3D model is designed with the help of the SolidWorks program as shown in Figure 5.

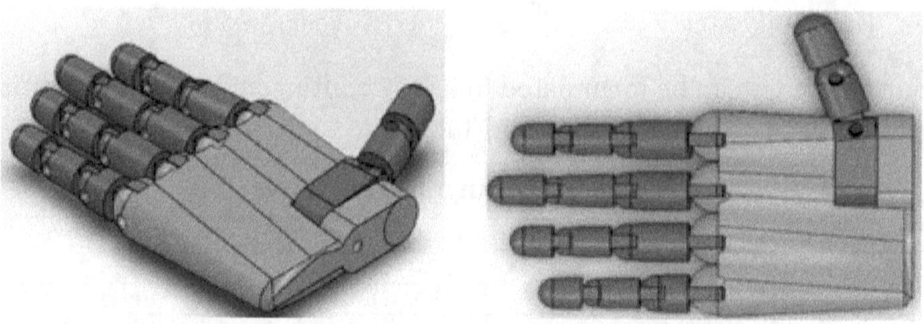

Figure 5. SolidWorks images of prosthetic hand.

4.2. Manufacturing of prosthetic hand via 3D printer

The prototype of the prosthetic hand was produced with the help of the EDISON 3D printer manufactured by 3D Design Company. The necessary adjustments for the production (e.g., resolution, amount of fullness, amount of support) were made using the Simplify 3D program, which was offered by the same company as the software program. After a hand of 16 parts was produced, it was assembled as shown in Figure 6.

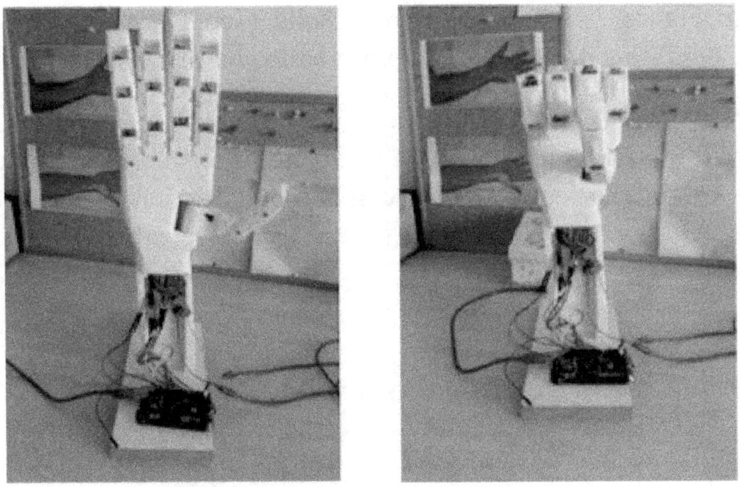

Figure 6. Prototype hand.

5. PROSTHETIC HAND SIMULATOR DESIGN

5.1. Mechanical design of prosthetic hand simulator via SimMechanics

SimMechanics used in the realization of simulations of mechanical systems [45, 46]. By transferring the 3D CAD model of the prosthetic hand developed in the SolidWorks program to the Matlab SimMechanics program, a chain structure containing each joint and link of the prosthetic hand was obtained as shown in Figure 7. Five fingers connected to the palm, three rotary hinges forming each finger, and three connecting links are arranged in series to form the hand SimMechanics model.

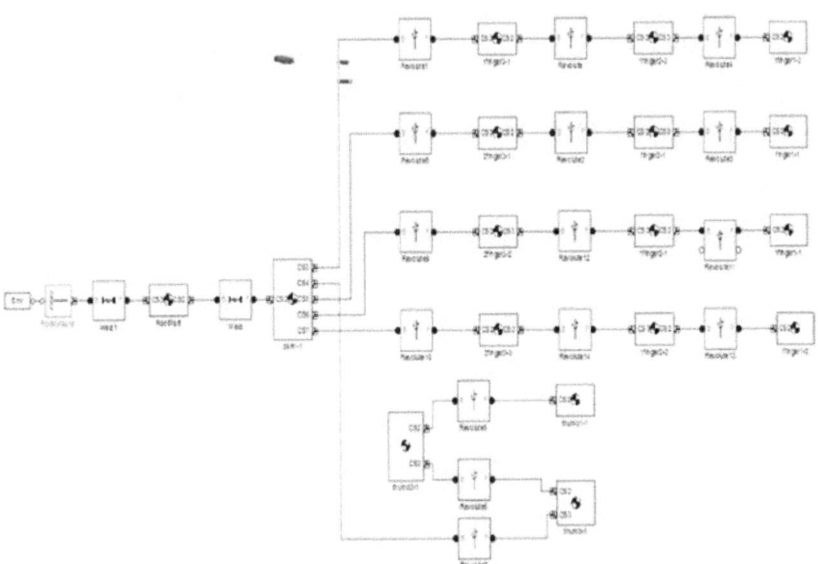

Figure 7. Prosthetic hand SimMechanics model.

As shown in Figure 7, when SolidWorks solid model is transferred to Matlab Program, a chain structure composed of revolute and link parts is obtained.

5.2. Modeling of the DC motor

In this study, it was decided to use a DC servo motor for movement of each joint in the prosthetic hand. The equivalent circuit of the DC servo motor is given in Figure 8 [47–49].

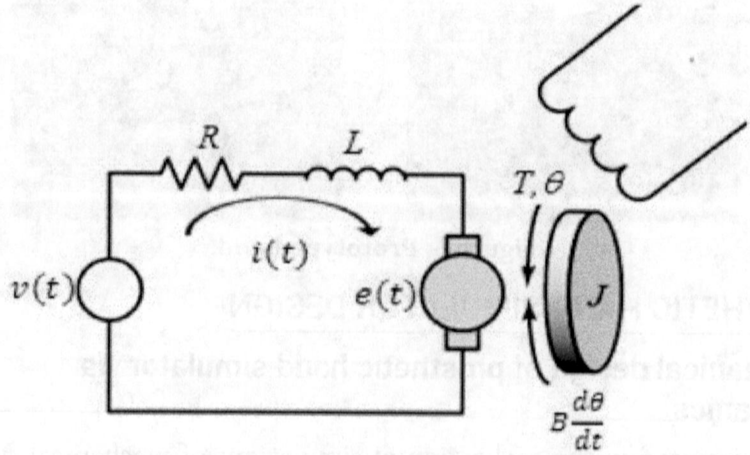

Figure 8. DC motor electrical and mechanical model.

Modeling equations of DC motor were expressed in terms of the Laplace variable s as Eqs. (11)–(13).

$$s(Js + B)\theta(s) = K_t I(s) \tag{11}$$

$$(Ls + R)I(s) = V(s) - K_e s\theta(s) \tag{12}$$

We arrive at the following open-loop transfer function by eliminating $I(s)$ between the two equations above, where the rotation is considered the output and the armature voltage is considered the input.

$$\frac{\theta(s)}{V(s)} = \frac{K}{s\left((Ls + R)(Js + b) + K^2\right)} \tag{13}$$

Using the mathematical model of the DC servo motor, the Matlab/Simulink model is constructed as shown in Figure 9.

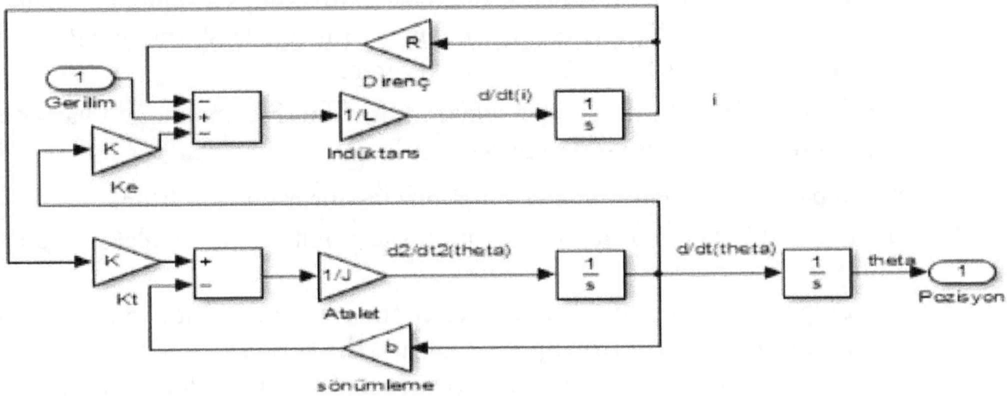

Figure 9. DC motor Matlab/Simulink model.

6. CONTROLLER DESIGN

Position of ultra-nano DC servomotors connected to joints is controlled using a PID controller. The controller's proportional gain coefficient (Kp), integral gain coefficient (Ki), and derivative gain (Kd) values are determined by Genetic Algorithm [11, 50–52] to ensure that the system quickly reaches a steady state without overshooting as shown in Table 8. The PID controller has an input-output relationship with input e (t) and output u (t) [53–55].

Table 8. PID parameters.

	K_p	K_i	K_d
All DC motors connected the each finger joints	0.42176	0.75724	0.0048566

$$u(t) = K_p \cdot e(t) + K_i \cdot \int_0^t e(\tau).d\tau + Kd.\frac{de(t)}{dt} \tag{14}$$

7. GRAPHICAL AND NUMERICAL RESULTS

Electromyography is used to measure EMG signals, which are extracted from the forearm muscles and classified with the help of four surface electrodes. The type of motion that one wishes to perform is the perceived and designed three-dimensional prosthetic hand simulator and the five-fingered and 15-jointed hand. These movements were made in real time on the prototype. Each joint of the prosthetic hand is moved with one ultra-nano servomotor, and the position control of the motors is provided by the designed PID.

The prosthetic hand was made with hand closure, hand opening, thumb-index touch, hand opening, thumb-middle touch, hand opening, thumb-ring touch, hand opening, thumb-pinkie touch, and hand opening movements. The hand opening movement is performed after the hand closing movement and touch movement.

1. EMG signals were taken from four channels, four groups of muscles simultaneously, as shown in Figures 10–13, and preprocessed. First, the signal amplitude was scaled from 0 to 10 V and then filtered.

As shown in Figures 14–17, the energy, maximum, effective, mean, and variance attribute values of the respective signals were calculated.

2. Motion pattern was determined by motion classification algorithm.
3. The specified type of motion information was input to the simulator and the prototype.
4. According to the recognized hand pattern, the reference joint angles in Table 9 were applied as the control input signal, and the closed loop position control of the DC servomotors was performed according to feedback information from sensors connected to the simulator joints.

Table 9. Reference value for each finger joints.

		Index			Middle			Ring			Pinkie			Thumb		
		θ_1	θ_2	θ_3	θ_1	θ_2	θ_3	θ_1	θ_2	θ_3	θ_1	θ_2	θ_3	θ_1	θ_2	θ_3
1	Motion 1	90	90	90	90	90	90	90	90	90	90	90	90	90	90	90
2	Motion 2	0	0	0	0	0	0	0	0	0	0	0	0	0	0	0
3	Motion 3	90	30	30	0	0	0	0	0	0	0	0	0	70	15	5
4	Motion 4	0	0	0	90	25	25	0	0	0	0	0	0	87	5	5
5	Motion 5	0	0	0	0	0	0	90	25	10	0	0	0	105	15	5
6	Motion 6	0	0	0	0	0	0	0	0	0	90	30	5	125	15	5

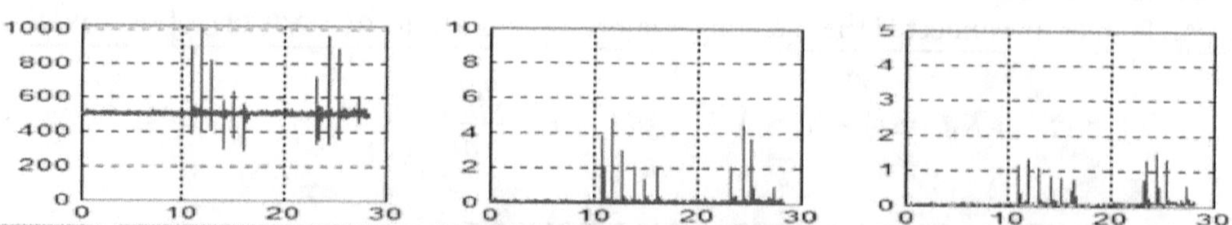

Figure 10. Preprocessing step graphics of EMG signal recorded Channel 1.

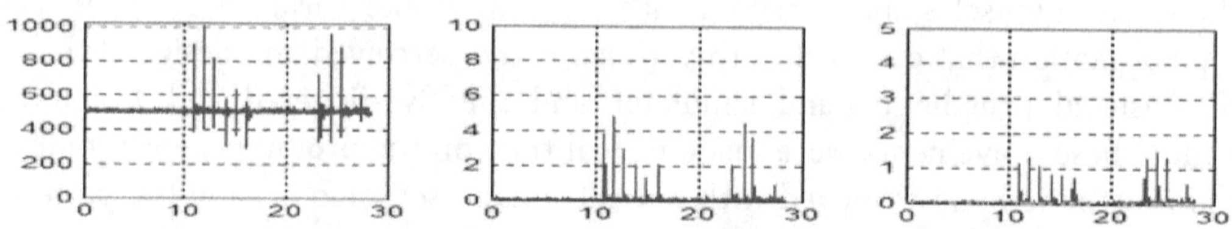

Figure 11. Preprocessing step graphics of EMG signal recorded Channel 2.

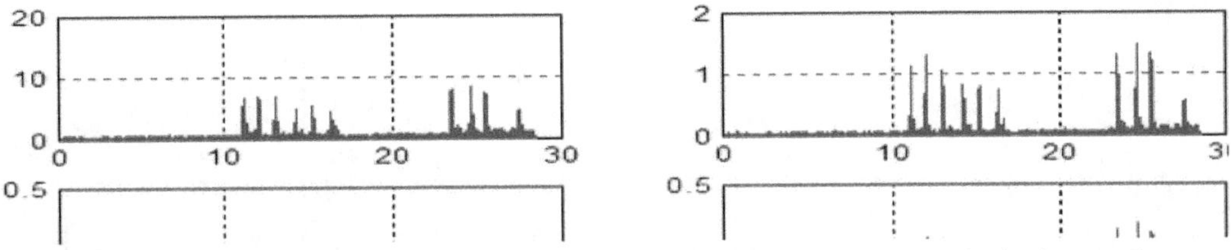

Figure 12. Preprocessing step graphics of EMG signal recorded Channel 3.

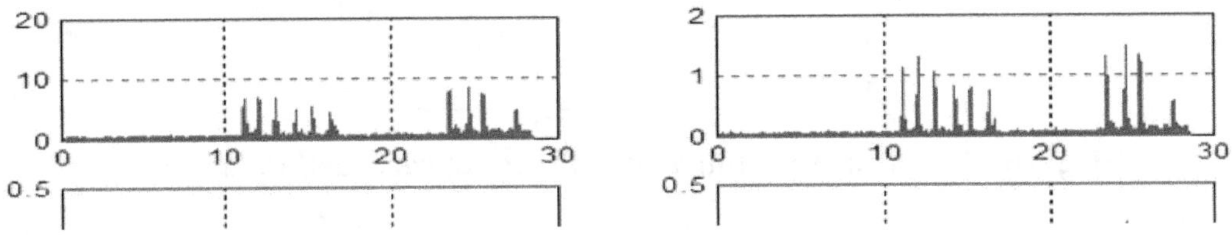

Figure 13. Preprocessing step graphics of EMG signal recorded Channel 4.

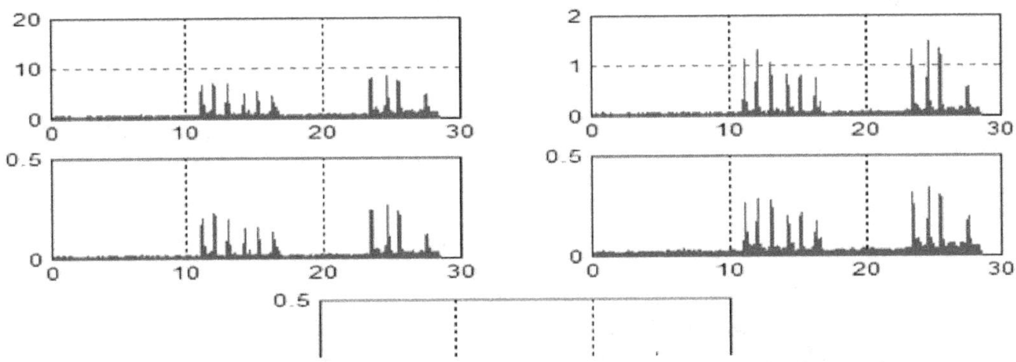

Figure 14. Features graphics of EMG signal recorded Channel 1.

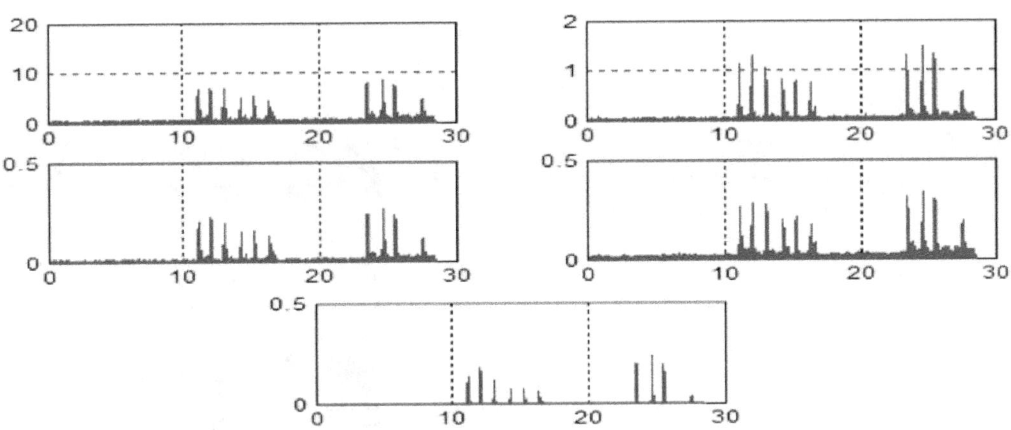

Figure 15. Features graphics of EMG signal recorded Channel 2.

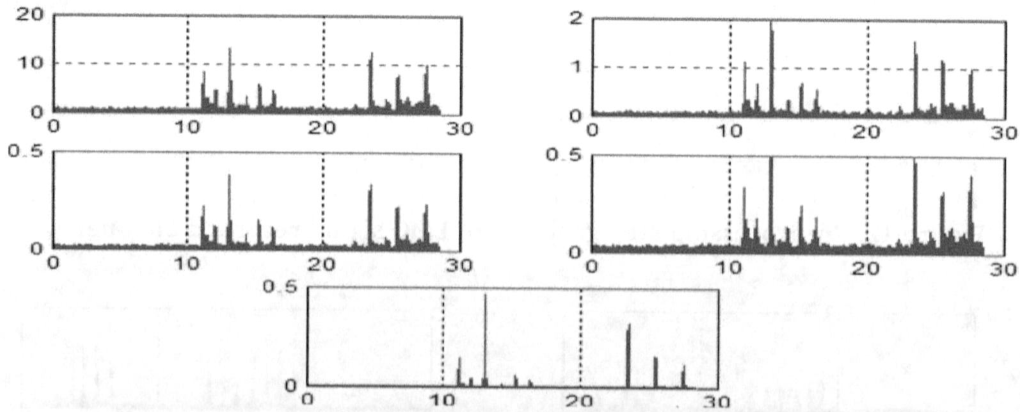

Figure 16. Features graphics of EMG signal recorded Channel 3.

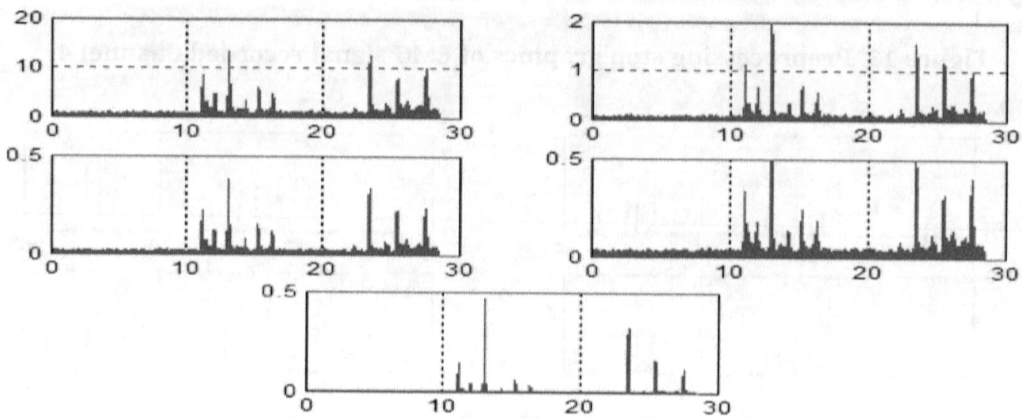

Figure 17. Features graphics of EMG signal recorded Channel 4.

Position control of the finger joints for six hand patterns was provided by the PID controllers as shown in Figures 18–23.

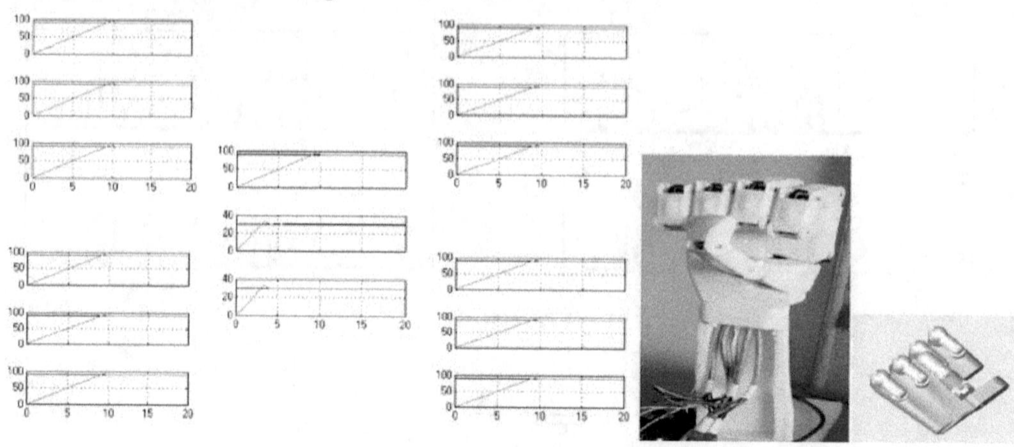

Figure 18. PID response graphics of five fingers for hand close and prosthetic hand photograph.

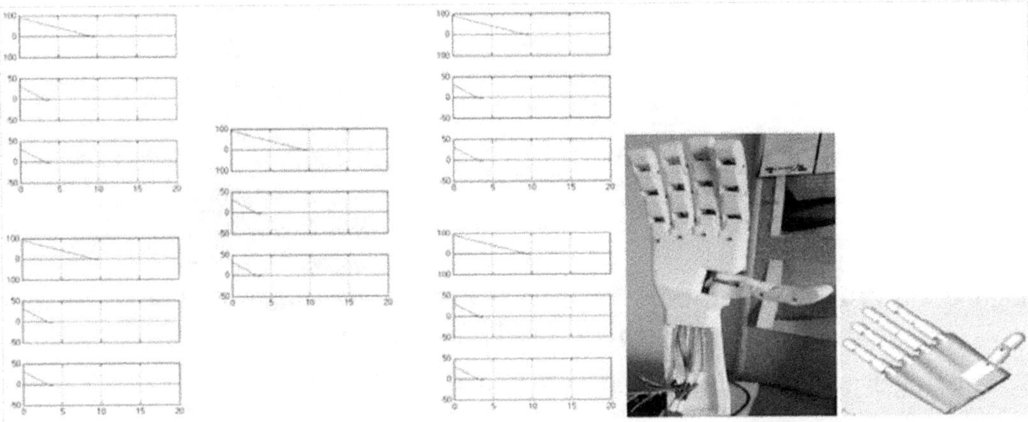

Figure 19. PID response graphics of five fingers for hand opening and prosthetic hand photograph.

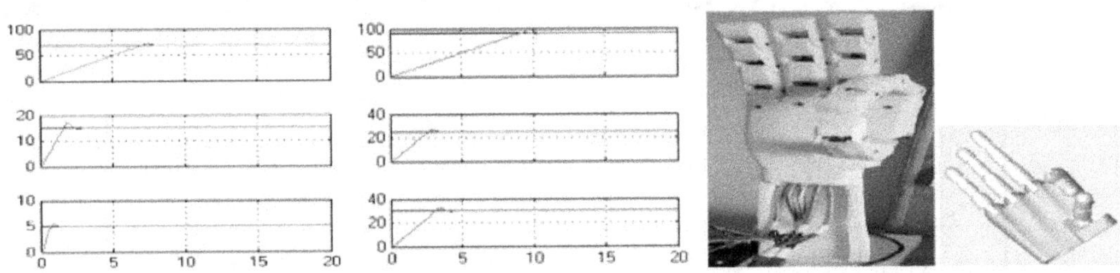

Figure 20. PID response graphics of five fingers for thumb-index touch and prosthetic hand photograph.

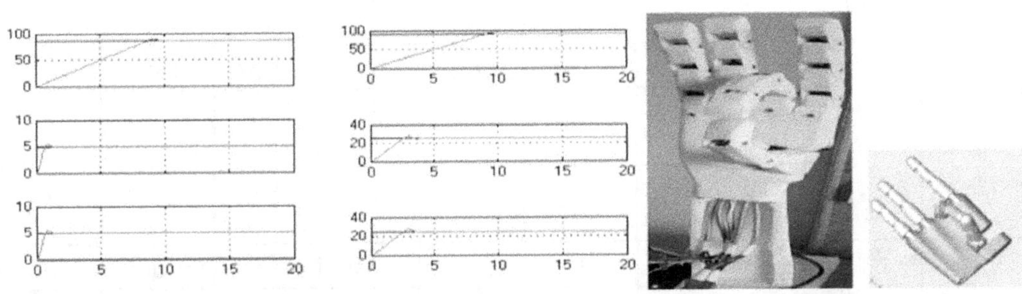

Figure 21. PID response graphics of five fingers for thumb-middle touch and prosthetic hand photograph.

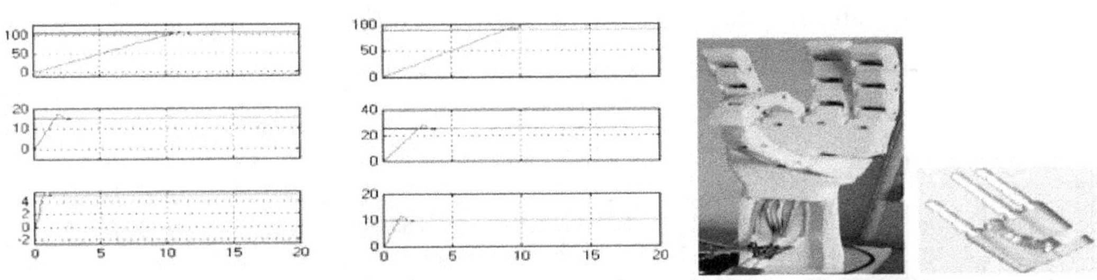

Figure 22. PID response graphics of five fingers for thumb-ring touch and prosthetic hand photograph.

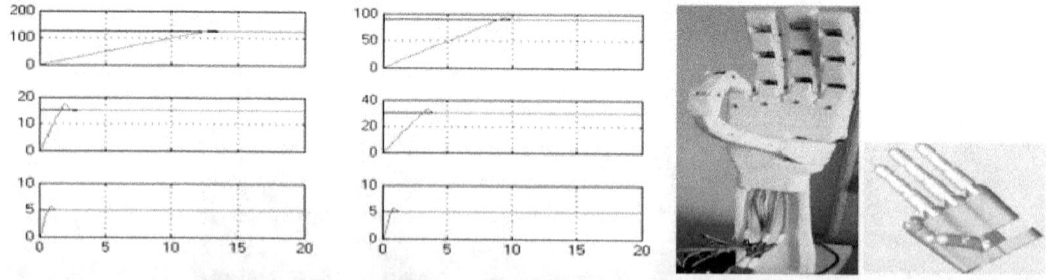

Figure 23. PID response graphics of five fingers for thumb-pinkie touch and prosthetic hand photograph.

For all finger joints, PID performance is shown in Table 10.

Table 10. PID performance value for each joint.

Finger	Joint no		Motion 1	Motion 2	Motion 3	Motion 4	Motion 5	Motion 6
Thumb finger	1	Overshoot (deg.)	2.835	0.2932	2.137	2.936	3.025	3.655
		Steady state time (s)	9.8084	13.413	8.8084	9.988	10.8084	12.8084
		Steady state error (deg.)	0.046	0.041	0.037	0.027	0.021	0.024
	2	Overshoot (deg.)	2.755	0.3265	0.652	0.252	0.652	0.652
		Steady state time (s)	4.415	2.883	1.952	0.752	1.952	1.952
		Steady state error (deg.)	0.0052	2.6e-3	0.0001	0.0001	0.0001	0.0001
	3	Overshoot (deg.)	2.754	0.2696	0.357	0.357	0.357	0.357
		Steady state time (s)	4.524	1.972	0.956	0.956	0.956	0.956
		Steady state error (deg.)	0.0053	1.5e-3	1.5e-3	1.5e-3	1.5e-3	1.5e-3
Index finger	1	Overshoot (deg.)	2.835	0.3299	2.835	0	0	0
		Steady state time (s)	9.915	9.71	9.915	0	0	0
		Steady state error (deg.)	0.045	10e-4	0.045	0	0	0
	2	Overshoot (deg.)	2.835	0.0368	2.349	0	0	0
		Steady state time (s)	9.915	4.555	7.0725	0	0	0
		Steady state error (deg.)	0.047	0.0183	0.0219	0	0	0

Finger	Joint no		Motion 1	Motion 2	Motion 3	Motion 4	Motion 5	Motion 6
	3	Overshoot (deg.)	2.835	0.348	0.377	0	0	0
		Steady state time (s)	9.915	4.535	7.429	0	0	0
		Steady state error (deg.)	0.047	0.0202	0.0255	0	0	0
Middle finger	1	Overshoot (deg.)	2.8356	0.3244	0	2.8368	0	0
		Steady state time (s)	10.5022	10.279	0	10.52	0	0
		Steady state error (deg.)	0.0474	1e-3	0	0.0475	0	0
	2	Overshoot (deg.)	2.8356	0.3244	0	2.812	0	0
		Steady state time (s)	10.5022	10.279	0	3.437	0	0
		Steady state error (deg.)	0.0474	1e-3	0	0.0036	0	0
	3	Overshoot (deg.)	2.8356	0.3244	0	2.7812	0	0
		Steady state time (s)	10.5022	10.279	0	3.9265	0	0
		Steady state error (deg.)	0.0474	1e-3	0	0.0036	0	0
Ring finger	1	Overshoot (deg.)	2.8356	0.3244	0	0	2.8368	0
		Steady state time (s)	9.922	9.907	0	0	9.914	0
		Steady state error (deg.)	0.047	1e-3	0	0	0.047	0
	2	Overshoot (deg.)	2.8356	0.3244	0	0	2.781	0
		Steady state time (s)	9.915	9.9075	0	0	3.412	0
		Steady state error (deg.)	0.047	1e-3	0	0	0.0035	0
	3	Overshoot (deg.)	2.8357	0.3244	0	0	2.545	0
		Steady state time (s)	9.9055	9.906	0	0	1.884	0
		Steady state error (deg.)	0.047	1e-3	0	0	0.0005	0

Finger	Joint no		Motion 1	Motion 2	Motion 3	Motion 4	Motion 5	Motion 6
Pinkie finger	1	Overshoot (deg.)	2.8356	0.3244	0	0	0	2.8368
		Steady state time (s)	9.9094	9.9122	0	0	0	9.29
		Steady state error (deg.)	0.0475	1e-3	0	0	0	0.0475
	2	Overshoot (deg.)	2.8357	0.3244	0	0	0	2.7883
		Steady state time (s)	9.9094	9.9122	0	0	0	4.8174
		Steady state error (deg.)	0.0475	1e-3	0	0	0	0.0052
	3	Overshoot (deg.)	2.8357	0.3244	0	0	0	2.636
		Steady state time (s)	9.9094	9.9122	0	0	0	1.3391
		Steady state error (deg.)	0.0475	1e-3	0	0	0	0

8. CONCISE SUMMARY

The primary determinant in enhancing the functioning of the prosthetic hand to replicate biological hand movements is the articulation of the fingers. The increased independence of finger movements enhances the prosthetic hand's mobility and its capacity to replicate the functionality of a human hand. This thesis enhances the functionality of the prosthetic hand through six distinct finger actions. Bioelectrical signals from the forearm muscles (flexor pollicis longus, flexor carpi radialis, brachioradialis, extensor carpi radialis, extensor digiti minimi, and extensor carpi ulnaris) of two distinct users were recorded using four groups of surface electrodes. Consequently, an extensive database of bioelectrical signals was established. The bioelectrical data were processed using a number of preprocessing and feature extraction techniques to determine the maximum, effective, mean, variance, and energy values of the EMG signals. An FL classification method was devised to establish an efficient cognitive interaction network, achieving a classification success rate of 90%. The detected bioelectrical signals were utilized in the constructed three-dimensional prosthesis portable simulator. The five-fingered and 15-jointed prosthetic hand prototypes produced with a 3D printer, and the positional control of the prosthetic finger joints was performed with the designed controllers. Each digit of the prosthetic hand was actuated by an ultra-nano DC motor, with the motor position controls governed by the engineered

PID system. Consequently, a cognitive interface and communication network were successfully constructed between the individual and the prosthetic hand.

AT A GLANCE

In recent years, researchers have undertaken numerous studies on the design and regulation of prosthetic devices that substitute for absent limbs. The functional capability of prosthetic hands that replicate biological hand functions improves with the amount of independent finger movements achievable. This study demonstrated that six distinct finger movements were transmitted to a prosthetic hand by bioelectrical signals, hence enhancing the usefulness of the prosthesis. Bioelectrical signals were obtained from four muscles using surface electromyography and surface electrodes. The captured bioelectrical signals underwent a sequence of preprocessing and feature extraction procedures. Fuzzy logic categorization methods were created to establish significant motion patterns and an efficient cognitive interaction network between humans and prosthetic hands. A prosthetic hand with five fingers and 15 joints was developed using SolidWorks, and a prototype was fabricated with a 3D printer. A prosthetic hand simulator was developed using Matlab/SimMechanics. Real-time pattern control of both the simulator and the prototype hand was accomplished. The position control of motors linked to each joint of the prosthetic hand was managed using a PID controller. Consequently, a proficient cognitive communication network was built between the user and the real-time pattern control of the prosthesis through bioelectrical signals.

REFERENCES

1. Boostani R, Moradi MH. Evaluation of the forearm EMG signal features for the control of a prosthetic hand. Institute of Physics Publishing Physiological Measurement. Physiological Measurement. 2003;24:309-319. PII: S0967-3334(03)38946-4
2. Nishikawa D, Yu W, Yokoi H, Kakazu Y. On-line supervising mechanism for learning data in surface electromyogram motion classifiers. Journal and Communication in Japan. 2002;33:14, pp. 2634-2643
3. Arieta AH, Katoh R, Yokoi H, Wenwei Y. Development of multi-DOF electromyography prosthetic system using the adaptive joint mechanism. Applied Bionics and Biomechanics. 2006;3:101-112
4. Zhao J, XIie Z, Jing L, Cai H, Hong L, Hırzınger G. EMG control for a five-fingered prosthetic hand based on wavelet transform and autoregressive model. International Conference on Mechatronics and Automation, China; 2006.

5. Carrozza MC, Cappiello G, Stellin G, Zaccone F, Vecchi F, Micera S, Dario P. On the development of a novel adaptive prosthetic hand with compliant joints: Experimental platform and EMG control. In: Proceedings of the IEEE/RSJ International Conference on Intelligent Robots and Systems; April 2005. pp. 3951-3956
6. Hocaoğlu E. Data acquisition and feature extraction for classification of prehensile SEMG signal for control of a multifunctional prosthetic hand [Master thesis]. Istanbul: Sabanci University, Mechanical Engineering; 2010.
7. Carozza M, Cappiello G, Stellin G, Zaccone F, Vecchi F, Micera S, Dario P., On the development of a novel adaptive prosthetic hand with compliant joints: Experimental platform and EMG control, Intelligent Robots and Systems, 2005. (IROS 2005). pp. 1271–1276, 2-6 August 2005, 10.1109/IROS.2005.1545585
8. Taylor Cl, Schwarz RJ. The anatomy and mechanics of the human hand. Artificial Limbs 2.2. 1955:22.
9. Rakibul H, Vepar SH, Hujijbert H. Modeling and control of the Barrett hand for grasping. In: 15th International Conference on Computer Modelling and Simulation; 10–12 April 2013; Cambridge University.
10. Hasan R, Rahideh A, Shaheed H. Modeling and interactional of the multifingered hand. In: Proceedings of the 19th International Conference on Automation & Computing. Brunel University, Uxbridge, UK; 2013.
11. Michalewicz, Z. Genetic Algorithms + Data Structures = Evolution Programs. Berlin Heidelberg, New York: Springer-Verlag; 1999.
12. Widhiada W, Douglas SS, Jenkinson ID, Gomm JB. Design and control of three fingers motion for dexterous assembly of compliant elements. International Journal of Engineering, Science and Technology. 2011;3(6):18–34
13. Al-Assaf Y, Al-Nashash H. Surface myoelectric classification for prostheses control. Journal of Medical Engineering & Technology. 2005;29(5):203–207
14. Parker P, Englehart K, Hudgins B. Myoelectric signal processing for control of powered limb prostheses. Journal of Electromyography and Kinesiology. 2006;16:541–548
15. Englehart K, Hudgins B. A robust, real-time control scheme for multifunction myoelectric control. IEEE Transactions on Biomedical Engineering. 2003;50(7):848–854
16. Hudgins B, Parker P, Scott RN. A new strategy for multifunction myoelectric control. IEEE Transactions on Biomedical Engineering. 1993;40(1):82–94

17. Asghari Oskoei M, Hu H. Myoelectric control systems - a survey. Biomedical Signal Processing and Control. 2007;4(4):275–294

18. Lamounier E, Soares A, Andrade A, Carrijo R. A virtual prosthesis control based on neural networks for EMG pattern classification. In: Proceedings of the Artificial Intelligence and Soft Computing; Canada; 2002

19. Chu J, Moon I, Kim S, Mun M. Control of multifunction myoelectric hand using a real time EMG pattern recognition. In: Proceedings of the IEEE/RSJ International Conference on Intelligent Robots and Systems; China; 2005. pp. 3957–3962

20. Mohammadreza AOI, Huosheng H. Review: Myoelectric control systems: A survey. Biomedical Signal Processing and Control. 2007;2:275–294

21. Zhao Z, Chen X, Zhang X, Yang J, Tu Y, Lantz V, Wang K. Study on Online Gesture SEMG Recognition, Advanced Intelligent Computing Theories and Applications. With Aspects of Theoretical and Methodological Issues. 2007:1257–1265

22. Mahdi K, Mehran J. A novel approach to recognize hand movements via sEMG patterns. Engineering in Medicine and Biology Society. In: 29th Annual International Conference of the IEEE; 2007.

23. Zhizeng L, Xiaoliang R, Yutao Z. Multi-pattern recognition of the forearm movement based on SEMG, Information Acquisition. In: Proceedings of the International Conference on IEEE; 21–25 June 2004. DOI: 10.1109/ICIA.2004.1373391

24. Roberto M. Electromyography Physiology, Engineering and Noninvasive Applications. IEEE Press, John Wiley & Sons Inc.; 2004, New Jersey, Canada

25. De Luca CJ. Electromyography. Encyclopedia of Medical Devices and Instrumentation, John Wiley Publisher; 2006. pp. 98–109, New York City, United States, DOI: 10.1002/0471732877.emd097

26. Hargrove L, Englehart K, Hudgins B. The effect of electrode displacements on pattern recognition based myoelectric control. In: IEEE Annual International Conference on Engineering in Medicine and Biology Society; 2006. pp. 2203–2206

27. De Luca CJ. The use of surface electromyography in biomechanics. Journal of Applied Biomechanics. 1997;1(2):135–163

28. SENIAM EMG protocol. Available from: http://www.seniam.org/(Download date: 21.03.2014)

29. Daud WMBW, Yahya AB, Horng CS, Sulaima MF, Sudirman R. Features extraction of electromyography signals in time domain on biceps brachii muscle. International Journal of Modeling and Optimization. 2013;3(6)

30. Yazıcı İ. EMG İşaretlerinin İşlenmesi Ve Sınıflandırılması [Master thesis]. Sakarya: Sakarya University, Institute of Science and Technology; 2008
31. Taşan D. Protez Denetimi İçin Elektromiyografi (EMG)de Örüntü Tanıma, Yüksek Lisans Tezi, Ege Universty, Institute of Science and Technology, İzmir, 2008
32. Akgün G, Demetgül M, Kaplanoğlu E. EMG Sinyallerinin Öznitelik Çıkarımı ve Geri Yayılımlı Yapay Sinir Ağı, Algoritması İle Sınıflandırılması, Otomatik Kontrol Ulusal Toplantısı, TOK2013, Malatya 26–28 Eylül 2013.
33. Chen WT, Wang Z, Ren X. Characterization of surface EMG signals using improved approximate entropy. Zhejiang University Science B. 2006;7(10): 844–848
34. Chu J, Moon I, Mun M. A real-time EMG pattern recognition system based on linear-non-linear feature projection for a multifunction myoelectric hand. IEEE Transactions on Biomedical Engineering. 2006;53:2232–2238
35. Chan F, Yong-Sheng Y, Lam F, Yuan-Ting Z, Parker P. Fuzzy EMG classification for prosthesis control. IEEE Transactions on Rehabilitation Engineering. 2000;8(3)
36. Yaraş B, Hüseynov R, Namazov M, ÇeliKKale İE, Şeker M. Fuzzy control and sliding mode fuzzy control of DC motor. Journal Of Engineering And Natural Sciences Mühendislik Ve Fen Bilimleri Dergisi, Sigma. 2014;32:97–108
37. Ajiboye A, Weir R. A heuristic fuzzy logic approach to EMG pattern recognition for multifunction prosthesis control. IEEE Transactions on Biomedical Engineering. 2005;52(11): 280–291
38. Karlik B, M.O., T., & M., A. A fuzzy clustering neural network architecture for multifunction upperlimb prosthesis. IEEE Transactions on Biomedical Engineering. 2003;50:1255–1261
39. Vuskovic M, Du SJ. Classification of prehensile EMG patterns with simplified fuzzy ARTMAP networks. In: Proceedings of the International Joint Conference on Neural Networks; 2002, 3, pp. 2539–2544
40. Khzeri M, Jahed M, Sadati N. Neuro-fuzzy surface EMG pattern recognition for multifunctional hand prosthesis control, Industrial Electronics. In: IEEE International Symposium; 2007
41. Zheng L, He X. Classification techniques in pattern recognition. In: WSCG, Conference Proceedings; 2005, ISBN 80-903100-8-7
42. Taşar B, Yakut O, Gülten A. Object detection and grip force control via force sensor for EMG based prosthesis hand, E205. In: International Conference on Electrical and Electronics Engineering (ICEEE); 27–28 Nisan 2015; Ankara

43. Taşar B, Gülten A, Yakut O. Modeling controlling and simulation of 15 DOF multifunctional prosthesis hand using SimMechanics, C201. In: International Conference on Automatic Control (ICOAC); 27–28 Nisan 2015; Ankara

44. Taşar B, Gülten A, Yakut O. Kinematic analysis of the human hand for desing prosthetic hand. In: 3rd International Congress on Natural and Engineering Sciences; September 9–13, 2015; Sarajevo, Bosnia and Herzegovina: International University of Sarajevo

45. Dung LT, Kang HJ, Ro YS. Robot manipulator modeling in Matlab SimMechanics with PD control and online Gravity compensation. IFOST Proceedings. 2010

46. Fedák V, Durovsky F, Üveges R. Analysis of robotic system motion in SimMechanics and MATLAB GUI Environment. MATLAB Applications for the Practical Engineer. Chapter 20. 2014

47. Bencsik AL. Appropriate mathematical model of DC servo motors applied in SCARA robots. Acta Polytechnica Hungarica. 2004;1(2):2004–2099

48. Matlan Tuturial, DC Motor Position Simulink Model. Available from:Dowload:http://ctms.engin.umich.edu/CTMS/index.php?example=MotorPosition&section=SimulinkModeling

49. Meier R. Modeling DC servo motors control systems, Tech Note, Milwaukee School of Engineering 1025 North Broadway, Milwaukee. https://faculty-web.msoe.edu/meier/ee3720/technotes/dcservo.pdf

50. Haupt Randly L, Haupt Sue E. Practical Genetic Algorithms. USA: A Wiley-Interscience Publication; 1998

51. Davis L. Handbook of Genetic Algorithms. New York, NY: Van Nostrand; 1991

52. Wook C, Ramakr Shna RS. Elitism-based compact genetic algorithms. IEEE Transactions on Evolutionary Computation. 2003;7(4):367–385

53. Astrom KJ, Hagglund T. PID Controllers: Theory, Design and Tuning. Instrument Society of America; 1995. p. 343, Research Triangle Park, North Carolina, 1995

54. Taguchi H, Araki M. Two-degree-of-freedom PID controllers. Proceedings of the IFAC Workshop on Digital Control: Past, Present and Future of PID Control; Elsevier; pp. 91–96. 5-7 April 2000, Terrassa, Spain

55. Bennett S. A History of Control Engineering, Peter Peregrinus; 1993. pp. 28–69, London, United Kingdom

17 Chapter

APPLICATION OF FUZZY LOGIC FOR THE CONTROL AND MONITORING OF CRITICAL MACHINE PARAMETERS IN A PROCESSING FIRM

1. INTRODUCTION: AN OVERVIEW

The beverage industry refers to the sector responsible for the production of drinks utilizing highly automated techniques that facilitate rapid beverage manufacturing. The beverage business is significantly impacted by profitability issues. The escalating expenses of new equipment and spare parts, persistent foreign currency constraints, and the necessity for enhanced competitiveness necessitate the implementation of more efficient maintenance systems. This leads to the optimal use of plant-installed capacity by enhancing reliability, uptime, quality, and asset longevity, all attained at the most favorable cost-benefit ratio. Emphasis should be placed on guaranteeing that appropriate maintenance is performed (executing the correct task) rather than just ensuring that maintenance is conducted accurately (executing the task correctly). At the processing company, process losses have historically constituted 1.5% of total losses, but escalated to 10% in 2011. This has generated concern, necessitating the exploration of novel maintenance management philosophies, particularly enhancing maintenance cost-effectiveness as a definitive method for augmenting total profitability [7].

1.1. Background

Currently, the plant uses run-to-failure maintenance as a strategy to take care of its machines. This approach is reported to be creating a high risk to workers, production and property. This is seen by the high rate of unscheduled events associated with breakdowns, emergency equipment, long working hours and high maintenance costs. This has lowered down the image of the company in terms of competition when compared the benchmarks with other beverage companies in developing nations. The progression towards a global economy has increased the base of competition for almost all companies. By competition, it is suggested that every organization out there is always targeting to keep a certain score. The record on a scorecard may be a ratio of more sales, increased revenue or a mounting customer base. Despite the benchmark used for the measurements, for a business to remain competitive, there is a basic company demand to improve at maintaining and taking care of machinery as well as meeting customers' requirements. The competitiveness of production corporations relies on the availability and productivity of their manufacturing equipment, which is possible if all manufacturing losses are recognized and eliminated so that the products can be sold on a marker at a minimum cost. The beverage industry encompasses the sector dedicated to the production of drinks through highly automated processes that enable swift manufacturing. The beverage industry is profoundly affected by profitability challenges. The rising costs of new equipment and spare parts, ongoing foreign currency limitations, and the need for improved competitiveness require the adoption of more effective maintenance systems. This results in the optimal utilization of plant-installed capacity by improving dependability, uptime, quality, and asset longevity, all achieved at the most advantageous cost-benefit ratio. Emphasis should be made on ensuring that proper maintenance is executed rather than merely verifying that maintenance is performed accurately. Historically, process losses at the processing company represented 1.5% of total losses, but increased to 10% in 2011. This has raised concerns, prompting the investigation of innovative maintenance management strategies, including improving maintenance cost-effectiveness as a key approach to increasing overall profitability [7].

1.2. Fuzzy logic

The rules can process simple arithmetic and logical operations with the help of Locator Identifier Separation Protocol (LISP), but a process control algorithm that is based on fuzzy logic is called fuzzy control [8]. The oldest and most common maintenance and repair strategy are "fix it when it breaks". The appeal of this approach is that no analysis or planning is required [9]. Early detection of faults in a machinery is a key parameter in control and maintenance to avoid failure of equipment [10].

2. CRITICAL MACHINERY

The criticality index is regularly used to decide the extent of the nature of assessment on a given machine by checking the machine's objective [11], excess (that is on the off chance that if the machine falls flat, is there a standby machine which can assume control), repairing expenses, idle sways, well-being, security and environmental concerns and other considerable features. The criticality record puts all machines into one of three classifications as follows:

Critical equipment: This is the fundamental machinery at the processing company. With basic equipment which is the core focus of the procedure, it is understood to require adequate online state checking to ceaselessly capture much information from the mechanism as could be expected paying little heed to expenditure and is frequently indicated by the plant protection. Estimations, for example, differential development, velocity, spiral relocation and shaft pivotal, dislodging and packaging vibration, temperatures, weights and loads, are taken where conceivable. Such qualities are regularly sustained into a hardware administration programming system, which is fit for slanting the recorded information and giving the administrators data, for example, executing information and even anticipating failures and giving conclusions of breakdowns beforehand.

Vital equipment: These are units that are essential to the procedure; however, in the event that there is a breakdown, the procedure still proceeds. Excess units (if accessible) would be in this domain. Analysing and regulation of these units are additionally crucial to keep up option strategies given that critical machinery falls flat.

Balance of industrial equipment or the general purpose: These are the components that complement the whole plant and are ordinarily assessed, utilizing a handheld information gatherer, as said beforehand, to occasionally build an idea of the strength of the machine.

2.1. Types of failure caused by not maintaining machinery

Breakdowns might be characterized by their seriousness in any of these four classes and are contingent upon the characterized breakdown impacts below:

1. Insignificant breakdown: Any breakdown that does not debase the general routine and adequacy of the framework past worthy breaking points [12].
2. Key breakdown: Any breakdown that will corrupt the framework implementation and viability past worthy points of confinement, however, they can be controlled [13].
3. Critical breakdown: Any breakdown that will corrupt the framework past adequate points of confinement and could make a security danger if a prompt restorative move is not made [14].
4. Catastrophic breakdown: Any breakdown that could bring about critical framework harm, for example, to block useful achievement, and bring about death and staff wounds [15].

2.2. Types of condition monitoring techniques

In line with the research area of interest, condition monitoring will be looked at since it is a branch of Condition Based Maintenance (CBM).

2.2.1. Noise and vibration

All machines vibrate, and when they are in good condition, their frequency spectrum has a characteristic form; any departure from this form indicates that something is wrong—fatigue, or wear, or ageing of something of some component. Parameters that are useful include amplitude, frequency and phase angle [16].

2.2.2. Amplitude

This gives an indication of the stress under which a piece of rotating machinery is working, in particular, it can give a measure of the eccentricity (out-of-roundness) of a rotor [17].

2.2.3. Frequency

Through the frequency spectrum, you can detect a fault in rotating machinery. Vibrations fall into two main classes:

Synchronous: Frequencies are in multiples or sub-multiples of the frequency of rotation, that is, they are harmonics or sub-harmonics of that frequency.

Asynchronous: These are not related to the rotation frequency; they can be the natural frequencies of various parts of the system, which can be identified [17].

2.2.4. Phase angle

It locates the high point in a rotor that is not perfectly circular and thus gauges its out-of-balance characteristics. Machines must be continuously monitored to denote their state. The following machines can do these:

1. Clearance recorder: recording the actual movements of the shaft which generate the vibrations.
2. Speed recorder: mounted externally to a machine, and it gives a strong signal at medium frequencies, depending on the temperature and the general environment.
3. Accelerometer: also installed to a machine, and it gives a strong signal at high frequencies [17].

2.2.5. Temperature

Temperature recording is a relatively simple matter at the industrial level. Change of temperature in rotating machinery is often a sign of deterioration and is therefore something to which close attention should be given. This is currently practised in thermography [18].

2.2.6. Tribology

An examination of the particles suspended in oil can give very valuable information. The amount of suspended material is an indicator of the state of deterioration of the machine; the composition can identify the source of the wear and thus the component that is failing. The necessary analysis can be done in the laboratory with the electron microscope. Basically, the monitoring techniques that were listed in Table 1 are almost the major ones currently used.

Table 1. Summary of condition monitoring techniques.

Type	Method
Visual	Eye
Temperature	Thermometer, thermocouple
Lubricant monitoring	Filtering, spectroscopy
Vibration	Signal frequency analysis
Crack	Di-penetrant analysis, radiography
Corrosion monitoring	Eye, corrosometer

3. PROBLEM BEHIND THE PROCESSING COMPANY (PC)

The reactive nature of the corrective maintenance strategy at the company has resulted in loss of trivial production time. Downtime due to breakdown is greatly affecting the production targets and delivery of juice products to customers which had resulted in loss of goodwill and trust from consumers. The processing company (PC) uses a number of raw materials for the production of the fruit juice concentrates: mangoes, guavas, oranges, tomatoes, lemons, grape fruits, granadilla and other continuous inputs such as municipality and borehole water, diesel, electricity and labour. The company has got equipment in place which were mainly supplied by a company in Italy. The mission of the production department is to safely produce quality products and maintain equipment at a lowest possible cost to meet sales and marketing demands.

3.1. Failure in equipment

Using records from previous seasons, equipment with high failure rates and high downtimes is identified. For easy interpretation and identification of this equipment, the information is sorted using Microsoft Excel and graphs are drawn. From the bar graph showing the relative contributions of equipment to the total downtime, equipment that contributed the most to the total downtime is identified. Breakdown of any of these equipment results in stoppage of the production; therefore, the production loss that is represented by a breakdown of each of these machines is equivalent to the production of the diffuser line, which is 60% of total production (300 TCH of 500 TCH). The following graph in Figure 1 shows the downtimes caused by the breakdown of each item during the last juice extraction season, and Figure 1 shows how extraction takes place.

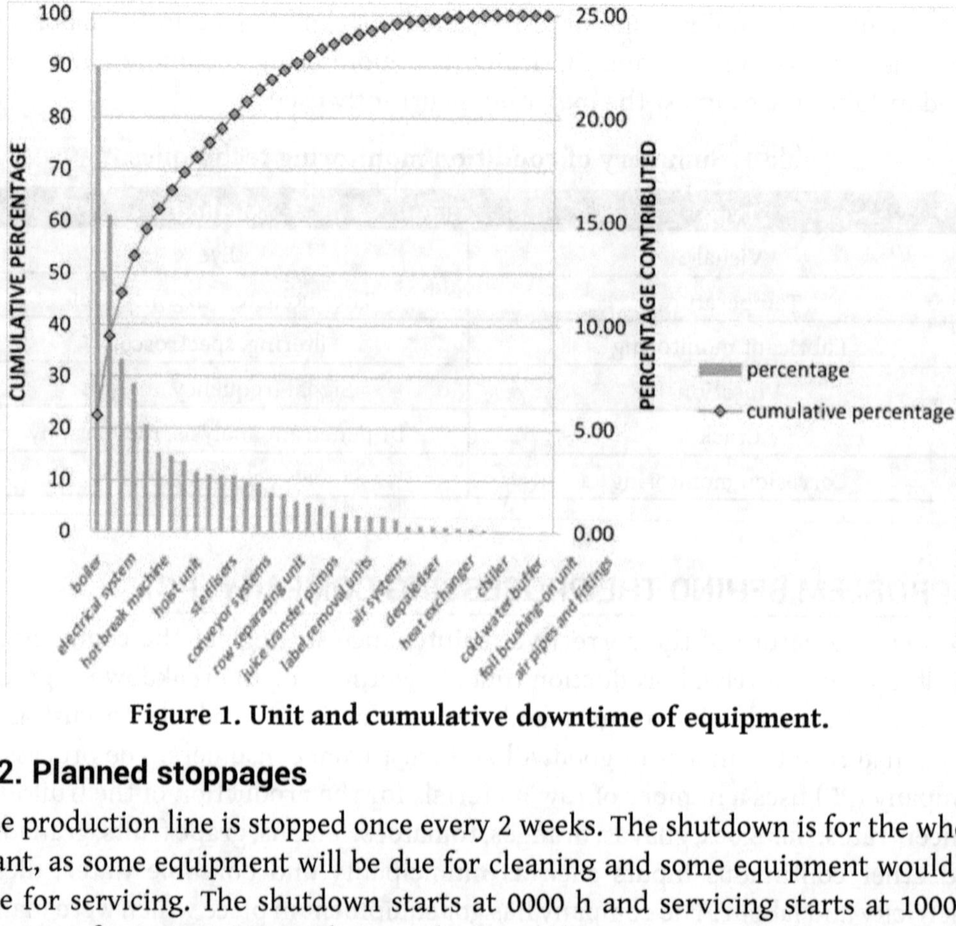

Figure 1. Unit and cumulative downtime of equipment.

3.2. Planned stoppages

The production line is stopped once every 2 weeks. The shutdown is for the whole plant, as some equipment will be due for cleaning and some equipment would be due for servicing. The shutdown starts at 0000 h and servicing starts at 1000 h. The time from 0000 to 1000 h is used to stop the processes, empty some vessels and allow some equipment to cool. This time is also used to plan some of the jobs, perform pre-task risk assessments, order and transport spares from the stores department and also to service some equipment. Full-fledged servicing kicks off at 1000 h and continues up to 1600 h. At 1600 h, the start-up process begins with steam being released into the vessels to warm them. Production is started around 1700 h when vessels are ready to receive juice, and hot imbibition water is available. The processes are then started, and the levels gradually increased. Full production levels are achieved by 1800 h. The total downtime for this planned stoppage is thus 18 h. A crushing season starts early April and ends in mid-December. On average, it has 36 weeks. The average number of hours for planned stoppages annually is thus:

$$\text{Total annual planned downtime} = \text{Number of stoppages} \times \text{Downtime per stoppage} \quad (1)$$

$$\text{Total annual planned downtime} = \frac{36}{2} \times 18 = 324 \text{ h.} \quad (2)$$

Figure 2 shows a comparison of planned stoppages and unplanned stoppages. This is a problem in this company with more stoppages. This has justified the need for my research, as shown in Figure 2.

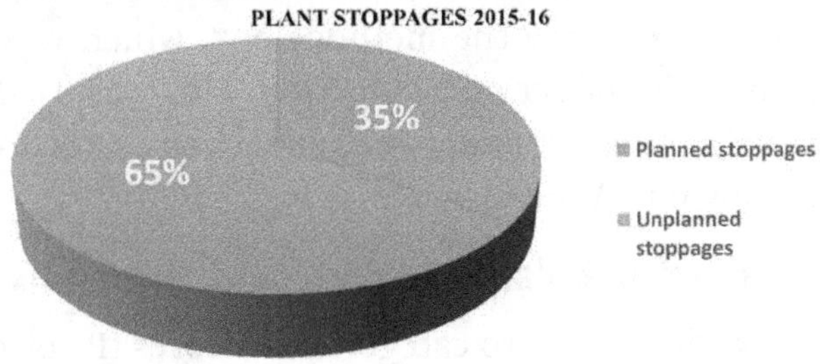

Figure 2. A pie chart for planned and unplanned stoppages.

Figure 2 is very critical as most of the time, the plant is always shutdown, which results in a great deficiency and loss of goodwill from the customers as the plant is failing to meet the targets.

3.3. Plant performance

From operating history, the availability performance for each item was evaluated, and the results are shown in Figure 3. All availability performance scores were above 90 so for clarity in comparing, the axis for availability performances was broken such that it starts at 90–100.The main contributor to downtime is planned downtime. This affects the availability of all equipment in the production line, but it is not all equipment that requires servicing every fortnight. Some of the equipment that requires servicing does not need 18 h. In view of this, maintenance schedules currently in use were analysed in order to identify equipment that really requires servicing fortnightly and has the least maintainability. Figure 3 shows the availability performance of the plant with planned downtime.

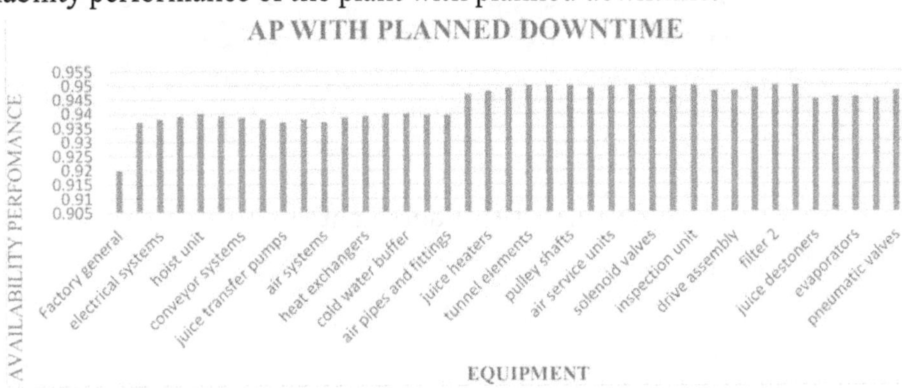

Figure 3. Equipment availability performance including planned downtime.

3.4. Equipment criticality analysis

This section now seeks to identify the equipment that is critical to mainly to production, health, safety, environment and operational costs. This classification of equipment according to risk level is done so as to prioritize equipment in terms of maintenance, work orders, costs and inspection. The criticality of equipment was analysed based on two sides: the effect of failure and its consequence. Equipment history data and interviews were the methodology, which was used to find out which equipment has major effect on production. This was used to give scores to various equipment using the risk analysis matrix balanced score card by Nowlan and Heap, which is shown in Appendix.

3.4.1. Risk analysis matrix balanced score card: Appendix

As shown in Appendix, there are two categories of scores (Priority and Equipment Score), which are used to calculate the Criticality Score. The Priority Score is obtained from multiplying the results of the three factors (Factor E, Factor F and Factor G). The Equipment Score is obtained from multiplying the results of the four factors (Factor A, Factor B, Factor C and Factor D). This is summarized below.

Criticality score (The Criticality Score is obtained by multiplying the Equipment Score by the Priority Score = **ES * PS**).

Equipment score (The equipment score (ES) is obtained by multiplying the results for the four factors = **Factor A * Factor B * Factor C * Factor D**). The factors A–D are shown in Appendices 1–4.

Priority score (The priority score is obtained by multiplying the results for the three factors **(= Factor E * Factor F * Factor G)** as shown in Table 2 and Figure 4.

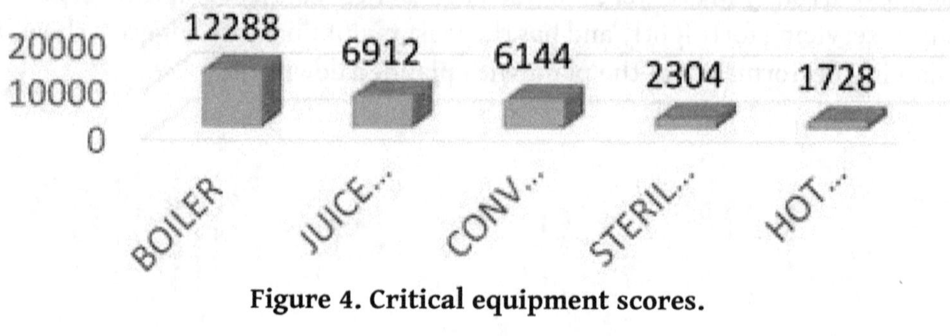

Figure 4. Critical equipment scores.

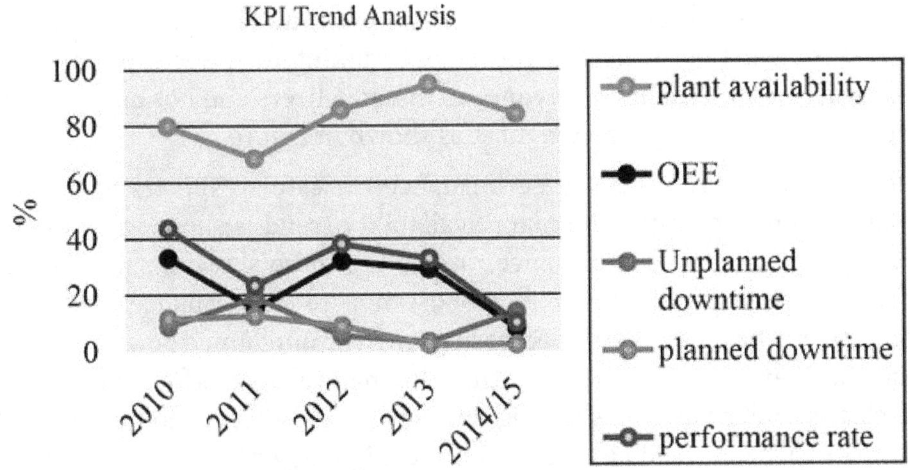

Figure 5. KPI trend analysis.

Table 2. Selection of critical machines.

Equipment	Factor A	Factor B	Factor C	Factor D	ES	Factor E	Factor F	Factor G	PS	CS
Boiler	4	4	3	4	**192**	4	4	4	64	**12,288**
Pumps and piping	4	4	3	3	**144**	4	3	4	48	**6912**
Plant conveyor Systems	4	4	3	4	**192**	4	2	4	32	**6144**
Sterilizers	4	4	3	2	**96**	2	4	3	24	**2304**
Hot break machine	4	4	3	2	**96**	2	3	3	18	**1728**
Evaporators	2	3	3	2	**36**	3	4	2	24	**864**
Heat exchangers	2	4	3	2	**48**	2	3	2	12	**576**
Super pulp creamers	4	4	3	2	**96**	2	2	1	4	**384**
Chillers	3	2	3	1	**18**	2	4	1	8	**144**
Aseptic filling machine	4	4	1	1	**16**	1	3	2	6	**96**
Sorting plant	4	3	1	2	**24**	1	1	2	2	**48**
Compressors	3	3	2	1	**18**	1	2	1	2	**36**
Washing line	2	2	2	1	**8**	1	2	1	2	**16**
Drier	1	1	1	1	**1**	1	1	1	1	**1**
CIP Plant	1	1	1	1	**1**	1	1	1	1	**1**

NB: The general details on the factors used are presented in Appendices 1–4. From Table 2, it is clear that criticality equipment ranking for PC is as follows: boiler, juice transfer pumps, plant conveyor systems, sterilizers and hot break machines, and the boiler is critically affected most as shown in Figure 5.

The graph above shows the Key Performance Indicators (KPI) trend in the year 2010–2015. From the graph, the plant availability trend is showing an increment over the years, and the performance rate seems to be showing the same trend; thus, it seems plant availability is proportional to performance rate. Planned downtime trend has shown to be decreasing with the unplanned downtime showing an increase over the years, meaning that the maintenance effort is opposing the world-class standard as it is showing an increase in reactive domination. OEE is showing a haphazard trend as it seems to be aligning with the performance rate. OEE depends mainly on the performance rate.

3.5. Boiler analysis

When failure occurs in any manufacturing process, it is critical to identify and analyse the root causes leading to that failure. Hybrid proactive maintenance strategy is condition-based maintenance complementing preventive maintenance (PM) strategy. PM uses mean time to failure (MTTF) as its pivot in the calculation of PM intervals. This gives calculations for MTTF on unrepairable equipment and mean time before failure (MTBF) on repairable equipment based on the assumption that the failure rate of each component is constant. Failure Modes Effects and Criticality Analysis (FMECA) analysis for the boiler is discussed in detail in a later chapter. The risk priority number chart is drawn with the highest scores corresponding to the most critical boiler component. The root cause failure analysis (RFCA) for the boiler machine is presented in the form of the Ishikawa diagram which was drawn using the Edraw max software as in Figure 6. The Ishikawa diagram is also called a fishbone diagram. The gearbox and feed check are reportedly failing almost every week. A Mamdani fuzzy logic controller is further presented in this chapter to monitor parameters such as moisture and dust effects on the feed check valve as well as oil level and torque control on the boiler gearbox. The fuzzy logic framework was determined among other counterfeit shrewd frameworks to be best fitting to comprehend the breakdown difficulties of the heater. A gearbox is continually sticking, and it is difficult to investigate the breakdown of fuzzy logic, and the reason is an instrument that was utilized for observation. On the risk priority side, it helps me knowing which segments to take and observe the most.

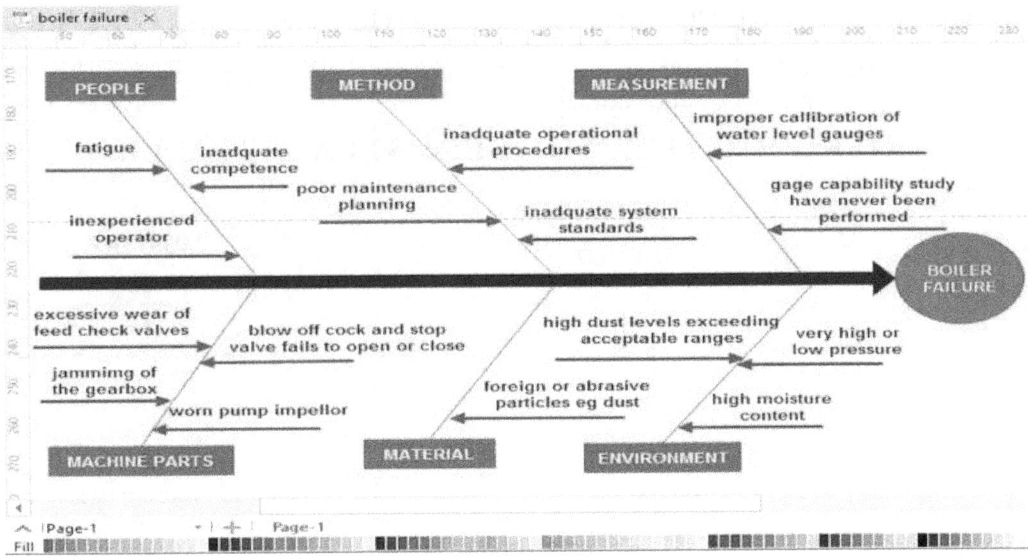

Figure 6. Ishikawa diagram for the root cause failure analysis of a steam boiler failure (Edraw Max).

In general, I will do main emphasis on the raw data analysis of the overall plant performance as shown in Figure 7. Critical equipment has been selected using the Nowlan and Heap procedure. The critical components for the plant were found to be the boiler, juice transfer pumps, conveyor system, sterilization unit and hot break machine. This is the equipment that stands to benefit the most from a change in maintenance policy. Because of the time constraint, suitable maintenance strategies will be produced for the boiler of these critical items, which show the greatest opportunity for improvement.

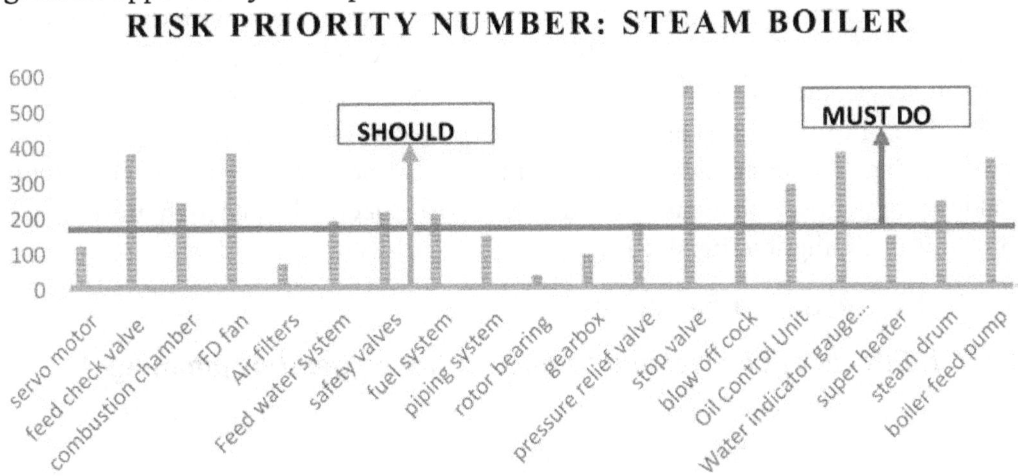

Figure 7. Risk priority number for steam boiler.

4. RESULTS OF THE BOILER FOR CASE STUDY USING FUZZY LOGIC

Analysis of the boiler is carried out in this section, and Figure 8 shows the risk priority number for the steam boiler.

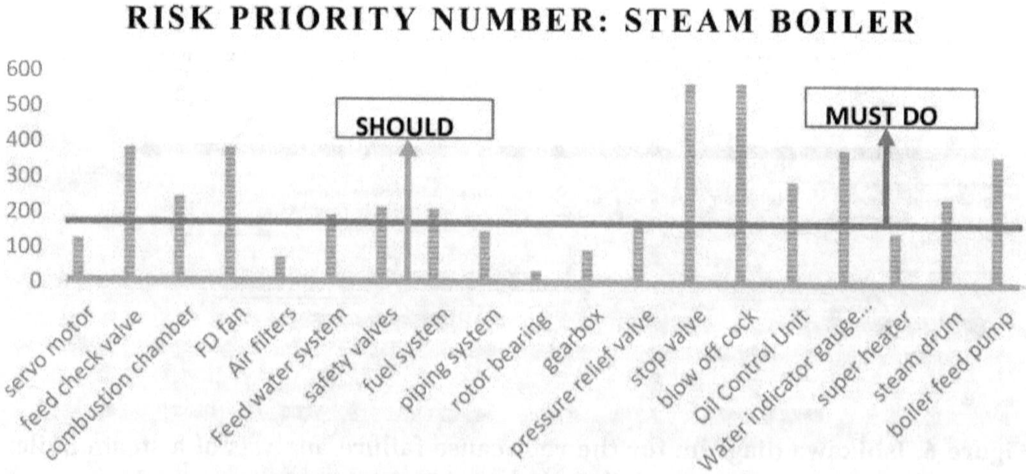

Figure 8. Risk priority number for steam boiler.

4.1. Fuzzy logic systems (using Matlab software)

The purpose of the gearbox in a boiler setup is to change the speed ratio between the motor and kicker. The gearbox is seen to be continually sticking, and it is difficult to investigate the underlying reason for the breakdowns along these lines; fuzzy rationale framework is created to screen the torque and oil level when the boiler container is stacked. The gearbox must be halted when terrible conditions exist in the evaporator as an integrated upkeep procedure to abstain from sticking, henceforth, canny fuzzy rationale control. From the interviews with the engineers and plant artisans, as well as maintenance of history log books, it was noted down that oil level and torque are the major factors which are contributing to the jamming of the gearbox whenever the boiler is being loaded. In this chapter, the oil level is being controlled and is to be modified with specific levels: high, acceptable and low. Controlled torque is to be modified with specific levels: high, normal and low. This is highlighted beneath.

4.1.1. Oil level and torque control

4.1.1.1. Effects of oil level and torque control on gearbox jamming

Moisture inside the valves may result from the compressed air passing through the valve. The effects of moisture on the valves include:

- Moisture may cause rusting in the moving parts of the valve.
- Increased rate of wear of the valve material may also result due to moisture. Moisture washes the lubrication away, which will result in the eventual

failure or malfunctioning of the valve.

- The industrial processes, which rely on the full functionality of the pneumatic control valves, may be jeopardized, and this usually results in costly breakdowns of the machine.
- Air- or gas-operated instruments may give inaccurate readings due to corrosion of the material and hence interrupting the plant processes.
- The rubber diaphragms inside the pneumatic valves can be stiffened and will eventually rupture due to the moisture flowing through them.

The following are the range of values of moisture, which are tolerated and some not tolerated inside the valve:

4.1.1.2. Oil level control

4.1.1.2.1. For a range of 0–100%

This is the aggregate scope of oil level either acknowledged or unaccepted in the gearbox.

- 0–40%

This alerts the artisans for refiling when oil is about to run out.

- 40–80%

At this level, it is acceptable, and the gearbox can continue operating.

- 80–100%

The level is fine and the gearbox has to keep running.

4.1.1.3. Torque control

4.1.1.3.1. For a range of 0–12,000 Nm

This is the total range for torque in the gearbox at any time (unaccepted or accepted).

- 0–2000 Nm

The range is very fine for starting the gearbox and is fine for running the boiler.

- 2000–8000 Nm

The torque is very fine for the gearbox to continue running.

- 8000–12,000 Nm

This range is very high, and it needs close monitoring, otherwise, stop the boiler.

4.1.1.4. Output control

4.1.1.4.1. For gearbox range of 0–100%

The gearbox has to run or stop, either of the 2, that is the meaning of 0 for stop and 100% for running that is the 0 or 1 for logic:

- 0 stop

At this range for the stop signal, the gearbox is to stop being controlled by torque and oil level

- 1 run

This involves the range 0–100%. All the conditions are being met (oil level and torque), and the gearbox can run effectively.

4.1.2. Simulation of the effects of oil level and torque on the gearbox using fuzzy logic

The rule base consists of a collection of expert rules, which are required to meet the control goals. These control rules can be developed from survey results, common sense, general principles and intuitive knowledge. The IF-THEN or *IF-THEN or IF-AND-THEN* rules are mainly going to be used in designing the controller as shown in Figure 9. The situation for which the rules are projected is given by the IF part. The fuzzy system reaction in this state will be given by the THEN part. Figure 9 below shows the fuzzy logic (Matlab software) screenshot for the two inputs (oil level and torque) and the output parameter (gearbox).

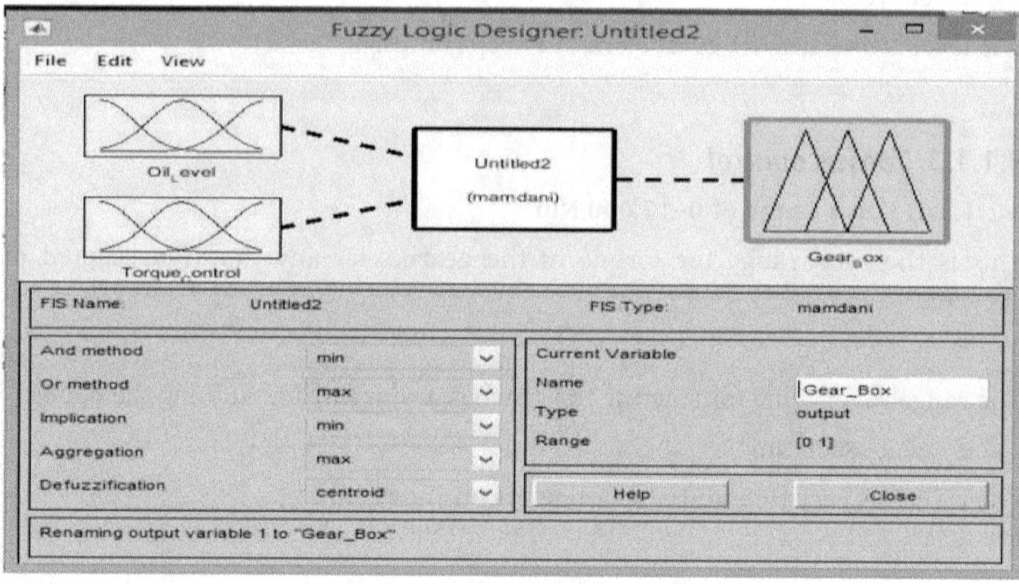

Figure 9. FIS editor for gearbox.

4.2. Control of gearbox jamming using fuzzy logic tool box

4.2.1. Membership function editor for oil level and torque monitoring

The membership function editor is the central concept of the fuzzy set, which has values ranging from 0 to 1 in the y axis. The ranges of control were done in Tables 3–5. Membership function editor is the stage whereby the ranges, as explained above, are inserted into the fuzzy logic Matlab software and the screenshot of the data is taken as shown in Figure 10. It helps to display all membership functions connected with the input and output for the entire Fuzzy Inference System (FIS). In this case, a triangular fuzzy set membership function is used for oil and torque monitoring. This is shown clearly in figures below with ranges for oil level monitoring varying from low, acceptable and high and the numerical values being added. For torque control, the corresponding numerical values range from low, normal and high. The gearbox is either run or stopped, depending on the oil level and torque conditions as in Table 5.

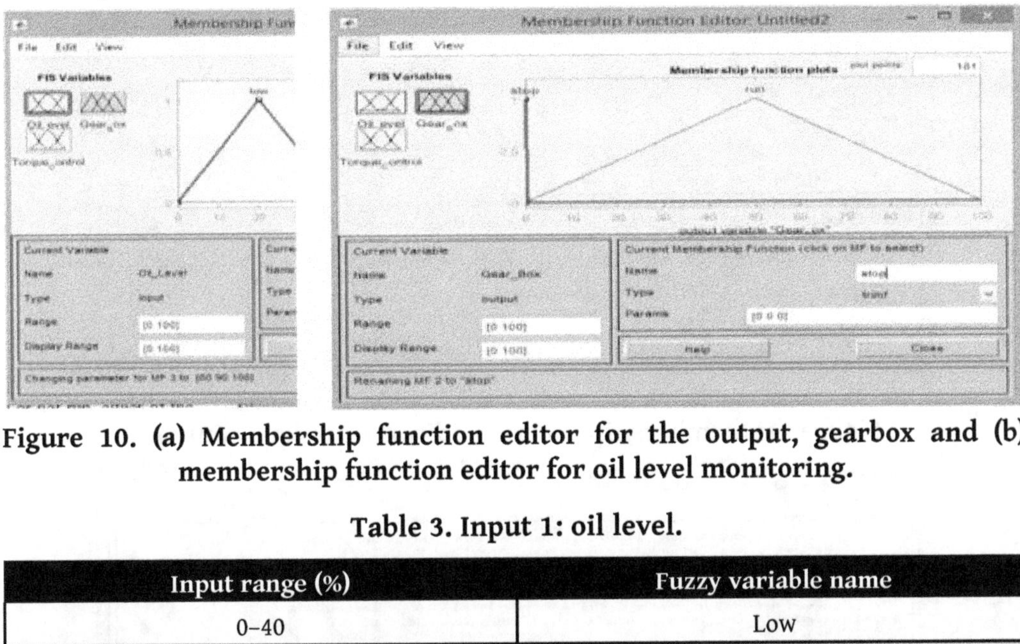

Figure 10. (a) Membership function editor for the output, gearbox and (b) membership function editor for oil level monitoring.

Table 3. Input 1: oil level.

Input range (%)	Fuzzy variable name
0–40	Low
40–80	Acceptable
80–100	High

Table 4. Input 2: torque control.

Crisp input range (%)	Fuzzy variable name
0–2000	Low
2000–8000	Normal
8000–12,000	High

Table 5. Output 1: gearbox.

Indicator	Symbol
Stop	0
Run	1

The membership functions editor for the output is shown in Figure 11, which is the gearbox that is plotted in the form of a triangle. Fuzzy logic is an artificial intelligence software which can store the output ranges in its memory and can learn the system to give solutions in what can be done. The Matlab software of fuzzy logic consists of the rule editor function which allows for the generation of the rules. This is done after inserting the range of values for oil level and torque monitoring and the gearbox outputs. The IF.....THEN......ELSE rules are being used for retrofitting as shown in Figure 12.

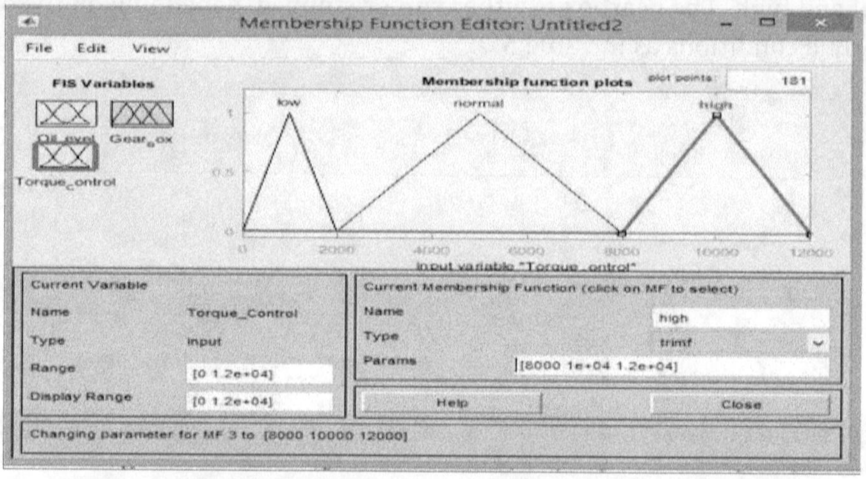

Figure 11. Membership function editor for torque control.

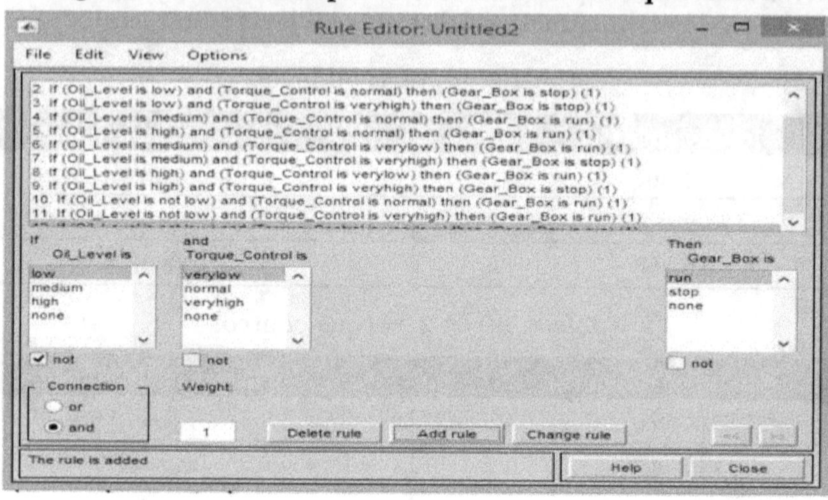

Figure 12. Rule editor for the fuzzy logic control.

After the rules are inserted, the overall results of combined effects of oil level and torque monitoring on gearbox monitoring are as discussed below. The rule viewer is an intelligent software, which is used to view the rules created using the rule editor. The rule viewer for the range values for oil level, torque and gearbox operation is discussed clearly in the figure below. However, the values obtained in the rule viewer are just optimum ranges and cannot be concluded as the best control values. Conclusions are only taken after the defuzzification process as described later in the chapter. The surface viewer shows a three-dimensional structure, which gives a conclusion to the range of values that are required to keep the gearbox functioning. This can be put in two-dimensional structure to highlight, in a simplified manner, the range of values recommended to keep the gearbox functioning as in Figure 13.

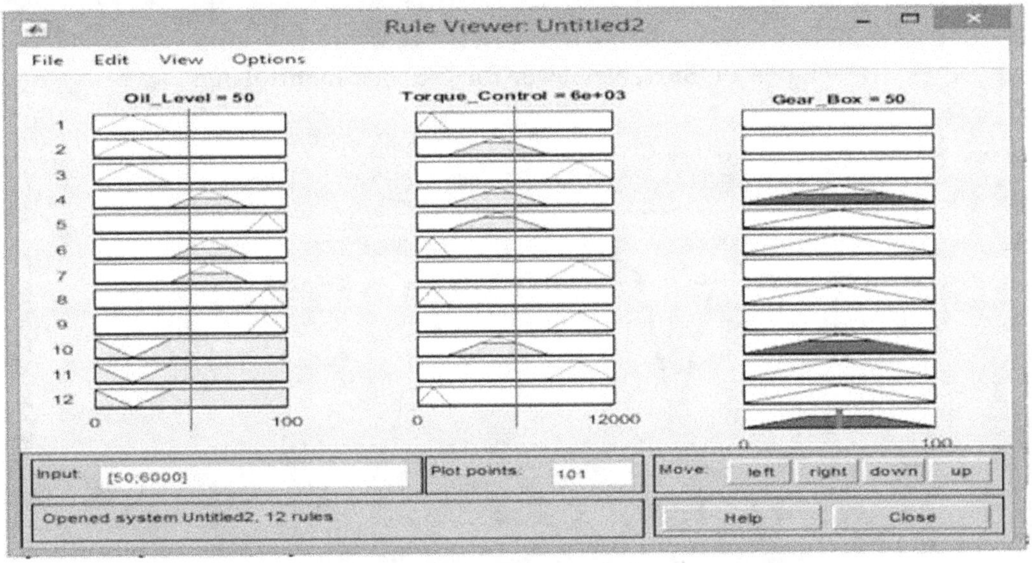

Figure 13. Rule viewer of the gearbox monitoring.

4.2.2. Analysis of the effects of oil level and torque on gearbox

4.2.2.1. Control of gearbox oil level

From the surface viewer of the gearbox against the oil level shown in Figures 14 and 15, it is seen that the gearbox is maintained and remains constant, at a 50% capacity if the oil level is 40%. If the oil level is at 40%, smooth running of the boiler is experienced. If the oil level by any means goes below 40%, stop the gearbox to avoid breakdown of the gearbox. It is seen that the fuzzy logic engine will instruct the gearbox to stop running without any human intervention so as to reduce the frequency of breakdown of the boiler machine through an effective gearbox operation. Figure 14 shows a three-dimensional, and Figure 15 is a two-dimensional image.

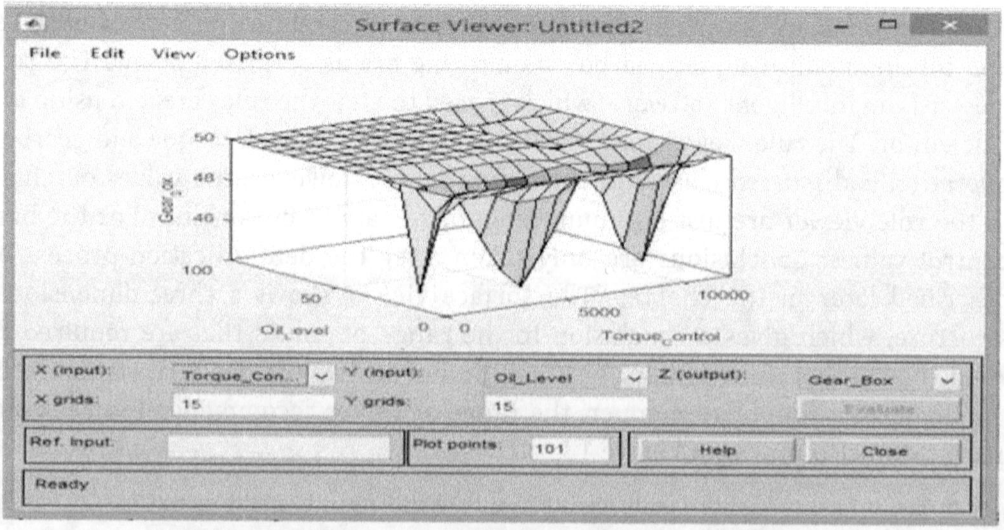

Figure 14. Surface viewer for gearbox monitoring.

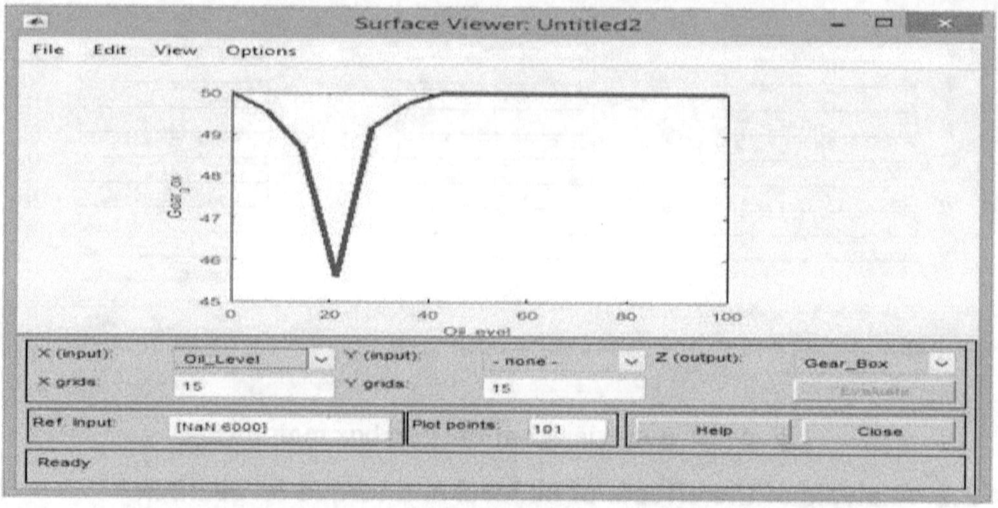

Figure 15. Surface viewer of oil level versus gearbox operation.

4.2.2.2. Control of gearbox torque

From the graph, Figure 14 of gearbox operation against torque control, it is shown that the gearbox should never be operated at torque greater than 8000 Nm as shown in Figure 16. The gearbox is seen to run smoothly at a 50% capacity if torque ranges from 0 to 8000 Nm. From this figure, it can be concluded that the required range for the gearbox to run smoothly is 0–8000 Nm of torque; otherwise, stop the gearbox and hence the boiler. Matlab software has stored this data in fuzzy logic and will automatically stop the gearbox operation and hence the boiler if the above conditions are not met.

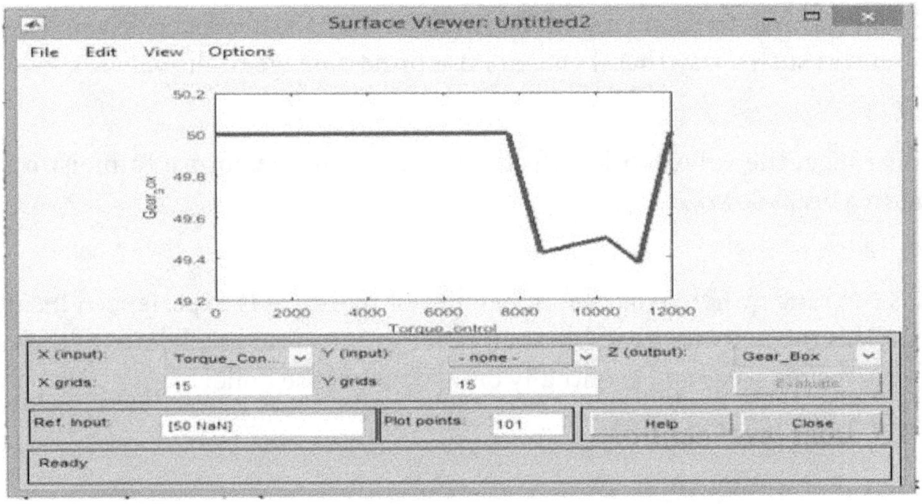

Figure 16. Surface viewer of gearbox control using torque.

4.2.3. Moisture and dust control on boiler feed check valves

4.2.3.1. Effects of moisture and dust levels on feed check valves

Moisture inside the valves may result from the compressed air passing through the valve. The effects of moisture on the valves include:

- Moisture may cause rusting in the moving parts of the valve.
- Increased rate of wear of the valve material may also result due to the moisture. The moisture washes the lubrication away which will result in the eventual failure or malfunctioning of the valve.
- The industrial processes which rely on the full functionality of the pneumatic control valves may be jeopardized, and this usually results in costly breakdowns of the machine.
- Air- or gas-operated instruments may give inaccurate readings due to corrosion of the material and hence, interrupting the plant processes.
- The rubber diaphragms inside the pneumatic valves can be stiffened and will eventually rupture due to the moisture flowing through them.

The following are the range of values of moisture which are tolerated and not tolerated inside the valve:

4.2.3.2. Moisture level control

4.2.3.2.1. For a range of 1–5%

This is the total range of moisture in the feed check valve at any time (either accepted or unaccepted).

- 0–1%

This defines that the amount of the moisture in the feed check valve is very minimum in such a way that it causes little or no damage to the valve.

- 1–3%

At this range, the valve works effectively but further exposure to moisture may result in a breakdown.

- 3–5%

This is the range when a maximum amount of moisture is experienced inside the valves. This range is unacceptable at all times as it leads to breakdown of the valves (rust, corrosion, wear and eventually breakdown of the boiler).

4.2.3.3. Dust level control

4.2.3.3.1. For a range of 0–1%

This is the total range of values of dust inside the feed check valve (accepted or unaccepted).

- 0–0.02%

At this range, the valves can run effectively without any damage.

- 0.02–0.5%

This amount of dust entering the valve may cause the valve to malfunction with further exposure. However, at this range, the feed check valves are seen to be operating effectively.

- 0.5–1%

These ranges of dust are not acceptable at all. The excessive exposure of dust inside the valve may result in scoring, wear and eventually total failure of the boiler machine.

4.2.3.4. Output control

The feed check valve has to either open or close, either of the two, and that is the meaning of 0 for close and 1 for open.

- 0 close

At this range, for close signal, the valve will close if the if the dust and moisture levels inside the valves are unacceptable.

- 1 open

This involves the acceptable range. All the conditions are being met (dust and moisture level), and the valve can effectively open at any angle between 0° and 90° depending on the amount of feed water to be pumped into the boiler.

4.2.4. Simulation of the effects of dust and moisture levels on feed check valves using fuzzy logic

The moisture and dust level variations in the boiler machine can be monitored using an artificial intelligence software. The inference engine of fuzzy logic can store two input parameters (moisture and dust level) and one output parameter (feed check valve operation) as shown by the print screen below from the laptop in Figure 17.

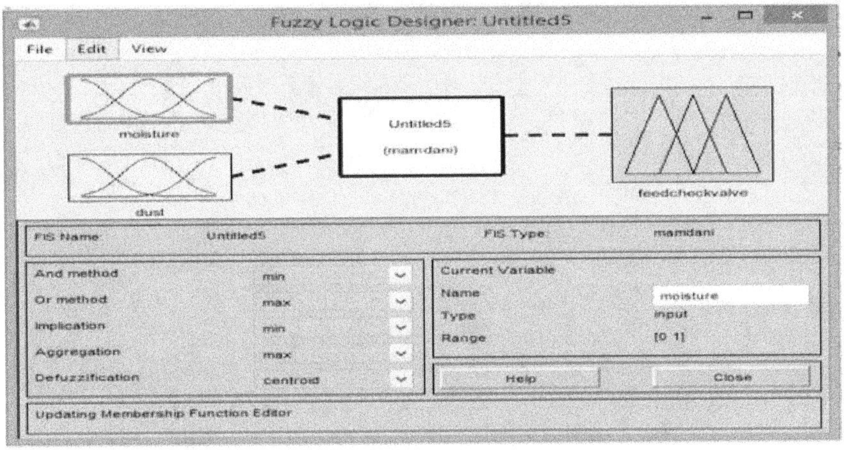

Figure 17. FIS Editor for feed check valve.

Control of feed check valve operation using fuzzy logic tool box is shown in Tables 6–8.

Table 6. Input 1: moisture level.

Input range (%)	Fuzzy variable name
0–1	Low
1–3	Medium
3–5	High

Table 7. Input 2: dust level.

Input range (%)	Fuzzy variable name
0–0.02	Acceptable
0.02–0.5	Average
0.5–1	Not acceptable

Table 8. Output 1: feed check valve operation.

Output range (%)	Fuzzy variable name
0	Closed
1	Open

4.2.5. Membership function editor for moisture and dust monitoring

This is shown clearly in Figures 18–20 with ranges for moisture level varying from low, acceptable and high and the numerical values being added. For dust control, the corresponding numerical values range from low, normal and high. The valve either closes or opens depending on moisture and dust levels. This is done to control the feed check valves so that they won#x2019;t operate in unfavourable conditions which might result in the failure of the boiler.

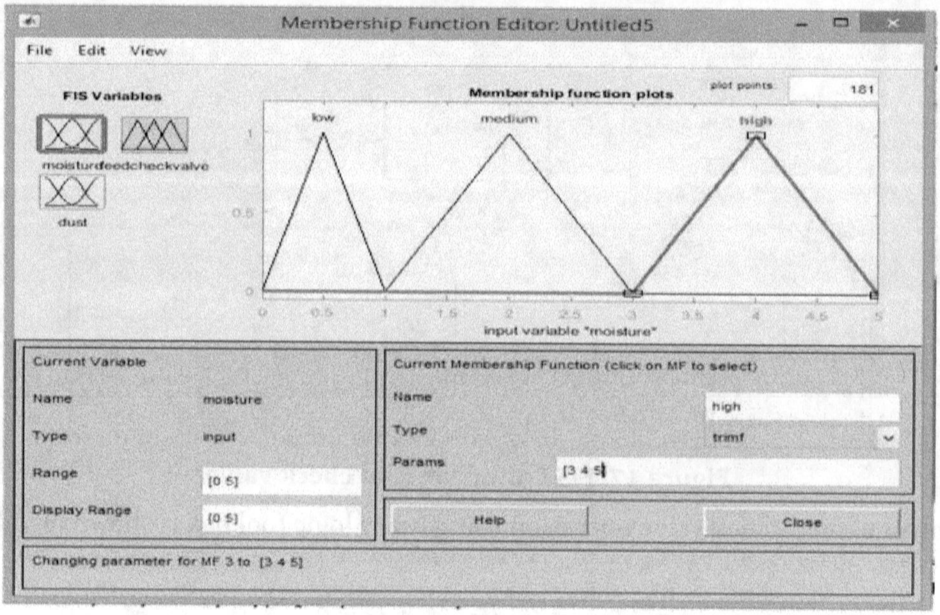

Figure 18. Membership function editor for moisture.

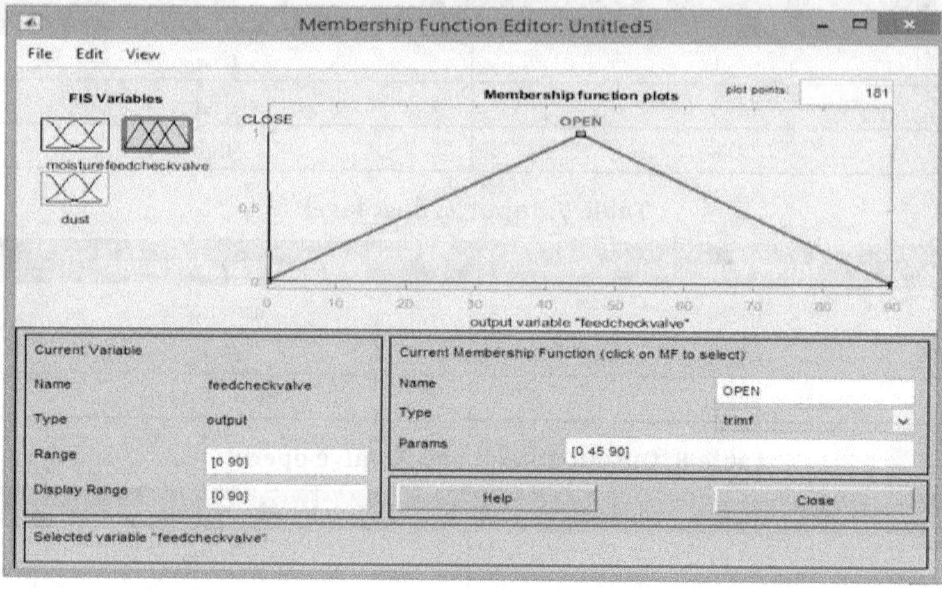

Figure 19. Membership function editor for the output and feed check valve.

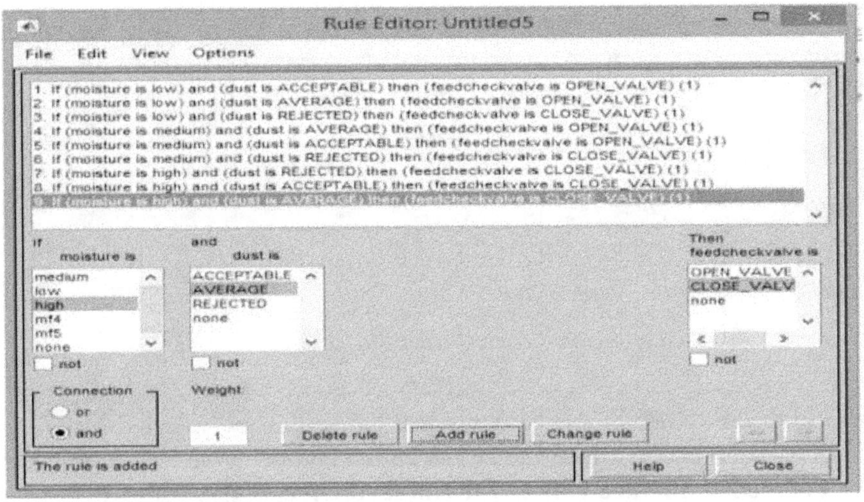

Figure 20. Rule editor for feed check valve.

The following rules were developed in Figure 20 and will instruct the controller what to do.

After the rules are inserted, the overall results of combined effects of moisture and dust levels on feed check valve are as discussed below. The rule viewer is an intelligent software which is used to view the rules created using the rule editor and is shown in Figure 21. The rule viewer for the range values for moisture level, dust and valve operation is discussed clearly in Figure 21. The surface viewer shows a three-dimensional layout in Figure 22, and the blue lines indicate the values of moisture level and dust level, which may result in the failure of the valve. The required values which should be maintained are shown by the yellow surfaces. However, the values obtained from the rule viewer are just ranges and not concluded as the best control values. Conclusions are only taken after the defuzzification process.

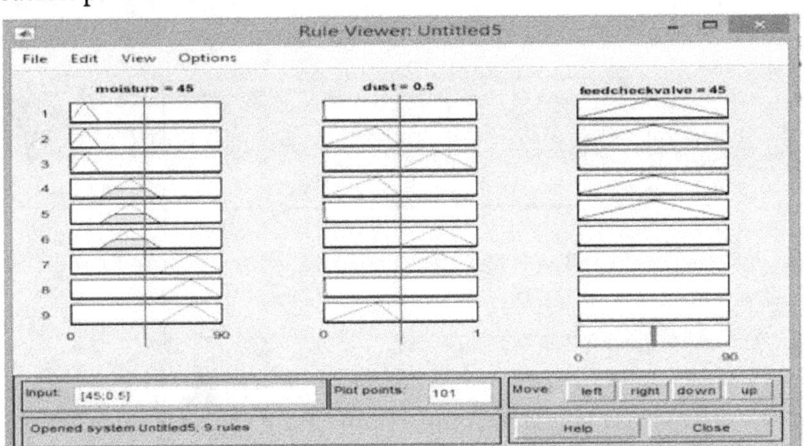

Figure 21. Rule viewer for the feed check valve monitoring.

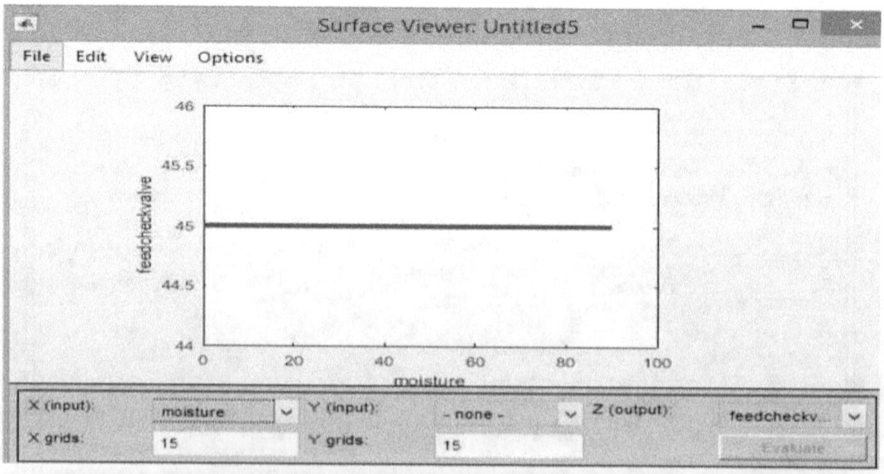

Figure 22. Surface viewer for the feed check valve monitoring.

4.2.6. Analysis of the effects dust and moisture on feed check valve

4.2.6.1. Control of feed check dust level

From the surface viewer of the feed check valve control against the amount of dust, it is seen that the valve remains constant at 45% corresponding to a dust range of 0–0.55%. The dust level should be maintained at this range. If the dust level is increased from 0.55 to 0.58%, there is a linear decrease in the feed check valve's opening angle from 45 to 0°. As the level of dust is further increased from 0.58 to 0.95%, the valve is kept constant at 0°. It is therefore concluded that dust level should remain between 0 and 0.55% so that the valves will run smoothly; otherwise, boiler breakdown will result. Fuzzy logic engine is an artificial intelligence system which will instruct the valves to immediately close as soon as the dust level exceeds 0.55%, as in Figure 23, without any human intervention.

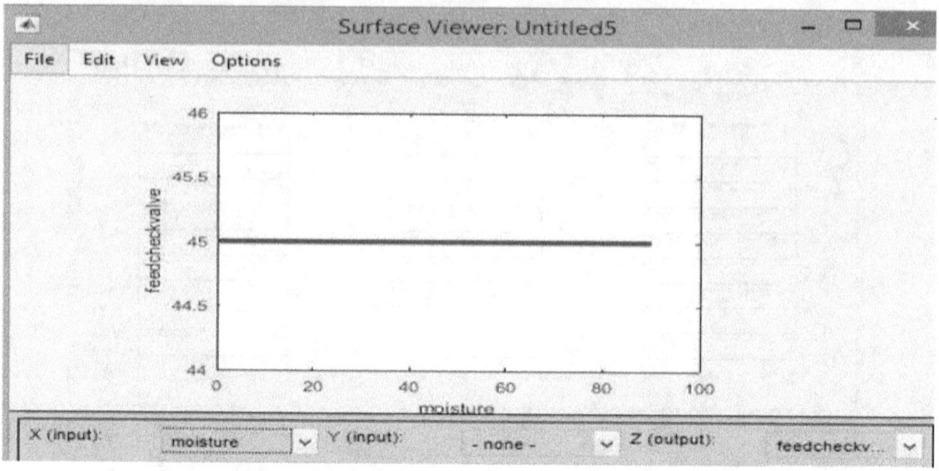

Figure 23. Surface viewer of gearbox control using dust level.

4.2.6.2. Control of moisture level

Figure 24 indicates that moisture level does not contribute much to the valve opening. From 0 to 85%, the feed check valves operate at a constant capacity of 45%. For smooth operation of the valves, moisture level should not exceed 85%; otherwise, fuzzy logic will set the valves to close automatically. This helps to prevent the breakdown of feed check valves.

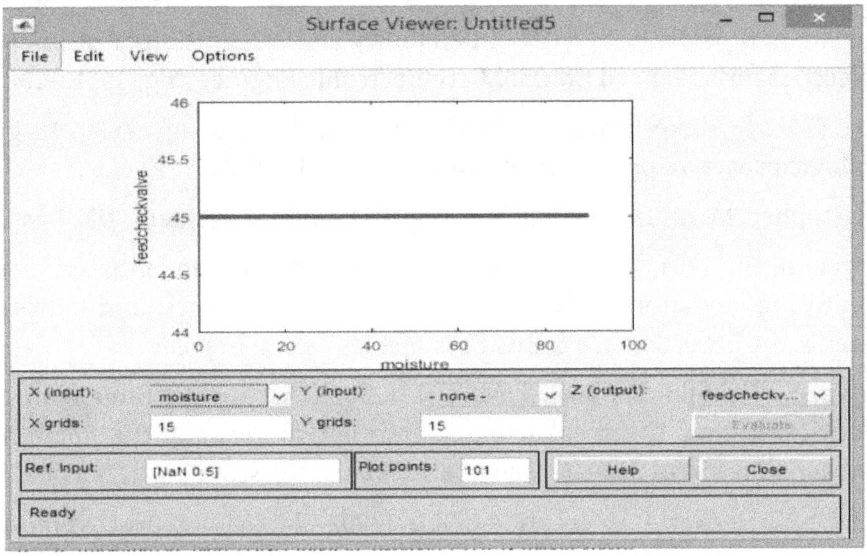

Figure 24. Surface viewer of feed check valve control using moisture.

5. CONCISE SUMMARY

The efficacy of any system is contingent upon the dedication of the administration. The design and implementation of a hybrid proactive approach necessitate a substantial initial investment for good outcomes.Given that systems are driven by individuals, worker participation is essential. Management must prioritize the allocation of resources for in-house training on the fuzzy logic system.

AT A GLANCE

The examined processing company determined that the boiler is the critical machine requiring artificial intelligence monitoring and control. The simulation was conducted using Matlab software, where the oil level, pressure, and temperature were modeled and regulated by a programmed logic controller (PLC), with a fuzzy logic controller serving as the primary control mechanism. The company specializes on fruit processing to manufacture juice.

REFERENCES

1. Bond, M., Meacham, T., Bhunnoo, R. & Benton, T. (2013). Food waste within global food systems. Swindon, UK: Global Food Security Programme. A Global Food Security Report.
2. Bose, B.K. (1994). Expert system, fuzzy logic, and neural network applications in power electronics and motion control. Proceedings of the IEEE, 82, 1303-1323.
3. Buchanan, D., & Lo, S.-H. (1997). Reliability and integration of ultra-thin gate dielectrics for advanced CMOS. Microelectronic Engineering, 36, 13-20.
4. Burt, D.N., & Pinkerton, R.L. (1996). A purchasing manager#x2019;s guide to strategic proactive procurement. New York: AMACOM.
5. Christopher, M. (2016). Logistics & supply chain management. UK: Pearson.
6. Isermann, R. (2011). Fault-diagnosis applications: model-based condition monitoring: actuators, drives, machinery, plants, sensors, and fault-tolerant systems. Springer Science & Business Media. Berlin, German.
7. Kropp, J., Zickfeld, K., & Eisenack, K. (2002). Assessment and management of critical events: the breakdown of marine fisheries and the north Atlantic thermohaline circulation. In The science of disasters. Springer. Berlin, German.
8. Liang, S.Y., Hecker, R.L., & Landers, R.G. (2002). Machining process monitoring and control: the state-of-the-art. In ASME 2002 International Mechanical Engineering Congress and Exposition. American Society of Mechanical Engineers, pp. 599-610. University of Missouri, United States of America.
9. Nagilla, S.R. (2015). Non-thermal reliability considerations for. Arlington: University of Texas.
10. Nwaobi, G.C. (2008). Energy power, digital infrastructure and e-learning platforms: African experience. Journal of Energy.
11. Okah-Avae. (1981). Condition monitoring - a maintenance management strategy for industrial machinery. Benin: University of Benin-City.
12. Papadopoulos, C., & Dimarogonas, A. (1992). Coupled vibration of cracked shafts. Journal of Vibration and Acoustics, 114, 461-467.
13. Pine, B.J. (1993). Mass customization: the new frontier in business competition. Harvard Business Press. Harvard.
14. Beverage Industry Environmental Roundtable. (2014). Beverage industry sector guidance for greenhouse gas emissions reporting. Beverage Industry Environmental Roundtable.

15. Scheffer, C., & Girdhar, P. (2004). Practical machinery vibration analysis and predictive maintenance. Elsevier. Amsterdam.
16. Seemann, K.W. (2000). Technacy education: towards holistic pedagogy and epistemology in general and indigenous/cross-cultural technology education. School of Tourism and Hospitality Management Papers, 9. United Kingdom.
17. Tavner, P.J., & Penman, J. (1987). Condition monitoring of electrical machines. Research Studies Press.
18. Zhang, Z. (2011). Energy and environmental policy in China: towards a low-carbon economy. Edward Elgar Publishing. Massachusetts, USA.

18 Chapter

APPLICATION OF FUZZY LOGIC IN THE DESIGN AND CONTROL OF NONLINEAR MIMO SYSTEMS

1. INTRODUCTION: AN OVERVIEW

The analytical description of several processes in technical practice is somewhat complex. This may result from their complexity, nonlinearity, transfer delays, intricate measurement of critical parameters, etc. Nevertheless, data regarding the efficacy of these processes can frequently be acquired through experimental means, utilizing appropriately selected measures or by observing their reactions to the operator's control actions. In such scenarios, fuzzy systems may consistently be seen as an alternative for system modeling and control. The implementation of fuzzy logic in the design of fuzzy controllers frequently presents a viable solution to control-related challenges across diverse industrial sectors, as these controllers serve as an effective means to attain superior quality attributes in controlled systems [1–9]. The drawback in this instance is the haphazard methodology employed in their synthesis and the comparatively rigorous assessment of their stability. The design of a fuzzy controller is fundamentally reliant on the fuzzification of its input range and the establishment of behavioral rules within that range. The operation of traditional fuzzy controllers was formulated based on language principles derived from experts. Nevertheless, this knowledge is not always readily accessible, particularly in instances of higher-order nonlinear systems [10, 11]. Consequently, recent years have seen a concentrated emphasis on the development of fuzzy control systems that do not rely on the pursuit of expert linguistic norms [12–17]. Control methods utilizing the fuzzy model of the controlled system exhibit several alterations contingent upon the specific application, while the quality of the fuzzy model itself is also critically important. Fuzzy modeling has been established as a nonlinear black-box modeling technique [11, 12, 18–20]. In the design of a black-box fuzzy system, it is essential to ascertain its qualitative features solely from experimentally obtained data, without knowledge of its structure or parameters. This frequently leads to issues with database inconsistency and inadequate coverage of the complete range of potential inputs, rendering the fuzzy model impractical for real-world applications. In constructing a black-box fuzzy model of a dynamic system, it is essential to employ an appropriate strategy for selecting qualitative attributes from the gathered database. The functional dependencies between inputs and outputs can be utilized to construct an appropriate nonparametric fuzzy model of the process for control design purposes [23–27].

Two typical examples for designing the control of nonlinear MIMO systems using fuzzy approach are presented in this chapter:

- The design of fuzzy PI torque controller of the PI type for a drive with an induction motor, whose parameters and rules are obtained by searching control input of such a vector which is optimal in terms of the selected criterion of optimality.
- The design of control for middle part of a continuous line for material processing by tension, where the continuous line presents a nonlinear MIMO system. Its control requires to ensure decoupled control of individual subsystems, because the output quality of the processed material depends directly on quality of the control. The controllers of the subsystems ensuring such decoupling usually are of complex structures and, when designing them by analytic methods, they are often unrealizable. When the continuous line is presented by a fuzzy model, it is possible to design simple controllers of the PI type ensuring high-quality dynamical properties of the controlled system.

2. DESIGN OF FUZZY TORQUE CONTROLLER FOR ASYNCHRONOUS MOTOR DRIVE

An asynchronous motor is a highly nonlinear fifth-order system, for which effective vector torque control is achieved by intricate mathematical transformations, resulting in a complex control structure [28, 29]. A fuzzy system has been employed for the construction of a torque controller for an asynchronous motor drive. The design of the fuzzy controller relies on recognizing the temporal sequence of the input signal to the controlled system, which will achieve the control objective according to the chosen optimality criterion. The fuzzy controller design method is distinguished by its simplicity, and the control quality is suitable for the specified drive.API-type discrete controller is generally described by the equation:

$$u_k = u_{k-1} + q_0 e_k + q_1 e_{k-1} \tag{1}$$

where u_k and u_{k-1} are values of controller output in the relevant sampling steps, e_k and e_{k-1} are values of the control error, and q_0 and q_1 are parameters of controller [28]. From these follows, it is possible structure, shown in Figure 1.

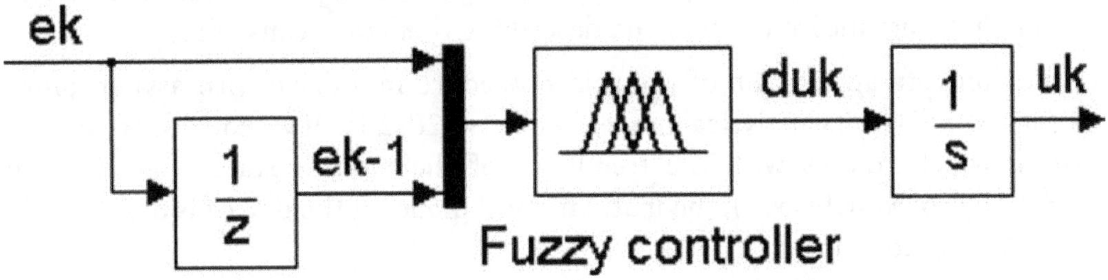

Figure 1. Block diagram of fuzzy controller.

A discrete fuzzy PI controller can be described, for example, by the following rules:

$$IF\ e_k\ is \ldots \wedge e_{k-1}\ is \ldots THEN\ du_k\ is \ldots \quad (2)$$

where quantities e_k, $e_{k-1,}$ and du_k are fuzzy variables that describe the relevant workspace of the fuzzy PI controller. The fuzzy controller design procedure consists of the following three steps:

Step 1. Finding the optimal sequence of input values

Fuzzy rules and fuzzification of the fuzzy PI controller workspace can be identified by means of relations expressed by triplets $[e_k e_{k-1} du_k]$ that have been obtained for its optimal behavior. This behavior is represented by the time sequence of input vector $d\mathbf{u}_{opt}$, at which the value of the optimality criterion is minimal. It is suitable to choose this criterion in the next form of the integral of the quadratic deviation of the system output y from the desired value w:

$$J(e) = \int (w(t) - y(t))^2\, dt \quad (3)$$

When obtaining the sequence of values of vector $d\mathbf{u}_{opt,}$ we apply to the drive input various sequences of input vector and evaluate the criterion value Eq. (3). This is a standard optimization task that can be solved, for example, by suitable geometric division of the drive workspace (which leads to rather high computing demands in the design process), or by applying the genetic algorithm method, which significantly speeds up the whole process of identification.

Step 2. Finding the database of optimal data

Having found the optimal input sequence $d\mathbf{u}_{opt}$ for the controlled drive, we then set up the database of triplets $[e_k e_{k-1} du_k]$, which describes relations between the inputs and the outputs of the optimal fuzzy PI controller.

Step 3. Designing the fuzzy controller from the optimal data database

From the obtained triplets $[e_k e_{k-1} du_k]$ of optimal data, it is possible to design a concrete fuzzy controller of various types using standard procedures of clustering the data into significant clusters and describing them by means of rules.

In the concrete application of the said procedure in a drive with asynchronous motor, we will use its analytical model (see Refs. [28, 29]). If we consider a rotating system which rotates with the frequency of the motor's stator field (usually marked by coordinates x, y), the mathematical model of the device is described by the following equations:

$$\begin{aligned} u_{sx} &= R_s i_{sx} + \frac{d\psi_{sx}}{dt} - \omega_s \psi_{sy} \\ u_{sy} &= R_s i_{sy} + \frac{d\psi_{sy}}{dt} - \omega_s \psi_{sx} \\ 0 &= R_r i_{rx} + \frac{d\psi_{rx}}{dt} - \omega_2 \psi_{ry} \\ 0 &= R_r i_{ry} + \frac{d\psi_{ry}}{dt} + \omega_2 \psi_{rx} \\ M_e &= \frac{3}{2} p \left(\psi_{ry} i_{rx} - \psi_{rx} i_{ry} \right) \end{aligned} \tag{4}$$

Used symbols:

i_{sx}, i_{sy} components of stator current space vector i_s

i_{rx}, i_{ry} components of rotor current space vector i_r

u_{sx}, u_{sy} components of stator voltage space vector u_s

ω_m the motor mechanical angular speed

ω_1 angular frequency of the stator voltage

p number of pole pairs ($p = 2$)

ω_2 slip angular speed $\omega_2 = \omega_1 - \omega_m$

R_s, R_r stator and rotor phase resistance

Ψ_s, Ψ_r stator and rotor magnetic flux

M_e electrical motor moment

AC drive parameters are given in the Appendix.

As a standard, asynchronous motors are supplied from static voltage frequency converters in which the stator frequency and voltage rate are U_1/ω_1.

When connected directly to the power supply network, the motor shows a large increase in torque and also in current (Figure 2).

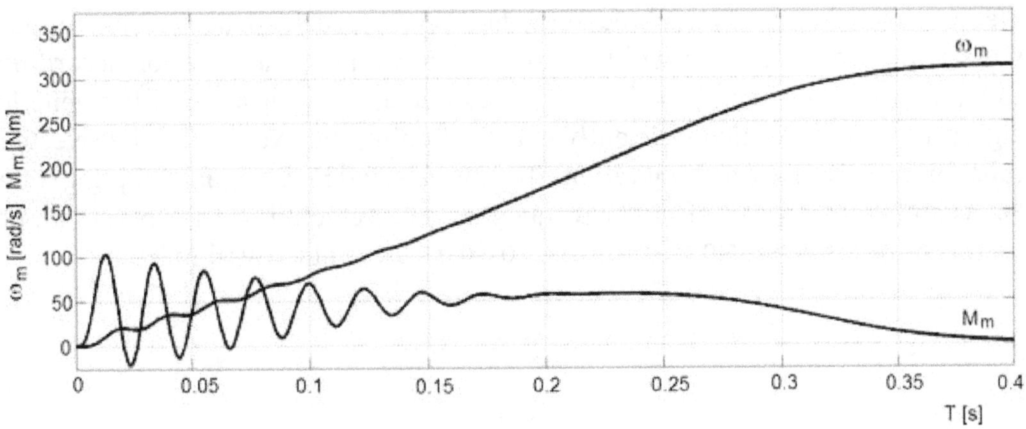

Figure 2. Direct connection of drive with AM.

Let the aim of the torque controller design be to adjust the slip ω_2 (i.e., the difference of $\omega_1 - \omega_m$) according to the desired torque value. The structure of the controlled system will be as shown in Figure 3.

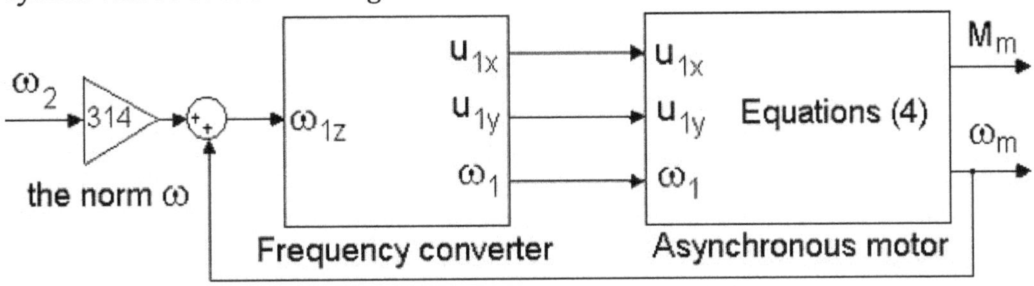

Figure 3. Structure of the controlled drive.

For finding the optimal input sequence $d\mathbf{u}_{opt}$, we shall use the diagram shown in Figure 4.

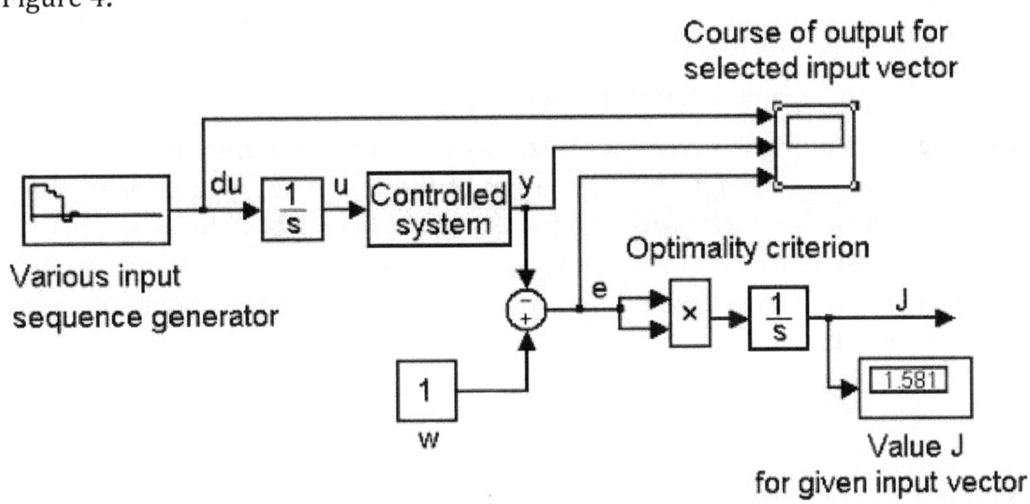

Figure 4. Block diagram for finding vector du$_{opt}$.

We search for the input signal sequence $d\mathbf{u}_{opt}$ through such changes of drive input voltage and frequency that will lead to minimal value of the criterion according to Eq. (3). This goal can be achieved, for example, by application of the genetic algorithm method, which efficiently enables finding the extreme of the selected function in a given space of mutations. Having selected input signal sampling time 50 ms and desired value of motor torque Mz = 30, we obtained the optimal input vector [0.165 0.145 0.155 0.12 0.105 0.095 0 0 0 0 0 0], as illustrated in Figure 5.

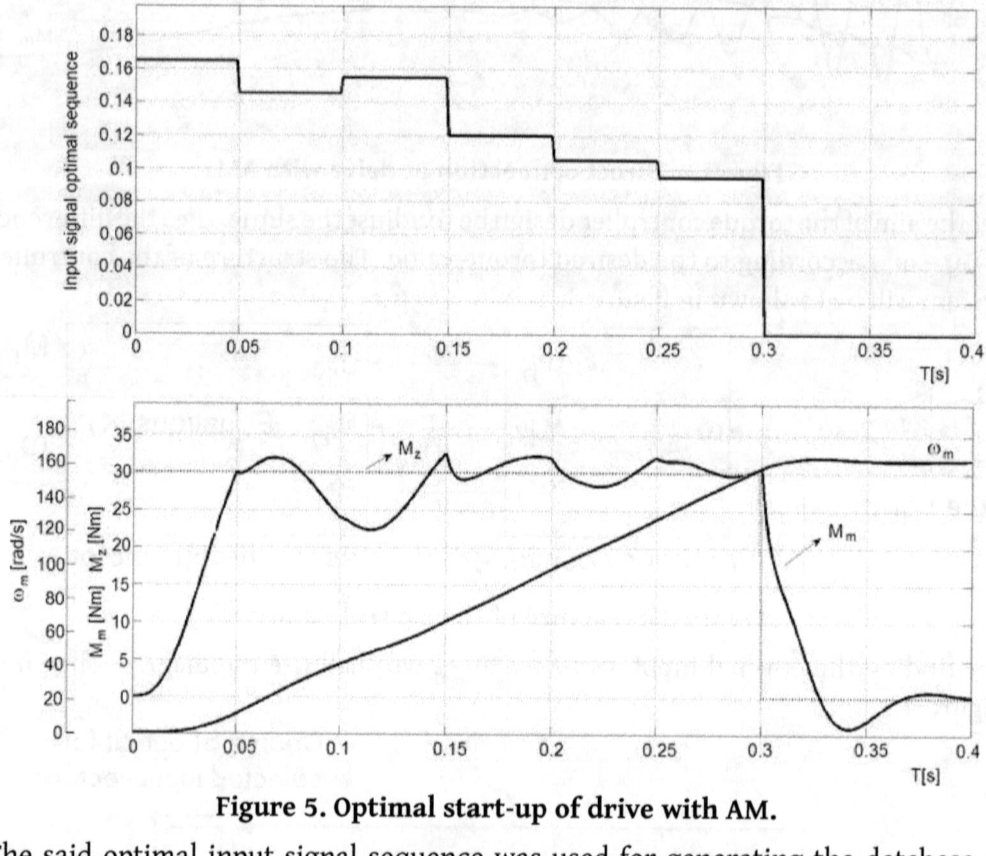

Figure 5. Optimal start-up of drive with AM.

The said optimal input signal sequence was used for generating the database of triplets $[\omega_{mk}\omega_{mk-1}\ (\omega_1-\omega_m)_k]$ for the design of the P-type fuzzy controller. Using the Anfisedit tool in the Matlab program, a static Sugeno-type fuzzy system was designed, based on the acquired database of measured data ($[\omega_{mk}\omega_{mk-1}\ (\omega_1-\omega_m)_k]$). This system has three zones for input data fuzzification, and its internal structure can be seen in Figure 6.

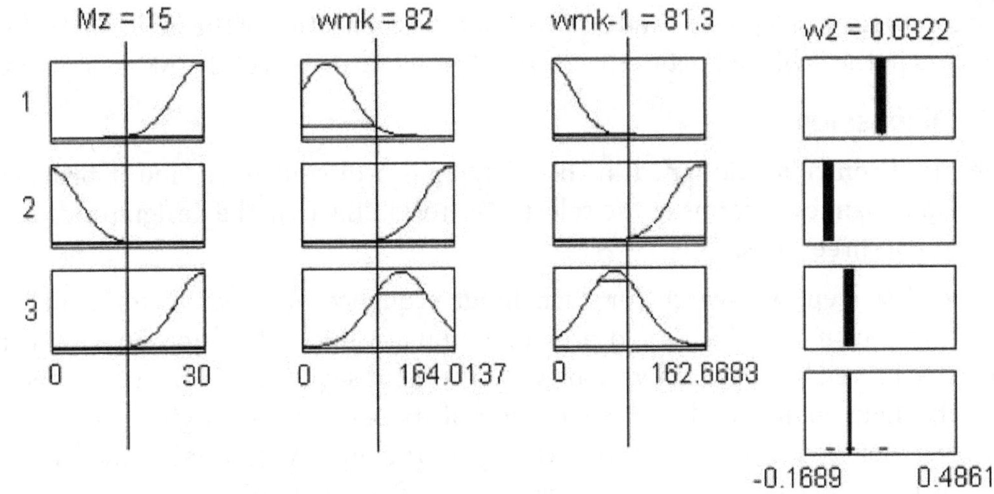

Figure 6. Structure of AM fuzzy controller.

The resulting torque control structure of AM with the designed fuzzy controller is shown in Figure 7.

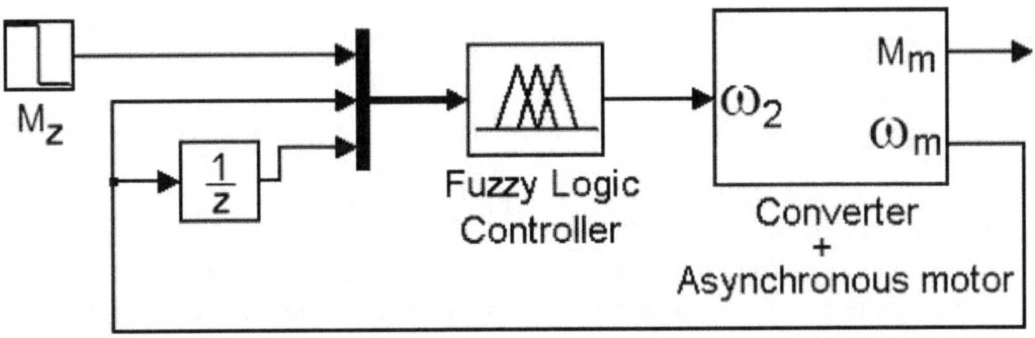

Figure 7. Structure of torque control in AM with fuzzy controller.

The start-up of the drive with desired torque Mz = 30 Nm using the designed fuzzy controller is shown in Figure 8.

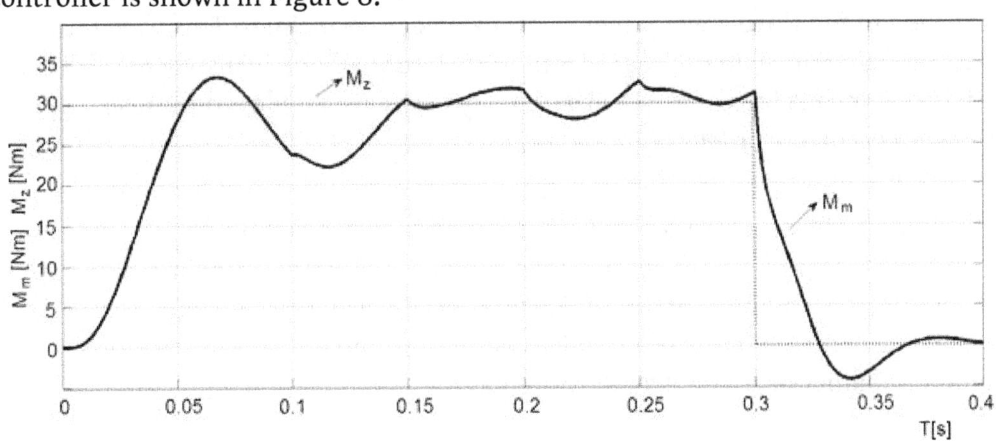

Figure 8. Optimal start-up of drive with AM with fuzzy controller.

The comparison of Figures 5 and 8 shows that the application of the fuzzy controller resulted in the achievement of the desired torque responses of the AM at start-up.

2.1. Discussion

The fuzzy controller designed in this chapter is a PI controller, and it provides optimal dynamics in terms of the selected criterion Eq. (3). The design procedure consists of three steps:

In the first step, we search for such input sequence of input vector **u** of the asynchronous motor that would provide minimal value of function Eq. (3). This can be achieved by modeling various vector **u** input sequences in the system with asynchronous motor, and by a simulation of its behavior for a given sequence. The selection of sampling time for a change of the input vector value and by that also of the number of its samples has to be defined according to the Shannon-Kotelnikov theorem. Obviously, with an increase in the number of samples also the time required for simulation and for optimal vector $\mathbf{u}_{opt}$ identification grows. As the whole process of optimal vector identification can be fully automated and the performance of computing means nowadays is sufficient, there is no essential problem to identify the optimal control input signal for a concrete drive with asynchronous motor.

In the second step, using the database obtained by application of the input vector $\mathbf{u}_{opt}$ to the asynchronous motor, the input/output relations of the optimal PI controller for the given controlled system are defined. Various input and output signals (e.g., depending on their actual measurability ...) can be chosen for setting up the suitable database. This makes it possible to adjust the controller design to a concrete drive with concrete technical capabilities.

The relations obtained in the second step of the design procedure, written down in the form of a suitable input and output signals database (mostly in a table), are described in the third step by means of fuzzy logic principles. This way a fuzzy PI controller similar to an optimal continuous PI controller is constructed. Standard computing means for working with fuzzy systems are employed in this step, such as the fuzzy toolbox in Matlab.

The whole procedure of fuzzy PI controller design was verified by simulation of its properties in the concrete control of a drive with asynchronous motor. The results of simulation experiments show that the controller, in spite of its simplicity and the uncomplicated computer oriented design procedure applied, enables considerable improvement in the control circuit dynamic properties also in case of strongly nonlinear higher order controlled systems.

3. MULTI-MOTOR DRIVE OPTIMAL CONTROL USING A FUZZY MODEL-BASED APPROACH

Typical representative of multi-motor drive is the middle part of the continuous line, where the individual working machines are coupled with each other through the material. It can be lines for processing continuous flows of material (e.g., sheet metal strips, tubes, processing lines in paper mills, and printing works) by material traction in the field of elastic or plastic deformation, which influences the material's mechanical properties. It means that the multiple motor drives are complex and coupled MIMO nonlinear systems. Therefore, due to the complexity of their mathematical models, which parameters are difficult to identify, the development of effective control systems is quite complicated task. This chapter presents the design of optimal control of continuous production line using a fuzzy model-based approach.

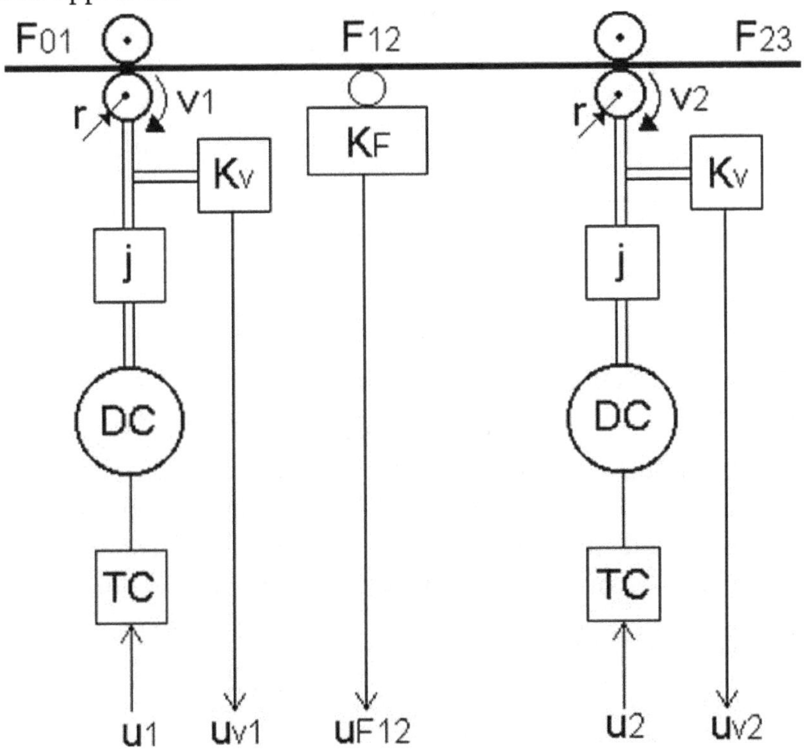

Figure 9. Structure of middle section of continuous line.

The structure of the middle part of the continuous line (further referred to as CL) is shown in Figure 9. The structure includes DC motors powered through static transistor converters TC. The working machines of the line are driven by the motors through gearbox *j*; v_1 and v_2 are machine rolls circumferential velocities, and F_{12} is the tension in the web of material between the two machines. The main line disturbances are tensions before and after the middle part of the considered line which are affecting the first and second drive (F_{01} and F_{23}). K_v is circumferential velocity sensors, K_F is tension sensor, u_{v1} and u_{v2} are outputs from velocity sensors, and u_{F12} is output of tension sensor. The controlling voltages u_1, u_2 of converters present the input variables of the system. The tension in the web of material F_{12} and the web of material velocity v_2 is the output variables (let us consider $y_1 = u_{F12}$ and $y_2 = u_{v2}$).

The described system with the mechanical coupling of two machines presents a third-order nonlinear MIMO system with two inputs and two outputs (Figure 10), the parameters of the system change depending on the mechanical properties of the material and on the speed of its motion. Defining precise parameters of this nonlinear system analytically presents a rather demanding task, and therefore, it is suitable to use for its description a fuzzy system (model) built only on basis of its measured input/output data.

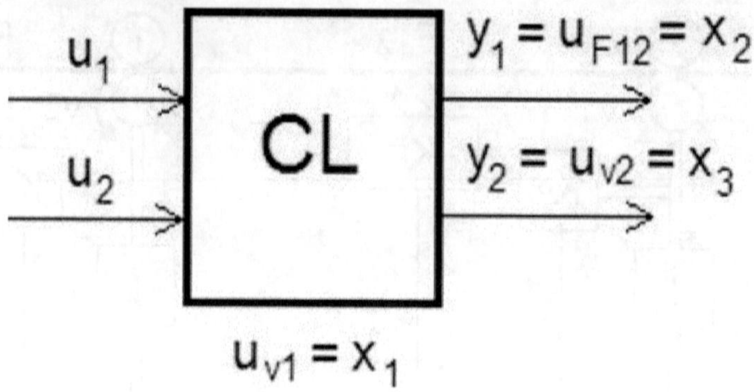

Figure 10. Middle section of continuous line as MIMO system.

Various fuzzy system structures consisting of static fuzzy subsystems and their dynamic parts can be found in the literature. In setting up the structure of the fuzzy model of a continuous line, we used its state description, where the given state of the system and the given input allow us to define the subsequent state, which can be expressed mathematically by the following equation:

$$\begin{aligned} x_{k+1} &= x_k + \Delta x_k \\ \Delta x_k &= f(u_{k-1}, y_{k-1}) \end{aligned} \tag{5}$$

where $\mathbf{u}$ is the model's input quantities vector, $\mathbf{x}$ is the state quantities vector, f is the searched for static vector function of the controlled system, and k is representing the sampling step.

Construction of the CL fuzzy model consists in determining the fuzzy approximation of this function on basis of the obtained CL inputs and outputs database. Considering the choice of CL input, state and output quantities presented in Figure 10, the structure of the proposed CL fuzzy model is shown in Figure 11.

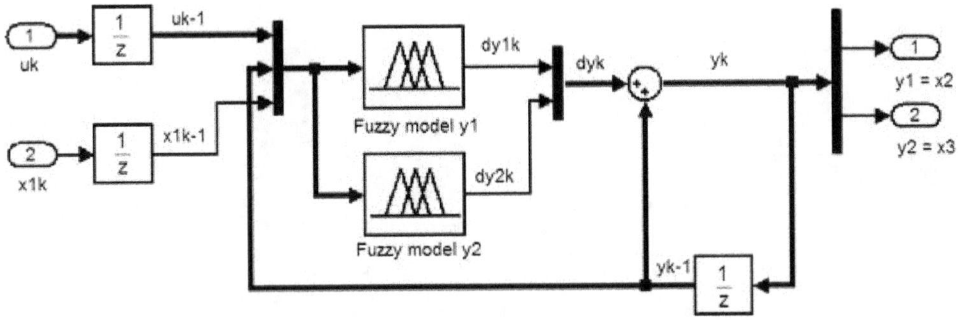

Figure 11. Structure of the discrete CL fuzzy model.

The whole design of CL optimal control consists of two steps:

Step 1. The design of the fuzzy model for the middle section of the continuous line.

The first step in the design of the fuzzy model for the middle section of the continuous line is the establishment of a consistent database from measured inputs and their corresponding outputs, which covers its entire assumed work space and describes the behavior of the modeled system. For establishing a consistent database, we can use, for example, the method of dividing the input range into *n*-levels and generating $n(n - 1)$ transient trajectories between them [12, 31, 32], or the method of exciting the system by a random input signal [10, 30] in case it is not possible (e.g., for operational reasons) to apply a pre-defined input signal at the system's input. Knowledge of the structure or of the parameters of the modeled system is not required in either of these methods. To define suitable sampling time *T* (according to the Shannon-Kotelnikov theorem) and approximate times for transitions for database measurement, we performed identification measurements on the physical model of the CL with input signals u_1 and u_2. Their value and performance are shown in Figure 12. Responses of the CL physical model output quantities to the input signals are illustrated in Figure 13.

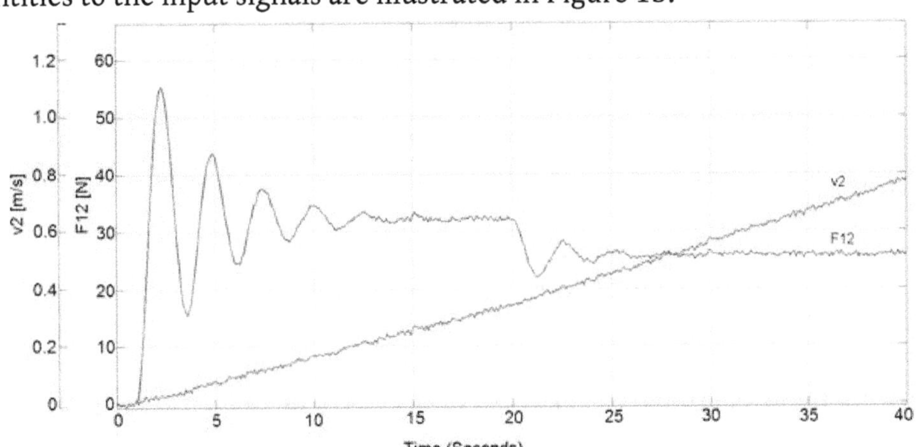

Figure 12. Input signals u_1 and u_2 for identification measurements.

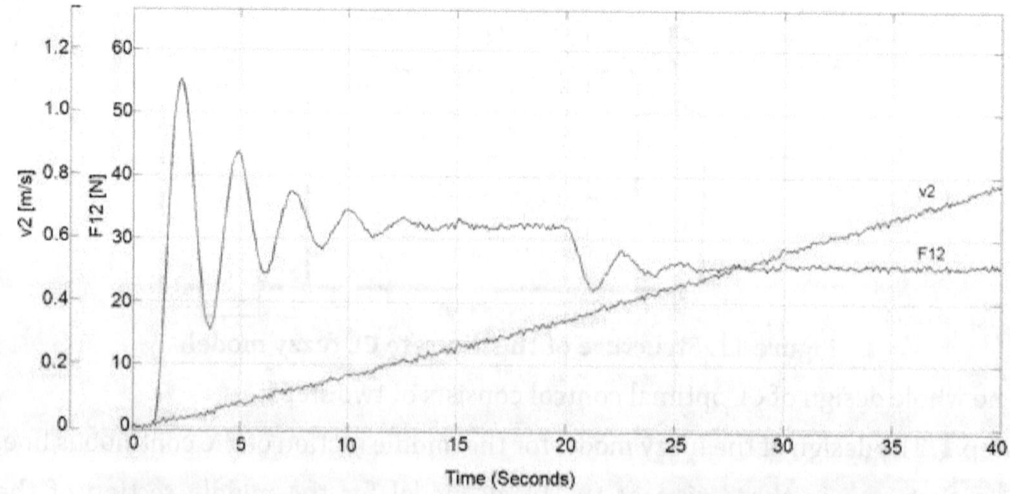

Figure 13. Identification response of CL middle section outputs.

The responses of the CL to inputs u_1 and u_2 (Figure 13) show that this system includes a fast tension subsystem and a slow speed subsystem. The range of input values for input u_1 (web tension) is assumed within the interval [-1, 1] and the range of input values for input u_2 (line speed) within the interval [-4, 4]. The database for CL fuzzy model set up will be generated so that line speed (input u_2) will increase in steps each 12 s, and each one second the faster (oscillating) part of the system will be excited by input u_1. The plot of input signals for generating the CL fuzzy model database is shown in Figure 14; Figure 15 shows the output quantities corresponding to these inputs.

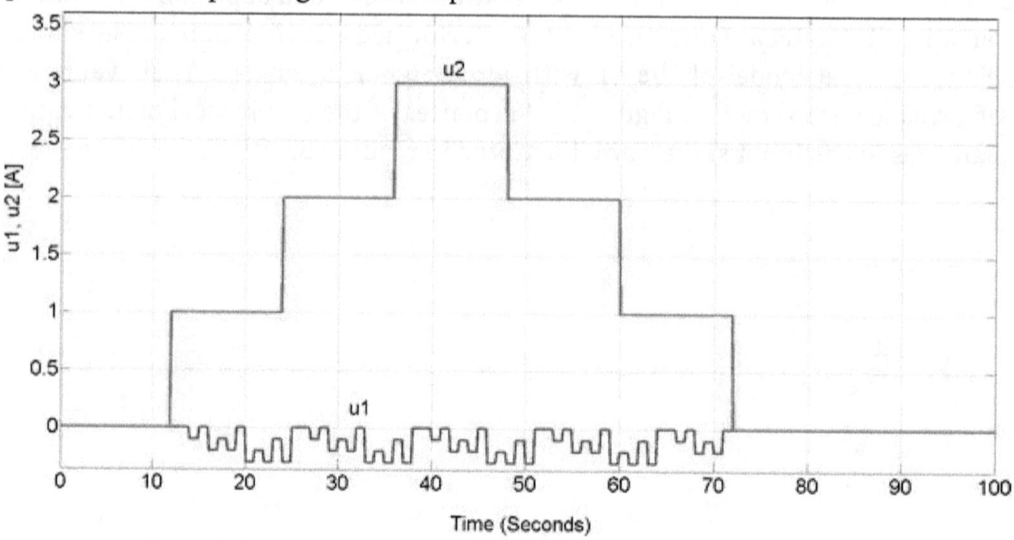

Figure 14. Identification of transitions for CL fuzzy model database generation.

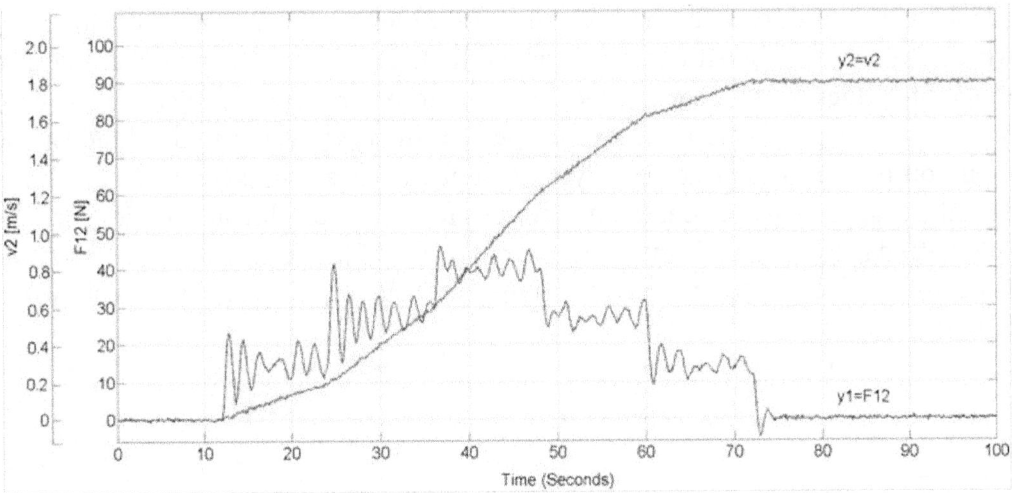

Figure 15. CL output variables corresponding identification transitions of inputs.

The database for CL fuzzy model was generated as demonstrated in Figure 16. With sampling time $T = 0.1$ s, we obtained a database with 1000 samples.

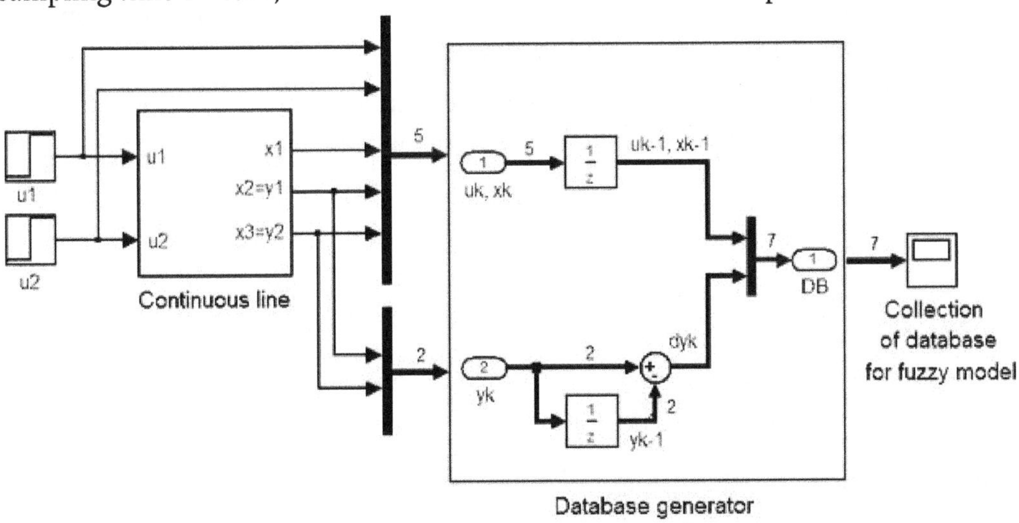

Figure 16. Generating database for CL fuzzy model.

This measured database can be used to search for two FIS structures of the given nonlinear system which best describe the measured relations between $[u_{1k-1}, u_{2k-1}, x_{1k-1}, x_{2k-1}, x_{3k-1}] \rightarrow dy_{1k}$, and $[u_{1k-1}, u_{2k-1}, x_{1k-1}, x_{2k-1}, x_{3k-1}] \rightarrow dy_{2k}$.

Using the measured database, the particular fuzzy model can be designed by standardly known procedures of cluster analysis and adaptive approaches to improve the quality of modeling and reduce development time. The fundamental features of cluster analysis are reduction of the number of fuzzy rules and provision of good initial rule parameters. For our purpose from the large number of methods for adaptive fuzzy networks development [33–36], we chose the adaptive neuro-fuzzy inference system (ANFIS) with subtractive clustering [14], which is a fast and robust data analysis method, having the following parameters: range of influence = 0.4, squash factor = 1.25, accept ratio = 0.4, reject ratio = 0.01. Subtractive clustering determines the optimal clusters [34] in a multi-dimensional input/output space that accurately represent the data [34, 37] and CL behavior. The ANFIS approach uses Gaussian functions for fuzzy sets, linear functions for the rule outputs, and Sugeno's inference mechanism [15]. The results were two static Sugeno-type fuzzy systems with two rules for each output quantity as is shown in Figure 17.

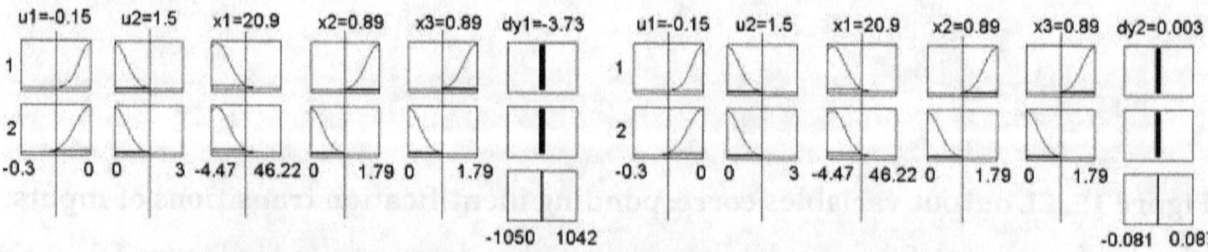

Figure 17. CL fuzzy model—SUGENO type with 2 rules.

The thus obtained fuzzy systems were implemented into the final continuous line fuzzy model structure, as illustrated in Figure 11.

To verify the correctness of the CL fuzzy model, randomly generated signals u_1 and u_2 were applied to its input, as demonstrated in Figure 18.

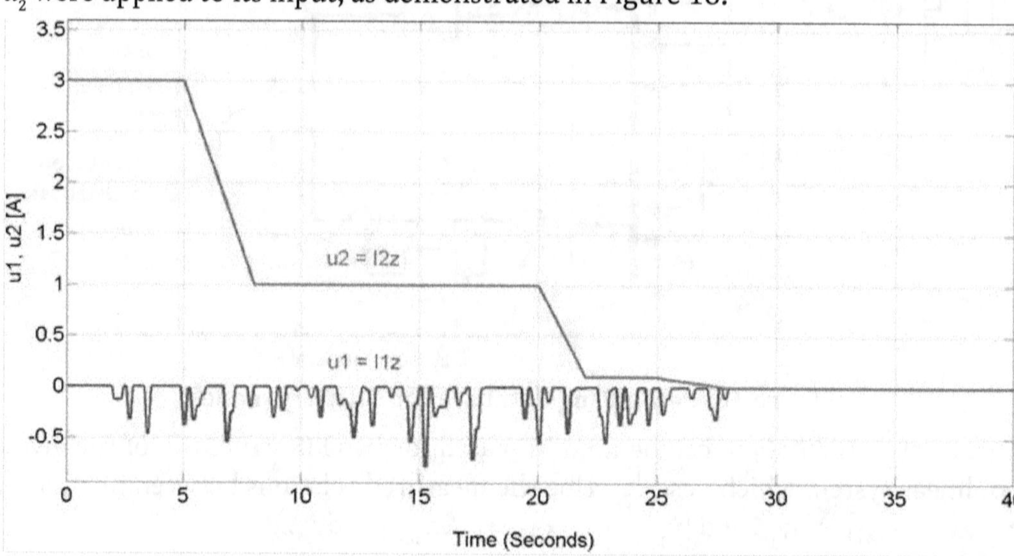

Figure 18. Performance of randomly generated signals u_1 and u_2.

The comparison of the fuzzy model outputs and CL physical model outputs for these inputs is shown in Figure 19.

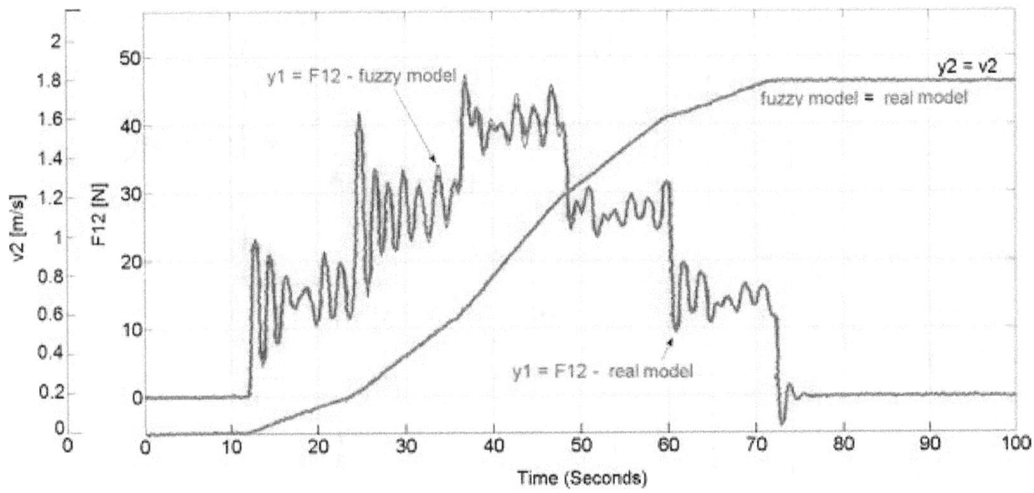

Figure 19. Performance of CL fuzzy model outputs y_1 and y_2 for randomly generated inputs u_1 and u_2.

The obtained results confirm that the designed fuzzy model very well approximates the performance of the continuous line also for randomly generated inputs and can be further used for the design of CL control.

Step 2. Design of optimal controller for middle section of continuous line.

The principal aim of CL control consists in achieving good dynamic control of tension in the material, with the speed of material movement being in accord with the pre-set CL speed. As it has been said above, this is in fact a nonlinear MIMO system an important feature of which is mutual influencing of the individual input and state quantities that can result in bad quality or even in the destruction of the material being processed. This fact makes the controller design methods and their subsequent resulting structures often very complex and presents an obstacle to their wider practical application in industry. Therefore, our aim was to design a simple CL controller that would ensure the desired dynamics in terms of the selected criterion for systems that are only described by input/output relationships, that is, on basis of their fuzzy model.

For control of middle section of CL (for which fuzzy model was designed), we chose the simplest control structure consisting of two standard PI controllers (one for tension control F_{12} and one for output velocity control v_2), as illustrated in Figure 20.

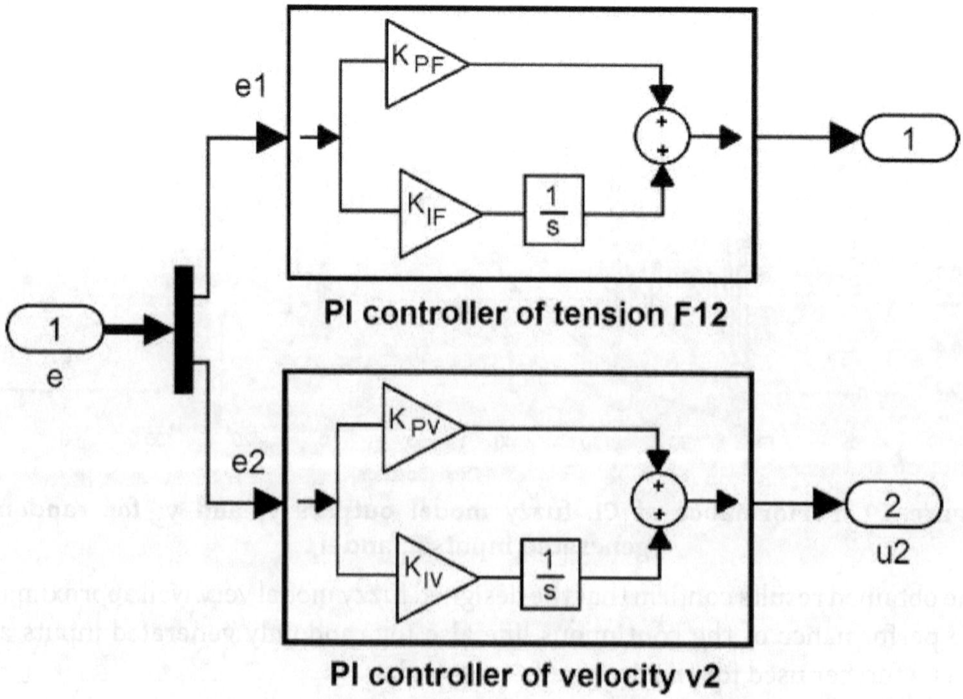

Figure 20. Continuous line PI controller structure.

Processing of material in a CL is usually carried out in operation cycles during which a required amount of prepared material is processed (e.g., a roll of paper, a sheet metal coil.) An operation cycle includes three stages—line start-up, line running at constant processing speed, and line delayed shut-off.

The objective of the optimization is to find such vector **K**[K_{PF}, K_{IF}, K_{PV}, K_{IV}] of the CL controller parameters for which the selected optimization criterion for a given CL operation cycle would be minimum. Most often this criterion is selected in quadratic form according to following equation

$$J(K) = \int \left(C_1 * e_1{}^2 + C_2 * e_2{}^2 \right) dt \quad (6)$$

where e_1 is the deviation between the desired and real tensile force of CL, e_2 is the deviation between the desired and real output velocity of the CL. Coefficients C_1 and C_2 determine the importance placed on the control errors of the particular outputs. In this case, we chose $C_1 = 5$ and $C_2 = 1$, which in terms of physics can be interpreted as larger emphasis put on the quality of regulation of error e_1 (tension in the strip of material which primarily determines its final quality). The optimal value of parameters of vector **K** is identified in the space of real values of gain of proportional and integrating elements of the particular controllers.

Let us note that what we are looking for is the extreme of the function of various variables, where the value of the criterial function for the individual vector **K** is determined by simulation on basis of the experimentally constructed CL fuzzy model according to Figure 21.

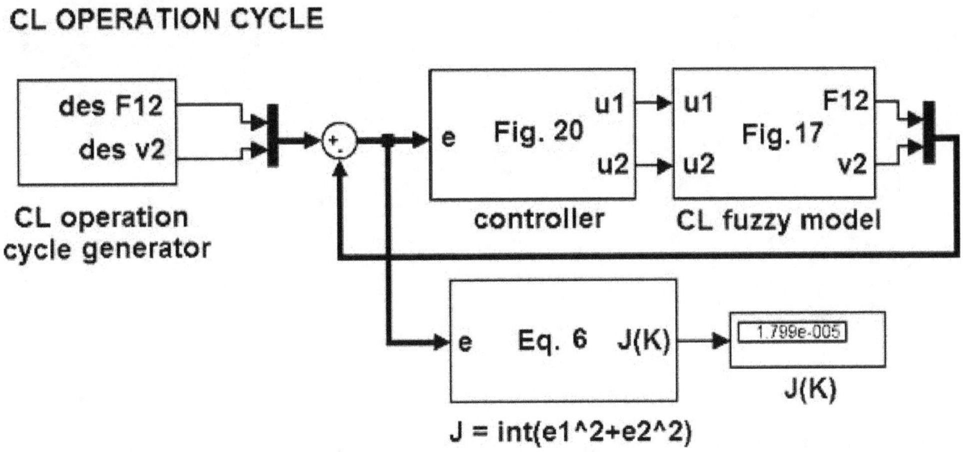

Figure 21. Simulation diagram for computing the value of criterial function J(K).

Several procedures can be applied for the purpose of optimization (e.g., genetic algorithm methods, and network charts). Thanks to today's availability of high-performance computing means, we chose the method of even geometrical division of the parameter space into equal intervals and of systematic searching within the whole range of the space. The advantage of this approach is that we can always identify the global minimum of the function Eq. (6), the disadvantage may be the time and computing demands in case vector **K** has several parameters, and the division of the space is more dense.

At the start of the optimization process, we determined the initial values of controller parameters $K_{PF0} = 1, K_{IF0} = 1, K_{PV0} = 1, K_{IV0} = 10$, and we divided the parameter space of individual gains of controllers by increments $\Delta K_{PF} = 1$, $\Delta K_{IF} = 5$, $\Delta K_{PV} = 1$, $\Delta K_{IV} = 5$. The maximum values of individual parameters were defined as $K_{PFmax} = 10$, $K_{IFmax} = 100$, $K_{PVmax} = 10$, $K_{IVmax} = 100$—this was based on physically standard values of proportional and integrating elements of PI controllers.

The value of the criterial function for initial values of vector $\mathbf{K} = \mathbf{K}_0$ and for the chosen CL operation cycle is equal to $J(\mathbf{K}_0) = 37.19$. The time responses of CL physical model output quantities corresponding to this initial PI controller parameter setting are illustrated in Figure 22.

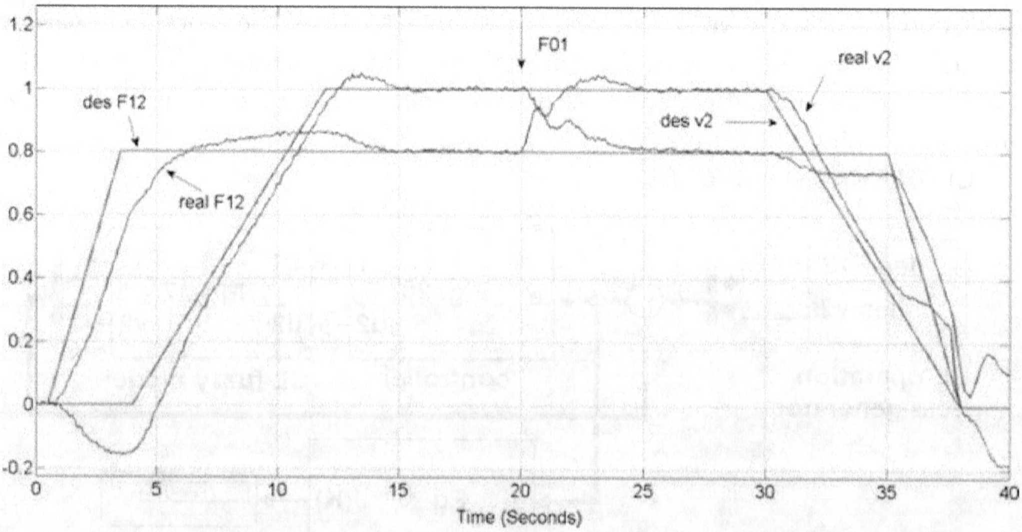

Figure 22. Time responses of CL outputs during operation cycle for K_0.

For finding optimal controller parameters, an m-file was created in Matlab program environment. At the end of the search process for optimal vector of CL controller parameters, the value of the optimization criterion was $J(\mathbf{K}_{opt}) = 0.3906$, which corresponds with optimal CL controller parameter values $\mathbf{K}_{opt}$ [9, 20, 7, 80]. Time responses of CL physical model output quantities for optimally set parameters of its controller are shown in Figure 23.

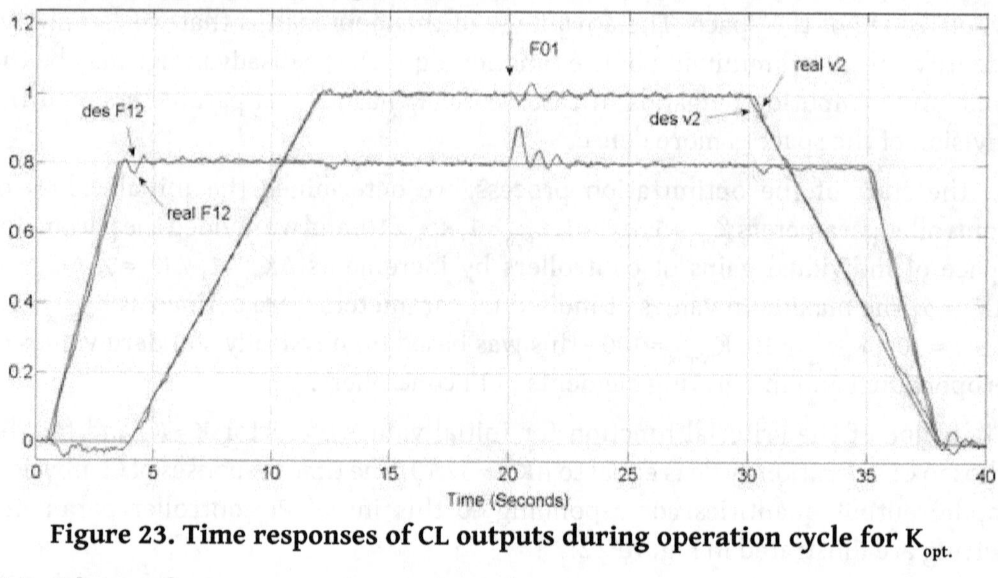

Figure 23. Time responses of CL outputs during operation cycle for K_{opt}.

3.1. Discussion

The proposed controller has been verified by experimental measurements on a real system which presented the physical model of the continuous line (the parameters of the CL physical model are specified in the Appendix). Figures 22 and 23 illustrate experimental results of the control of the continuous line middle section for selected operational cycle. In industrial practice, the required tension in the strip of material is set the first and then the line starts up to the desired operational speed.

Figure 22 shows the selected CL operation cycle in which first the desired value of tension in the strip of material is set to 0.8 N and at time 4 s the line starts up to reach the operational speed. At time 20 s, failure F_{01} = 0.7*FN occurs at entry of the line middle section (caused e.g., by a change in material thickness—material weld). The Figure shows that at initial values of parameters of CL speed and tension PI controllers, the deviation in tension from the desired value is up to 10%, the speed deviation is up to 25%, and at certain moments, the line also runs in the opposite direction. Autonomy and invariance in terms of failures are poor in dynamic states. On the contrary, when we set optimal values of parameters of CL speed and tension PI controllers, we can see—as illustrated in Figure 23—that tension in the line is maintained also in dynamic states within the range of 2% (which ensures high material processing quality during the whole operation cycle), and line speed only briefly falls outside the desired value by approx. 8% at a moment of influence by an external step disturbance. Optimal setting of the CL PI controller parameters therefore ensures good quality dynamics, autonomy and invariance of the controlled system against failures.

For the control of continuous line tension and velocity, a very simple control structure with two PI controllers was designed. We looked for four optimal parameters in the structure, such that would best satisfy the chosen quadratic optimality criterion for the given operation cycle of the line.

The quality of the designed controllers depends on the quality of the constructed fuzzy model which very well approximates the performance of the modeled system and can be further employed in the design of various CL control structures and also in the identification of non-measurable additive disturbances influencing the system, principally in real time.

The quality of the proposed controller depends on a large extent on a good quality of the nonlinear system fuzzy model which is constructed in the first step of the design procedure The model is constructed only on basis of suitably measured relations between the system's inputs and outputs, without the necessity of preliminary knowledge of its internal structure and parameters. The fuzzy model design is based on the basic idea of dynamic system description in state space.

The quality of the proposed controller depends on a large extent on a good quality of the nonlinear system fuzzy model which is constructed in the first step of the design procedure (see Section 4). The model is constructed only on basis of suitably measured relations between the system's inputs and outputs, without the necessity of preliminary knowledge of its internal structure and parameters. The fuzzy model design is based on the basic idea of dynamic system description in state space.

With this method, no principal limitations for the investigated system's nonlinearities are defined, and therefore, there is good reason to assume that the presented method will find wide use in multi-motor drives in steel industry, paper-making, printing and textile industries, in the production of synthetic fibers and foils in the chemical industry and in other industries.

AT A GLANCE

Conventional analytical methods frequently prove inadequate or entirely ineffectual for the design of nonlinear control systems with poorly defined parameters. The application of fuzzy logic concepts offers a viable method for controlling systems, applicable for both modeling and control design. The benefit of employing this method is in its simplicity and the straightforward development of the algorithm, facilitating the creation of controllers and the modeling of the characteristics of the constructed structures through the utilization of computer technology. The simplicity of the proposed structure, often utilizing PI controllers, and the ability to ascertain their parameters without elaborate mathematical descriptions represent a significant advantage of the used technology. This chapter provides two representative examples of designing control for nonlinear multi-input multi-output (MIMO) systems within the domain of mechatronic systems, utilizing fuzzy logic principles.

REFERENCES

1. Bai Y., Zhuang Q., Roth Z.S. Fuzzy logic control to suppress noises and coupling effects in a laser tracking. IEEE Transactions on Fuzzy Systems. 2005;13(1):113-121.
2. Baturone I., Moreno-Velo F.J., Sanchez-Solano S., Ollero A. Automatic design of fuzzy controllers for car-like autonomous robots. IEEE Transactions on Fuzzy Systems. 2004;12(4):447-465.
3. Chang C.Y., Chen B.S. Intelligent robust tracking controls for holonic and nonholonomic mechanical systems using only position measurements. IEEE Transactions on Fuzzy Systems. 2005;13(4):491-507.
4. Kim E., Lee S. Output feedback tracking control of MIMO systems using a fuzzy disturbance observer and its application to the speed control of a PM synchronous motor. IEEE Transactions on Fuzzy Systems. 2005;13(6):725-741.
5. Mousavi S.A., Hashemipour M., Sadeghi M., Petrofsky J.S., Prowse M.A. A fuzzy logic control system for the rotary dental instruments. Iranian Journal of Science & Technology, Transaction B: Engineering. 2010;34(No. B5):539-551.

6. Amaral da Silva J., Luiz de Oliveira Serra G. A new methodology for fuzzy model based robust control for nonlinear systems. In: IEEE International Conference on Fuzzy Systems (FUZZ-IEEE); 2012. pp. 1-8. doi:10.1109/FUZZ-IEEE.2012.6251329
7. Beirami H., Zerafat M.M. Self-tuning of an interval type-2 fuzzy PID controller for a heat exchanger system. Iranian Journal of Science & Technology Transactions of Mechanical Engineering. 2015;39(No. M1):113-129.
8. Fedor P., Perduková D., Ferková Ž. Optimal input vector based fuzzy controller rules design. Advances in Intelligent Systems and Computing. 2013;189:371-380.
9. Fedor P., Perduková D. A simple fuzzy controller structure. Acta Electrotechnica et Informatica. 2005;5(4):53-56.
10. Babuška R. Fuzzy modeling for control. Boston: MA: Kluwer; 1998.
11. .Fedor P., Perduková D. Fuzzy model of a nonlinear mechatronic system. Procedia Engineering. 2014;96:91-100.
12. Feng G. A survey on analysis and design of model-based fuzzy control systems. IEEE Transactions on Fuzzy Systems. 2006;14(5):676-697.
13. Qi P., Liu Ch., Ataka A., Lam H.K., Althoefer K. Kinematic control of continuum manipulators using a fuzzy-model-based approach. IEEE Transactions on Industrial Electronics. 2016;63(9):5022-5035.
14. Kiriakidis K. Fuzzy model-based control of complex systems. IEEE Transactions on Fuzzy Systems. 1998;6(4):517-529.
15. Wu Ch., Wang J., Li H., Liang H. Fuzzy-model-based control for nonlinear networked systems with random packet losses. In: 5th International Conference on Information Science and Technology (ICIST); 2015. pp. 455-460. doi:10.1109/ICIST.2015.7289015
16. Fedor P., Perduková D. Model based fuzzy control applied to a real nonlinear mechanical system. Iranian Journal of Science and Technology, Transactions of Mechanical Engineering. 2016;40(2):113-124.
17. Li B., Fan X., Zhang D., Jiang G. Modeling and fuzzy sliding decoupling control of looper multivariable system. In: Chinese Control and Decision Conference (CCDC); 2016. pp. 2467-2472. doi:10.1109/CCDC.2016.7531400
18. Sjoberg J., Zhang Q., Ljung L., Benveniste A., Delyon B., Glorennec P., Hjalmarsson H., Juditsky A. Nonlinear black-box modeling in system identification: a unified overview. Automatica. 1995;31:1691-1724.
19. Tečec Z., Petrović I., Matuško J. A Takagi-Sugeno fuzzy model of synchronous generator unit for power system stability application. AUTOMATIKA: Journal for Control, Measurement, Electronics. Computing and Communications.

2010;51(2):12-137.

20. Fedor P., Perduková D. Fuzzy model for middle section of continuous line. International Journal of Engineering Research in Africa. 2015;18:75-84.
21. Liu X., Xioung Z., Chen L., Zhu Z. A new Takagi-Sugeno fuzzy approach of process modeling and fault detection. In: 35th Chinese Control Conference – CCC; 2016. pp. 7126-7130. doi:10.1109/ChiCC.2016.7554483
22. Johansen T.A. Fuzzy model based control: Stability robustness and performance issues. IEEE Transactions on Fuzzy Systems. 1994;2(1):221-233.
23. Girovský P. Fuzzy control of synchronous motor with permanent magnet. Acta Electrotechnica et Informatica. 2016;16(4):17-20.
24. Tang K.S., Man K.F., Chen G., Kwong S. An optimal fuzzy PID controller. IEEE Transactions Industrial Electronics. 2001;48(4):757-765.
25. Tang W., Rui Z., Jiang H., Gao H. Coagulation control fuzzy modeling based on cluster analysis and optimization technique. In: 35th Chinese Control Conference (CCC); 2016. pp. 2154-2159. doi:10.1109/ChiCC.2016.7553687
26. Tao C.W., Taur J.S. Robust fuzzy control for a plant with fuzzy linear model. IEEE Transactions on Fuzzy Systems. 2005;13(1):30-41.
27. Zeng K., Zhang N.Y., Xu W.L. A comparative study on sufficient conditions for Tagaki-Sugeno fuzzy sytems as universal approximators. IEEE Transactions on Fuzzy Systems. 2000;8(6):773-780.
28. Vo H.H., Brandstetter P., Dong C.S.T. MRAS observers for speed estimation of induction motor with direct torque and flux control. Lecture Notes in Electrical Engineering. 2016;371:325-335.
29. Vittek J., Butko P., Ftorek B., Makyš P., Gorel L. Energy near-optimal control strategies for industrial and traction drives with a.c. motors. Mathematical Problems in Engineering. 2017;2017:22 pages/ID 1857186. doi:10.1155/2017/1857186
30. Tseng Ch.S., Chen B.S., Uang H.J. Fuzzy tracking control design for nonlinear dynamic systems via TS fuzzy model. IEEE Transactions on Fuzzy Systems. 2001;9(3):381-392.
31. Zadeh L.A. Outline of a New Approach to the Analysis of Complex Systems and Decision Processes. IEEE Transactions on Systems, Man, and Cybernetics. 1973;1:28-44.
32. Brandstetter P., Dobrovsky M., Kuchar M. Implementation of Genetic Algorithm in Control Structure of Induction Motor AC Drive. Advances in Electrical and Computer Engineering. 2014;14(4):15-20.

33. Jang J.S.R. ANFIS: Adaptive-Network-Based Fuzzy Inference System. IEEE Trans. Syst., Man, and Cyber.. 1993;23(3):665-685.

34. Chiu S.C. Extracting Fuzzy Rules from Data for Function Approximation and Pattern Classification, Fuzzy Information Engineering: a guided tour of applications. New York: John Wiley & sons; 1997.

35. Takagi T., Sugeno M. Fuzzy identification of systems and its application to modeling and control. IEEE Trans. Syst., Man and Cyber. 1985;15:116-132

36. Yager R.R. Essentials of Fuzzy Modeling and Control . New York: Wiely; 1994.

37. Hüseyin A., Faruk O. ANFIS & PID controller design and comparison for overhead cranes. Indian Journal of Engineering & Materials Sciences. 2011;18:191-203.

19 Chapter

VIBRATION SUPPRESSION CONTROLLER FOR MULTI MASS RESONANCE SYSTEMS UTILIZING FUZZY CONTROL

1. INTRODUCTION: AN OVERVIEW

Recently, motor drive systems, comprising several motors, shafts, gears, and loads, are extensively employed in industrial sectors. These mechanical systems are designed to meet the demands for rapid reaction, weight reduction, miniaturization, and high precision in many industrial applications. Therefore, in the industrial sector, the system is regarded as a multi-mass resonance system, comprising multiple inertial moments, torsional shafts, and gear couplings. The first-order approximation of the multi-mass resonance model is the two-mass resonance model. For instance, several control methods, which are PID control (Proportional plus Integral plus Derivative Control) with a resonance ratio control using the disturbance observer, coefficient diagram method (CDM), full state feedback control with the state observer, the pole placement method, fractional order PIDk control, and $H\infty$ control method, are effective to control for two-mass resonance system [1–3]. Ikeda et al. [4] elucidated the efficacy of the controller design strategy employing the pole placement method for the two-mass position control system. In recent years, the resonance system necessitates greater precision and enhanced response speed in control. Consequently, it is imperative to address a higher-order model of the resonance system. The electric vehicle's drive train is comprised of a four-mass system. Similarly, the ball screw drive stage is generally a four-mass system. The thermal power generation system, consisting of various turbines and generators, is represented as a twelve-mass resonance system. Consequently, numerous vibration suppression control techniques for three-mass resonance systems or more have been presented [5, 6]. A modified-IPD speed controller utilizing the Taguchi Method has been proposed in References [7, 8]. Simultaneously, the state equations and parameters of the controlled object are necessary for the construction of control systems. References [9, 10] previously introduced a controller gain tuning methodology for a vibration suppression-type speed controller utilizing fictitious reference iterative tuning (FRIT) for single-input multi-variable control systems, without requiring knowledge of the system state equations and parameters. In contrast, a fuzzy control system may be assumed as one technique for overcoming these challenges. A fuzzy control system employing fuzzy inference exemplifies a non-mathematical control technique developed from experience and intuition. Numerous applications have integrated fuzzy control systems into motor drive systems [11–14]. This chapter presents a vibration suppression controller utilizing fuzzy inference. The control system comprises a speed fuzzy controller and a proportional-integral (PI) current controller to achieve accurate speed and torque response in a two or three inertial resonance system. In the control system, just the state variables on the motor side are employed to regulate the resonance system. This chapter also addresses the suggested control system as a digital control system. The proposed control system is an enhanced version designed to implement the previously proposed control system for the simulation model referenced in [13, 14] onto actual experimental equipment. The fuzzy controller comprises three scaling factors, while the PI current controller has two controller gains. This chapter employs a differential evolution method (DE) to ascertain these five controller parameters [13–18]. DE, suggested by Price and Storn, is an evolutionary optimization strategy. Utilizing DE facilitates the swift and efficient identification of appropriate controller parameters. Finally, the validity of the controller design, its robustness, and the control efficacy of the suggested technique were confirmed using simulations and experiments utilizing the test setup.

2. MULTI-MASS VIBRATION SUPPRESSION CONTROL SYSTEM

2.1. 2-mass model

Figure 1 shows the two-mass resonance model. The model is configured of two rigid inertial masses with a torsional shaft, where ω_M, T_{dis}, ω_L, T_{in}, J_M, J_L, K_s, and T_L denote the motor angular speed, the torsional torque, the load angular speed, the input torque, the inertia of motor, the inertia of load, the shaft torsional stiffness, and the load torque, respectively.

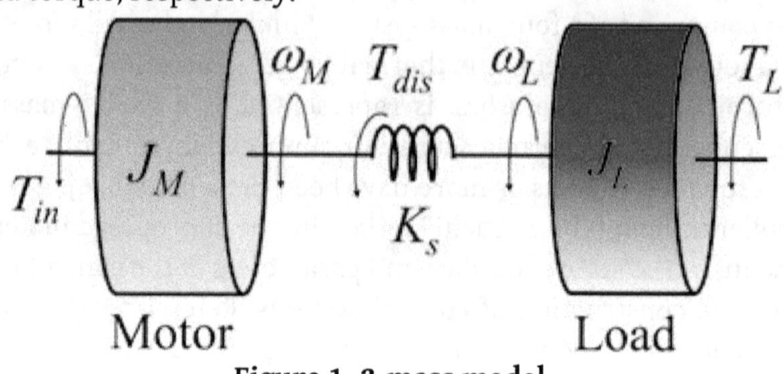

Figure 1. 2-mass model.

If all the state variables can be observed by several sensors and all the system parameters are known or identified, it is easy to construct the optimal control system. However, in general, it is difficult to measure the state variables of the load side due to constraints on scarce measurement environment and sensor installation location. Therefore, in this chapter, we use only the motor side variables. Furthermore, we contemplate for the current minor control in order to compensate torque response. Eq. (1) shows the continuous state equation of two-mass resonance model, where the viscous friction is not considered.

$$\frac{d}{dt}\begin{pmatrix}\omega_M\\ \omega_L\\ T_{dis}\end{pmatrix}=\begin{pmatrix}0 & 0 & -\frac{1}{J_M}\\ 0 & 0 & \frac{1}{J_L}\\ K_s & -K_s & 0\end{pmatrix}\begin{pmatrix}\omega_M\\ \omega_L\\ T_{dis}\end{pmatrix}+\begin{pmatrix}\frac{1}{J_M}\\ 0\\ 0\end{pmatrix}T_{in}+\begin{pmatrix}0\\ -\frac{1}{J_L}\\ 0\end{pmatrix}T_L \tag{1}$$

Eq. (2) shows the transfer function of two-mass model, which input signal is T_{in} and output signal is ω_M.

$$\frac{\omega_M}{T_{in}}=\frac{s^2+\omega_a^2}{J_M s(s^2+\omega_r^2)} \tag{2}$$

where ω_r is a resonance frequency and ω_a is an anti-resonance frequency. Here, we use the DC servo motor as the driving motor. Eq. (3) is the voltage equation of dc servo motor, where R_a is the armature resistance, L_a is the armature inductance, K_e is the back-emf constant, and K_0 is the converter gains of the DC power supply. Input torque is calculated by $T_{in} = K_t i_a$, where K_t is the torque constant.

$$L_a\frac{di_a}{dt}+R_a i_a = K_0 u_c - \omega_M \tag{3}$$

Figure 2 is indicative of the block diagram of the two-mass resonance system.

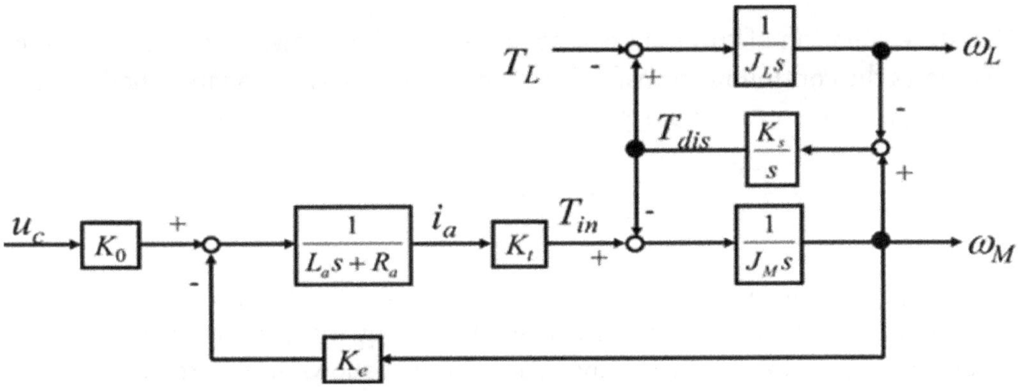

Figure 2. Block diagram of two-mass resonance model.

The inertia ratio R of two-mass model is given by Eq. (4), where J_{Mn} and J_{Ln} represent the nominal values of the motor and load inertias, respectively.

$$R = \frac{J_{Ln}}{J_{Mn}} \tag{4}$$

2.2. Three-mass model

Similarly to two-mass resonance model, Figure 3 reveals the three-mass model. The model consists of three rigid inertias and two shafts. Here, J_c and J_L are the load 1 inertia moment and the load 2 inertia moment, respectively. Furthermore, ω_c, ω_L, T_{dis1}, T_{dis2}, K_{s1}, and K_{s2} denote load 1 angular speed, load 2 angular speed, shaft 1 torsional torque, shaft 2 torsional torque, the shaft 1 stiffness, and the shaft 2 stiffness, respectively.

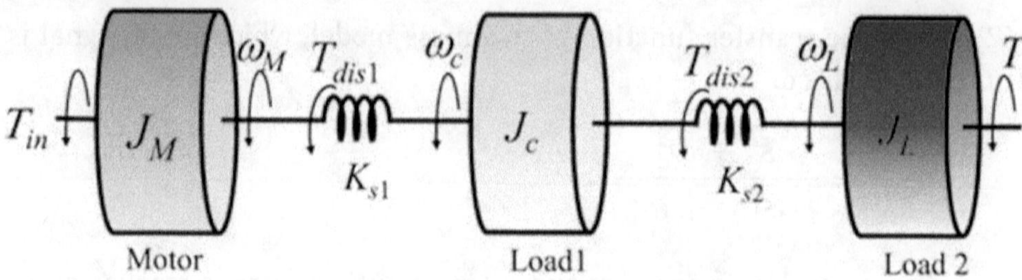

Figure 3. Three-mass model.

$$\frac{d}{dt}\begin{pmatrix}\omega_M\\ \omega_M\\ \omega_M\\ T_{dis}\\ T_{dis}\end{pmatrix}\begin{pmatrix}\omega_M\\ \omega_L\\ T_{dis}\end{pmatrix} = \begin{pmatrix} 0 & 0 & 0 & -\frac{1}{J_M} & 0 \\ 0 & 0 & 0 & \frac{1}{J_c} & -\frac{1}{J_c} \\ 0 & 0 & 0 & 0 & \frac{1}{J_L} \\ K_{s1} & -K_{s1} & 0 & 0 & 0 \\ 0 & K_{s2} & -K_{s2} & 0 & 0 \end{pmatrix}\begin{pmatrix}\omega_M\\ \omega_M\\ \omega_M\\ T_{dis}\\ T_{dis}\end{pmatrix} + \begin{pmatrix}\frac{1}{J_M}\\ 0\\ 0\\ 0\\ 0\end{pmatrix} T_{in} + \begin{pmatrix}0\\ 0\\ -\frac{1}{J_L}\\ 0\\ 0\end{pmatrix} T_L \tag{5}$$

The state equation of three-mass resonance model is shown in Eq. (5). Then, Eq. (6) shows the continuous transfer function of three-mass resonance model, which input signal is T_{in} and output signal is ω_M.

$$\frac{\omega_M}{T_{in}} = \frac{(s^2+\omega_{a1}^2)(s^2+\omega_{a2}^2)}{J_M s(s^2+\omega_{r1}^2)(s^2+\omega_{r2}^2)} \tag{6}$$

In this equation, ω indicates the angular frequency, where ω_{r1}, ω_{r2}, ω_{a1}, and ω_{a2} are the resonance frequencies, and anti-resonance frequency, respectively. Then, the block diagram realized by using above equations is shown in Figure 4.

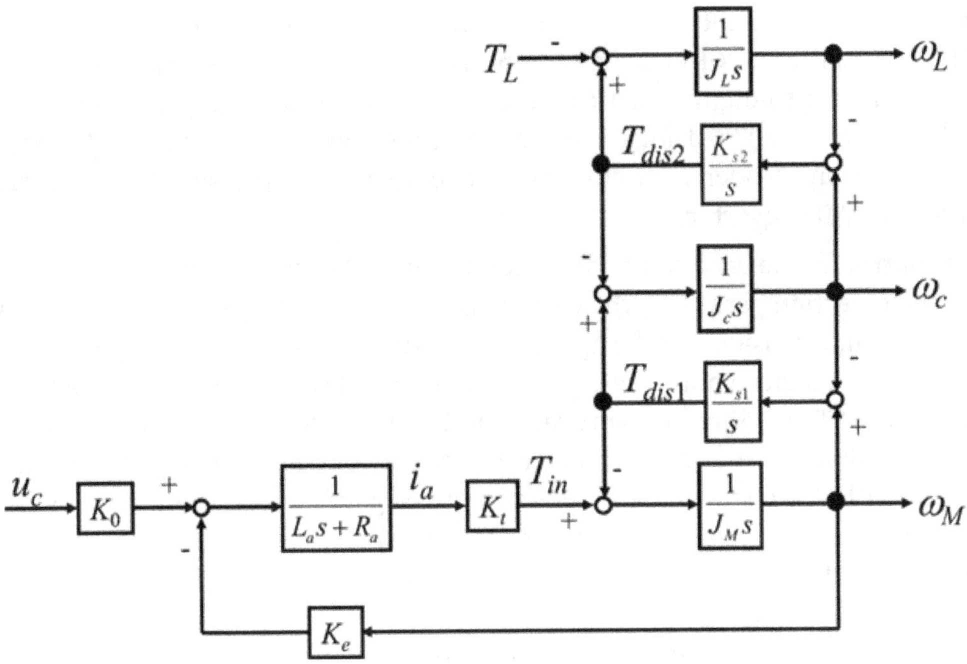

Figure 4. Block diagram of three-mass model.

2.3. Experimental set up

This chapter confirms the effectiveness and performance of the proposed method by experiments using the experimental equipment.

Figure 5 is the appearance of the experimental system constructed in this research. The two-mass resonance system is simulated by utilizing the dc servo motor and the dc generator with a finite rigid coupling. The controller is realized on a digital signal processor, which calculates the PWM signal to a four-quadrant dc chopper.

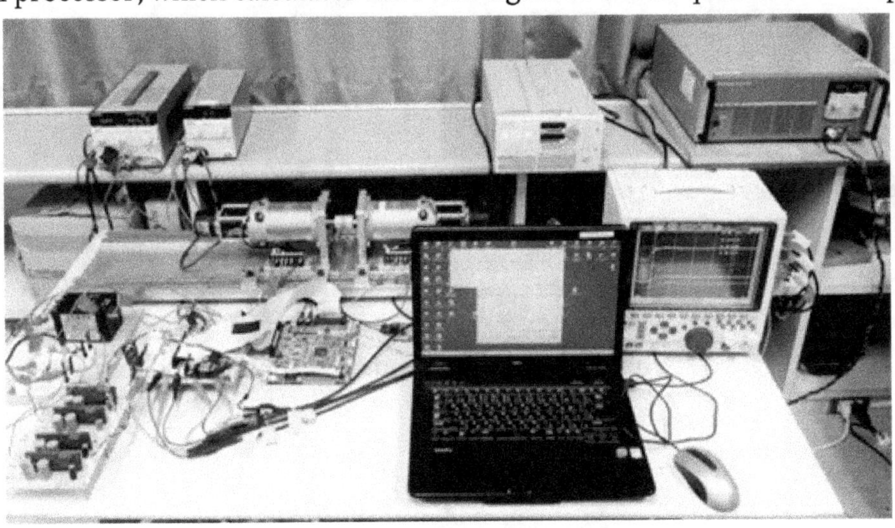

Figure 5. Experimental apparatus.

The DSP board (PE-PRO/F28335 Starter Kit, Myway Plus Corp.) consists of the DSP (TMS320F28335PGFA), a digital input/output (I/O), ABZ counters for encoder signals, analog-to-digital (AD) converters and digital-to-analog (DA) converters [19]. The motor and load angles and angular speeds are detected using 5000 pulses-per-revolution encoders. The current of dc servo motor is measured by the current sensor and AD converter.

The control frequency and the detection frequency of the encoder are both 1 ms, and the detection period for the current is 10 μsec. The design language used was C. Then, while considering the application of the system to specific apparatus, we constructed a digital control system that contains a discrete controller. In addition, we used MATLAB/Simulink software for the proposed off-line tuning process based on simulation and constructed the fuzzy control system as a continuous system [20]. The disturbance is added to the dc generator as the torque by using the electric load device on constant current mode. Figures 6 and 7 show the apparatus of the two-mass model and three-mass model used in the experimental set up, respectively. Figure 8 shows the experimental system configuration. For reference, the nominal parameters of the experimental two-mass model and three-mass model are given in Tables 1 and 2, respectively.

Figure 6. Photograph of two-mass resonance model.

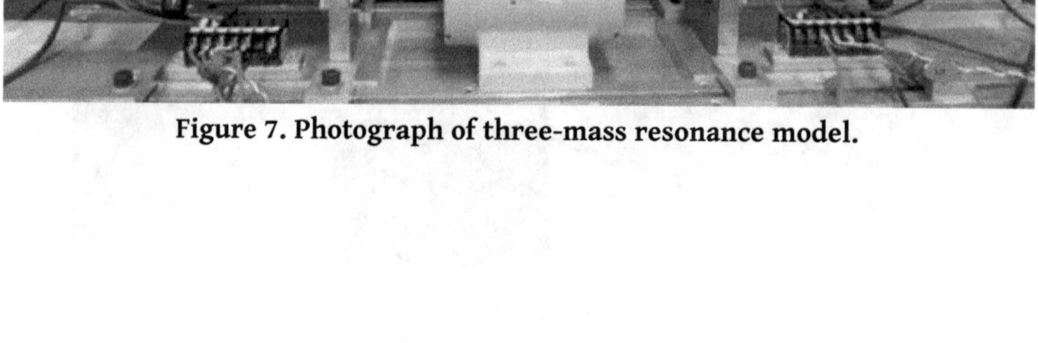

Figure 7. Photograph of three-mass resonance model.

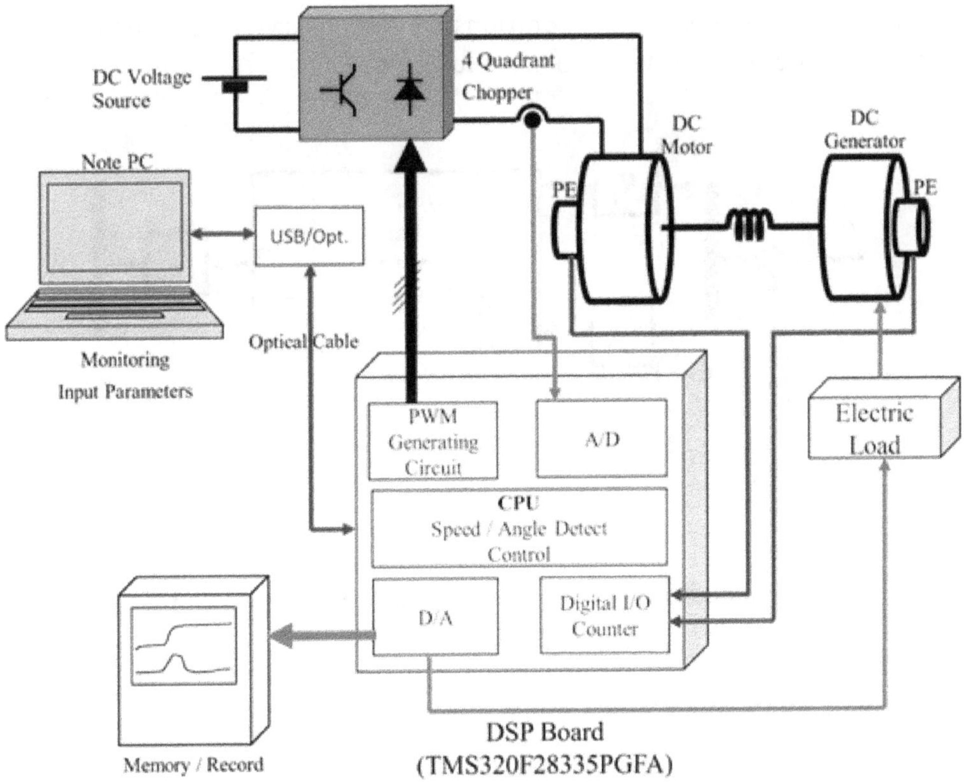

Figure 8. Configuration of experimental system (two-mass resonance model).

Table 1. Nominal parameters of two-mass experimental model.

Symbol	Parameter	Value
J_{Mn}	Motor inertia	2.744×10^{-4} (kgm^2)
J_{Ln}	Load inertia	2.940×10^{-4} (kgm^2)
K_{sn}	Shaft stiffness	18.5 (Nm/rad)

Table 2. Nominal parameters of three-mass experimental model.

Symbol	Parameter	Value
J_{Mn}	Motor inertia	2.744×10^{-4} (kgm^2)
J_{cn}	Load 1 inertia	1.112×10^{-4} (kgm^2)
J_{Ln}	Load 2 inertia	2.940×10^{-4} (kgm^2)
K_{s1n}	Shaft stiffness 1	18.5 (Nm/rad)
K_{s2n}	Shaft stiffness 2	18.5 (Nm/rad)

Figure 9 shows an example of experimental result using two-mass model. These step waves are the motor and load angular speeds with direct current voltage input. Similarly, Figure 10 shows an example of experimental result using three-mass model, which are the motor and load angular speeds with same above condition.

In these figures, the resonance vibrations can be observed. The purpose of this research is to suppress these resonance vibrations.

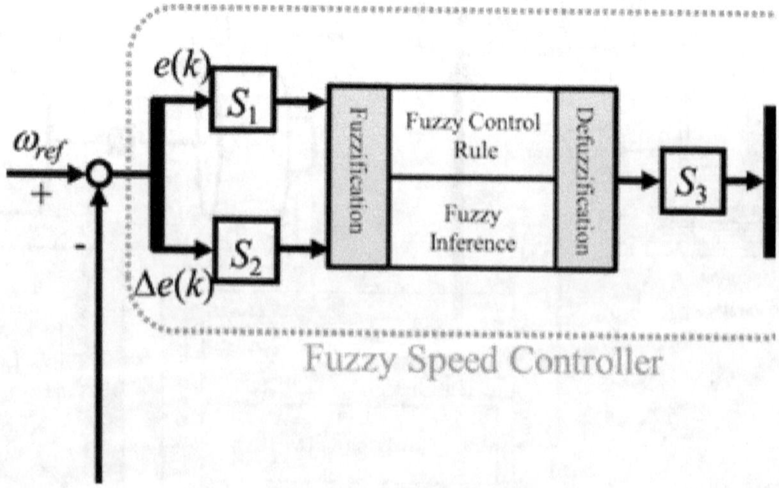

Figure 9. Angular speeds (ω_M and ω_L) of the step responses to a DC voltage input (two-mass model).

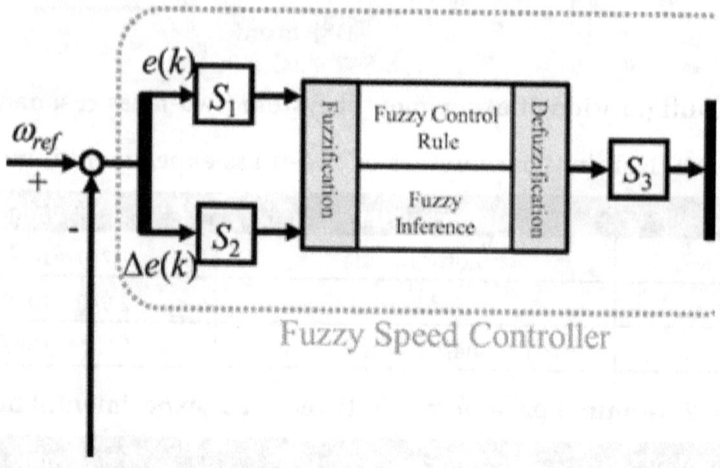

Figure 10. Angular speeds (ω_M and ω_L) of the step responses to a DC voltage input (three-mass model).

3. PROPOSED FUZZY CONTROL SYSTEM

3.1. Fuzzy speed controller

Fuzzy controller, which is executed by the fuzzy set and the fuzzy inference, can control for nonlinear systems or uncertain model. Figure 11 indicates the proposed fuzzy speed controller in this chapter. The speed controller is based on fuzzy control. The current controller is typical PI controller. Furthermore, the load side state variables are not utilized for control, where S_1, S_2 and S_3 are the parameters to determine the scale of the membership function, which are called scaling factors or scaling coefficient. K_{pc} and K_{ic} are the current PI controller gains. Eq. (7) shows the transfer function of current PI controller. Additionally, this chapter uses the discrete control system.

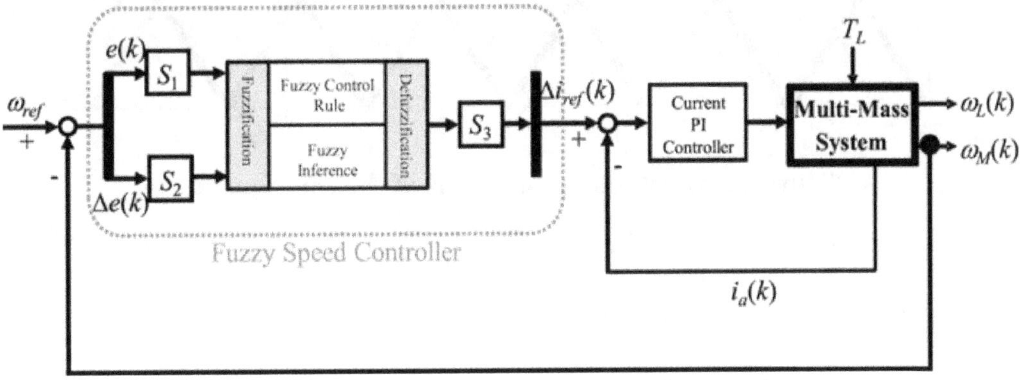

Figure 11. Block diagram of the proposed control system.

$$u_c(k) = \left(K_{pc} + \frac{1}{s}K_{ic}\right)e(k) \tag{7}$$

Figure 12 is indicative of the membership function for the premise variables. This membership function is a shape of triangle with a dense center. Figure 13 indicates the membership function, which is formed uniformly triangle for the consequent variable. Here, the s denotes the scaling factor. PB, PM, PS, ZE, NS, NM, and NB are the linguistic variables of the fuzzy control where, PB indicates positive big, PM indicates positive medium, PS indicates positive small, ZE indicates zero, NS indicates negative small, NM indicates negative medium, and NB indicates negative big, respectively. The premise variables are $e\omega_M(k)$ and $\Delta e\omega_M(k)$.

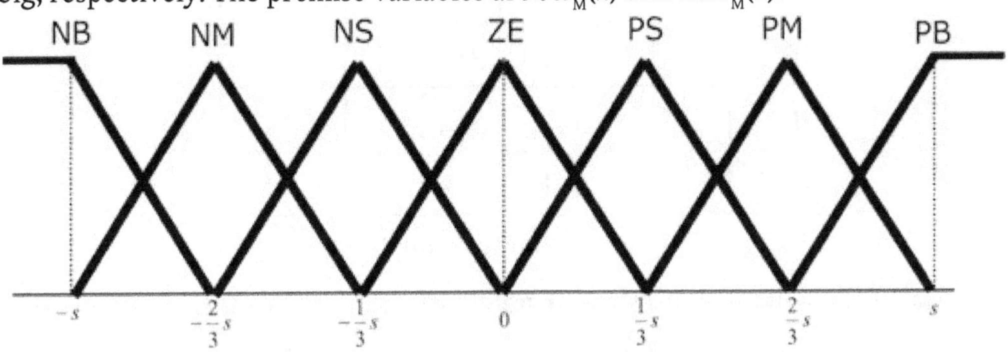

Figure 12. Membership functions of the antecedence.

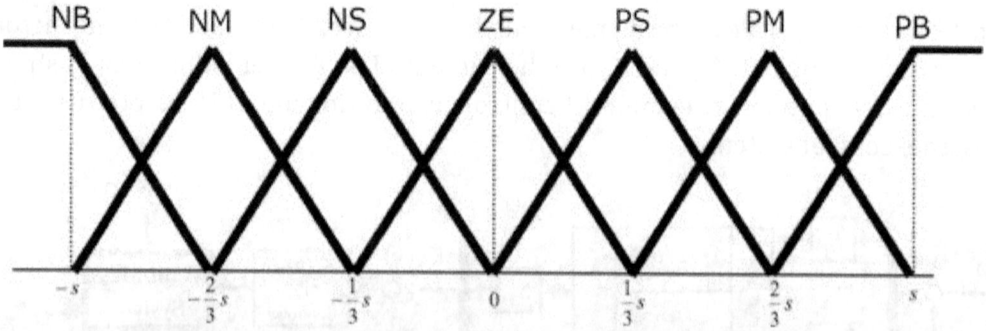

Figure 13. Membership functions of the consequence.

$$e_{\omega M}(k) = \omega_{ref} - \omega_M(k) \tag{8}$$

$$\Delta e_{\omega M}(k) = e_{\omega M}(k) - e_{\omega M}(k-1) \tag{9}$$

Then, the consequence variable is the variation width of the current input $\Delta i_{ref}(k)$. Therefore, the proposed fuzzy controller is nearly same as the proportional-derivative (PD) type controller.

Figure 14 is indicative of the fuzzy rule table. The rule is included the rising correction of the angular speed response.

e \ Δe	NB	NM	NS	ZE	PS	PM	PB
NB				NB	NM		
NM				NM			
NS				NS	ZE		PM
ZE	NB	NM	NS	ZE	PS	PM	PB
PS	NM		ZE	PS			
PM				PM			
PB			PM	PB			

Figure 14. Control rule table.

3.2. Design method of controller parameters by differential evolution

In this chapter, five parameters (S_1, S_2, S_3, K_{pc} and K_{ic}) of the proposed controller have to be designed. However, it is difficult to determine them by trial and error or some. Therefore, this chapter proposes the differential evolution (DE) to search the optimal controller parameters. Here, DE is one of evolutionary optimized solution search methods. DE is the optimization method-based multi-point search method. In particular, basic GA expresses parameter by binary coding, whereas DE uses the parameters by real variable vector. The DE design is conducted by the initial population, the mutation, the crossover, and the selection. The design flow of DE is shown in Figure 15. In this chapter, DE/rand/1/bin design strategy is used for the determination of five controller parameters.

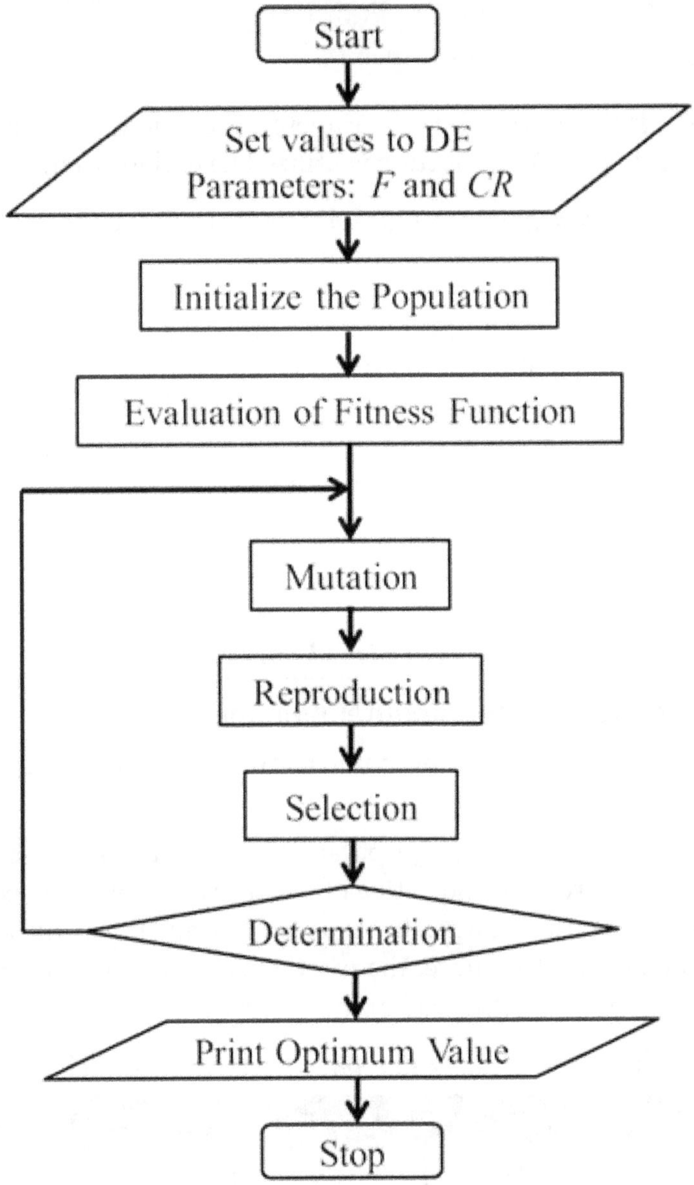

Figure 15. Flow of DE algorithm.

where D is the number of design parameter vectors, NP is the number of members in each population. Each parameter vector is represented by the parameter vector (target vector) $\mathbf{x}_{i,G}$, where G denotes one generation. The mutation vector $\mathbf{v}_{i,G}$ is calculated by Eq. (10). From this equation, F indicates the step width (scaling factor) of DE design, and CR indicates of the crossover rate, where r_1, r_2, and r_3 are different values.

$$v_{i,G+1} = x_{r1,G} + F(x_{r1,G} - x_{r3,G}), \quad r_1 \neq r_2 \neq r_3 \neq i \tag{10}$$

$$u_{j,G+1} = \begin{cases} v_{j,G+1} & \text{rand} \leq CR \text{ or } j = \text{ST} \\ x_{j,G} & \text{rand} > CR \text{ or } j \neq \text{ST} \end{cases} \tag{11}$$

In Eq. (11), $\mathbf{u}_{j,G+1}$ is the vector of trial parameter, the rand is random value, and ST indicates the start point. The selection is utilized next algorithm,

$$x_{i,G+1} = \begin{cases} u_{i,G+1} & \text{if } y(u_{i,G+1}) > y(x_{i,G}) \text{ for maximaization problems} \\ x_{i,G} & \text{otherwise} \end{cases} \tag{12}$$

As previously described, the proposed method uses five control parameters (S_1, S_2, S_3, K_{pc}, and K_{ic}). The population size is 2000, the order of each vector is 20, and the coefficient of membership function F is 0.5. Moreover, the rate of crossover CR is 0.9. Then, the performance index function is shown in Eq. (13). Meanwhile, this chapter utilizes the inverse of y as a fitness function.

$$y = \int_0^{\infty} t\sqrt{(\omega_{ref} - \omega_L)^2}\,dt \tag{13}$$

4. SIMULATION AND EXPERIMENTAL RESULTS

4.1. Verification results of computer simulation

Next, the simulation results of the proposed method are demonstrated by computer simulation.

Table 3 shows the results of design parameter using the proposed method for two-mass model. Figure 16 is indicative of the transition of the maximum fitness function. In this simulation design, the step response and the disturbance response have been evaluated. Furthermore, the inertia ratio R is 1.07, and the stiffness of shaft K_{sn} has been set to 18.5 Nm/rad in the simulation design.

Table 3. Results of design parameter calculated by DE.

S_1	S_2	S_3	K_{pc}	K_{ic}
8.486	0.4802	0.4001	4.678	1.0×10^{-6}

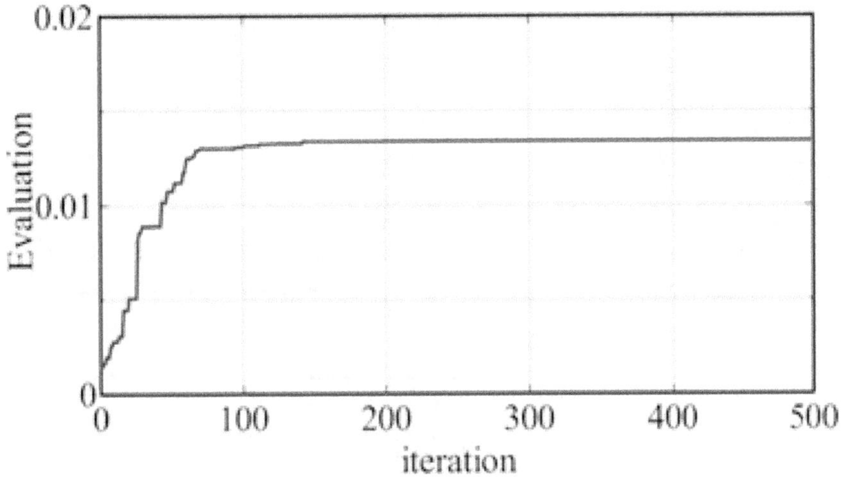

Figure 16. Convergence of index function y.

Figures 17 and 18 show the step responses that were obtained for the motor and load angular speeds, and armature current when using the proposed method. In this chapter, ω_{ref} is 30 rad/s, the DC voltage input is 25 V, and the disturbance input T_L is changed from 0 to 20% at t = 0.3 s. As shown by these figures, good waves are observed for the reference-following, vibration suppression, and the disturbance performance. Figure 19 is indicative of the search process of the S_1 vector. Similarly, Figures 20–23 show the transition of the S_2 vector, S_3 vector, K_{pc} vector and K_{ic} vector, respectively. In particular, from Figure 23 and Table 3, K_{ic} is 1.0×10^{-6} of the design limitation value. Therefore, integral gain of the current PI controller can be omitted for this control object.

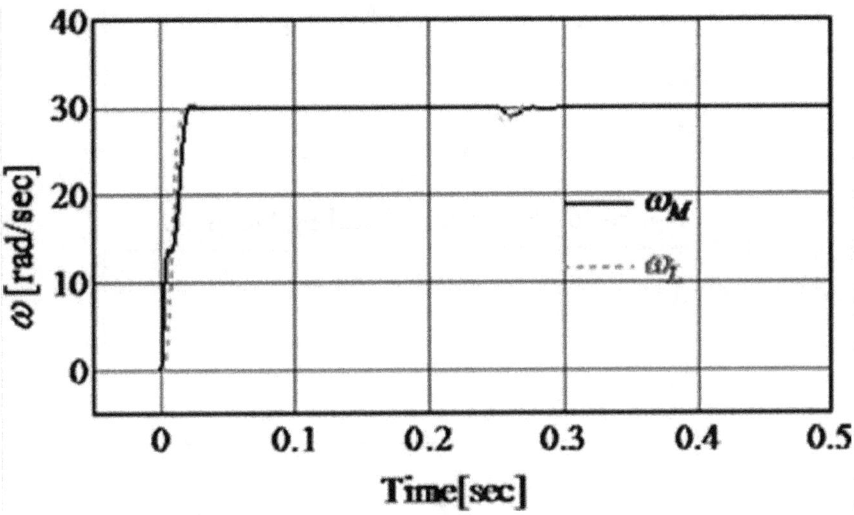

Figure 17. Simulation results ω_M and ω_L (two-mass, R = 1.07, K_{sn} = 18.5 Nm/rad).

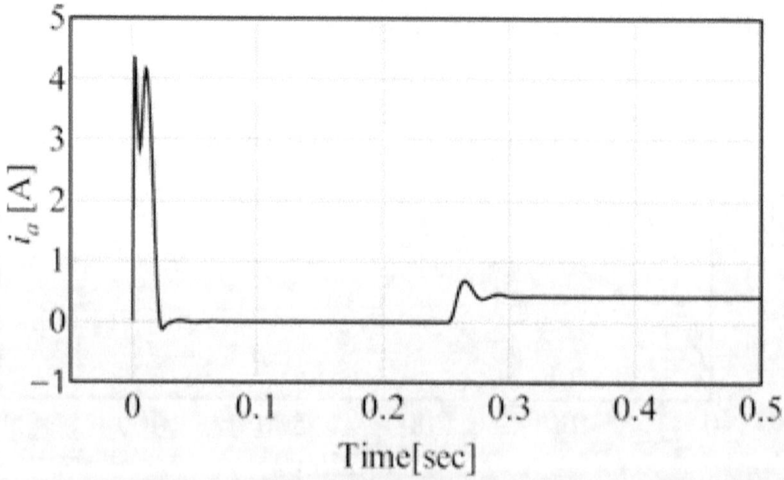

Figure 18. Simulation results i_a (two-mass, R = 1.07, K_{sn} = 18.5 Nm/rad).

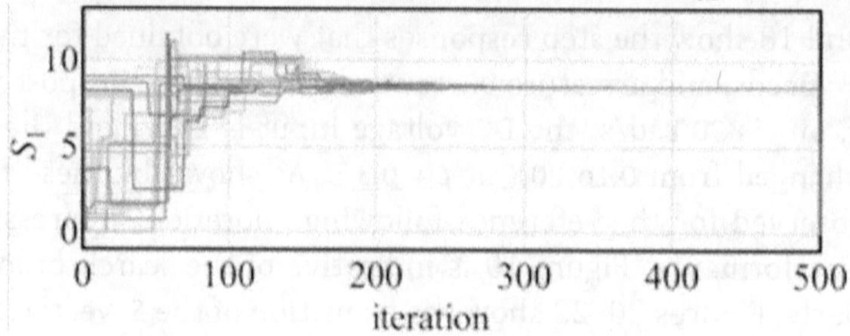

Figure 19. Transition of scaling factor S_1.

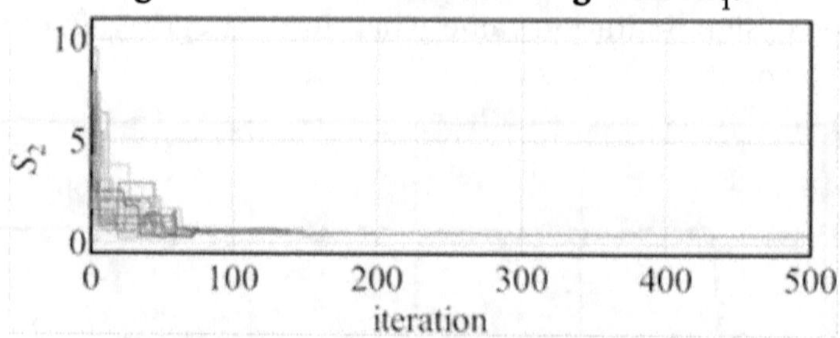

Figure 20. Transition of scaling factor S_2.

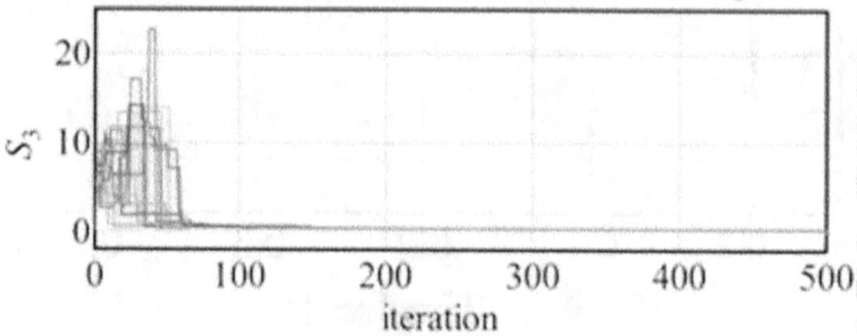

Figure 21. Transition of scaling factor S_3.

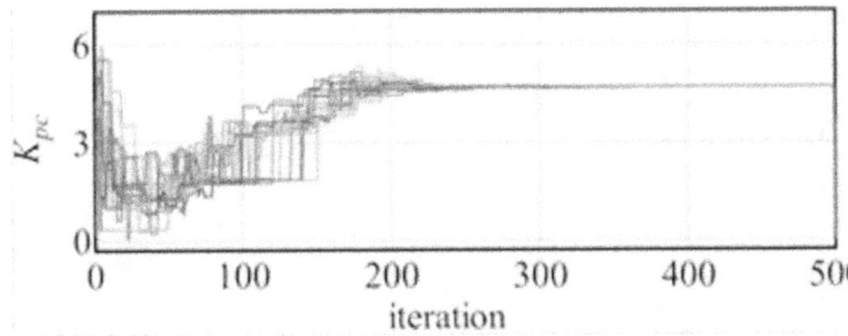

Figure 22. Transition of current proportional gain K_{pc}.

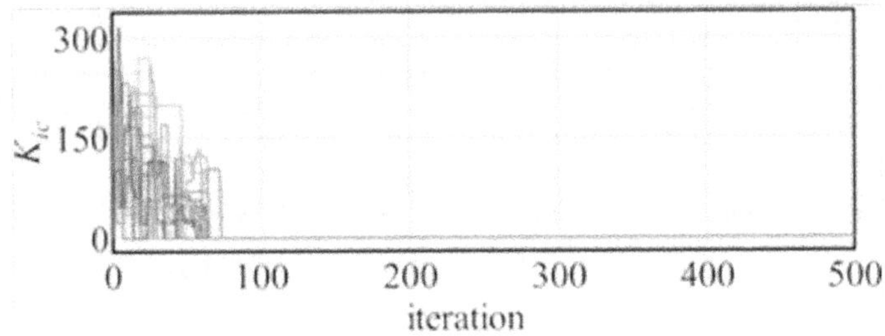

Figure 23. Transition of current integral gain K_{ic}.

4.2. Experimental results

4.2.1. 2-mass model

Next, the experimental results by using the proposed method are illustrated in this section. Figures 24 and 25 show the experimental results of two-mass model using the proposed method, where the condition (R = 1.07, K_{sn} = 18.5 Nm/rad) is same as the above simulation results shown in Figures 17 and 18. From these figures, it is observed that the resonance vibrations between the motor and the load angular speed (ω_M and ω_L) have been suppressed very well. Furthermore, after inputting disturbance, it can be seen that the angular speeds immediately have followed the reference speed ω_{ref} without resonance vibrations. Hence, the validity of the control system, which consists of the proposed method, can be confirmed.

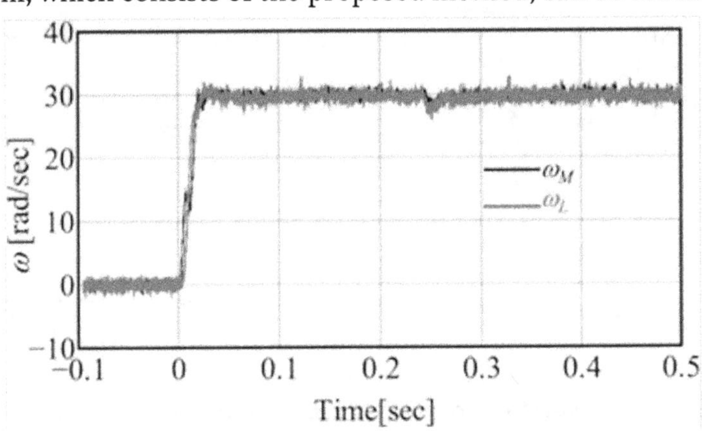

Figure 24. Experimental results for ω_M and ω_L obtained using the proposed method (two-mass, R = 1.07, K_{sn} = 18.5 Nm/rad).

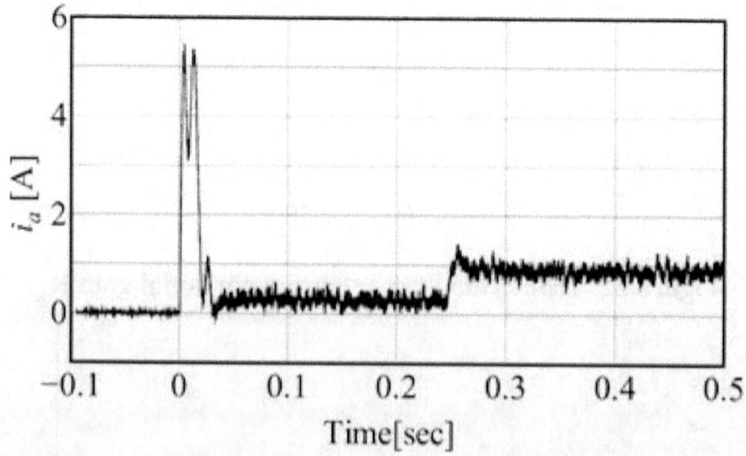

Figure 25. Experimental results for i_a obtained using the proposed method (two-mass, R = 1.07, K_s = 18.5 Nm/rad).

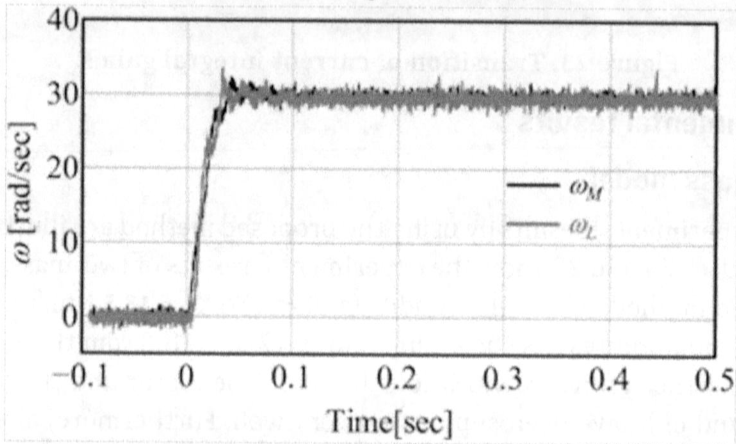

Figure 26. Robustness verification results (two-mass, R = 0.42, K_{sn} = 18.5 Nm/rad).

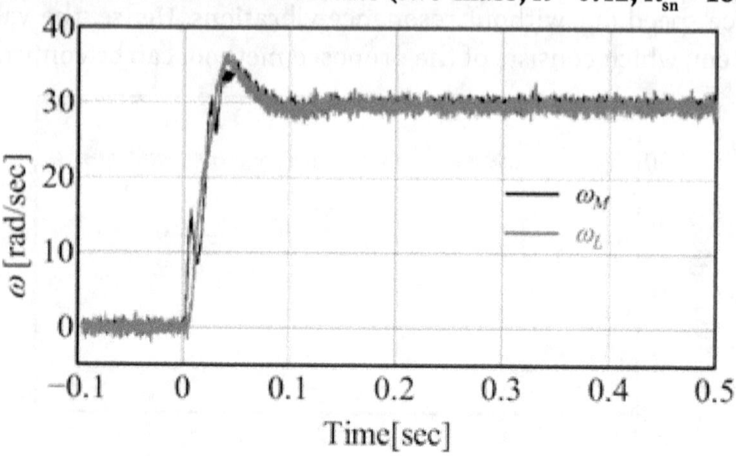

Figure 27. Robustness verification results (two-mass, R = 2.67, K_{sn} = 18.5 Nm/rad).

5. EFFECTS OF PARAMETER VARIATION

Next, it is described the effectiveness of robustness by the proposed design method. This section evaluates the robustness to variations in the ratio of inertia and the stiffness of the rigid shaft based on a nominal value.

Figures 26 and 27 show the experimental results of the motor and load angular speeds obtained for the inertia ratio variation when using the same controller gains that were designed using the proposed method, when R = [0.42, 2.65], where the disturbance torque input was skipped. From these figure, although it can be observed some overshoot and resonance vibration, the good results can be confirmed that were obtained for the design condition.

Figure 28 shows the experimental results of the motor and load angular speeds obtained for the stiffness of shaft variation using the same controller gains, when K_{sn} = 70.7. From this figure, it can be seen some resonance vibrations. However, the vibrations rapidly have been suppressed well.

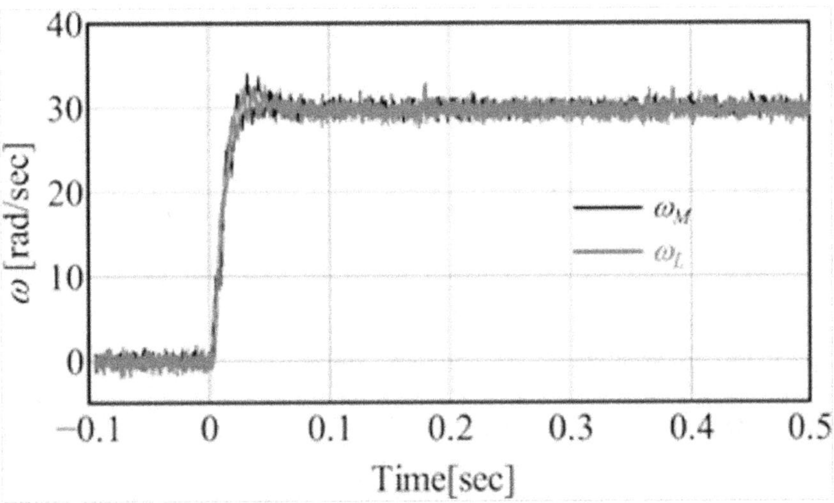

Figure 28. Robustness verification results (two-mass, R = 1.07, K_{sn} = 70.7 Nm/rad).

Similarly, Figure 29 shows the experimental results when K_{sn} = 3.1. As can be seen, the motor and load angular speeds oscillated and overshot. Therefore, if the stiffness of shaft of the experimental model is less than the design value, the settling time to suppress the resonance vibration becomes longer, although the proposed control system is not unstable. In addition, Figure 30 shows the experimental result when the control parameter redesigned with the stiffness of shaft K_{sn} as the nominal value of experimental model. Good responses can be observed in this figure.

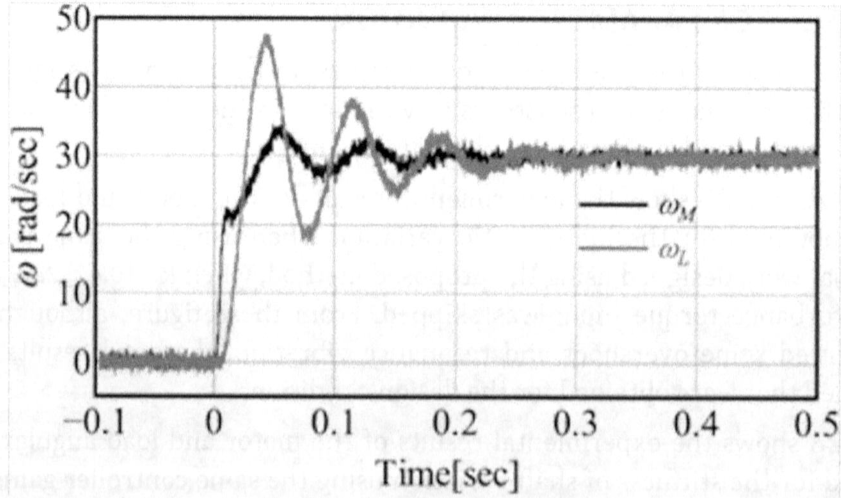

Figure 29. Robustness verification results (two-mass, R = 1.07, K_{sn} = 3.1 Nm/rad).

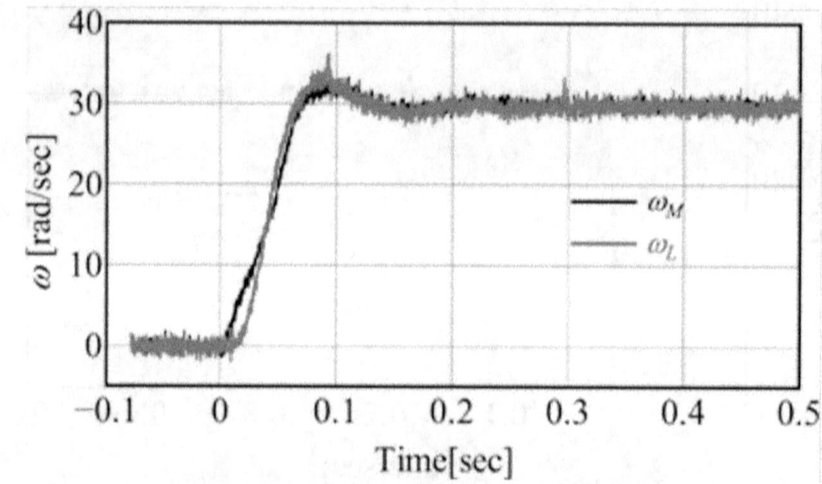

Figure 30. Experimental results for ω_M and ω_L redesigned using the proposed method (two-mass, R = 1.07, K_{sn} = 3.1 Nm/rad).

Furthermore, the proposed fuzzy control system is applied to a three-mass resonance model. Figure 31 shows the experimental results of the motor and load angular speeds when using the same controller gains designed for two-mass model (R = 1.07, K_{sn} = 18.5 Nm/rad, where the nominal parameters of the three-mass experimental setup are $J_{Mn} = 2.774 \times 10^{-4}$ kgm^2, $J_{Ln} = 2.940 \times 10^{-4}$ kgm^2, K_{s1n} = 18.5 Nm/rad, K_{s2n} = 18.5 Nm/rad. From this figure, the effectiveness of the proposed method can be confirmed in a similar manner to the two-mass model case.

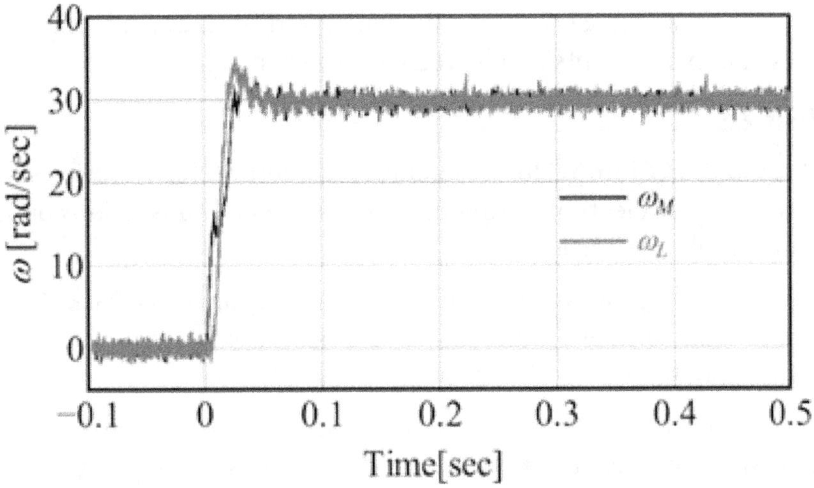

Figure 31. Robustness verification results (three-mass, J_{Mn} = 2.774 × 10^{-4} kgm^2, J_{cn} = 1.112 × 10^{-4} kgm^2, J_{Ln} = 2.940 × 10^{-4} kgm^2, K_{s1n} = 18.5 Nm/rad, K_{s2n} = 18.5 Nm/rad).

6. CONCISE SUMMARY

This chapter presents a speed control method designed to mitigate resonance vibrations in multi-inertial models, namely in two-mass and three-mass systems. The controller has been constructed with the digital fuzzy controller for speed control and the digital PI controller for current control. In the control system, only motor side state variables have been used for controlling the resonance system. This chapter employed the DE to ascertain these five controller parameters. The validity of the controller design, robustness, and control efficacy of the suggested method have been confirmed through simulations and tests utilizing the test setup.

AT A GLANCE

The regulation of vibration suppression in mechanical systems is a critical technique for achieving high precision, rapid response, and energy efficiency. The mechanical system is typically represented as a multi-mass resonance system, with vibration suppression control implemented. This chapter introduces an innovative controller design methodology for the speed control system aimed at mitigating resonance vibrations in both two-mass and three-mass resonance systems. The target systems comprise a motor, finite rigid shafts, and loads. The control system consists of a speed fuzzy controller and a proportional-integral (PI) current controller to realize exact speed and torque response. To implement the experimental system, it is regarded as a digital control system. This chapter use differential evolution (DE) to identify five optimal controller parameters: three scaling factors for the fuzzy controller and two gains for the PI current controller. This chapter conclusively demonstrated the efficacy of suppressing resonant vibrations and the robustness of the suggested strategy through computer simulations and experiments conducted with the test rig.

REFERENCES

1. Hori Y, Sawada H, Chun Y. Slow resonance ratio control for vibration suppression and disturbance rejection in torsional system. IEEE Transactions on Industrial Electronics. 1999;46(1):162–168
2. Kwang-Ho Yoon, Jong-Kwang Lee, Ki-Ho Kim, Byung-Suk Park, Ji-Sup Yoon. Hybrid robust controller design for a two mass system with disturbance compensation. In: Proc. ICCAS 2008, pp. 1367–1372
3. Szabat K, Kowalska TO. Vibration suppression in a two-mass drives system using PI speed controller and additional feedbacks – comparative study. IEEE Transactions on Industrial Electronics, 2007;54(2):1193–1206
4. Ikeda H, Hanamoto T, Tsuji T, Tanaka Y. Position control of 2-mass systems with speed minor loop designed by pole placement method. IEEJ Transactions on Industry Applications. 1999;119-D(4):544–545
5. Zhang G, Furusho J. Control of three-inertia system by PI/PID control. Transactions of IEE Japan. 1999;119-D(11):1386–1392
6. Eker I, Vural M. Experimental online identification of a three-mass mechanical system. Proceedings of 2003 IEEE Conference. 2003;1:60–65
7. Ikeda H, Hanamoto T, Tsuji T. Design of multi-inertia digital speed control system using Taguchi method. In: Proceedings of ICEM 2008, Paper ID 1167, PB.3.9; 2008. pp. 1–6
8. Ikeda H, Hanamoto T, Tsuji T, Tomizuka M. Design of vibration suppression controller for 3-inertia systems using Taguchi method. In: Proceedings of SPEEDAM 2006, Mechatronic Systems, S10-19 to S10-25; 2006.
9. Ikeda H, Hanamoto T. Design of vibration suppression controller for 2-inertia system by fictitious reference iterative tuning. In: Proceedings of ICEE 2015, ICEE15A-123; 2015. p. 6
10. Ikeda H, Ajishi H, Hanamoto T. Application of fictitious reference iterative tuning to vibration suppression controller for 2-inertia resonance system. In: Proceedings of IECON 2015, TS-48, YF-008451; 2015. pp. 1825–1830
11. Malhotra R, Kaur T, Deol GS. DC motor control using fuzzy logic controller. International Journal of AEST. 2011;8(2):291–296
12. Chakravorty J, Sharma R. Fuzzy logic based method of speed control of DC motor. International Journal of ETAE. 2013;3(4):64–66

13. Ikeda H, Hanamoto T. Fuzzy controller of multi-inertia resonance system designed by differential evolution. In: Proceedings of ICEMS 2013, MC-1883; 2013. pp. 2291–2295
14. Ikeda H, Hanamoto T. Fuzzy controller of three-inertia resonance system designed by differential evolution. Journal of International Conference on Electrical Machines and Systems. 2014;3(2);184–189
15. Yadav JS, Patidar NP, Singhai J. Controller design of discrete system by order reduction technique employing differential evolution optimization algorithm. International Journal of IMS. 2010;6(1):43–49
16. Yamaguchi S. An automatic control parameters tuning method for differential evolution. Transactions of IEE Japan. 2008;128-C(11):1696–1703
17. Brest J, et al. Self-adapting control parameters in differential evolution: A comparative study on numerical benchmark problems. IEEE Transactions on Evolutionary Computation. 2006;10(6):646–657
18. Ikeda H, Hanamoto T. Design of m-IPD controller of multi-inertia system using differential evolution. In: Proceedings of IPEC-Hiroshima 2014 – ECCE ASIA, 21J1-2; 2014. pp. 2476–2482
19. Myway Plus Corporation, https://www.mayway.co.jp/, Yokohama, Japan.
20. The Mathworks, https://www.mathworks.com, Massachusetts, U.S.A.

20 Chapter

DESIGN AND STABILITY ASSESSMENT OF A FUZZY-BASED ADAPTIVE CONTROLLER FOR WASTEWATER TREATMENT FACILITIES

1. INTRODUCTION: AN OVERVIEW

This chapter addresses the issue of tracking the dissolved oxygen reference trajectory in an aerobic reactor for nutrient removal, utilizing a direct model reference adaptive controller at the activated sludge wastewater treatment plant (WWTP). The reference trajectory is supplied in real-time by the higher control layer of the comprehensive control system. The controller architecture employs several time scales for the internal dynamics of dissolved oxygen and for disturbance inputs. This chapter presents two types of adaptive control: Direct Model Reference Adaptive Control (DMRAC) and a fuzzy logic variant of DMRAC with a two-level control system.

2. ADAPTIVE CONTROL

Adaptive control fundamentally consists of two primary types. The first is termed model reference adaptive control (MRAC) mode, while the second is designated self-tuning mode. An adaptive control characteristic is that the control parameters are flexible and are changed in real-time based on the system's signals.

2.1. Model reference adaptive control

A model reference adaptive control can be divided into four parts, such as plant, reference model, control law and controller. A plant includes unknown parameters. The reference model is described as control system output. The closed-loop control law is adjusting mechanism for adjustable control parameters. The controller updates the adjustable parameter with time-varying control system.

The plant is supposed to exist with known system structure, but plant parameters are unknown in the real. The structure of the dynamic equation is known with some unknown parameters in the nonlinear plants. The number of poles/zero are supposed to be known with unknown location poles/zero.

The reference model is used in order to obtain assignation ideal response of adaptive control system to control output. For the adaptive control system, that mean it is supply the ideal response by adjusting ideal plant parameters in the adaptation mechanism. To design the adaptive control system, first step is choice of the reference model. It is needed to meet two following clauses.

- The reference model should satisfy the performance of adaptive control system such as rise time, overshoot and settling time.
- The ideal plant parameter should be implemented by the adaptive control system.

The controllers are composition of several adjustable parameters. This implies that the controllers are distribution signal to each adjustable parameter with online update. The controller needs to have good tracking performance which means it can achieve tracking convergence behaviour. To design controller, two conditions need to be considered.

- If the plant parameters are known, then the plant out should track model reference trajectory by relevant controller parameters.
- If the plant parameters are unknown, then the plant out should track model reference trajectory by adjusting the controller parameters.

The linearly parameterized that mean is the control law with linear term of the adjustable parameters. To guarantee stability and tracking performance, adaptive control design is used by linear parameterization of the controller.

Adjusting parameters in adaptive control law is call adaptation mechanism. In the MRAC systems, the adaptation law is used in order to search the plant parameters, therefore the plant out can track set-point (model reference) with good performance by adaptive controller. The difference between ideal adaptive control parameter and real plant parameter is call tracking error. The tracking error converge to zero that implies that adaptive control system is stable.

2.2. Self-tuning model reference adaptive control

The control parameters estimate plant unknown parameters in control system. If a plant parameter is unknown, then a parameter estimator provides estimation values to those plant unknown parameters. If a plant parameter is known, then control parameters would transmit plant parameters by on-line update on model reference. The estimator provides estimation control parameters with on-line update from model reference, it is called self-tuning controller. The self-tuning controller is estimation unknown parameter in the plant at the same time.

The self-turning MRAC manipulate processes:

- The estimator transfer estimated plant parameters to controller; therefore, it can compute the plant corresponding unknown parameters at the same time. The plant estimation parameters depend on the past plant input and output.
- Computes a control input and rely on control parameters and measured signal, and this control input rely on new plant output.
- The close-loop parameters and plant input are updated on-line with time-varying adaptive control system.

The estimation parameter can be taken from an ideal parameters and real parameters by plant input/plant output data that are updated on-line with time-varying adaptive control system. The error dynamic is described as the difference between ideal plant parameters and real plant parameters; this implies that if tracking errors converge to zero by adjusting parameters adaptation then plant output complete tracking reference model. It is purpose of self-turning adaptive control design.

The self-turning control includes two types of adaptive controllers, one is called Indirect Model Reference Adaptive Control (IMRAC) and the another one is called Direct Model Reference Adaptive Control. The plant unknown parameters are provided by adaptive controller estimation of those plant parameters. If the estimation plant parameters need transfer into controller parameters,furthermore control law parameters can influent plant unknown parameters. This implies that the control parameters can adjust plant unknown parameters with standard estimation approach. It is called IMRAC. On the other hand, if it does not need transfer process, this method is called DMRAC.

2.3. Direct model reference control design

The property of adaptive control is used for plant with unknown parameters; therefore, choosing the adaptive control law is more implicated in controller design. Since we mention before, adaptive control law produce controller parameters. Also the stability analysis for control system need to be considered in controller design. In this chapter, we used Lyapunov theory to analyse control system whether stable or unstable. The process of adaptive control design includes three steps. The first step is choosing control law (include plant variable parameters). The second step is choosing adaptation law. The final step is stability analyses to guarantee convergence of control system.

3. DMRAC WITH INPUT SATURATION APPLY ON WWTP

3.1. Introduction

An activated sludge wastewater treatment plant (WWTP) is a complex nonlinear system due to multiple time scale and unmeasurable state variables. In addition, it has time-varying input disturbances and saturation during the WWTP operation; hence, the hierarchical structures which were considered in Refs. [1, 2]. The two-level controller of tracking prescribed a concentration of the dissolved oxygen (DO) trajectory, while the reference of concentration dissolved oxygen (DO^{ref}) was developed in Refs. [3, 4]. The activated sludge plant contained two main components, such as bioreactor and settler as illustrated in Figure 1.

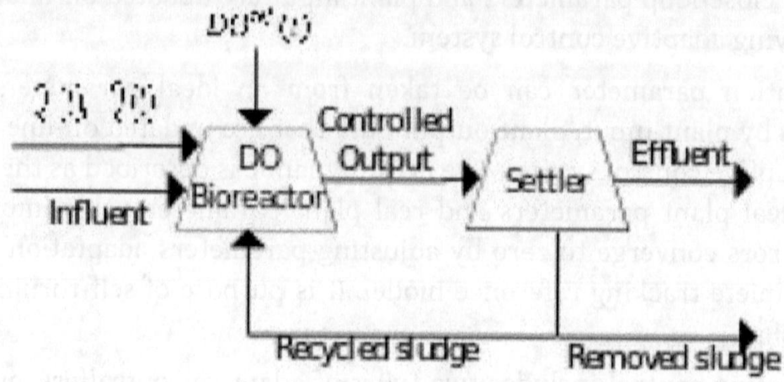

Figure 1. Structure of WWTP for nutrient removal.

The microorganism produced the biomass to nutrient removal in the bioreactor. The concentration of dissolved oxygen control is an important state parameter that feeds the microorganisms. The concentration of DO control was considered in Ref. [5]. The upper level controller produced airflow $Q_{air}{}^{ref}(t)$ into the aerobic biological reactor zone. The lower level controller produced the concentration of DO to track the $Q_{air}{}^{ref}(t)$ set-point trajectory. The airflow is the control input, and the concentration of DO is the control output.

The dissolved oxygen reference trajectory $DO^{ref}(t)$ set-point was optimized by the upper control layer which was the medium control layer in overall WWTP [1]. The clean water came out from the settler after being separated from the biomass and sludge. The concentration of substrate and biomass were unmeasurable state variables; hence, they were not able to be on-line updated. The upper layer control with input saturation was presented in Ref. [6]. The saturation function was assumed considering that aeration system controller was ideal and thus the airflow was equal to airflow reference. However, the physical modelling of the wastewater treatment plant was used to design the controller. In this chapter, we consider the upper level controller with input saturation by designing the new direct model reference adaptive control (DMRAC).

3.2. Problem statement

A mathematical model of the WWTP is based on the mass balance equations, which are illustrated in Figure 1. Hence, they represent the plant variables to produce the model in the state-space format [8]:

$$\frac{dX}{dt} = \mu(t)X(t) - D(t)(1+r)X + rD(t)X_r(t) \tag{1}$$

$$\frac{dS}{dt} = -\frac{\mu(t)X(t)}{Y} - D(t)(1+r)S(t) + D(t)S_{in}(t) \tag{2}$$

$$\frac{dDO}{dt} = -\frac{K_0\mu(t)X(t)}{Y} - D(t)(1+r)DO(t) + k_{La}(Q_{air}(t))(DO_{max} - DO(t)) + D(t)DO_{in}(t) \tag{3}$$

$$\frac{dX_r}{dt} = D(t)(1+r)X(t) - D(t)(\beta + r)X_r(t) \tag{4}$$

$$\mu(t) = \mu_{max}\frac{S(t)}{K_s + S(t)} \times \frac{DO(t)}{K_{DO} + DO(t)} \tag{5}$$

where

$$D(t) = \frac{Q_{in}}{V_a}; \quad r = \frac{Q_r}{Q_{in}}; \quad \beta = \frac{Q_w}{Q_{in}} \tag{6}$$

$X(t)$, $S(t)$, DO_{max}, $X_r(t)$, $D(t)$, $S_{in}(t)$, $DO_{in}(t)$, Y, $\mu(t)$, μ_{max}, K_S, K_{DO}, $Q_{air}(t)$, K_0, r, β, $Q_{in}(t)$, $Q_r(t)$, $Q_w(t)$, V_a are biomass concentration, substrate concentration, maximum dissolved concentration, recycled biomass concentration, dilution rate, substrate concentration in the influent, dissolved oxygen concentration in the influent, biomass yield factor, biomass growth rate, maximum specific growth rate, affinity constant, saturation constant, aeration rate, model constant, recycled sludge rate, removed sludge rate, influent flow rate, effluent flow rate, recycled flow rate, waste flow rate and aerator volume, respectively.

The function $k_{La}(Q_{air}(t))$ is the oxygen transfer, which depends on the aeration actuating system and sludge conditions [4]. In this chapter, it is assumed that

$$k_{La}(t) = \alpha Q_{air}(t) + \delta \tag{7}$$

where α and δ are two known constant values relating to oxygen transfer.

As only the DO output is considered in this chapter, the model is sufficiently accurate. Otherwise, more detailed model, for example, the ASM3, than it should be utilized as it has been done in Ref. [7]. In Figure 2, the detailed structure of the activated sludge WWTP for nutrient removal with the airflow actuator is illustrated. The actuator dynamics are described by a complex hybrid model. The output of aeration control system airflow output $Q_{air}(t)$ needs to follow the reference airflow input $Q_{air}^{ref}(t)$, which was described in Ref. [8]. The plant input with time-varying disturbances are dissolved oxygen concentration in the fluent $DO_{in}(t)$, the substrate concentration in the influent $S_{in}(t)$ and influent flow rate $Q_{in}(t)$. The controller needs to have high performance to enables the airflow output to track the reference airflow input.

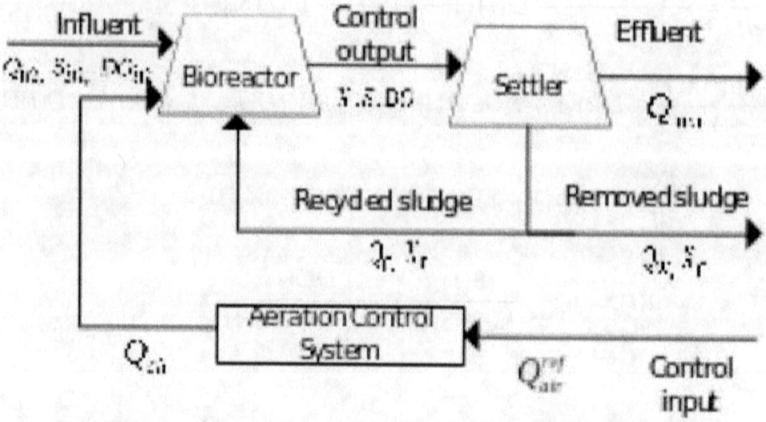

Figure 2. Structure of wastewater treatment plant.

The structure of *DO* control system for nutrient removal with input saturation at activated sludge WWTP is illustrated in Figure 3.

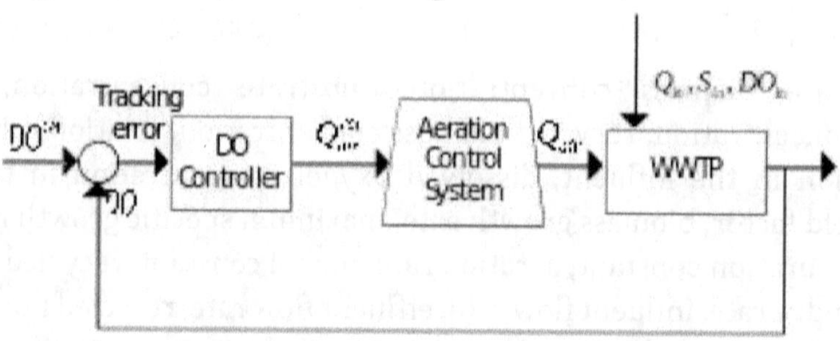

Figure 3. Structure of DO control system with input saturation.

The aeration controller designed in this chapter considers the aeration control system as the input $Q_{air}(t)$ with input saturation to achieve $Q_{air}^{ref}(t)=Q_{air}(t)$. This is the main difference in comparison with Ref. [8]. If the plant dynamics have several serially coupled reactors, the decentralized controller needs to consider the input saturation [6]. Previous papers [1–3] considered the two-level controller to remove the nutrient in the activated sludge WWTP. The upper layer controller generated the expected $Q_{air}^{ref}(t)$ into each of the aerobic biological reactor zones. The input of the lower layer controller was $Q_{air}(t)$, which needs to track prescribed upper layer output $Q_{air}^{ref}(t)$ for each of the reactor zones. If the upper layer controller had an input saturation condition, it influenced global control system stability and performance. As the plant dynamics have very high order and nonlinear dynamics as in Eqs. (1)–(5). The fixed parameter linear controller could not continue to keep the expected performance under full range of operating conditions. This was verified in Ref. [4] by using fixed parameter PI controller in low layer control. The upper layer controller used a fuzzy supervised controller. It obtained the expected performance. In practice, if the disturbance of the input becomes large, fast varying and with saturation input, the PI controller becomes very complex. The DMRAC with input saturation in upper layer control is considered in this chapter, which is not based on previous papers [7]. The $D(t)$ of the DMRAC input-output model rearranges the state-space model from Eqs. (1)–(4). As the state variable are not measureable, the unknown quantities in this input-output model will integrate into one term known as respiration. The parameter adaptation laws of the adaptive controller enable the respiration to be estimated indirectly and automatically.

3.3. DMRAC design

The direct state-space model of WWTP is represented in Eqs. (1)–(5). The dynamics are uncertain and nonlinear. The state variables $X(t)$, $X(t)$, $S(t)$ are unmeasurable. The state variables $Q_{\text{air}}(t), Q_{\text{in}}(t), Q_{\text{w}}(t), DO(t)$ are measured by on-line updates. To design the direct model reference adaptive controller, we shall derive dissolved oxygen dynamics model in input-output format of first-order and with input disturbance and input saturation. We shall rewrite Eq. (3) by substituting Eqs. (5) and (7). The term $(t)DO_{\text{in}}(t)$ can be neglected in state-space model, since it is very small in comparison with other state variables. The dissolved oxygen input-output model (DOIOM) is derived as follows:

$$\frac{dDO}{dt} = -a_p(t)DO(t) - c_p(t)f(DO(t)) + b_p(t)Q_{\text{air}}(t) + d_p \tag{8}$$

where $a_p(t)$,$c_p(t)$,$b_p(t)$,d_p are DOIOM parameters and

$$\begin{aligned} a_p(t) &= \frac{Q_{\text{in}}(1+r)}{V_a} + \delta \\ c_p(t) &= \frac{K_0 X(t)}{Y} \frac{\mu_{\max} S(t)}{(K_S + S(t))} \\ f(DO(t)) &= \frac{DO(t)}{(K_{DO} + DO(t))} \\ b_p(t) &= \alpha(DO_{\max} - DO(t)) \\ d_p &= \delta DO_{\max} \end{aligned} \tag{9}$$

The parameters $a(t)$ and $c_p(t)$ are slowly varying and unknown. The parameters $a(t)$ is dependent on upper control layer which operates in the time scale and is slower in comparison with $DO(t)$ control time scale (6) and (9). The parameter $c(t)$ is dependent on X and S, and is slower in comparison with $DO(t)$ (1) and (15). The parameter $b(t)$ is dependent on the fast internal dynamics of $DO(t)$ time scale (9). Hence, $D(t)$ is fast varying and known. The parameter d_p is slowly varying and known. The model reference dynamics (MRD) generate achieved $D(t)$ dynamics.

The $D(t)$ tracks the prescribed dissolved oxygen trajectory $DO_{ref}(t)$ by the controller. The MRD equation is as follows:

$$\frac{dDO_{m,\mathrm{ref}}}{dt} = -a_m DO_{m,\mathrm{ref}}(t) + b_m DO^{\mathrm{ref}}(t) \tag{10}$$

where $DO_{m.ref}(t)$ is the reference dynamics output and the parameters a_m and b_m are constant.

The $D(t)$ dynamics SISO input-output model with input saturation yields is as follows:

$$\frac{dDO}{dt} = -a_p(t)DO(t) - c_p(t)f(DO(t)) + b_p(t)W(t) + d_p \tag{11}$$

where $W(t)$ is assumed saturation control input with constraint (SCIC) $W(t)$=saturation($Q_{air}(t)$),

$$W(t) = \begin{cases} Q^{L}_{air}(t), & \text{if} \quad Q_{air}(t) > Q^{L}_{air}(t) \\ Q_{air}(t), & \text{if} \quad Q^{L}_{air}(t) \leq Q_{air}(t) \leq Q^{U}_{air}(t) \\ Q^{U}_{air}(t), & \text{if} \quad Q_{air}(t) < Q^{U}_{air}(t) \end{cases} \tag{12}$$

where $Q^{L}_{air}(t)$ and $Q^{U}_{air}(t)$ are actuator lower and upper constant bounds. If under the input saturation condition, the filter tracking error (t) is increasing, then global stability will be unstable for the control system. The model reference adaptive control law without input saturation was proposed in Ref. [5]. This motivates us to develop a new control law in comparison with Ref. [8] by explicitly considering the influence of the actuator input saturation nonlinearity.

The filter tracking error is applied as follows:

$$n(t) = e(t) - \lambda(t) \tag{13}$$

where $e(t)$ represents the difference between $DO(t)$ and $DO_{m,ref}(t)$ with on-line update.

$$e(t) = DO(t) - DO_{\mathrm{m,\,ref}}(t) \tag{14}$$

We define the auxiliary signal as follows:

$$\frac{d\lambda}{dt} = b_p \Delta Q_{\mathrm{air}}(t) - \Phi\lambda(t) \tag{15}$$

where Φ is small position constant parameter. The parameter $\Delta Q_{\mathrm{air}}(t)$ is the difference between SCIC $W(t)$ and control input $Qair(t)$.

The affine MRAC law is applied as follows:

$$\begin{aligned} W(t) = & \frac{1}{b_p} a_{DO}(t) DO(t) + \frac{1}{b_p} a_f(t) f(DO(t)) \\ & + \frac{1}{b_p} a_{DO^{\mathrm{ref}}}(t) DO^{\mathrm{ref}}(t) - \frac{1}{b_p} d_p \\ & - \frac{1}{b_p} \Phi(t)(DO(t) - DO_{\mathrm{m,\,ref}}(t)) \end{aligned} \tag{16}$$

The MRD in Eq. (10) has linear dynamics. The terms $\frac{1}{b_p} a_f(t) f(DO(t))$ and $-\frac{1}{b_p} d_p$ in Eq. (16) can be cancelled by closed-loop with an impact of the nonlinear and additive terms in Eq. (8). The control input saturation is described by the last term in Eq. (16) which is retained in DOIOM. The fifth term in Eq. (16) is updated on-line. The parameters $a_{\mathrm{DO}}(t)$, $a_{\mathrm{f}}(t)$ and $a_{\mathrm{DO}}{}^{\mathrm{ref}}(t)$ are updated by adaptive control law. The MRD is achieved in the closed-loop for ideal parameter. Closing the loop by Eq. (16) yields:

$$\begin{aligned} \frac{dDO}{dt} = & -(a_p(t) - a_{DO}(t)) DO(t) \\ & - (c_p(t) - a_f(t)) f(DO(t)) \\ & + a_{DO^{\mathrm{ref}}}(t) DO^{\mathrm{ref}}(t) \\ & - \Phi(DO(t) - DO_{\mathrm{m}}(t)) \end{aligned} \tag{17}$$

and

$$-(a_{\mathrm{p}}(t) - \hat{a}_{\mathrm{DO}}(t)) = -a_{\mathrm{m}} \tag{18}$$

$$-(c_{\mathrm{p}}(t) - \hat{a}_{\mathrm{f}}(t)) = 0 \tag{19}$$

$$\hat{a}_{DO^{\mathrm{ref}}}(t) = b_{\mathrm{m}} \tag{20}$$

$$-\Phi(DO(t) - DO_{\mathrm{m}}(t)) = -\Phi e(t) \tag{21}$$

where $\hat{a}_{\mathrm{DO}}(t)$, $\hat{a}_f(t)$ and $\hat{a}_{\mathrm{DO^{ref}}}(t)$ are the ideal parameters, which can now be obtained as follows:

$$\hat{a}_{\mathrm{DO}}(t) = -a_m + a_p(t) \tag{22}$$

$$\hat{a}_{\mathrm{f}}(t) = c_{\mathrm{p}}(t) \tag{23}$$

$$\hat{a}_{\mathrm{DO^{ref}}}(t) = b_{\mathrm{m}} \tag{24}$$

The parameter adaption laws which can achieve stability for a DMRAC system with SISO-controlled plant were derived in Ref. [6]. It was a first-order dynamic system composed of the mixed linear uncertainty in constant but not time-varying parameters and additive structured nonlinear. Applying these laws to Eq. (8) yields:

$$\frac{da_{\mathrm{DO}}}{dt} = -\gamma_1 e(t) DO(t) \tag{25}$$

$$\frac{da_f}{dt} = -\gamma_2 e(t) f(DO(t)) \tag{26}$$

$$\frac{da_{\mathrm{DO^{ref}}}}{dt} = -\gamma_3 e(t) DO^{\mathrm{ref}}(t) \tag{27}$$

γ_1, γ_2 and γ_3 are small enough positive constants representing the parameter adaptation gains which are used to control the parameter adaptation rates. In order to guarantee the stability of the closed-loop system, these rates shall be harmonized with the process variable rates. The DMRAC structure is presented in Figure 4.

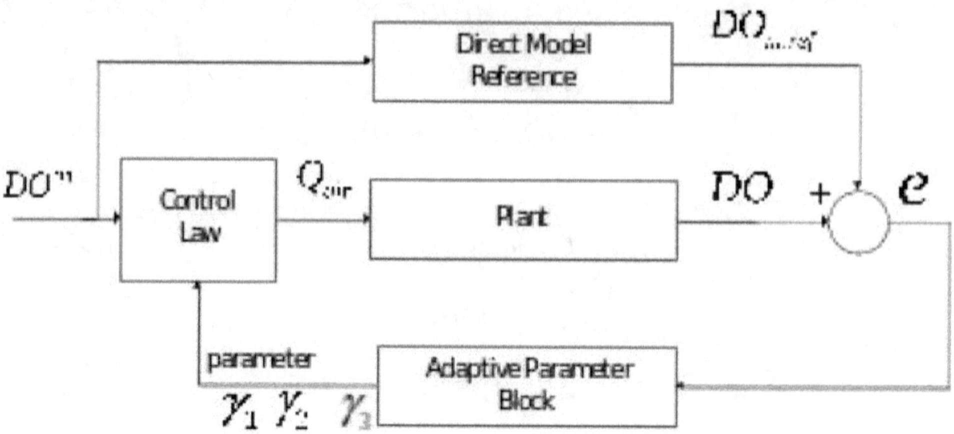

Figure 4. DMRAC structure.

3.4. Stability analysis

The estimated parameters $a_{DO}(t)$, $a_f(t)$ and $a_{DO}{}^{ref}(t)$ are updated on-line by the adaptation laws (25)-(27). The error between estimated parameter and ideal parameters are denoted as $\Delta a_{DO}(t)$, $\Delta a_f(t)$ and $\Delta a_{DO}{}^{ref}(t)$:

$$\Delta a_{DO}(t) = a_{DO}(t) - \hat{a}_{DO}(t) \tag{28}$$

$$\Delta a_f(t) = a_f(t) - \hat{a}_f(t) \tag{29}$$

$$\Delta a_{DO^{ref}}(t) = a_{DO^{ref}}(t) - \hat{a}_{DO^{ref}}(t) \tag{30}$$

Considering the following Lyapunov function:

$$V(t) = \frac{1}{2}n^2(t) + \frac{1}{2}\Delta a_{DO}{}^2(t) + \frac{1}{2}\Delta a_f{}^2(t) + \frac{1}{2}\Delta a_{DO^{ref}}{}^2(t) \tag{31}$$

Hence,

$$\begin{aligned} \frac{dV(t)}{dt} &= n(t)n(t) + (a_{DO}(t) - \hat{a}_{DO}(t))(a_{DO}(t)) - a_{DO}(t)) = \frac{1}{\gamma 1} \\ &+ (a_f(t) - a_f(t))(\hat{a}_f(t) - a_f(t))\frac{1}{\gamma_2} \\ &+ (a_{DO^{ref}}(t)) - \hat{a}_{DO^{ref}}(t))(a_{DO^{ref}}(t) - a_{DO^{ref}}(t))\frac{1}{\gamma_3} \end{aligned} \tag{32}$$

It follows from Eqs. (13), (15), (10) and (17) that

$$\begin{aligned} n(t) = (a_{DO}(t) - \hat{a}_{DO}(t)DO(t)) \\ + (a_f(t) - a_f(t))f(DO(t)) \\ + (a_{DO^{ref}}(t)) - \hat{a}_{DO^{ref}}(t))DO_{m,ref}(t) \\ - \Phi n(t) \end{aligned} \tag{33}$$

Applying Eqs. (33), (25) (26) and (27) into (32), yields:

$$\frac{dV(t)}{dt} = -\Phi n^2(t) \tag{34}$$

Summarizing the result of the Lyapunov function (RLF) with input saturation closed-loop $D(t)$ dynamic system, it can be seen that RLF progressively approaches zero. If the RLF approaches zero, then filter tracking error approaches zero, when time approaches infinity, Φ is small position constant and $n^2(t)$ is positive variable. To find the bounded saturation control input by limiting error between control output and dissolved oxygen trajectory reference with auxiliary signal, yields:

$$\text{limit}\{e(t) - \Lambda(t)\} \leq 0 \tag{35}$$

If time approaches infinity, the (t) approaches zero. To confirm whether the auxiliary signal is negative or positive when time goes to infinity by considering the following Lyapunov function, yields:

$$V_\Lambda(t) = \frac{1}{2}\Lambda(t) \tag{36}$$

Hence

$$\frac{dV_\Lambda(t)}{dt} = \Lambda(t)\Lambda(t) \tag{37}$$

Applying Eq. (12) into Eq. (37), yields:

$$\frac{dV_\Lambda(t)}{dt} = \Lambda(t)b_p(t)\Delta Q_{air}(t) - \Lambda(t)\Lambda(t)\Phi \tag{38}$$

$$\frac{dV_\Lambda(t)}{dt} = \Lambda(t)$$

Assume term

$$b_p(t)\Delta Q_{air}(t) = \Delta Q_{air}^{plant}(t) \tag{39}$$

It follows Eqs. (38) and (39) so that

$$\frac{dV_\Lambda(t)}{dt} = -\Lambda^2(t)\Phi + \frac{1}{2}\Lambda^2(t) + \frac{1}{2}\Delta Q_{air}^{plant^2}(t) \tag{40}$$

Now we assume

$$\Phi = \frac{1}{2}a_0, \quad a_0 < 0, \tag{41}$$

where a_0 is small positive constant value. Applying Eq. (41) into Eq. (40) yields:

$$\frac{dV_\lambda(t)}{dt} = -2V_\lambda(t)a_0 + \frac{1}{2}\Delta Q_{air}^{plant^2}(t) \tag{42}$$

The second term $\frac{1}{2}\Delta Q_{air}^{plant^2}(t)$ is bounded. To find bound of first term by integral, yields:

$$V_\lambda(t) = \frac{\Delta Q_{air}^{plant^2}(t)}{4a_0} + (V_{\lambda.0}(t) + \frac{\Delta Q_{air}^{plant^2}}{4a_0})e(t)^{-2a_0} \tag{43}$$

where the $V_{\lambda.0}(t)$ is the initial value of the $V\lambda(t)$. As time approaches infinity and a_0 is large enough for a positive value, then second term is equal to zero in (37).

$$V_\lambda(t) = \frac{\Delta Q_{air}^{plant^2}(t)}{4a_0} \tag{44}$$

It follows from Eqs. (36) and (44) and the limitation $V\lambda(t)$ as negative or zero by squared.

$$V_\lambda{}^2(t) = \frac{\Delta Q_{air}^{plant^2}(t)}{4a_0}$$

$$\frac{1}{2}\lambda^2(t) = \frac{\Delta Q_{air}^{plant^2}(t)}{4a_0} \tag{45}$$

$$\lambda(t) = \sqrt{\frac{\Delta Q_{air}^{plant^2}(t)}{2a_0}}$$

It follows from Eqs. (35) and (41) that

$$e(t) \leq \sqrt{\frac{\Delta Q_{air}^{plant^2}(t)}{2a_0}} \tag{46}$$

If the value of a_0 is large enough, then the tracking error (t) is closer to zero. The control system will be more stable. Finally, the standard application of the Barbalat's lemma allows concluding the adaptive control system that achieves the asymptotic tracking of $DO_{ref}(t)$ under-bounded $a_{DO}(t)$, $\hat{a}_{\hat{D}\hat{O}}(t)$, $a_f(t)$, $a_f(t)$, $a_{DO^{ref}}(t)$, $a_{DO^{ref}}(t)$ if (a) the parameter in adaptive control law are close enough to the set-point in initial condition; (b) the parameter adaptation rates are positive small enough and (c) the saturation input is small enough, the control parameters bounded are stabilized

3.5. Simulation results

The simulation data are based on the real record. We assumed WWTP without disturbance, the very good $D(t)$ tracking model reference $DO_{ref}(t)$ performance has been shown in Figure 5. The real plant contained some disturbances such as effluent flow rate, recycled flow rate and waste flow rate. The controller that we have designed still indicated perfect tracking performance in Figure 6.

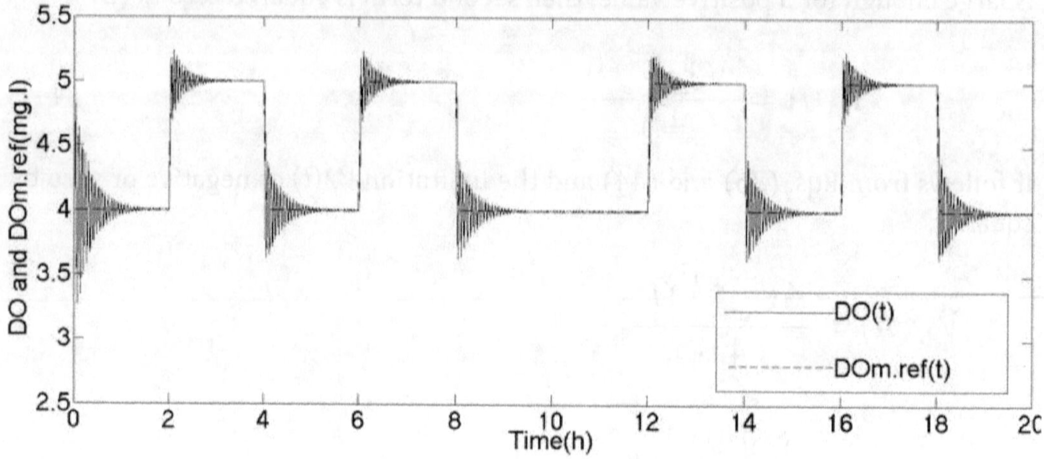

Figure 5. DO and DO^{ref} with input saturation.

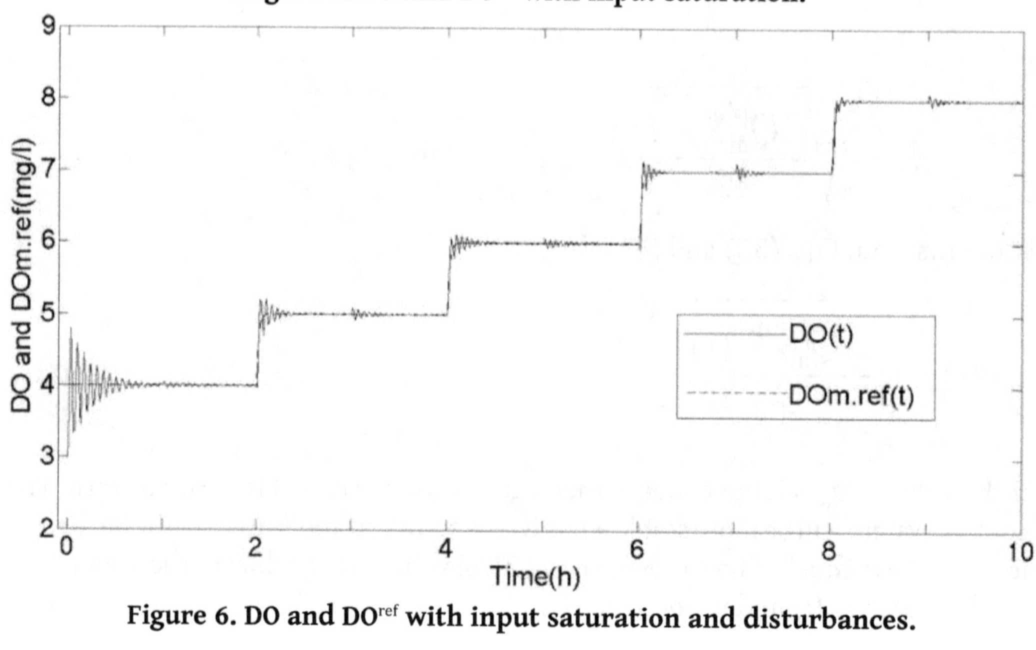

Figure 6. DO and DO^{ref} with input saturation and disturbances.

4. FUZZY SUPERVISOR BASED ON MULTIPLE DMRAC

4.1. Introduction

In this section, we consider that fuzzy control are based on multiple DMRAC. The fuzzy control represents upper level control and DMRAC represents lower level control. More detail information are described in the next section.

4.2. Problem statement

In Ref. [4] is descried two-level controller tracking previously set-point of DO trajectory in several serially coupled reactors for the nutrient removal served by one actuator system with several air blower at WWTP. The upper level control delivers airflow into each bioreactors to be bioreactor set-point trajectory close to ideal trajectory. The lower level control is used for the concentration of DO trajectory flowing the set-point. The structure of WWTP with coupled reactors is illustrated in Figure 7. The structure of WWTP is different from Section 5 that contains two bioreactors. The capacity limit is that total airflow $Q_{air}^{max}(t)$ should be small or equal to sum of all the adaptive control signals $Q_{air.k}(t)$.

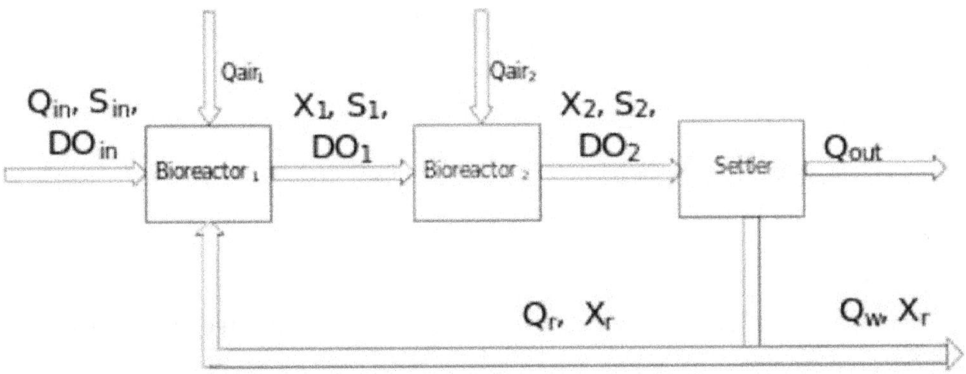

Figure 7. The structure of wastewater treatment plant for nutrient removal with coupled reactors.

$$\sum_{k=1}^{q} Q_{air.k}(t) \le Q_{air}^{max}(t) \tag{47}$$

where k is the number of bioreactor.

4.3. Controller design

4.3.1. Lower DMRAC design

As mentioned in Section 3, the process of DMRAC design is explained in detail; therefore, in this section we provide essential equations. The state-space format is same with last section for single reactor as in Eqs. (1)–(6).

- The dissolved oxygen input-output model (DOIOM) with coupled bioreactors is derived as follows:

$$\frac{dDO_i}{dt} = -a_{p.i}(t)DO_i(t) - c_{p.i}(t)f(DO_i(t)) + b_{p.i}(t)Q_{air.i}(t) + d_{p.i} \quad (48)$$

where $a_{p.i}(t), c_{p.i}(t), b_{p.i}(t), d_{p.i}$ are DOIOM parameters and

$$\begin{aligned} a_{p.i} &= \frac{Q_{in}(1+r_1)}{V_{a.i}} + \delta_i \\ c_{p.i} &= \frac{K_{0.i}X_i(t)}{Y_i}\frac{\mu_{max.i}S_i(t)}{K_{s.i}+S_i(t)} \\ b_{p.i} &= \alpha_i(DO_{max.i} - DO_i(t)) \\ d_{p.i} &= \delta_i DO_{max.i} \end{aligned} \quad (49)$$

where i=1,2

- The plant parameters status are exactly same with single reactor. The model reference dynamics equation is set as:

$$\frac{dDO_{m.ref.j}}{dt} = -a_{m.j}DO_{m.ref.j}(t) + b_{m.j}DO_j^{ref}(t) \quad (50)$$

where $j = 1, 2$

- The affine model reference adaptive control law is applied as follows:

$$Q_{air.k}(t) = a_{DO.k}(t)DO_k(t) + a_{f.k}(t)f(DO_k(t)) + a_{DO_k^{ref}}(t)DO_k^{ref}(t) - \frac{\delta_k DO_{max.k}}{b_{p.k}(t)} \quad (51)$$

where $k = 1, 2$

- Model reference adaptive control law is used as:

$$\frac{da_{DO_n}}{dt} = -\gamma_{zone.z.1}e_l(t)DO_i(t) \quad (52)$$

$$\frac{da_{fn}}{dt} = -\gamma_{zone.z.2}e_l(t)f(DO_i(t)) \quad (53)$$

where $n = 1, 2$; $z = 1, 2$; $l = 1, 2$; $i = 1, 2$.

4.3.2. Fuzzy supervisor design

The purpose of fuzzy supervisor is to divide total airflow $Q_{air}^{max}(t)$ into two lower control signal, but those should satisfy capacity limit (47). Each of the bioreactors airflow restrict lower control output by MRAC. This implies that if fuzzy supervisor delivers airflow big enough then bioreactor output is more close to set-point trajectories (model reference). The error dynamic described each bioreactors output approaching uniform level. The fuzzy supervisor is designed as following:

4.3.2.1. Step 1: Fuzzification

Linguistic variable is at lower level for each DMRAC error dynamics. Those error dynamics are divided into three types such as small, medium and big by percentage of lower level error dynamics (54). Membership function used in this chapter are Sigmoidal condition (55) and Gauss condition (56).

$$V(t) = e_i(t)^{-1} g \sum_{i=2}^{q} e_i(t) \tag{54}$$

where $V(t)$ is percentage of lower level error dynamics for each airflow.

$e(t)$ is DMRAC error dynamics for each bioreactors.

$$\mu(v(t)) = \frac{1}{1 + \exp\left(-a(v(t)) - c\right)} \tag{55}$$

where $a,\ c$ are membership function shape parameters.

$$\mu(v(t)) = \begin{cases} \exp\left(-(v(t)) - c_1\right)^2/\sigma_1^2) \\ 1; \\ \exp\left(-(v(t)) - c_2\right)^2/\sigma_2^2) \end{cases} \tag{56}$$

where a, σ are membership function shape parameters.

4.3.2.2. Step 2: Fuzzy rule

4.3.2.2.1. First rule

If error dynamics is small and sum of level airflow is greater than total airflow,

Then bioreactor receives corresponding percentage of total airflow.

If (t) is small and $\sum_{i=1}^{q} Q_{air.i}(t) \geq Q_{air}^{max}(t)$

Then

$$Q_{air.i.2}^{supervisor}(t) = \frac{Q_{air}^{max}(t)}{\sum_{i=1}^{q} Q_{air.i}(t)} \times Q_{air.i}\, g10\% \tag{57}$$

4.3.2.2.2. Second rule

If error dynamic medium and sum of level airflow is greater than total airflow,
Then bioreactor receives corresponding percentage of total airflow.

If (t) is medium and $\sum_{i=1}^{q} Q_{air.i}(t) \geq Q_{air}^{max}(t)$
Then

$$Q_{air.1.2}^{supervisor}(t) = \frac{Q_{air}^{max}(t)}{\sum_{i=1}^{q} Q_{air.i}(t)} \times Q_{air.i}\, g30\% \tag{58}$$

4.3.2.2.3. Third rule

If error dynamic is big and sum of level airflow is greater than total airflow,
Then bioreactor receive corresponding percentage of total airflow.

If (t) is big and $\sum_{i=1}^{q} Q_{air.i}(t) \geq Q_{air}^{max}(t)$
Then

$$Q_{air}^{supervisor}(t) = \frac{Q_{air}^{max}(t)}{\sum_{i=1}^{q} Q_{air.i}(t)} \times Q_{air.i}\, g60\% \tag{59}$$

4.3.2.3. Step 3: Defuzzification

Each of the bioreactors obtain airflow by a fuzzy value.

5. CONCISE SUMMARY

In this chapter, we studied two alternative adaptive control. The initial adaptive control is implemented in a wastewater treatment plant with saturated control inputs. The second adaptive control elucidates the operation of upper-level fuzzy control, which is predicated on the lower-level DMRC implemented in the coupling bioreactors of the wastewater treatment plant (WWTP).

AT A GLANCE

This chapter discusses the design and stability evaluations of a direct model reference control system for a wastewater treatment plant. The objectives of controller design encompass input saturation management and a two-level control system featuring fuzzy supervisory control. The wastewater treatment plant is a highly uncertain nonlinear system with unknown parameters; hence, controller design occurs under these conditions.

REFERENCES

1. Brdys MA, Grochowski M, Gminski T, Konarczak K, Drewa M. Hierarchical predictive control of integrated wastewater treatment system. Control Engineering Practice. 2008;16(6):751–767.

2. Piotrowski R, Brdys MA, Konarczak K, Duzinkiewicz K, Chotkowski W. Hierarchical dissolved oxygen control for activated sludge processes. Control Engineering Practice. 2008;16:114–131.

3. Brdys MA, Diaz-Maiquez. Application of fuzzy model predictive control to the dissolved oxygen concentration tracking in an activated sludge process. In: Proceedings of the 15th IFAC World Congress; 21–26 July; Barcelona.

4. Brdys MA, Chotkowski W, Duzinkiewicz K, Konarczak K, Piotrowski R. Two-level dissolved oxygen control for activated sludge processes. In: Proceedings of the 15th IFAC World Congress; 21–26 July 2002; Barcelona.

5. Olsson G, Newell R. Wastewater Treatment System. Modelling, Diagnosis and Control. London, UK: IWA Publishing; 1999.

6. Zubowicz T, Brdys MA, Piotrowski R. Intelligent PI controller and its application to dissolved oxygen tracking problem. Journal of Automation Mobile Robotics & Intelligent Systems. 2010;4(3):16–24.

7. Li M, Brdys MA. Direct model reference adaptive control of nutrient removal at activated sludge wastewater treatment plant. In: The 20th International Conference on Methods and Models in Automation and Robotics; 24–27 August 2015; Miedzyzdroje, Poland.

8. Duzinkiewicz K, Brdys MA, Kurek W, Piotrowski R. Genetic hybrid predictive controller for optimized dissolved-oxygen tracking at lower control level. IEEE Transaction on Control System Technology. 2009;17:1183–1192.

21
Chapter

A FRAMEWORK FOR ASSESSING SOIL SUSCEPTIBILITY TO EROSION UTILIZING REMOTE SENSING DATA AND A FUZZY LOGIC SYSTEM

1. INTRODUCTION: AN OVERVIEW

Soil is regarded as a nonrenewable resource, vital for food security and a sustainable future. It is defined as the uppermost layer of the Earth's crust, composed of mineral particles, organic matter, water, air, and living organisms. Desertification linked to soil erosion processes is the primary focus for soil protection and preservation. Furthermore, desertification and land degradation refer to the decline in ecosystem productivity and are linked to the loss of vegetation cover. The United Nations initiative to combat desertification is essential since it addresses hunger and poverty, fosters stability, and enhances resilience to climate change in some of the world's most susceptible regions. Soil erosion, water retention capacity, salinity, sodicity, and nutrient depletion are prevalent markers employed to classify land degradation [1]. The formation of soil is a protracted process, with the development of a 30 cm layer requiring between 1,000 and 10,000 years. This landscape, shaped by soil degradation and the protracted process of development, offers few options for soil protection, which is vital for humanity. Soil vulnerability refers to the susceptibility of one or more ecological activities of the soil system to damage, including biomass production, filtration, buffering, transformation, genetic reserves, and the preservation of flora and animals. In other words, it is the susceptibility of soil to degradation processes. Soil erosion, driven by climate change, is the most significant degrading process globally associated with desertification. Climate change is a primary catalyst for ecosystem transformation in a desertified environment [6]. Vegetation is essential for soil protection. Furthermore, soil erosion is linked to biomass production, which plays a significant role in the future energy landscape [7]. Numerous variables influence soil erosion.

Nonetheless, vegetation and geography significantly influence soil vulnerability. The primary source of organic matter for soil, it safeguards against erosion caused by rain and wind. The second one assesses and enables the quantification of soil losses resulting from erosive processes and transport. Soils characterized by minimal vegetation cover and steep gradients are more susceptible to degradation processes. Consequently, it is essential to understand these aspects to assess soil vulnerability, so safeguarding it, preventing erosion, and preserving resources for sustainable future utilization. Numerous nations, including Cyprus, the Greek Islands, Spain, and Italy, have soil erosion, and there are significant challenges in accessing or acquiring soil data due to orographic conditions, accessibility issues, and other environmental concerns. It is estimated that around 115 million hectares, constituting 12% of Europe's total land area, are affected by erosion [9]. Furthermore, it is projected that in the Mediterranean region, which is particularly susceptible to climate change impacts, water erosion may result in a loss of 20 to 40 tons per hectare of soil following a single cloudburst, with extreme cases potentially exceeding 100 tons per hectare. Consequently, it is essential to employ approaches that enhance the generation and aggregation of spatial information. These techniques are encompassed under digital soil mapping (DSM), which consists of numerical models that infer the geographical and temporal fluctuations of soil types and attributes based on soil observations, knowledge, and associated environmental variables [12]. An overview of soil conditions and characteristics is essential for decision-makers and to enhance public access to soil data. Land and water resources are fundamental to agriculture and rural development, inherently connected to global issues of food insecurity, poverty, climate change adaptation and mitigation, as well as the degradation and depletion of natural resources impacting the livelihoods of millions of rural inhabitants worldwide. [13]. The application of straightforward models by administration for decision-making is crucial for resource conservation and the welfare of individuals. In this context, models utilizing remote sensing data and mathematical frameworks, such as those founded on fuzzy logic systems, can be advantageous. Models derived from fuzzy logic systems are associated with fuzzy set theory, which is fundamental to fuzzy logic. Zadeh established the theoretical foundation of this theory and articulated the notion of fuzzy sets as follows: "A fuzzy set A in X is defined by a membership function fA(x), which assigns to each point in X a real number within the interval [0, 1], where the value of fA(x) at x signifies the 'grade of membership' of x in A." The closer the value of fA(x) is to one, the greater the membership grade of x in A. A notable characteristic of fuzzy sets is their efficacy in addressing data uncertainty. A rigid classification of environmental and soil components is susceptible to subjectivity and the negligence of uncertainties, such as measurement errors, decimal choices, and data omissions. Fuzzy sets are employed for classifications in which the categories lack clearly defined boundaries [15]. This methodology is readily applicable to the analysis of environmental phenomena, including soil nutrient depletion, atmospheric pollution dispersion, and forest productivity, owing to the significant spatial variability and potential discontinuities of these phenomena. Fuzzy theory offers a robust mathematical foundation for comprehending decision-making challenges and formulating decision rules in multi-criteria assessment and integration. A diverse array of fuzzy logic methodologies has been established to enhance the concept of fuzzy sets. A fuzzy measure denotes any function that exhibits monotonicity concerning set membership [17]. The range of applicable functions is extensive. A fuzzy function is, in essence, a generalization of the classical function notion. A classical function f is a mapping from its domain D into a space S; the range of f, denoted f(D), is a subset of S. The various characteristics of the classical notion of a function may be seen as ambiguous [18]. Fuzzy measurements encompass Bayesian probabilities, beliefs and plausibilities from Dempster-Shafer theory, and membership grades of fuzzy sets, establishing a framework for techniques in uncertainty studies [19]. This approach integrates fuzzy sets with soil erosion to enhance decision-making through straightforward criteria applicable by administration. Desertification, climatic change, and soil fertility must be taken into account. This methodology is founded on the necessity for global comprehension of local processes and the capabilities of computers and remote sensing tools. Due to these environmental, social, and economic factors, a decision-making tool is essential. Decision theory pertains to the rationale employed in selecting among alternatives. The process of selecting among various possibilities is facilitated by a decision support system (DSS). The examination of decision support systems is a practical field that utilizes knowledge and particularly theories from various other areas [21]. The understanding of soil science and mathematical modeling facilitates the development of an efficient Decision Support System (DSS). A prevalent decision support system is the multi-criteria evaluation (MCE) grounded in the multi-criteria analysis (MCA) of the analytical hierarchy process (AHP) [22, 23]. The AHP analysis is founded on three fundamental concepts as outlined by Zorpas and Saranti [22]: the configuration of preferences, the decomposition of the problem into subproblems, and the pairwise evaluation of criteria/subcriteria against the provided alternative scenarios. The MCA [22, 23] delineates four primary stages: first, the identification of the problem (including potential subproblems) and the establishment of a hierarchical framework; second, the pairwise comparison of decision elements to generate normalized absolute scales, which are subsequently utilized as priorities; third, the management of the established priorities to address the problem; and finally, the assessment and evaluation of alternative scenarios to resolve the specific issue. This methodology can be employed by decision-makers, engineers, consultants, researchers, and federal and local authorities to assess or propose specific solutions to particular problems [21]. The MCE examines the integration of information from several criteria to create a unified assessment index utilizing discrete or continuous elements [21–23]. In environmental sciences, theme information aggregation methods are employed during the criteria combination process. Two aggregation approach lineages have been extensively employed for MEC [24]: (1) Boolean overlay, wherein all criteria are evaluated against suitability levels to generate Boolean maps utilizing logical operators such as union, intersection, or negation. and (2) weighted linear combination (WLC), in which continuous criteria are standardized—typically through a straightforward linear transformation—to a uniform numeric range and subsequently aggregated by weighted averaging. Both aggregation criteria exhibit considerable inflexibility due to their intrinsic logical aggregation (types of operators and characteristics). Aggregation operations on fuzzy sets are methods that aggregate many fuzzy sets according to specific rules and operators to yield a singular fuzzy set [25]. Fuzzy measurements offer a theoretical foundation for investigating and enhancing the comprehension of multi-criteria evaluation procedures and the development of novel aggregation operators [24]. Two traditions of logical operators have been extensively utilized for decades: (1) MIN and MAX operators for Boolean overlay and fuzzy membership aggregation, and (2) the averaging operator for weighted linear combination. Jiang and Eastman [24] proposed employing weighted linear combinations as a fuzzy set membership operator alongside the MIN and MAX operators within the context of fuzzy measurements. This is a highly adaptable method for fuzzy set aggregations. This chapter presents a soil vulnerability model pertaining to soil erosion, utilizing fuzzy set aggregation and simplified functions for analysis.

2. A FUZZY LOGIC MODEL TO EVALUATE SOIL VULNERABILITY

Soil erosion is regarded as the paramount degrading process in this context, with soil susceptibility determined by the principal criteria influencing this phenomenon. This model utilizes two categories of information: (1) topographic characteristics and (2) vegetation. Remote sensing methodologies enable the acquisition of data from any location on Earth. Digital elevation models (DEMs), which delineate the topography of any place, can be generated using data acquired by NASA's Shuttle Radar Topography Mission and various other platforms. Remote sensing techniques can also provide vegetation status data for the analysis. Data acquired from Landsat missions [27] or other programs enable the calculation of vegetation indices such as the normalized difference vegetation index (NDVI).

The analytical capabilities derived from the use of DEM are enormous, ranging from basic topographic feature estimation, to flood simulations, and others. Moreover, remote sensing has a great potential to obtain imagery from optical, thermal, and microwave spectral regions across wide regions of the planet with a great temporal frequency, enough accuracy, and open source. Image processing methods of remotely-sensed data could extract valuable and specialized information of selected targets (e.g., soils, vegetation, waters, etc.). An added value of image processing techniques is the capability to analyze temporal series of data which can be useful to study the vulnerability and changes occurred in soils along time.

The proposed model is based on a set of three initial inputs, selected considering the previous indications (Figure 1):

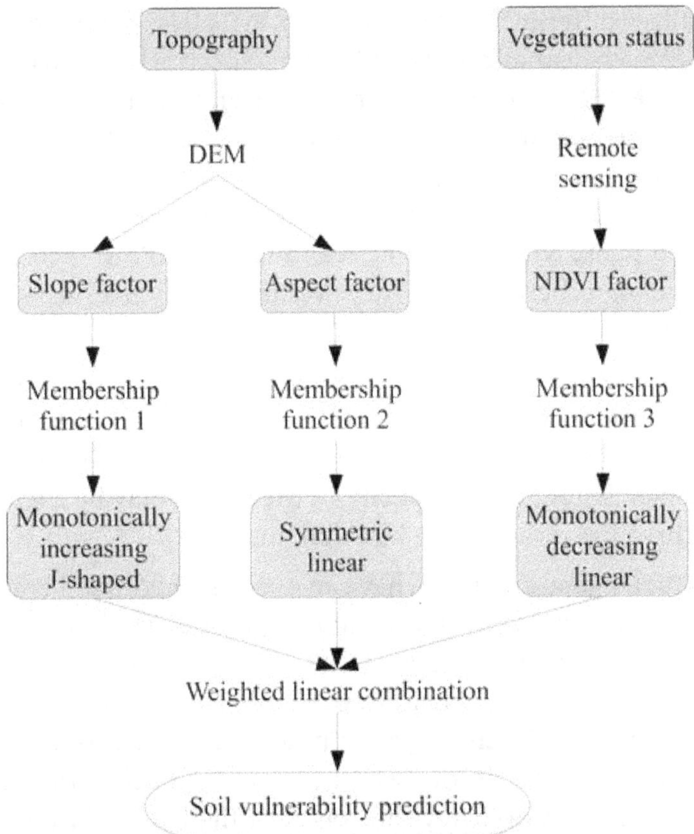

Figure 1. Flowchart of the fuzzy logic model to soil vulnerability evaluation.

- Slope: This topographic parameter could be derived from a DEM, and it is defined as the variation of altitude between two points in relation with the distance that separate them. Slope parameter reports information about the roughness of the territory and plays a key role in soil erosion because it could increase or reduce the effect of water [28]. The universal soil loss equation (USLE), and derived equations from this, considers slope parameter as a key factor for superficial soil water erosion [29].

- Aspect: This topographic parameter could be derived from a DEM. It reports information about the geographic orientation of the slopes. As a general rule (in the north hemisphere), a slope closer to a north aspect presents lower temperature and higher moisture levels (lower incident radiation levels). This parameter could suggest information about soil moisture, water availability by plants, etc. This general rule could be dismissed in special local situations (e.g., for places where wet maritime winds impact against slopes closed to a south aspect). This factor is of great importance in plant distribution and growth and soil development and properties [30].
- Normalized difference vegetation index (NDVI) is computed as a normalized ratio between near infrared (NIR) and red spectral bands of remotely-sensed imagery. NDVI values range between -1 and +1. This vegetation index is strongly related with several vegetation parameters such as changes in biomass and chlorophyll content [31, 32]. NDVI is related with other vegetation parameters too [33] such as leaf water content, CO_2 net flux, absorbed photosynthetically active radiation (APAR), and leaf area index (LAI), among others. A high NDVI value (usually over 0.3–0.5) indicates a vigorous and dense vegetal coverage.

Slope, aspect, and NDVI are the parameters computed in our model and could be extracted for wide regions from remote sensing data. Those parameters are computed following the flowchart as Figure 1 presents, with an easy function that can be implemented. The simplicity of the model may be the limiting factor because only three inputs are taken into account.

However, this tool is good enough to have an overview of the soil erosion and could be a good tool for regional soil vulnerability assessment and long-term monitoring with a low economic cost. A near real-time monitoring of soil vulnerability, depending on the availability of data from satellite, could be performed and used by land managers and scientist. The main advantage in this sense is that field campaign to check soil, vegetation status, slope, and other filed parameters is initially not necessary. Moreover, the centered field work in determined areas reduces the cost of large field campaigns of wide territories.

The proposed model considers input parameters as fuzzy sets (factors) to be standardized by the definition of individual membership functions. A great variety of fuzzy set membership functions have been developed.

Based on Eastman [16], fuzzy set membership functions can be defined by three parameters:

- Shape of the function that provides three possible options: (1) monotonically increasing which begins at 0 and then rise and stay at 1, (2) monotonically decreasing which is opposite to the previous one, and (3) symmetric which firstly rise and then fall again

- Control points that govern the shape of the curve
- Type of function with three possible options: (1) sigmoidal produced by a cosine function, (2) J-shaped produced by a sine function, and (3) linear produced by a linear function

The definition of these parameters provides a great number of possibilities to define the effect of individual factors in the observed phenomena. The shape and type of membership functions should be defined according to the knowledge of environmental phenomena. Weighted linear combination is used for the standardized aggregation factors. This method allows the possibility to assign equal or differential weights to factors in function of experts' knowledge about their relative importance for the studied phenomena. Finally, it is important to mention that this simple model approach could be completed with other factors to be considered in the aggregation process or in other stages of the model. Equilibrium must be found between model simplicity and valuable obtained data to soil conservation.

3. SOIL VULNERABILITY UNDER HIGH SLOPE CHANGES: A CASE STUDY

A test of soil vulnerability evaluation of this model after a fire event was done, considering that the area to test the model has many changes in slope and is close to the sea, affected from marine breeze and moisture. The selected area is located on the south-east coast of Spain (Alicante province) in the area of "La Granadella" (38.73°N, 0.19°E). This test site supported a fire event from 26 to 30 on August 2000. The topography is characterized by a highly rough relief (cliffs above the sea of more than 150 m) and a wet Mediterranean climate (700–1000 mm/year of rainfall). This is an interesting test area which combines a wet climate and a complex and precipitous relief and exhibits a great recurrence of fires.

The hypothesis to verify with the fuzzy logic model is that vegetation regeneration has a primary role in soil vulnerability reduction. The potential advantage for the use of fuzzy sets with respect to other methods of factor combination must be owed to their great flexibility derived from the defined membership functions.

Vegetation status information was obtained from two satellite scenes acquired by the multispectral advanced spaceborne thermal emission and reflection radiometer (ASTER): (http://asterweb.jpl.nasa.gov/) sensor. The first scene was acquired 4th October in 2000, some days after the forest fire, while the second scene was acquired 7th June in 2003, being a reasonable period for a substantial evolution in the landscape.

ASTER sensor is composed of three subsystems in function of their spectral and spatial resolution characteristics: visible near-infrared radiometer (VNIR) with 15 m of spatial resolution and stereoscopic capability, short-wave infrared radiometer (SWIR) with 30 m of spatial resolution, and thermal infrared radiometer (TIR) with 90 m of spatial resolution [34]. Only VNIR system bands were employed for our analyses. Both scenes were in origin preprocessed to an ASTER high-level product denominated AST_07 which contain data of surface reflectance [35].

Topographic information was obtained from a high-resolution digital elevation model (Figure 2) that was computed using the previous vector cartography (scale 1:10,000). Triangulated irregular network (TIN) polygons were computed with the nodes of the vector cartography and used to develop the raster DEM. The spatial resolution of the DEM was adjusted to ASTER-VNIR imagery (15 m). Satellite images were geometrically corrected using the high-resolution DEM and additional cartography in order to minimize the positional errors among the different sources of information. The root-mean-square error (RMSE) of the geometric correction was less than half a pixel for both ASTER scenes.

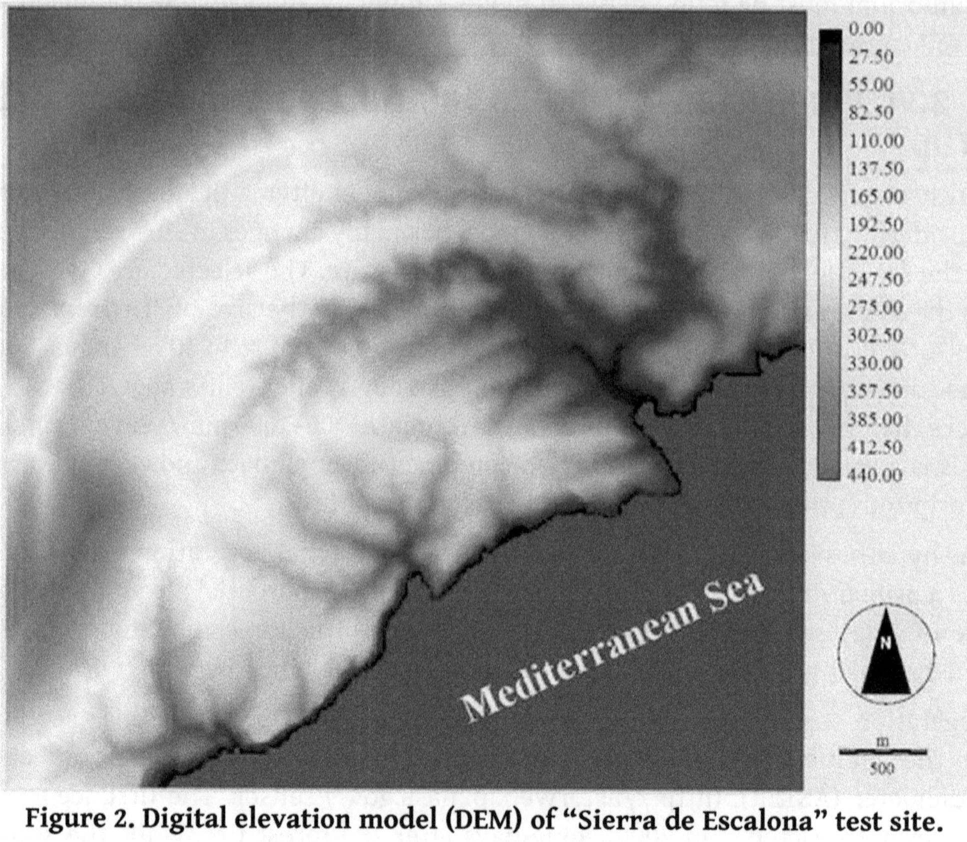

Figure 2. Digital elevation model (DEM) of "Sierra de Escalona" test site.

Topographic parameters (slope and aspect) were computed using the DEM. Both parameters were quantized as degrees. Normalized difference vegetation index is derived from a spectral band transformation between near-infrared and red bands. The original NDVI formulation has been attributed to Rouse et al. [36] and their research with early Landsat images. The original NDVI formulation has been adapted to subsequent sensors whose spectral characteristics are different among them. The following equation is the adaptation of NDVI for ASTER spectral bands (Eq. (1)):

$$NDVI = \frac{\rho_{VNIR3} - \rho_{VNIR2}}{\rho_{VNIR3} + \rho_{VNIR2}} \quad (1)$$

where ρ_{VNIR2} is the spectral reflectance for the second band of the ASTER-VNIR subsystem and ρ_{VNIR3} is the spectral reflectance for the third band of the same subsystem. The computation of NDVI is more suitable with surface reflectance data (like AST_07 high-level ASTER product) in order to minimize differential wavelength-dependent atmospheric disturbances. NDVI was computed in the same way for the 2000 and 2003 scenes (Figure 3 (a) and (b)).

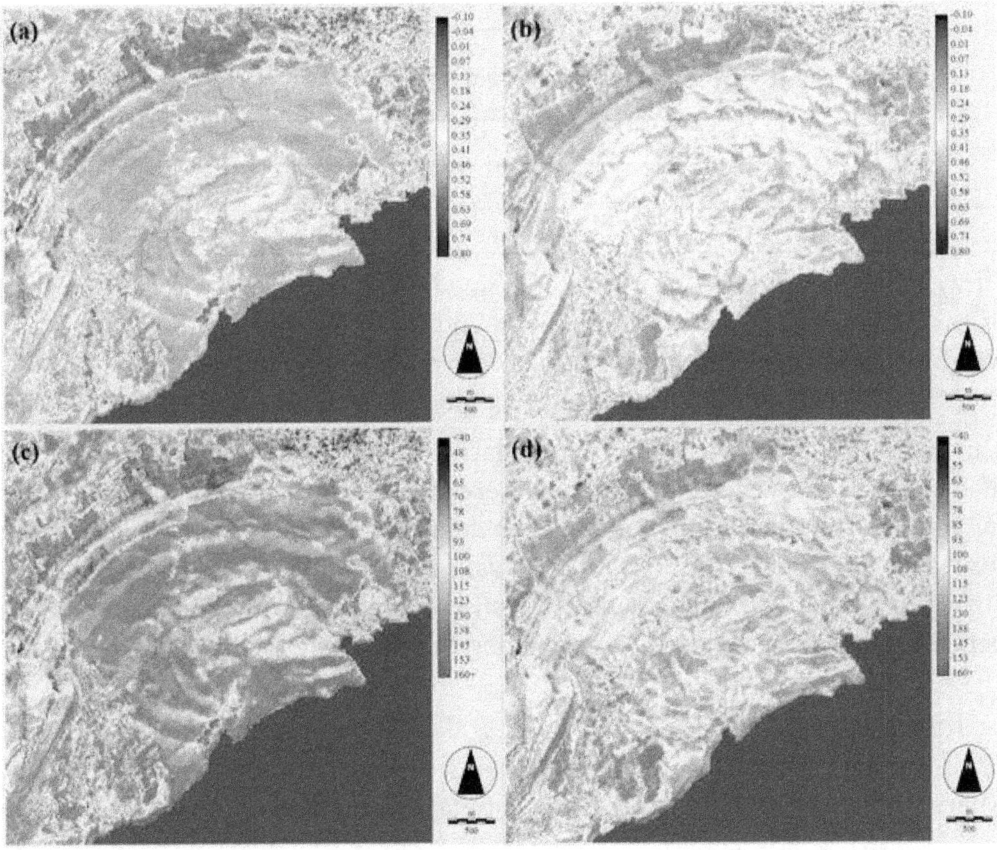

Figure 3. NDVI estimations for 2000 (a) and 2003 (b). Soil vulnerability estimations for 2000 (c) and 2003 (d). Soil vulnerability results area in 8 bits of quantization.

Soil vulnerability estimations were computed for both 2000 and 2003 scenarios. Individual fuzzy set membership functions are characterized by their shape and type and were defined for each of the considered soil vulnerability factors (i.e., slope, aspect, and NDVI). In this sense, Table 1 provides a synthesis of model parameters used with the fuzzy logic model. The proposed model employed the following membership function for input variables: (1) NDVI, a monotonically decreasing linear function; (2) slope, a monotonically increasing J-shaped function; and (3) aspect, a symmetric linear function. Membership functions required the definition of several control points (CP) for a potential model generalization.

Table 1. Model parameters for the soil vulnerability factors slope, aspect, and NDVI.

Factors		Slope	Aspect	NDVI
Membership functions	Shape	Monotonically increasing	Symmetric	Monotonically decreasing
	CP 1	0	0	0
	CP 2	90	180	1
	CP 3	No	180	No
	CP 4	No	360	No
	Type	J-shaped	Linear	Linear
Weights		0.35	0.05	0.6

Fuzzy membership functions were combined weighting the relative importance of each one using a linear system. Jiang and Eastman [24] suggested the use of weighted linear combinations as a fuzzy set membership operator together with the MIN and MAX operators, as a very flexible approach for fuzzy set aggregations.

This weighting process considers the relative importance of each variable in the modeling process. Vegetation and slope were considered the most important factors for the soil vulnerability model. Further details of the definition of membership functions and model calibration are available in Melendez-Pastor et al. [37]. Finally, a temporal change analysis of soil vulnerability was computed using the percentage of change procedure employing the following formulation (Eq. (2)):

$$\text{Temporal change}\,(\%) = \frac{(t_2 - t_1)}{t_1} \cdot 100 \qquad (2)$$

where t_1 and t_2 are the soil vulnerability estimations for 2000 and 2003, respectively.

The application of the fuzzy logic model for this study area covers two environmental facts, an (almost) invariant one associated to the slope and orientation (which are mainly determined by the geomorphology of the landscape) and a highly changing parameter as the vegetation through NDVI. Landscape topography greatly affects soil profile formation [30, 38], while vegetation status and dynamics are highly related with the quality of the soils [39]. The DEM analysis revealed that the study area has a mean slope value of 16 degree, with a maximum slope value of 78 degrees. Relief is mainly configured on a NE-SW direction with preferential slope orientation to the SE direction. NDVI mean values varied from 0.31 (standard deviation of 0.12) on 2000 to 0.38 (standard deviation of 0.09) on 2003.

Soil vulnerability simulations (Figure 3 (c) and (d)) revealed the severe impact of the forest fire on vegetation. The model marks the importance of vegetation dynamics, the relation with the presence of soil although both are limited to the position as the model indicates (slope and aspect). A characteristic pattern of soil vulnerability mitigation by intense vegetation regeneration could be advertised along the valleys where NDVI has increased faster since 2000. The temporal change estimation of soil vulnerability simulations (Figure 4) remarks the local high postfire ecosystem regeneration. Change rates are up to a 30% of less soil vulnerability within burned area. Soil vulnerability change map also highlighted areas where soil vulnerability has increased. Those changes correspond to land use conversions to urban (change rates up to a 30% of more soil vulnerability) and vegetation status variations.

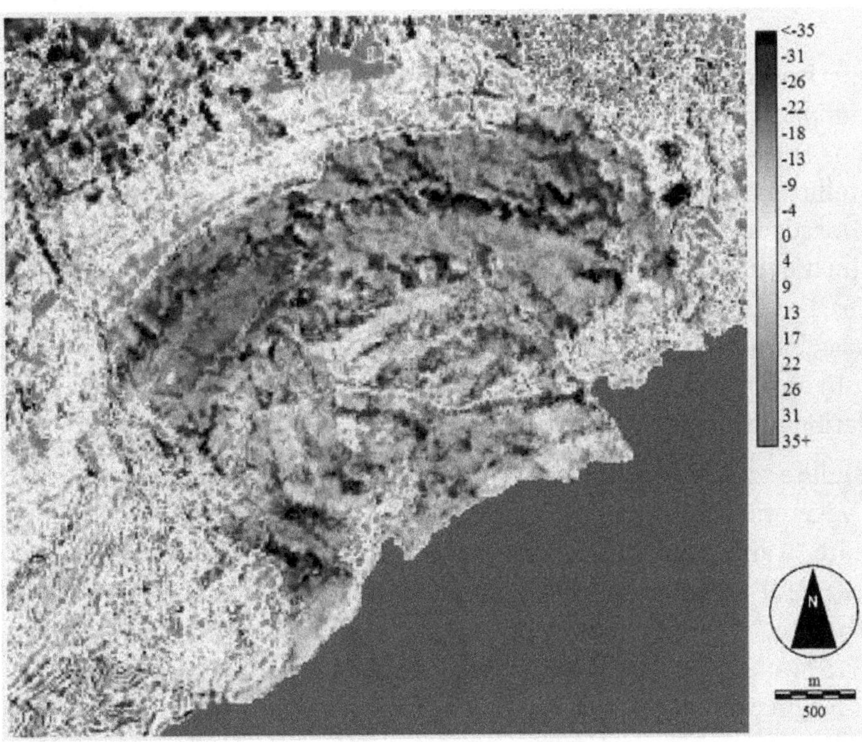

Figure 4. Temporal change (%) of soil vulnerability estimation. Green pixels correspond to less vulnerable areas on 2003.

4. SOIL VULNERABILITY AND VEGETATION AS A LIMITING FACTOR

The proposed model was also engaged in a research of soil vulnerability evaluation in a semiarid area in the south of Alicante province (Spain). This study area is located in a portion of the "Sierra de Escalona" (37.97°N, 0.86°W), an area with altitudes ranging from 80 to 340 m.a.s.l., a semiarid Mediterranean climate with hot summers and scarce precipitations (less than 250–300 mm), and fragile soils severely affected by erosion and land degradation processes. Dominant land cover classes are intensive agriculture (citrus and almond trees) at low to medium slopes, xerophytic shrubs in abandoned fields and marginal areas, and sparse *Pinus halepensis* stands at the highest slopes. A large reservoir fed by a transbasin diversion is located in the north of the study area. This is an interesting test area which combines a semiarid climate, intensive exploitation of soil resources by agriculture, and high erosion rates resulting transport of sediments and nutrients to the reservoir.

The hypothesis to verify with this fuzzy logic model application is that soil vulnerability is enhanced during drought periods when vegetation status is less protective against land degradation drivers. The engagement of satellite images allowed the estimation of soil vulnerability changes within a drought period and between hydrologic years with different precipitation patterns (drought vs. normal year).

A comparison of the changes regarding soil vulnerability between late spring/ early summer and late summer for different hydrologic years was included in this research. The selections of the dates were based on meteorological information and satellite image availability. Meteorological information for the nearby Pilar de la Horadada town meteorological stations was obtained from the Spanish Agroclimatic information System for Irrigation (Ministry of Agriculture and Fisheries, Food and Environment). Precipitation data indicated that the 2000 water year was characterized by a severe drought (169.8 mm), while 2003 was a more regular hydrologic year (253.0 mm). This information was the starting point for the compilation of vegetation and topography datasets.

Four satellite scenes acquired by the multispectral ASTER sensor were employed to obtain vegetation information. Two scenes were acquired for the 2000 hydrologic year (June 30 and August 1) and two other images for the 2003 water year (May 22 and August 10). The first images correspond to the end of spring and early summer, while the second scenes correspond with the end of summer. Only VNIR system bands (15 m of spatial resolution) were employed for our analyses. Both scenes were in origin preprocessed to the ASTER high-level product of surface reflectance (AST_07).

Topographic information was obtained from a high-resolution digital elevation model (Figure 5) that was computed using the previous vector cartography (scale 1:10,000). Triangulated irregular network (TIN) polygons were computed with the nodes of the vector cartography and used to develop the raster DEM in the same way as the previous case study. The spatial resolution of the DEM was adjusted to ASTER-VNIR imagery (15 m). Satellite images were geometrically corrected using the high-resolution DEM and additional cartography in order to minimize the positional errors among the different sources of information. The root-mean-square error (RMSE) of the geometric correction was less than half a pixel for both ASTER scenes.

Figure 5. Digital elevation model (DEM) of "La Granadella" test site.

Slope and aspect were computed using the DEM and quantized as degrees. Normalized difference vegetation indices were computed with the surface reflectance data spectral band transformation between near-infrared and red bands according to Eq. (1). Soil vulnerability estimations were computed for the four dates of the ASTER scenes (Table 2). The proposed model employed the following membership function and weights for input variables: (1) NDVI, a monotonically decreasing linear function with a weight value of 0.4, (2) slope, a monotonically increasing J-shaped function with a weight value of 0.4, and (3) aspect, a symmetric linear function with a weight value of 0.2. Membership function control points were the same of the previous case study. Further details of the definition of membership functions and model calibration are available in Melendez-Pastor et al. [40]. Finally, a temporal change analysis of soil vulnerability was computed using the percentage of change procedure employing the formulation of Eq. (2).

Temporal change analyses were done between different seasons for the same year estimations (E01 vs. E02 and E31 vs. E32) and between the same seasons for different years (E01 vs. E31 and E02 vs. E32).

Table 2. Soil vulnerability estimations and employed ASTER scenes. Hydrologic year accumulated precipitation for Pilar de la Horadada meteorological station is shown.

Estimation	Year	Season	ASTER scene	Accumulated precipitation
E01	2000	Spring summer	06-30-2000	167.2
E02	2000	Late summer	08-01-2000	167.8
E31	2003	Spring summer	05-22-2003	246.4
E32	2003	Late summer	08-10-2003	250.0

NDVI estimations highlighted the critical importance of hydrologic year accumulated precipitation for the maintenance of vegetation status for nonirrigated areas. In Figure 6 (a) and (b), the largest portion of the study area corresponds with nonirrigated crops, sclerophyllous vegetation, and some coniferous forest areas. They had very low NDVI values, while irrigated areas and some wet ravines exhibited quite high NDVI values (0.4–0.6). These NDVI images correspond with an intense drought water year with less than 168 mm of accumulated precipitation.

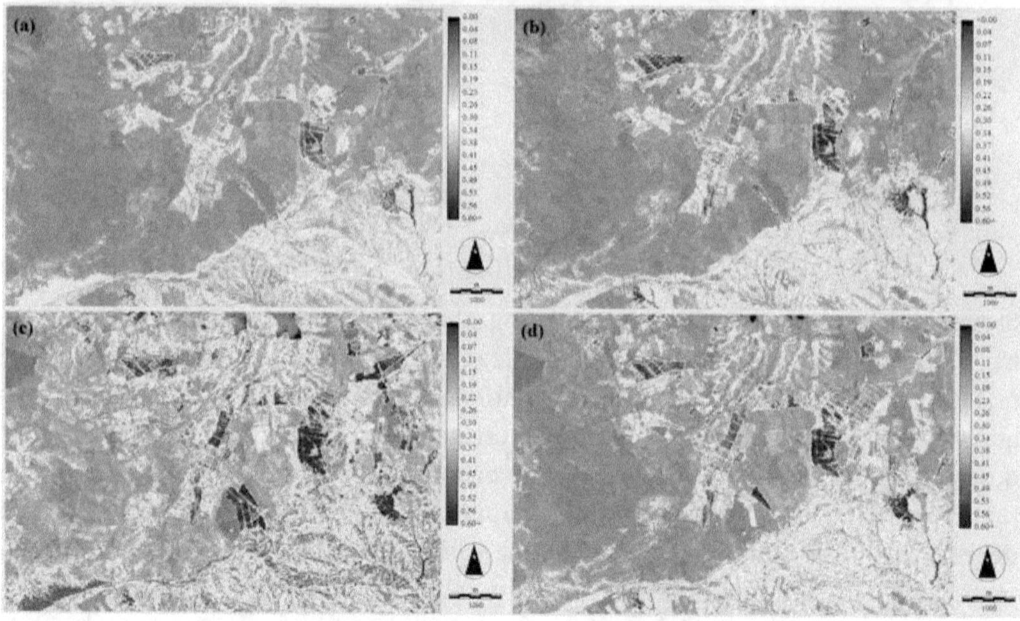

Figure 6. NDVI images for the estimations E01 (a), E02 (b), E31 (c), and E32 (d).

Climatic conditions for the hydrologic year 2003 were very different with almost 50% more precipitation. Greener areas in Figure 6 (c) and (d) correspond with the irrigated crops and also with some coniferous forest and dense sclerophyllous vegetation areas. More intense NDVI changes for irrigated crops were associated with the harvest of seasonal crops.

High values of soil vulnerability estimations were obtained for almost the whole study area in all the scenes (Figure 7), due to the abrupt orography and the lack of vigorous vegetation cover. Southern slopes with high slope values are shown as the most vulnerable areas since its vegetation cover, and edaphic development is very scarce. The lowest soil vulnerability estimations were obtained for some crop fields with permanent crops in flat areas.

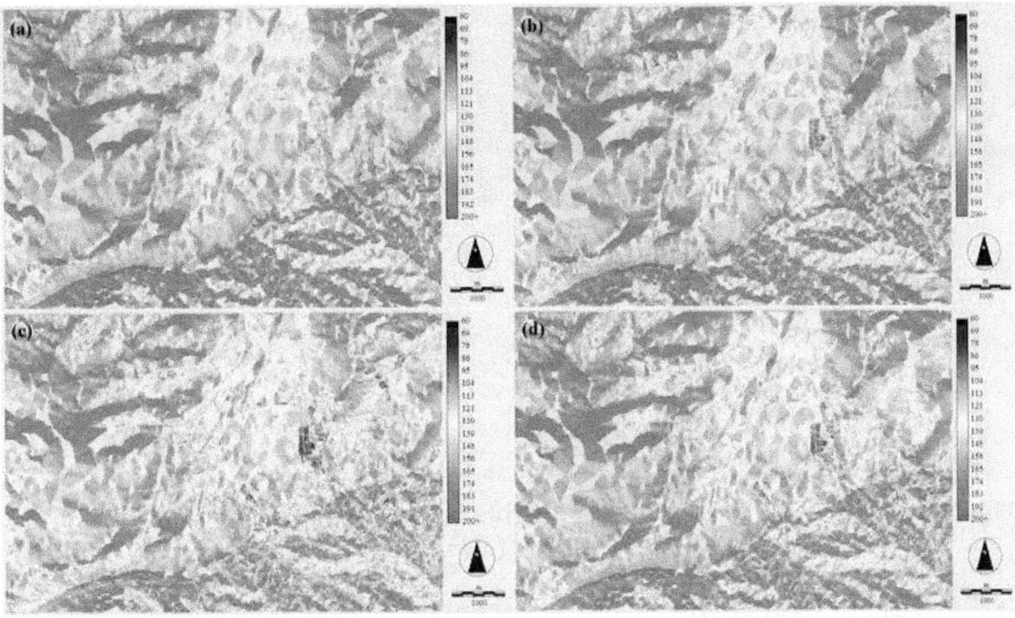

Figure 7. Soil vulnerability estimations E01 (a), E02 (b), E31 (c), and E32 (d).

Temporal change analyses between different seasons for the same year estimations were done. Figure 8 (a) corresponds with the change in 2000 (E01 vs. E02), and Figure 8 (b) is for the change in 2003 (E31 vs. E32). The most remarkable changes were associated with the irrigated crops. Temporal crops had been collected within the period of time of the change analysis, and the soil vulnerability increased by the elimination of the vegetation cover. On the other hand, temporal change between the same seasons but for different years (2000 vs. 2003) was analyzed. Figure 8 (c) corresponds with the change analysis for late spring-early summer (E01 vs. E31), while Figure 8 (d) corresponds with the change analysis for late summer (E02 vs. E32). The intense effect of drought mitigation in soil vulnerability values was estimated. This effect was more intense for the change analysis of late spring-early summer when very few areas exhibited an increase of soil vulnerability estimations. Figure 8 (c) indicated a large increase of soil vulnerability in areas of the center and east of the image, which are produced by crop replacements. Two higher soil vulnerability areas in the north of the image are part of the reservoir basin without a soil profile. Late summer temporal changes were less evident for the nonirrigated areas by the absence of precipitations in 2000 and 2003. Soil vulnerability reductions were associated to irrigated areas less affected by the typical summer drought of the study area.

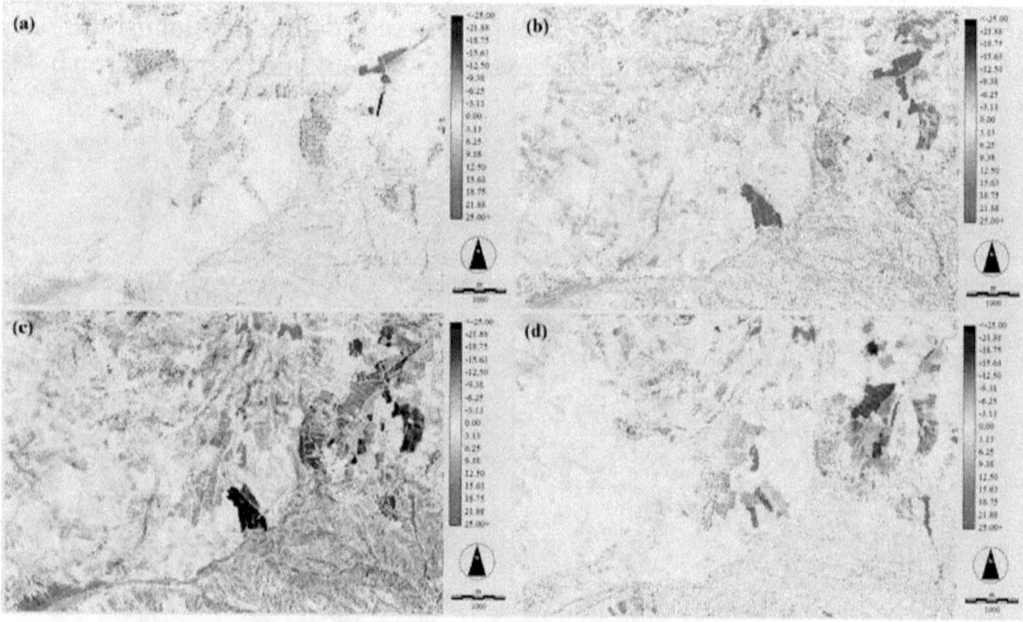

Figure 8. Temporal change (%) of soil vulnerability estimation E01 vs. E02 (a), E31 vs. E32 (b), E01 vs. E31 (c), and E02 vs. E32 (d).

5. CONCISE SUMMARY

The primary benefit of the suggested approach is in its facilitation of straightforward data collecting and calculation, which may be employed to assess soil vulnerability and erosion. The studies conducted exemplify the substantial potential of the proposed methodology and its heightened sensitivity in assessing the current condition and identifying temporal variations in soil vulnerability, a dynamic critical parameter crucial for soil conservation. Our research confirmed the efficacy of this straightforward method for updating satellite images using a fuzzy logic model technique. This approach's usefulness in extensive and sparsely inhabited regions with minimal field data may enhance land management techniques and facilitate a more comprehensive understanding of soil degradation processes. Decision support systems for local and regional authorities and land management serve as instruments that assist decision-makers in formulating appropriate responses for environmental, societal, and sustainable futures.

AT A GLANCE

Soil vulnerability refers to the susceptibility of one or more ecological services within the soil system to damage. The notion is intricate, necessitating the identification of several environmental elements and land management across diverse temporal and spatial dimensions. The utilization of geospatial information with robust updating capabilities may serve as an effective instrument for evaluating prospective alterations in soil sensitivity across extensive regions. This chapter discusses the implementation of two land degradation case studies that are straightforward, synoptic, and appropriate for a continuous monitoring approach utilizing fuzzy logic. The model integrates topographical and vegetative data to evaluate soil susceptibility to land degradation. Topographic parameters were extracted from digital elevation models (DEM), while vegetation state information was gathered via the calculation of the normalized difference vegetation index (NDVI) using satellite imagery. This spectral index offers significance and is revised for each scene, providing evidence regarding biomass, soil productivity, and vegetation density or stress (e.g., forest fires, droughts). Modeled output maps are appropriate for temporal change analysis, facilitating the identification of the impacts of land management methods, soil and vegetation regeneration, or climatic influences.

REFERENCES

1. Doula MK, Sarris A. Soil Environment. In: Poulopoulos SG, Inglezakis JV editors. Environment and Development: Basic Principles, Human Activities and Environmental Implications. Netherlands: Elsevier; 2016. pp. 213-216.

2. D'Odorico P, Ravi S. Land degradation and environmental change. In: Sivanpillai R, Shroder JF Jr editors. Biological and Environmental Hazards, Risks, and Disasters. Netherlands: Elsevier. 2016. pp. 219-227.

3. FAO. Soil Is A Non-Renewable Resource. FAO fact sheet. Job Number I4373E/1/2015; 2015. 4 p. Available from: http://www.fao.org/documents/card/en/c/ec28fc04-3d38-4e35-8d9b-e4427e20a4f7/ [Accessed: 2017-04-03]

4. Häberli R, Luscher C, Praplan Chastonay B, Wyss C. L'affaire sol. Pour une politique raisonnée de l'utilisation du sol. Rapport final du programme national de recherche "Utilisation du sol en Suisse" (PNR 22), Switzerland: Georg Editeur SA; 1991. 192 p.

5. Blum WEH. The challenge of soil protection in Europe. Environmental Conservation. 1990;17:72-74.

6. D'Odorico P, Bhattachan A, Davis KF, Ravi S, Runyan CW. Global desertification: Drivers and feedbacks. Advances in Water Resources. 2013;51:326-344.

7. Kort J, Collins M, Ditsch D. A review of soil erosion potential associated with biomass crops. Biomass and Bioenergy. 1998;14(4):351-359.
8. Department of Agriculture RSA. Soil erosion. Resource Centre, Directorate Agricultural Information Services. South Africa: Department of Agriculture; 2008. 7 p.
9. COM/2006/0231. Communication from the Commission to the Council, the European Parliament, the European Economic and Social Committee and the Committee of the Regions—Thematic Strategy for Soil Protection [SEC(2006)620] [SEC(2006)1165]. Available from: http://eur-lex.europa.eu/legal-content/EN/TXT/?uri=CELEX:52006DC0231 [Accessed: 2017-04-03]
10. Franchis L, Ibańez F, Fox T. Plan Blue Papers-2, Threats to Soil in Mediterranean Countries-Document Review. France: UNEP; 2003. 78 p.
11. COM 2006/0232. Proposal for a Directive of the European Parliament and of the Council establishing a framework for the protection of soil and amending Directive 2004/35/EC. Available from: http://eur-lex.europa.eu/legal-content/EN/TXT/?uri=CELEX%3A52006PC0232 [Accessed: 2017-04-03]
12. Lagachierie P. Digital soil mapping: A state of the art. In: Hartemink AE, McBratney A, Mendonça-Santos ML editors. Digital Soil Mapping with Limited Data. Netherlands: Springer; 2008. pp. 3-14.
13. FAO. The State of the World's Land and Water Resources for Food and Agriculture (SOLAW)—Managing Systems at Risk. Italy and United Kingdom: Food and Agriculture Organization of the United Nations and Earthscan; 2011. 308 p.
14. Zadeh LA. Fuzzy sets. Information and Control. 1965;8:338-353.
15. Urbanski JA. The use of fuzzy sets in the evaluation of the environment of coastal waters. International Journal of Geographical Information Sciences. 1999;13(7):723-730.
16. Eastman JR. IDRISI Kilimanjaro. Guide to GIS and Image Processing. USA: Clark University; 2003. 328 p.
17. Dubois D, Prade H. A class of fuzzy measures based on triangular norms. International Journal of General Systems. 1982;8:43-61.
18. Zimmermann HJ. Fuzzy set theory. WIREs Computational Statistics. 2010;2:317-332.
19. Bonissone PP, Decker K. Selecting uncertainty calculi and granularity: An experiment in trading-off precision and complexity. In: Kanal LN, Lemmer JF editors. Uncertainty in Artificial Intelligence. Netherlands: Elsevier Science; 1986. pp. 217-247.

20. Power DJ. Decision support systems: A historical overview. In: Burstein F, Holsapple CW editors. Handbook on Decision Support Systems 1. USA: Springer; 2008. pp. 121-140.

21. Department for Communities and Local Government. Multi-Criteria Analysis: A Manual. UK: Communities and Local Government Publications; 2009. 168 p.

22. Zorpas AA, Saranti A. Multi-criteria analysis of sustainable environmental clean technologies for the treatment of Winery's wastewater. International Journal of Global Environmental. 2016;15:151-168.

23. Zorpas AA, Pociovălisteanu DM, Georgiadou L, Voukkali I. Environmental and technical evaluation of the use of alternative fuels through multi criteria analysis model. Progress of Industrial Ecology. 2016;10(1):3-15.

24. Jiang H, Eastman JR. Application of fuzzy measures in multi-criteria evaluation in GIS. International Journal of Geographical Information Sciences. 2000;14(2):173-184.

25. Sasikala KR, Petrou M. Generalized fuzzy aggregation in estimating the risk of desertification of a burned forest. Fuzzy Sets and Systems. 2001;118:121-137.

26. Buckley JJ. The multiple judge, multiple criteria ranking problem: A fuzzy set approach. Fuzzy Sets and Systems. 1984;13:25-37.

27. Jet Propulsion Laboratory. Shuttle Radar Topography Mission. Available from: http://www2.jpl.nasa.gov/srtm/[Accessed: 2017-01-25].

28. USGS. Landsat Missions. Available from: https://landsat.usgs.gov/[Accesed: 2017-01-25].

29. Antolin C (coord.). El suelo como recurso natural en la Comunidad Valenciana. Spain: Conselleria d'Obres Públiques, Urbanismo i Transports de Valencia; 1998. 187 p.

30. Wischmeier WH, Smith DD. Predicting Rainfall Erosion Losses from Cropland East of the Rocky Mountains. US: USDA, U.S. Government Printing Office, Agric. Handbook 282; 1965. 47 p.

31. Carter BJ, Ciolkosz EJ. Slope gradient and aspect effects on soils developed from sandstone in Pennsylvania. Geoderma. 1991;49(3-4):199-213.

32. Jensen JR. Remote Sensing of the Environment: An Earth Resources Perspective. (2nd edition). USA: Pearson; 2007. 592 p.

33. Chuvieco E. Teledetección Ambiental. Madrid, Spain: Ariel; 2010. 528 p.

34. ERSDAC ASTER Level 1 Data Products Specifications (GDS Version). Version 1.3. Japan: Earth Remote Sensing Data Analysis Center (ERSDAC); 2001. 126 p.

35. Abrams M, Hook S, Ramachandran B. ASTER User Handbook. Version 2; USA: Jet Propulsion Laboratory-NASA/California Institute of Technology; 2002. 135 p.

36. Rouse JW, Hass RH, Schell JA, Deering DW. Monitoring vegetation systems in the great plains with ERTS. In Proceedings, Third Earth Resources Technology Satellite-1 Symposium. NASA SP-351. USA: NASA; 1974; pp. 3010-3017.

37. Melendez-Pastor I, Navarro-Pedreño J, Koch M, Gómez I, Hernández EI. Evaluation of land degradation after forest fire with a fuzzy logic model. Environmental Engineering and Management Journal. 2013;12:2087-2096.

38. Jenny H. Factors of Soil Formation. USA: McGraw-Hill; 1941.

39. Morgan RPC. Erosión y conservación del suelo. Spain: Mundi-Prensa; 1997. 348 p.

40. Melendez-Pastor I, Córdoba Sola P, Navarro-Pedreño J, Gómez I. Evaluación de la vulnerabilidad a la degradación por erosión en suelos mediante un modelo de lógica borrosa. Revista de Cięncias Agrarias. 2010;33:171-181.

INDEX

www.ingramcontent.com/pod-product-compliance
Lightning Source LLC
LaVergne TN
LVHW062107200926
848484LV00007B/1614